ACOUSTIC HIGH-FREQUENCY DIFFRACTION THEORY

ACOUSTIC HIGH-FREQUENCY DIFFRACTION THEORY

FRÉDÉRIC MOLINET

MOMENTUM PRESS

MOMENTUM PRESS, LLC, NEW YORK

Acoustic High-Frequency Diffraction Theory

First published in 2011 by
Momentum Press®, LLC
222 East 46th Street, New York, N.Y. 10017
www.momentumpress.net

ISBN-13: 978-1-60650-100-9 (hard back, case bound)
ISBN-10: 1-60650-100-3 (hard back, case bound)

ISBN-13: 978-1-60650-102-3 (e-book)
ISBN-10: 1-60650-102-X (e-book)

Cover Design by Jonathan Pennell
Interior Design by Scribe, Inc. (www.scribenet.com)

First Edition: May 2011

10 9 8 7 6 5 4 3 2 1

Printed in Taiwan

To my wife Odile who typed the manuscript for the strength of her love

CONTENTS

List of Figures		xvii
List of Tables		xxix
About the Author		xxxi
Preface		xxxiii
Chapter 1	Introduction to the Geometrical Theory of Diffraction	1
	1.1. General Overview and Basic Concepts	1
	1.2. The Fermat Principle	3
	1.2.1. Conditions for a Path to Be a Ray	4
	1.2.2. Application of Conditions (1.5) and (1.6) to Specific Problems	6
	1.2.2.1. Segments in a Fluid with Constant Sound Velocity	6
	1.2.2.2. Reflection from a Smooth Surface	7
	1.2.2.3. Transmission through a Smooth Interface between Two Different Homogeneous Fluids	8
	1.2.2.4. Excitation of Elastic Waves at a Smooth Interface between a Homogeneous Fluid and an Isotropic Homogeneous Elastic Body	9
	1.2.2.5. Diffraction by an Edge in a Homogeneous Fluid	11
	1.2.2.6. Surface Rays	12

1.3. Extension of Fermat's Principle to Surface Waves 12

1.3.1. Planar Interface 13

1.3.2. Curved Interface 20

1.4. Fundamentals of Asymptotic Expansions 22

1.4.1. Asymptotic Sequence 23

1.4.2. Compatible Asymptotic Sequence 25

1.4.3. Properties of an Asymptotic Expansion 27

1.5. Asymptotic Solution of the Wave Equation in a Source-Free, Unbounded Medium 28

1.5.1. Derivation of the Asymptotic Expansion of the Solution 28

1.5.2. Resolution of the Eikonal Equation 33

1.5.3. Properties of the Characteristic Curves 34

1.5.4. Resolution of the Transport Equation 35

1.6. Acoustic Field Reflected by a Smooth, Rigid, Soft, or Impedance Surface 40

1.7. Reflected and Transmitted Waves at a Smooth Interface between a Fluid and an Elastic Medium 45

1.8. Acoustic Field Diffracted by the Edge of an Impenetrable Wedge 53

1.9. Acoustic Field in the Shadow Zone of a Smooth Convex Object 63

References 70

Chapter 2 Canonical Problems and Nonuniform Asymptotic Theory of Acoustic Wave Diffraction 71

2.1. Introduction 71

2.2. The Wedge 72

2.2.1. Hard and Soft Straight Wedge at Normal and Oblique Incidence 72

2.2.1.1. Line Source Excitation 72

2.2.1.2. Plane Wave Incidence 89

2.2.2. Impedance Wedge at Normal and Oblique Incidence: Maliuzhinets's Solution 92

2.2.2.1. Analytical Details for the Justification of the General Form of the Solution 93

2.2.2.2. Derivation of the Solution 101

2.2.2.3. Asymptotic Evaluation of the Solution 110

2.2.2.4. Generalization to Oblique Incidence 122

2.2.3. Elastic Wedge-Shaped Shell 125

2.2.3.1. Derivation of the Solution 125

2.2.3.2. Asymptotic Evaluation of the Solution 132

2.2.3.3. Generalization of the Elastic Wedge-Shaped Shell Solution to Oblique Incidence 137

2.2.3.4. Typical Examples 138

2.3. The Circular Cylinder 140

2.3.1. Circular Cylinder with Hard, Soft, or Impedance Boundary Conditions 140

2.3.1.1. General Solution to the Problem 140

2.3.1.2. Asymptotic Expansion and Physical Interpretation of the Solution 148

2.3.1.3. Expression of the Diffraction Coefficient 162

2.3.1.4. Noncircular Convex Cylinder 165

2.3.1.5. Field at an Observation Point Located on the Surface of a Circular Cylinder 166

2.3.1.6. Position of the Poles in the Complex ν-Plane 169

2.3.1.7. Oblique Incidence 171

2.3.2. Three-Dimensional Convex Surface with Hard, Soft, or Impedance Boundary Conditions 176

2.3.3. Hollow Elastic Circular Cylinder 178
2.3.3.1. General Solution for Normal Incidence 178
2.3.3.2. Watson Transformation 182
2.3.3.3. Asymptotic Expansion and Physical Interpretation of the Solution 186
2.3.3.4. Field Diffracted in the Shadow Region by Creeping Waves 192
2.3.3.5. Noncircular Convex Cylindrical Shell 195
2.3.3.6. Oblique Incidence for Creeping Waves 195
2.3.3.7. Elastic Surface Waves 197
2.3.3.8. Oblique Incidence for Elastic Surface Waves 204
2.3.3.9. Noncircular Cylindrical Shell at Normal and Oblique Incidence 207
2.3.4. Three-Dimensional Convex Shell 209
2.3.4.1. General Solution for Creeping Waves 209
2.3.4.2. General Solution for Elastic Surface Waves 211
2.4. Concave Surface 211
2.4.1. Introduction 211
2.4.2. Solution of the Canonical Problem of a Line Source Parallel to the Generatrix of a Concave Circular Cylinder 212
2.4.2.1. Derivation of the Solution 212
2.4.2.2. Extraction of the GA Contributions 220
2.4.2.3. Expression of the Remainder Integral 228
2.4.3. Extension of the Solution to More Generalized Situations 233
References 235
Chapter 3 Uniform Asymptotic Theory of Acoustic Wave Diffraction 239
3.1. Introduction 239
3.2. The Three-Dimensional Convex Wedge 240

3.2.1. Uniform Solution for the Acoustic Field Scattered by a Convex Three-Dimensional Hard or Soft Wedge 240

3.2.1.1. Uniform Asymptotic Solution of a Hard or Soft Wedge with Planar Faces Submitted to a Planar Sound Wave 242

3.2.1.2. Uniform Asymptotic Solutions for a Soft or Hard Wedge with Curved Faces 267

3.2.2. Uniform Asymptotic Solutions for the Acoustic Field Scattered by a Convex Impedance Wedge 272

3.2.2.1. Uniform Asymptotic Solution of an Impedance Wedge with Planar Faces 273

3.2.2.2. Uniform Asymptotic Solution for a General Three-Dimensional Impedance Wedge 290

3.2.3. Uniform Asymptotic Solutions for the Acoustic Field Scattered by a Convex Wedge-Shaped Elastic Shell 294

3.2.3.1. Uniform Asymptotic Solution for a Wedge-Shaped Shell with Planar Faces 295

3.2.3.2. Uniform Asymptotic Solution for a Three-Dimensional Wedge-Shaped Shell 296

3.3. The Three-Dimensional Smooth Convex Surface 297

3.3.1. Uniform Asymptotic Solution through the Shadow Boundary of a Smooth Convex Two-Dimensional Surface 297

3.3.1.1. Uniform Asymptotic Solution for a Circular Impedance Cylinder 298

3.3.1.2. Uniform Asymptotic Solution for a Smooth Convex Impedance Cylinder 312

3.3.1.3. Field in the Boundary Layer of a Smooth Convex Impedance Cylinder 314

3.3.2. Uniform Asymptotic Solution through the Shadow Boundary of a Smooth Convex Three-Dimensional Surface 315

3.4. The Three-Dimensional Smooth Convex Shell 319

3.4.1. Uniform Solution for a Hollow Elastic Cylinder 320

3.4.1.1. Uniform Solution for the GA Field Associated with the Creeping Wave Field 321
3.4.1.2. Uniform Solution for the Reflected Field Associated with the Elastic Surface Wave Field 324
3.4.1.3. Oblique Incidence 334
3.4.1.4. Noncircular Cylindrical Shell 335
3.4.2. Uniform Solution for a Three-Dimensional Convex Shell 336
3.4.2.1. Uniform Solution for Creeping Waves 336
3.4.2.2. Uniform Solution for Elastic Surface Waves 338
3.5. Numerical Results 339
3.5.1. Results Concerning the Wedge 339
3.5.2. Results Concerning the Circular Cylinder 341
References 345
Chapter 4 Wave Field Near a Caustic 349
4.1. Introduction 349
4.2. Techniques for Calculating the Field on a Caustic and in Its Neighborhood 351
4.2.1. Observation Point Located Close to a Regular Caustic 351
4.2.1.1. Canonical Problem for a Regular Caustic 354
4.2.1.2. Uniform Asymptotic Expansion for a Regular Three-Dimensional Caustic 358
4.2.1.3. Caustic of Rays Reflected by a Smooth Surface 361
4.2.1.4. Caustic of Rays Diffracted by an Edge 362
4.2.1.5. Caustic of Rays Diffracted by a Smooth Surface (Creeping Rays) 363
4.2.1.6. Comments on the Solutions for a Regular Caustic 365
4.2.2. Observation Point Located Close to a Line Cusp of a Caustic 366
References 367

Chapter 5 Hybrid Diffraction Coefficients 369

5.1. Introduction 369

5.2. Edge-Excited and Edge-Diffracted Creeping Waves 369

5.2.1. Spectral Representation of the Fock Field on a Smooth Convex Cylindrical Surface 371

5.2.2. Hybrid Diffraction Coefficients for Creeping Waves on a Curved Wedge 373

5.2.2.1. Two-Dimensional Wedge 373

5.2.2.2. Three-Dimensional Wedge 378

5.2.2.3. Solution Valid at Grazing Incidence and Grazing Observation 382

5.3. Edge-Excited and Edge-Diffracted Surface Waves 384

5.3.1. Impedance Wedge 384

5.3.2. Elastic Wedge-Shaped Shell 388

5.3.3. Elastic Surface Wave on a Curved Wedge 390

5.4. Edge-Excited and Edge-Diffracted Whispering Gallery Waves 392

References 394

Appendix A A Brief Presentation of the Governing Equations for Wave Processes in Fluids 397

A.1. Wave Propagation in Ideal Fluids 397

A.2. Sound Waves 398

A.3. Boundary Conditions 400

A.3.1. Absolutely Rigid Boundary 400

A.3.2. Absolutely Soft Boundary 401

A.3.3. Interface between Two Fluids at Rest 401

A.4. Harmonic Waves 401

A.5. Reflection at a Boundary 402

A.5.1. Absolutely Rigid Boundary 403

A.5.2. Absolutely Soft Boundary 404

A.5.3. Impedance Boundary 404

A.6. Reflection and Refraction at the Interface of Two Homogeneous Fluids 404

References 406

Appendix B A Brief Presentation of the Governing Equations of Linearized Elasticity 409

B.1. Deformation 409

B.2. Linear Momentum and Stress Tensor 410

B.3. Hooke's Law 413

B.4. Waves in an Elastic Medium 414

B.5. Boundary Conditions at Interfaces 415

B.6. Elastic Waves in Homogeneous Isotropic Solids 415

B.7. Harmonic Waves 418

B.8. Reflection and Refraction at a Plane Interface 419

B.9. Reflection on an Elastic Plate Separating a Fluid from Vacuum 425

B.9.1. Application of the Boundary Conditions 426

B.9.2. Solution of the System of Linear Equations 428

References 430

Appendix C Surface Waves 431

C.1. Introduction 431

C.2. Rayleigh Waves 432

C.3. Surface Waves at Fluid–Solid Interfaces 434

C.4. Leaky Waves 438

References 443

Appendix D General Formulas for the Principal Radii of Curvature of the Reflected Wave Front on a Three-Dimensional Surface 445

References 446

Appendix E Symmetric Form of the Maliuzhinets Diffraction Coefficient 447

Appendix F Elements of the Determinant of the Boundary Conditions for a Circular Elastic Shell in a Fluid 453

F.1. Normal Incidence 453

F.2. Oblique Incidence 456

Index 459

FIGURES

Figure 1.1. Constraints for edge diffraction. 4

Figure 1.2. Constraints for the diffraction by a smooth surface at an observation point in the region not reached by direct or reflected sound rays. 4

Figure 1.3. General path between the two extremities of a ray. 5

Figure 1.4. Reflection from a smooth surface. 7

Figure 1.5. Transmission through a smooth interface between two fluids. 9

Figure 1.6. Excitation of elastic waves at a smooth interface between a fluid and an elastic solid. 10

Figure 1.7. Keller's cone. 11

Figure 1.8. Configuration of the ray path involving a surface ray. 13

Figure 1.9. Schematic representation of the complex ray path joining Q to P. 14

Figure 1.10. Real ray joining Q to P and having a path on the interface. 16

Figure 1.11. Incident and radiated inhomogeneous local plane waves. 18

Figure 1.12. Excitation of a surface wave at the characteristic angle θ' and continuous shading of rays. 19

Figure 1.13. Schematic representation of the complex ray path joining the source Q to the observation point P. 20

Figure 1.14. Geometrical representation of the real ray path joining the source Q to the observation point P and having an arc on the surface S. The generalized Rayleigh wave is excited on the convex surface S at the critical angle θ_1' and radiates off the surface into the fluid under the local critical angle along the geodesic $Q_1'Q_2'$. 22

Figure 1.15. Geometrical representation of the real ray path joining the source Q to the observation P and having an arc on the surface S. This ray path corresponds to a Scholte-Stoneley wave or a creeping wave propagating on a convex surface S. 23

Figure 1.16. Typical curve giving the variation of the error Δ as a function of the number of terms N of an asymptotic expansion. 28

Figure 1.17. Wave front, rays, and caustic. 35

Figure 1.18. Ray tube. 36

Figure 1.19. Curvilinear (ray) coordinate system. 37

Figure 1.20. Astigmatic pencil of rays. 40

Figure 1.21. Incident ray field intercepted by a smooth body. 41

Figure 1.22. Incident and reflected rays. 44

Figure 1.23. Incident, reflected, and refracted rays at a curved interface between a fluid and an elastic solid. 48

Figure 1.24. Orientation of the displacement vectors. 49

Figure 1.25. General shape of a curved wedge. 54

Figure 1.26. Astigmatic pencil of diffracted rays. 55

Figure 1.27. Keller's cone and unit vectors $\mathbf{s}^i$, $\mathbf{s}^d$, $\mathbf{t}$. 56

Figure 1.28. Straight wedge tangent to the curved wedge at the diffraction point Q. 58

Figure 1.29. Curved wedge: Local coordinate system and caustic boundary layer B. 59

Figure 1.30. Diffracted rays emanating from the virtual caustic point O in the plane formed by O and the tangent $\mathbf{t}$. 62

Figure 1.31. Creeping rays. 64

Figure 1.32. Distance along the surface wave front between adjacent rays. 67

Figure 2.1. The two-dimensional wedge and line source in the circular cylindrical coordinate system. 73

Figure 2.2. The contours L and L' for the integral representation of the Bessel function of the first kind $I_q(x)$. 79

Figure 2.3. Sommerfeld contours. 81

Figure 2.4. SDP Γ and Γ' in the complex ξ-plane ($\xi = u + iv$). 84

Figure 2.5. Geometry for the ray coordinate system for oblique incidence. 91

Figure 2.6. Geometry of the diffraction by a wedge. 93

Figure 2.7. Sommerfeld contour $\gamma = L - L'$. 94

Figure 2.8. Half-space $y > 0$. 95

Figure 2.9. Integration contour in the complex α-plane and the domain in which $I_m(k \sin \alpha) > 0$. 95

Figure 2.10. Contour C_φ. 96

Figure 2.11. Regions in the complex α-plane where $I_m(\cos \alpha) < 0$. 97

Figure 2.12. Graphical view of the contours C_φ^-, C_φ^+ and C^-, C^+. 98

Figure 2.13. Deformation of the contours C^+ and C^-. 100

Figure 2.14. Sommerfeld contour $\gamma = \Delta + \Delta' = L - L'$. 101

Figure 2.15. The double loop path γ and the SDP Γ and Γ'. 110

Figure 2.16. Location of the poles of $R(\alpha)$ in the complex α-plane for a passive surface. 113

Figure 2.17. Surface waves for a pure imaginary impedance. 117

Figure 2.18. Surface waves for $R_e Z^{\pm} > 0$. 118

Figure 2.19. Oblique incidence on a straight wedge: (a) coordinate system and angle β_0 and (b) projections of $\mathbf{k}^i$ and $\mathbf{k}^r$ and angles φ_0 and φ. 123

Figure 2.20. Wedge-shaped shell and coordinate system. 125

Figure 2.21. Position of the pole and integration path. 131

Figure 2.22. Poles of the reflection coefficient (a) in the complex θ-plane and (b) in the complex α-plane. 135

Figure 2.23. Lines of constant amplitude and the limits of existence of the generalized Lamb wave. 136

Figure 2.24. Radiation diagram of a generalized Lamb wave. 137

Figure 2.25. Experimental setup: r_1 = 30 cm, r_{23} = 20 cm. 139

Figure 2.26. Generalized Lamb waves excited at the edge of a shell of brass shaped like a right-angle wedge, surrounded by water: Experimental results. 140

Figure 2.27. Generalized Lamb waves excited at the edge of a shell of brass shaped like a right angle, surrounded by water: Theoretical results. 141

Figure 2.28. Contour C for the Watson transformation. 143

Figure 2.29. Contour of integration D and the Stokes lines. 145

Figure 2.30. Regions of validity of the asymptotic approximations of (a) $H_\nu^{(1)}(kz)$ and (b) $H_\nu^{(2)}(kz)$, in the complex ν-plane. 146

Figure 2.31. Representation of the angles ϕ_1 and ϕ_2. 149

Figure 2.32. Representation of the dominant creeping rays. 151

Figure 2.33. Deformation of the contour C_O into a contour C_S passing through the saddle points ν_R and ν_D. 153

Figure 2.34. Graphical interpretation of the condition represented by equation (2.289). 154

Figure 2.35. Reflected ray and reflection angle θ_R. 160

Figure 2.36. Creeping rays reaching a point Q on the surface of a circular cylinder. 168

Figure 2.37. Trajectories of the zeros of equation (2.346) for (a) purely imaginary values of Z and (b) complex values $Z = |Z|e^{i\theta}$, when $|Z|$ varies from zero to infinity. 170

Figure 2.38. Geometrical parameters for oblique incidence. 172

Figure 2.39. Cylinder geometry. 179

Figure 2.40. Original integration contour for the Watson transformation. 183

Figure 2.41. Modified contour for the Watson transformation. 184

Figure 2.42. Final integration contours for the Watson transformation. 185

Figure 2.43. Ray geometry for the reflected field. 191

Figure 2.44. Geometrical interpretation of the creeping wave solution. 193

Figure 2.45. Geometrical representation of the elastic surface rays. 200

Figure 2.46. Schematic representation of an elastic surface wave shading a local inhomogeneous plane wave. 203

Figure 2.47. Geometrical representation of the rays corresponding to transverse and oblique incidence. 205

Figure 2.48. Line source within a circular cavity bounded by planar absorbers. 213

Figure 2.49. Path of integration in the complex μ-plane. 216

Figure 2.50. Path of integration and poles in the complex ν-plane. 216

Figure 2.51. Contour deformation and branch cut. 217

Figure 2.52. Contour deformation of C' into C'' and C'''. 218

Figure 2.53. Deformation of the integration path through the stationary points v_1 and v_2. 218

Figure 2.54. Integration path C_1. 223

Figure 2.55. Graphical interpretation of the saddle point condition for ϕ_1. 225

Figure 2.56. Graphical interpretation of the saddle point conditions corresponding to (a) ϕ_2 and (b) ϕ_3. 226

Figure 2.57. Graphical interpretation of the saddle point condition corresponding to ϕ_4. 226

Figure 2.58. Caustic of singly reflected rays. 227

Figure 2.59. Reflected ray caustics. 228

Figure 2.60. Path deformations in the complex v- and ξ-planes. 230

Figure 2.61. Source and observation point on a concave hard circular cylinder modulus of $G_h(\sigma, 0, 0)$ versus σ, $ka = 8$. 232

Figure 2.62. Source and observation point on a concave hard circular cylinder modulus of $G_h(\sigma, 0, 0)$ versus σ, $ka = 80$. 233

Figure 3.1. Intersection of (a) a line caustic boundary layer and (b) a surface caustic boundary layer with a shadow boundary layer. 240

Figure 3.2. Region of validity of the nonuniform solution for a wedge. 241

Figure 3.3. Geometry of the scattering from a wedge with plane wave incidence, where (a) is the coordinate system and (b) is the projection of incident and diffracted rays on a plane perpendicular to the edge. 243

Figure 3.4. Canonical transformation. 244

Figure 3.5. Branch cut in the complex X-plane uniquely defining Kouyoumjian's transition function for a complex argument. 247

Figure 3.6. Extension of the field reflected on a straight wedge outside its physical domain. 265

Figure 3.7. Critical situation for line source or point source illuminations of a wedge with planar faces. 267

Figure 3.8. Extension of the field reflected on a curved wedge outside its physical domain and the transition zone of the diffracted field, where (a) represents virtual caustic and (b) represents real caustic. 271

Figure 3.9. Wedge-shaped shell with convex faces. 294

Figure 3.10. Boundary layers in the diffraction by a smooth convex impenetrable surface. 297

Figure 3.11. Integration paths for the Airy functions. 300

Figure 3.12. Contour of integration C and SDP Γ. 303

Figure 3.13. Geometrical parameters in the shadowed part of the transition region. 305

Figure 3.14. Reflected and pseudo-ray paths. 309

Figure 3.15. Shadow boundaries on a circular hollow cylinder excited by a plane wave. 320

Figure 3.16. Geometrical parameters appearing in the formula for the field diffracted in the transition region by the creeping wave $Q_1Q_1'Q$. 322

Figure 3.17. Pseudo-ray path $QQ'P$. 323

Figure 3.18. Deformation of C_s into the SDP in the vicinity of ν_R and ν_D. 325

Figure 3.19. Elastic surface ray $Q_1Q_1'M_1$ and pseudo-elastic surface ray $Q_1Q_2'M_2$. 332

Figure 3.20. Positions of the transition regions: (1) creeping rays, (2) elastic surface rays, (3) GA rays and elastic surface rays. 336

Figure 3.21. Diffraction by a right-angle hard wedge. 339

Figure 3.22. Diffraction by a right-angle soft wedge. The dashed curve gives the total field. 340

Figure 3.23. Diffraction by a right-angle impedance wedge with a surface impedance $Z = 1/\sqrt{2}$. 341

Figure 3.24. Diffraction by a right-angle impedance wedge (surface impedance $Z = \sqrt{2}$). 342

Figure 3.25. Elastic right-angle wedge–shaped shell: Definition of the incident and observation angles. 342

Figure 3.26. Diffraction by an elastic right-angle wedge–shaped shell. The solid curve corresponds to the sum of the reflected and diffracted fields. 343

Figure 3.27. Diffraction of a plane wave by a circular cylindrical shell made of stainless steel, immerged in water. This is the nonuniform solution for the amplitude of the sum of the contributions of the creeping waves, elastic surface waves, and reflected waves. Each graduation corresponds to 10 dB: $a = 1$ cm, $ka = 8.48$, $b/a = 0.97$, and $r = 10\,a$. 344

Figure 3.28. Diffraction of a plane wave by a circular cylindrical shell made of stainless steel, immerged in water. This is the uniform solution for the amplitude of the sum of the contributions of the creeping waves, elastic surface waves, and reflected waves. Each graduation corresponds to 10 dB: $a = 1$ cm, $ka = 8.48$, $b/a = 0.97$, and $r = 10\,a$. 345

Figure 3.29. Diffraction of a plane wave by a circular cylindrical shell made of stainless steel, immerged in water. This is the uniform solution for the amplitude of the sum of the contributions of the creeping waves, elastic surface waves, and reflected waves. Each graduation corresponds to 10 dB: a = 1.5 cm, ka = 30, b/a = 0.97, and r = 10 a. 346

Figure 3.30. Diffraction of a plane wave by a circular cylindrical shell made of stainless steel, immerged in water. This is the uniform solution for the amplitude of the sum of the contributions of the creeping waves, elastic surface waves, and reflected waves. Each graduation corresponds to 10 dB: a = 1.5 cm, ka = 40, b/a = 0.97, and r = 10 a. 347

Figure 4.1. Virtual caustic inside a cylindrical body struck by parallel rays. 350

Figure 4.2. Caustic of rays emanating from a point source S and diffracted by a circular edge. 350

Figure 4.3. Real and virtual caustic of parallel rays reflected by a convex-concave rotational symmetric reflector. 351

Figure 4.4. Degenerated caustics: (a) and (b) illustrate an axial caustic and (c) is a point caustic at infinity. 352

Figure 4.5. At an observation point close to a real caustic, the field passes two rays tangent to the caustic. 353

Figure 4.6. Rays tangent to a circular caustic. 356

Figure 4.7. Airy integral paths; $l_m t^3 > 0$ in the shadowed regions. 357

Figure 4.8. Involute C_1 of C. 359

Figure 4.9. Paths followed by two creeping rays passing through a point near a real caustic. 360

Figure 4.10. Geometry of the rays near a cusp of a caustic. 367

Figure 5.1. Hybrid diffraction phenomena: (a) an edge-diffracted creeping wave and (b) an edge-excited creeping wave. 370

Figure 5.2. Geometry for the representation of the Fock field on a smooth convex surface. 371

Figure 5.3. Creeping wave diffracted into a space wave. 374

Figure 5.4. Pseudo-ray path. 377

Figure 5.5. Oblique incidence of a plane wave on a two-dimensional wedge. 379

Figure 5.6. Grazing incidence and observation on a wedge. 383

Figure 5.7. Schematic representation of the complex ray path and reflected wave front. 385

Figure 5.8. Contour C and steepest descent path (SDP) Γ in the complex ξ-plane. 386

Figure 5.9. Edge-excited surface wave at Q_0 diffracted into a space wave at Q_1. 388

Figure 5.10. Rectangular waveguide immerged in water. 390

Figure 5.11. Diffraction at the edge B of a Scholte-Stoneley wave excited at A. 391

Figure 5.12. Circular cylindrical curved wedge-shaped shell. 391

Figure 5.13. Wedge in a cylindrical concave shell. 393

Figure A.1. Reflection of a sound wave at a boundary. 403

Figure A.2. Reflection and refraction of a plane sound wave at the interface between two media. 405

Figure B.1. Incident, reflected, and refracted waves. 420

Figure B.2. Coordinate system and geometrical and acoustic parameters. 425

Figure C.1. Intrinsic representation of an inhomogeneous plane wave. 440

Figure C.2. The generalized Rayleigh wave as a set of three inhomogeneous plane waves. 441

Figure C.3. Geometrical representation of two inhomogeneous plane waves in the fluid corresponding to the same value of ξ. 442

Figure C.4. Pseudo-generalized Rayleigh wave. 443

TABLES

Table 2.1. Physical parameters 138

Table 2.2. θ_p in degrees 139

ABOUT THE AUTHOR

Frédéric Molinet was born in Wengelsbach, France, in 1934. He received an engineering degree from Ecole Centrale de Lyon in 1959, a doctorate (3e cycle) in theoretical nuclear physics from the University of Strasbourg in 1963, and a docteur ès siences from the University of Paris in 1971. From 1964 to 1971, he was involved in research in plasma physics at the Institut Henri Poincaré in Paris. In 1971 he joined the Laboratoire Central de Télécommunications, where he lead the Department of Theoretical Studies and Modeling. In 1980 he founded his own company, MOTHESIM (Modelisation, Optimisation, Théorie, Simulation Mathématique), where he continues as manager. The activities of MOTHESIM are focused on numerical and asymptotic solutions of electromagnetic and acoustic radiation and scattering problems. Since 1990 he has been teaching a course on high-frequency techniques at the Ecole Supérieure d'Electricité (Supélec) in Paris.

Dr. Molinet is the coauthor of the two books *Asymptotic Methods in Electromagnetics* (published by Springer in 1997) and *Asymptotic and Hybrid Methods in Electromagnetics* in the IEE Electromagnetic Waves Series (published by the Institution of Engineering and Technology [IET] in 2005). He is a former president of the Waves and Fields chapter of the Société des Electriciens, Electroniciens et des Radioélectriciens (SEE), a member of the Institute of Electrical and Electronics Engineers (IEEE), and a member of the International Union of Radio Science (URSI) Commissions B and E. From 2004 to 2007, he was the chairman of Commission B of the National Radio Science Committee of URSI. In 1990 he was the recipient of the Science and Defense Award of the French Ministry of Defense.

PREFACE

This book was written for researchers and engineers working with aerial and underwater acoustics. It examines the interactions of acoustic waves with obstacles that may be rigid, soft, elastic, or characterized by an impedance boundary condition. The approach is founded on asymptotic high-frequency diffraction methods based on the concept of rays. Despite the progress in the field of numerical methods for diffraction problems, ray methods remain the most useful approximate methods for analyzing wave motions. Ray methods provide considerable physical insight into diffraction mechanisms and allow for the analytic treatment of objects that are still too large in terms of wavelength to be solved in the realm of numerical methods.

Rays were originally defined in the field of optics as the path along which light travels. Later on, ray methods were developed for sound propagation in air and water. Subsequently they were extended to the electromagnetic field and more recently to elastic wave motion in elastic media.

In the ray theory, it is assumed that acoustic waves propagate along rays and that the interactions of such rays with obstacles follow simple geometrical rules. From the mathematical point of view, ray theory provides an expansion having asymptotic validity. It is an approximate solution that becomes more and more exact as the frequency becomes larger. The principal term of the ray expansion gives the geometrical optics (GO) field.

In fact, the ray method is a formal technique for constructing asymptotic expansions with respect to the inverse powers of the wave number, k, of solutions to the boundary value problems when k is large. This technique has been applied intensively in acoustics, in electromagnetic theory, and to a lesser extent, in elastic wave theory. In all these domains, the technique permits an analytic treatment of wave propagation in an unbounded media and its reflection at (or transmission through) a plane or curved smooth interface. However, construction of a high-frequency asymptotic solution by the ray method is possible only if the field of rays is regular. This condition, which will be mathematically described in chapter 1, is normally satisfied away from caustics and shadow boundaries.

A significant advance in ray theory was its extension to diffraction problems, known as the geometrical theory of diffraction (GTD). This development was mainly due to J. B. Keller, who introduced the concept of diffracted rays. The extension consists of a generalization of Fermat's principle to include any extremal path, satisfying constraints inferred by the environment. For example, these rays may change direction at a sharp edge (i.e., edge diffraction) or follow the surface of a smooth object (i.e., creeping rays).

A further extension of the ray theory, also due to J. B. Keller and his co-workers, concerns the excitation and propagation of scalar and elastic surface waves. A surface wave is a wave that travels along a surface near the location where the wave motion is confined. On boundaries of elastic solids, these waves are known as Rayleigh waves. The geometrical theory of surface waves, which is an extension of GO, involves complex rays that travel from the source to the surface, then along the surface, and finally from the surface to the point of observation, which may be in a solid or in a fluid surrounding it. As in the case of waves that travel in an unbounded medium (i.e., space waves), surface waves may also be diffracted by a sharp edge or by a higher-order singularity of the surface.

All these extensions of the ray theory may be written in the form of a short-wavelength asymptotic expansion of the exact diffracted or surface wave field. Keller's theory gives the leading term of these expansions. The excitation coefficients and diffraction coefficients that enter into the theory are determined from the solutions of canonical problems, which are problems that can be solved exactly, usually by the method of separation of variables. These serve as prototypes for the local behavior of the solution to a more general problem. The purpose of this book is to (1) state the basic ideas of the ray method and its generalization to diffraction problems, (2) discuss the necessary methods of applied mathematics used in asymptotic expansions, and (3) present a number of basic formulas that can be used as a toolbox for solving practical diffraction problems.

The book comprises five chapters: In chapter 1, we give an introduction to the GTD. We start with a brief discussion of the relationship between ray representations and the asymptotic terms of the solution of the wave equation. We examine how this duality can be used to give a unified presentation of the ray method and its extensions to diffraction problems. The chapter then presents the generalized Fermat principle and its extension to complex rays, followed by some fundamentals on asymptotic expansions. We proceed with a treatment of a series of wave interaction problems frequently encountered by engineers and researchers in practical projects. For each problem, we derive the solution for a general incident doubly curved free scalar wave interacting with a three-dimensional body or interface. When diffraction phenomena occur, the solution is expressed in terms of diffraction coefficients. In chapter 1, the reader can make use of these coefficients in a general diffraction problem without going through all the details of the solutions, which are provided in chapter 2. The specific values of the coefficients are established in chapter 2 from the asymptotic solutions of canonical problems.

As mentioned previously, in chapter 2 we give a synthesis of the present knowledge on the solutions of some canonical problems. Those problems include the diffraction of a planar scalar wave by a solid wedge or a wedge-shaped elastic shell and by a solid circular cylinder or a circular elastic shell at normal and oblique incidences. In addition to the exact high-frequency solutions, we also consider some heuristic high-frequency solutions that may be useful in practical applications. Special emphasis is devoted to the canonical problem of a concave circular cylindrical surface submitted to the sound emission of an acoustic line source, which plays an important role in the computation of the asymptotic surface displacement generated by edge-excited whispering gallery waves. Some experimental results are also provided and compared to high-frequency solutions for a straight elastic wedge-shaped shell and a circular elastic shell.

The formulas derived in chapters 1 and 2 are nonuniform near reflection and shadow boundaries. In chapter 3, we give them uniform validity. For this purpose, we introduce the uniform theory of diffraction (UTD) for wedge-shaped structures. It comprises two steps: The first is a uniform solution for the canonical problem. It involves some transition functions depending on geometrical parameters. In a second step, this uniform solution is taken over to the original, non-canonical problem, and the geometrical parameters are adapted in such a way that the total field is continuous through the reflection and shadow boundaries. The details concerning the uniform asymptotic evaluation of the exact or approximate solution of the canonical problem are given first. Then the uniform canonical solutions for a straight wedge and for a circular cylinder with different boundary conditions are adapted to a curved two-dimensional or three-dimensional wedge and to a general convex smooth three-dimensional surface. Uniform diffraction coefficients are derived that can be incorporated in the formulas established in chapter 1 for an incident doubly curved scalar wave.

The classical and uniform GTD presented in chapters 1 and 3 breaks down at caustics. In three dimensions, the wave front has two principal radii of curvature. Therefore, the caustic constituted by the locus of the centers of curvature of the wave front comprises two sheets. These sheets are joined together along a cusp line. In chapter 4, corrections are presented for the field near a smooth caustic, and some comments are given for solving the problem of a cusped caustic. We do not present corrections for singular points on a cusp line, which may also occur.

In chapter 5, we consider diffraction problems involving a combination of two types of diffraction phenomena, such as diffraction of a creeping wave or a surface wave by an edge discontinuity or creeping wave and surface wave launching at discontinuities in a curved surface. Such diffraction processes are described by hybrid diffraction coefficients that are useful for solving a number of diffraction problems important for applications, especially in situations where no reflections occur.

We have attempted to make the presentation self-contained by adding two brief appendices (A and B) on the governing equations of wave processes in ideal fluids and in elastic media. Four other appendixes (C through F) provide derivations and general formulas for selected concepts used in the chapters.

Chapter 1

INTRODUCTION TO THE GEOMETRICAL THEORY OF DIFFRACTION

1.1. GENERAL OVERVIEW AND BASIC CONCEPTS

Although the concept of rays has been known since antiquity, its relationship to the wave equation and especially to the Helmholtz equation $\Delta u + k^2 u = 0$ for harmonic waves was only outlined at the beginning of the 20th century by Sommerfeld and Runge [1.1]. The mathematical tool that allowed this breakthrough was the asymptotic expansion or, in short, "the asymptotics" of a function of a parameter α, for large values of this parameter, introduced two decades before by Poincaré [1.2]. Sommerfeld and Runge showed that a direct relationship exists between the ray representations of geometrical optics (GO) or geometrical acoustics (GA) and the asymptotics of the solution of the Helmholtz equation for k large, where k is the wave number. This result has provided the basis of a formal technique called the ray method, or ray theory, for constructing an asymptotic expansion of the solution of a scattering problem involving smooth objects, known as the Luneburg-Kline expansion [1.3]. Such an expansion has asymptotic validity. It is an approximate solution that becomes more and more accurate as the frequency becomes larger. In acoustics, the principal term of the ray expansion gives the GA solution.

A significant advance in ray theory was its extension to diffraction problems by J. B. Keller in the mid-1950s [1.4–1.7]. Keller's theory, which is known as the geometrical theory of diffraction (GTD), is founded on the following postulates [1.8]:

1. The diffracted field propagates along rays that are determined by a generalization of Fermat's principle to include points on the boundary surface in the ray trajectory.

2. Diffraction, like reflection and transmission, is a local phenomenon at high frequencies (i.e., it depends only on the nature of the boundary surface and the incident field in the immediate neighborhood of the point of diffraction).
3. The diffracted wave propagates along its ray so that power is conserved in a tube of rays, and the phase delay along the ray path equals the product of the wave number of the medium and the distance.

Diffracted rays are initiated at points on the boundary surface where the incident field is discontinuous (i.e., at points on the surface where there is a shadow or reflection boundary of the incident field like edges, vertices, or points at which the incident ray is tangent to a smooth convex surface). The amplitudes of the diffracted waves are derived from the asymptotic evaluation (as the wave number tends to infinity) of known, exact solutions to scattering from simple shapes, referred to as the canonical problems of GTD. A canonical problem is the simplest boundary value problem having the same local geometry as the original problem, near the point of diffraction, that can be solved exactly, usually by the method of separation of variables. Keller's GTD gives the principle term of the asymptotic expansion of the diffracted field.

Another way to understand GTD is to view it as an asymptotic $(k \to \infty)$ theory of the solution of the Helmholtz equation (i.e., as a mathematical discipline). In Russian literature [1.9], Keller's theory is known as the model or etalon problem method. It is based on the principle that similar ray geometry leads to similar asymptotic formulas (as $k \to \infty$) for wave fields. It is the simplest problem in which the field of rays has the same singularities as in the original problem (e.g., caustics, shadow boundaries). The simplest model problem, like the canonical problem, can be solved exactly, and an asymptotic expansion is obtained that describes the field in the region of interest. The analytical expression for the field, found in the investigation of the model problem, is taken over to the original problem. This procedure permits us to start the resolution of the original problem by stating an asymptotic expansion of the solution, usually called an *ansatz,* having the same analytical form and especially the same asymptotic sequence as that of the model problem. The coefficients of the asymptotic series are determined recursively by substituting this analytical expression into the Helmholtz equation and boundary conditions that correspond to the original problem. This strategy, which—with rigorous mathematical proofs—allows easy derivation of the specific GTD laws from the model problem solutions, will be adopted in our presentation. It also has the advantage of being easily extended to complex rays and to the construction of a geometrical theory for surface waves and their diffraction by singularities of the surface.

It is important to mention that the asymptotic character of the series obtained by the model problem method has only been proved in special cases, especially for two-dimensional problems. A rigorous mathematical proof (at the level of theorems) that the formulas derived by the model problem method are asymptotic expansions of exact solutions of the given boundary value problems does not exist in the general case. However, it has now been checked in a variety of theoretical and experimental circumstances and can be reckoned to be fully confirmed by the physics of short-wave propagation.

In summary, the construction of GTD solutions of a diffraction problem comprises three main steps:

1. Determine the ray paths by a generalization of Fermat's principle.
2. Represent the solution in the form of an asymptotic series—the analytical form of which is given by the solution of a canonical or model problem.
3. Calculate the field amplitudes.

We will develop each step and derive the general form of GA and GTD solutions of typical diffraction problems encountered in practical applications, including the diffraction of scalar and elastic surface waves.

1.2. THE FERMAT PRINCIPLE

The Fermat principle from GO can be easily transposed to GA. In optics, a lossless medium is defined by its refractive index $n(\mathbf{r}) = \frac{k(\mathbf{r})}{k_o}$, where k is the wave number and $\vec{r}$ is the position vector. The wave number k is related to the speed of light $\upsilon(\mathbf{r})$ in the medium by $k = \frac{\omega}{\upsilon}$, where $\omega = 2\pi f$ is the angular frequency (f being the frequency of the wave). The wave number $k_o = \frac{\omega}{c}$ corresponds to a wave at the same frequency propagating in a vacuum (c being the speed of light in a vacuum). In a lossless medium, n is independent of the frequency and completely characterizes the medium for the propagation of optical waves. In acoustics, the vocabulary "refractive index" is not used. However, for the statement of the Fermat principle in GA, it is convenient to introduce the same concept, but now $k_o = \frac{\omega}{\upsilon_o}$ is related to some fictitious homogeneous medium with constant sound speed υ_o. Keeping this definition in mind, let T be any curve joining the points A and B in a medium at rest with refractive index $n(\mathbf{r})$. Then the curvilinear integral

$$L(T) = \int_T n(\vec{r})\,ds \qquad (1.1)$$

is known as the optical (acoustical) path length. In general, the optical (acoustical) path length will depend not only on the end points A and B but also on the actual track T traced between them.

In 1654 the French mathematician Fermat postulated that regardless of the kind of reflection or refraction a ray encounters, it travels from one point to another in such a way that the time taken is minimum. By comparison with the equation of a particle in the theory of mechanics, Fermat's principle was later put in the following equivalent form by Hamilton between 1824 and 1844.

A ray between A and B is defined as a trajectory satisfying the Fermat principle, which selects it from all the curves from A to B as the one rendering the integral from A to B stationary with respect to infinitesimal variations of the path.

It is important to recognize that Fermat's principle does not require the optical (acoustical) path length to be a minimum. It is sufficient that this path length be stationary under local deformations of the path. This point is very important; otherwise, significant rays may be lost.

The generalized Fermat principle consists of a generalization of the concept of rays to include any extremal path satisfying constraints inferred from the environment. Such a generalization is well known for a reflected ray on a smooth surface, where the ray is treated as an extremal path constraint having a point on the smooth surface. It has also been applied to the refraction of a ray into a different media through a smooth interface, in which case the ray is subject to have a point on the interface.

In studying the diffraction by a wedge, Keller introduced diffracted rays that are defined by an extremum path that is subject to the constraint that it must include a point somewhere on the edge (fig. 1.1). He also extended the principle to include diffraction by smooth convex surfaces in the region not reached by direct or reflected sound rays. In this case, the ray is defined by an extremum path that is subject to the constraint that it must include an arc on the surface (fig. 1.2).

1.2.1. CONDITIONS FOR A PATH TO BE A RAY

Since our main objective is to treat the interaction of sound waves with obstacles that are large compared to the wavelength but small compared to the distance where the sound speed varies

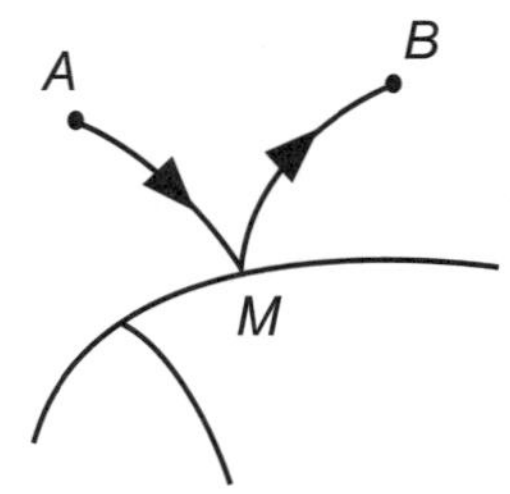

Figure 1.1. Constraints for edge diffraction.

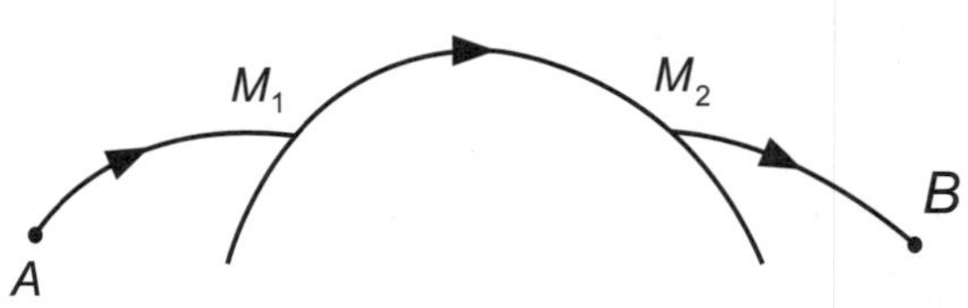

Figure 1.2. Constraints for the diffraction by a smooth surface at an observation point in the region not reached by direct or reflected sound rays.

significantly, we make the assumption throughout this book that the sound speed is constant in the fluid surrounding the obstacle. Under this assumption, a generalization of the Fermat principle including all types of interactions can be carried out in the following way: Let us consider a path T connecting the points M_o and M_{N+1}. Let T be composed of N regular segments T_i and the connection points be M_i, $i = 1, \ldots, N$, where either the direction or the curvature of T changes (fig. 1.3).

The segments T_i may either reside in the space outside the object or be located on the surface. The points M_i, in turn, are located either on regular parts of the surface, on edges, or on the tips. N represents the total number of interactions of the path T with the object.

Let us designate $\mathbf{t}_i$ as the unit vector of the tangent at M_i to the segment T_i, and let $\mathbf{t}'_i$ be the unit vector of the tangent at M_i to the segment T_{i-1}, where $\mathbf{t}_i$ is not equal to $\mathbf{t}'_i$ in general. Since the speed of the sound is constant, the acoustical path is simply the length of the path T multiplied by a constant, which for our purpose can be omitted so that we can write

$$L(T) = \int ds. \tag{1.2}$$

This is a functional expression defined on the entire set of paths, compatible with the connection on the surface that links the points M_o to M_{N+1} in such a way that the intermediate points M_i located on the surface, edge, or tip remain there, and the segments T_i on the object remain on the surface of the object.

According to Fermat's principle, T is a ray if and only if the length of T is stationary for all paths satisfying the connections on the surface. Applying the calculus of variations to express the infinitesimal variation $\delta(L(T))$ of the length of the path T when each point of it undergoes an incremental displacement δM compatible with the connections on the surface, we obtain, for an object located in the same media,

$$\delta(L(T)) = \int \delta(ds) = \int \mathbf{t} \cdot d(\delta \mathbf{M}), \tag{1.3}$$

since

$$\delta(ds) = \delta\left(\sqrt{(d\mathbf{M})^2}\right) = \frac{d\mathbf{M}}{ds} \cdot \delta(d\mathbf{M}) = \mathbf{t} \cdot \delta(d\mathbf{M}) = \mathbf{t} \cdot d(\delta \mathbf{M}),$$

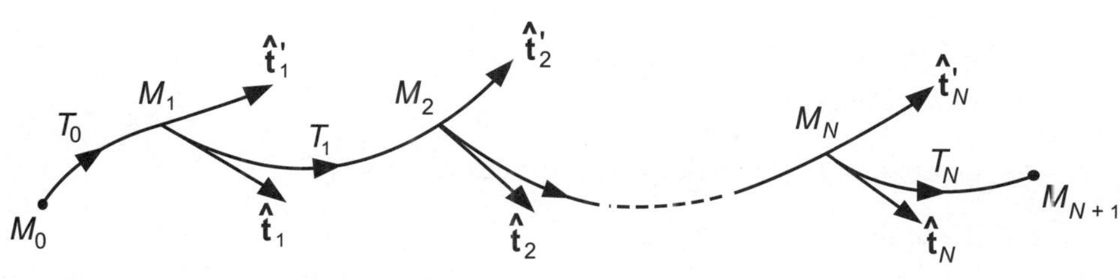

Figure 1.3. General path between the two extremities of a ray.

where $\mathbf{t} = \dfrac{d\mathbf{M}}{ds}$ is the unit vector of the tangent to T at point M and where we have used the fact that d and δ are two independent operators that can be commuted $\delta d = d\delta$.

Integrating by parts, (1.3) becomes

$$\delta\big(L(T)\big) = \sum_{i=1}^{N} \left(\mathbf{t}'_i - \mathbf{t}_i\right) \cdot \delta \mathbf{M}_i - \sum_{i=0}^{N} \int_{T_i} \delta \mathbf{M} \cdot d\mathbf{t}. \tag{1.4}$$

We can now define a ray as follows: T is a ray if $\delta(L(T)) = 0$, for all $\delta\mathbf{M}$ compatible with the constraints imposed on the ray segments. Applying this definition to (1.4) leads to the following two types of conditions:

1. $N + 1$ conditions characterizing the $N + 1$ elementary segments T_i,

$$\int_{T_i} \delta \mathbf{M} \cdot d\mathbf{t} = 0 \tag{1.5}$$

2. N conditions of transition associated with the diffraction points,

$$(\mathbf{t}'_i - \mathbf{t}_i) \cdot \delta\mathbf{M}_i = 0 \tag{1.6}$$

Until now we have supposed that the ray remains in the same medium. When the ray crosses at some connection point $\mathbf{M}_q$, the interface between two media with constant but different velocities υ_1 and υ_2, the condition (1.5) still applies. However, the condition (1.6) at $\mathbf{M}_q$ is modified and replaced by

$$\left(\upsilon_1^{-1}\mathbf{t}'_q - \upsilon_2^{-1}\mathbf{t}_q\right) \cdot \delta \mathbf{M} = 0. \tag{1.7}$$

At all other points $\mathbf{M}_i$ connecting two paths T_i and T_{i+1} lying in the same medium, condition (1.6) holds.

1.2.2. APPLICATION OF CONDITIONS (1.5) AND (1.6) TO SPECIFIC PROBLEMS

1.2.2.1. Segments in a Fluid with Constant Sound Velocity

In this case, $\delta\mathbf{M}$ is arbitrary and has three degrees of freedom. According to (1.5), $d\mathbf{t} = 0$; hence $\mathbf{t} = \mathbf{t}_i = \mathbf{t}'_i$ = constant, and as a result (1.6) is also satisfied. This leads us to the following important result: in a fluid with constant sound velocity, the rays are straight lines.

1.2.2.2. Reflection from a Smooth Surface

We consider a homogeneous fluid having a smooth interface with another medium (e.g., fluid, rigid or soft solid, or elastic solid). Since in a homogeneous fluid the velocity of the sound is constant, both the incoming and outgoing rays corresponding to a reflection point M are straight lines. Let us denote the unit vector along the incident ray as $\mathbf{t}'_i = \mathbf{i}$ and the corresponding unit vector along the reflected ray as $\mathbf{t}_i = \mathbf{r}$. Since the reflection point is on the surface S, $\delta\mathbf{M}$ must be a vector in the tangent plane P at the reflection point M. From (1.6), it appears that $\mathbf{i} - \mathbf{r}$ must be normal to S:

$$\mathbf{i} - \mathbf{r} = \lambda \mathbf{n}, \tag{1.8}$$

where λ is a scalar and $\mathbf{n}$ is the unit vector normal to S at M, pointing outward. Let us denote by θ_i, the incident angle (fig. 1.4). Thus

$$\mathbf{i} \cdot \mathbf{n} = -\cos\theta_i. \tag{1.9}$$

Writing (1.8) in the form

$$\mathbf{r} = \mathbf{i} - \lambda \mathbf{n}$$

and using the fact that $\mathbf{r}$ is a unit vector, we obtain

$$\lambda(\lambda + 2\cos\theta_i) = 0. \tag{1.10}$$

Hence for nongrazing incidence and observation ($\lambda \neq 0$), we have

$$\lambda = -2\cos\theta_i,$$

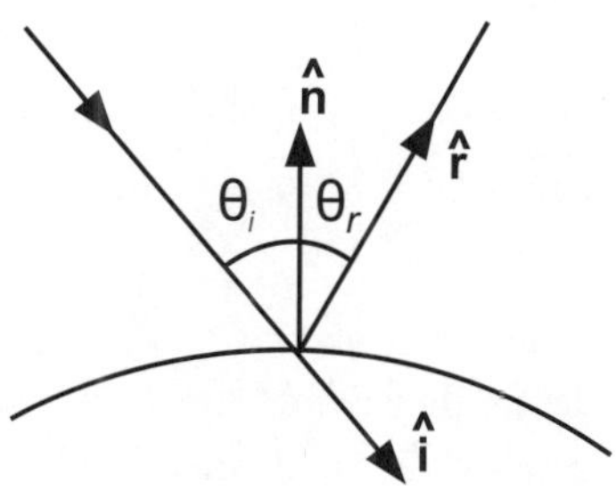

Figure 1.4. Reflection from a smooth surface.

and according to (1.8),

$$\mathbf{r} = \mathbf{i} + 2\cos\theta_i\,\mathbf{n}. \tag{1.11}$$

Equation (1.11) represents the law of reflection. It is often stated as follows: the reflected ray is in the plane of incidence, which is defined by the surface normal and the propagation vector of the incident ray, and the angle of reflection is equal to the angle of incidence. It can be shown that (1.11) is also valid at grazing incidence, since for $\theta_i = \frac{\pi}{2}$ it gives $\mathbf{r} = \mathbf{i}$ in accordance with the solution $\lambda = 0$ of (1.10).

1.2.2.3. Transmission through a Smooth Interface between Two Different Homogeneous Fluids

We consider a ray coming from fluid 1 with sound velocity υ_1 and penetrating fluid 2 with sound velocity υ_2. The ray penetrating fluid 2 is known as the refracted ray. An example is the interface between air and water.

Let us denote by $\mathbf{M}_q$ the refraction point and by $\mathbf{t}'_q = \mathbf{i}$ and $\mathbf{t}_q = \mathbf{t}$ the unit vectors along the incident ray in medium 1 and along the refracted ray in medium 2, respectively. Since the refraction point $\mathbf{M}_q$ is constrained to be on the interface S, it appears from (1.7) that

$$\upsilon_1^{-1}\mathbf{i} - \upsilon_2^{-1}\mathbf{t} = \lambda\mathbf{n}, \tag{1.12}$$

where λ is a scalar and $\mathbf{n}$ is the unit vector normal to S at $\mathbf{M}_q$, pointing toward medium 1.

In view of (1.12), vector $\mathbf{t}$ is in the plane of incidence defined by $\mathbf{i}$ and $\mathbf{n}$. Moreover, if we denote the incident angle as θ_i and the refracted angle as θ_t (fig. 1.5), then by multiplying both terms of (1.12) by the unit vector $\mathbf{b}$ tangent to the interface S and lying in the plane of incidence, we get

$$\upsilon_1^{-1}\sin\theta_i = \upsilon_2^{-1}\sin\theta_r. \tag{1.13}$$

Equation (1.13) is known as the Snell-Descartes law.

Equations (1.12) and (1.13) define the law of refraction, which can be stated as follows: the refracted ray is in the plane of incidence that is defined by the normal to the interface and the propagation vector of the incident ray, and the angle of refraction is related to the angle of incidence by the law (1.13).

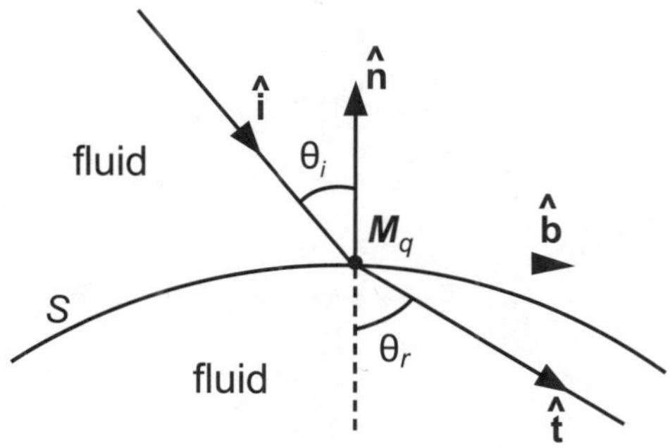

Figure 1.5. Transmission through a smooth interface between two fluids.

1.2.2.4. Excitation of Elastic Waves at a Smooth Interface between a Homogeneous Fluid and an Isotropic Homogeneous Elastic Body

We consider a ray coming from a fluid (medium 1) with sound velocity υ_1 exciting at a smooth interface S with an isotropic homogeneous elastic medium and two rays penetrating into the elastic medium (medium 2)—one corresponding to a longitudinal wave with velocity $\upsilon_{2\ell}$ and another corresponding to a transverse wave with velocity υ_{2t}.

The transverse waves are also called shear waves (see appendix B). Applying the condition (1.7), we need to consider two ray paths with a common connection point on the surface. Let us denote this connection point by $\mathbf{M}_q$ and the unit vector along the incident ray in medium 1 by $\mathbf{t}'_q = \mathbf{i}$. Likewise, let $\mathbf{t}_{q1} = \mathbf{l}$ and $\mathbf{t}_{q2} = \mathbf{t}$ be the unit vectors along the rays corresponding to the longitudinal wave and the transverse wave, respectively. Since $\mathbf{M}_q$ is constrained to be on the surface S, it follows from (1.7) that

$$\upsilon_1^{-1}\,\mathbf{i} - \upsilon_{2\ell}^{-1}\,\mathbf{l} = \lambda\,\mathbf{n}, \tag{1.14}$$

$$\upsilon_1^{-1}\mathbf{i} - \upsilon_{2t}^{-1}\,\mathbf{t} = \mu\,\mathbf{n}, \tag{1.15}$$

where λ and μ are scalars and $\mathbf{n}$ is the unit vector normal to S at $\mathbf{M}_q$, pointing toward medium 1.

In view of (1.14) and (1.15), $\boldsymbol{\ell}$ and $\mathbf{t}$ are in the plane of incidence defined by $\mathbf{i}$ and $\mathbf{n}$. Furthermore, if we denote the incident angle and the refraction angles of the longitudinal and transverse waves by θ_i, θ_ℓ, and θ_t, respectively, and if we multiply both terms of (1.14) and (1.15) by the unit vector $\mathbf{b}$, tangent to the interface S and lying in the plane of incidence, then we obtain the following expression of the Snell-Descartes law (fig. 1.6):

$$\upsilon_1^{-1}\sin\theta_i = \upsilon_{2\ell}^{-1}\sin\theta_\ell = \upsilon_{2t}^{-1}\sin\theta_t. \tag{1.16}$$

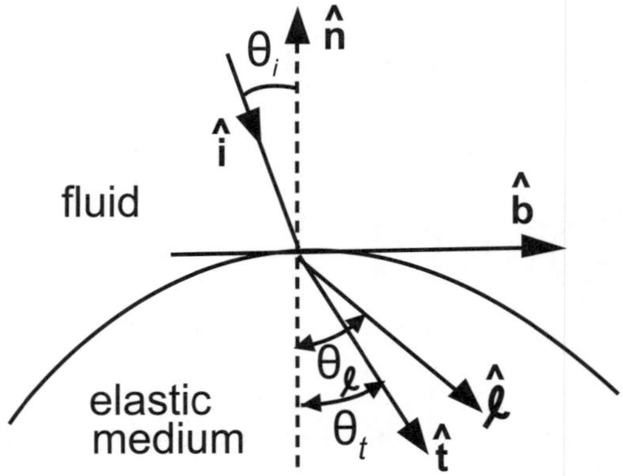

Figure 1.6. Excitation of elastic waves at a smooth interface between a fluid and an elastic solid.

We know from the theory of elastic waves that the velocity of longitudinal waves is always greater than that of transverse waves ($\upsilon_{2\ell} > \upsilon_{2t}$). Moreover, in most cases of interest, the sound velocity in the fluid υ_1 is less than that of the longitudinal waves in the solid $\upsilon_{2\ell}$. The sound velocity in the fluid can also be less than the velocity of the transverse waves υ_{2t}. Let us first consider the case where $\upsilon_{2t} < \upsilon_1 < \upsilon_{2\ell}$. From (1.16) we see that for $\sin\theta_i > \frac{\upsilon_1}{\upsilon_{2\ell}}$ the angle θ_ℓ is complex, while the angle θ_t is real for all θ_i. Thus the longitudinal wave in the solid will be an inhomogeneous wave that travels along the boundary and decays while departing from it.

Now consider the case $\upsilon_1 < \upsilon_{2t} < \upsilon_{2\ell}$. It follows from (1.16) that θ_ℓ and θ_t are real angles for $0 \le \sin\theta_i < \frac{\upsilon_1}{\upsilon_{2\ell}}$. At $\frac{\upsilon_1}{\upsilon_{2\ell}} < \sin\theta_i < \frac{\upsilon_1}{\upsilon_{2t}}$, the angle θ_t will be real, whereas θ_ℓ will be complex. For $\sin\theta_i > \frac{\upsilon_1}{\upsilon_{2t}}$, angles θ_ℓ and θ_t are both complex. This means that both longitudinal and transverse waves in the solid are inhomogeneous and propagate along the boundary. The angles $\theta_i = \arcsin\left(\frac{\upsilon_1}{\upsilon_{2\ell}}\right)$ and $\theta_i = \arcsin\left(\frac{\upsilon_1}{\upsilon_{2t}}\right)$ are critical to longitudinal and transverse waves, respectively.

1.2.2.5. Diffraction by an Edge in a Homogeneous Fluid

We consider a homogeneous fluid surrounding a wedge-shaped interface with a rigid or soft solid or an elastic solid and an incident ray diffracted in the fluid by the edge of the wedge. The sound velocity being constant, both the incoming and outgoing rays corresponding to a diffraction point M on the edge are straight lines. Let us denote by $\mathbf{t}'_i = \mathbf{i}$ the unit vector along the incident ray and by $\mathbf{t}_i = \mathbf{d}$ the corresponding unit vector along the diffracted ray. Since the diffraction point M is on the edge, $\delta\mathbf{M}$ is along the tangent $\mathbf{s}$ to the diffracting edge. Then from (1.6), we have

$$(\mathbf{i} - \mathbf{d}) \cdot \mathbf{s} = 0. \tag{1.17}$$

Equation (1.17) defines a cone of diffracted rays, having the tangent to the edge as its axis. The semiangle of this cone β is the angle between $\mathbf{i}$ and $\mathbf{s}$. The cone of diffraction is called Keller's cone (fig. 1.7).

Equation (1.17) states that all diffracted rays originating from a diffraction point on the edge must reside on the Keller cone. This is the law of diffraction from an edge. This law is valid for any line discontinuity. It applies also to a curvature discontinuity or a discontinuity of higher-order derivatives of the surface.

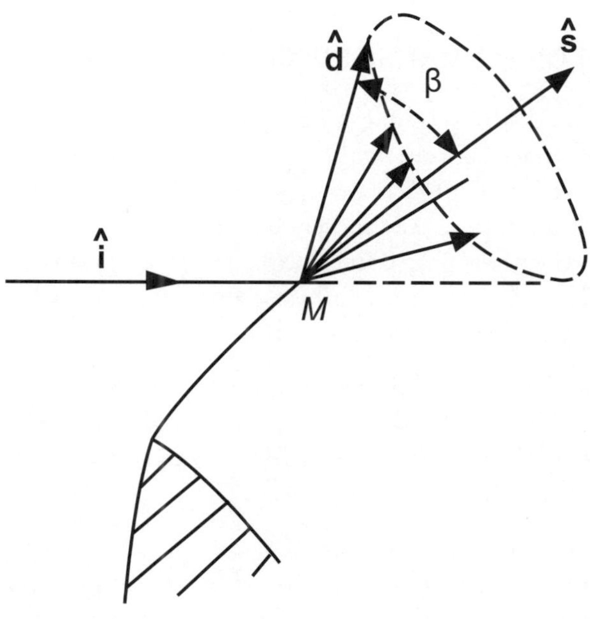

Figure 1.7. Keller's cone.

1.2.2.6. Surface Rays

If the elementary segment $\mathbf{T}_i$ of a ray lies on a surface, then $\delta\mathbf{M}$ lies in the plane tangential to the surface, and consequently $\mathbf{n} \cdot \delta\mathbf{M} = 0$, where $\mathbf{n}$ is the normal to the surface at M. However, since in the tangent plane to the surface at M, $\delta\mathbf{M}$ takes an arbitrary value, the condition (1.5) can only be satisfied if $\delta\mathbf{M} \cdot d\mathbf{t} = 0$. Hence

$$\frac{d\mathbf{t}}{ds} = \lambda\,\mathbf{n}. \tag{1.18}$$

Since according to Frenet's formula,

$$\frac{d\mathbf{t}}{ds} = \frac{\mathbf{h}}{R(s)}, \tag{1.19}$$

where $\mathbf{h}$ is the principal normal to the surface ray, we see that $\mathbf{h}$ and $\mathbf{n}$ coincide. This property is characteristic of the geodesics on a surface. It leads to the following rule governing the propagation of surface rays in an inhomogeneous fluid, along a smooth convex, or concave boundary with a solid: the surface rays follow the geodesics of the surface. In the case of a convex surface, the surface rays are also referred to as creeping rays. This denomination was introduced by Franz [1.10] for electromagnetic surface waves. Creeping waves are sometimes also called Franz waves in scientific literature.

1.3. EXTENSION OF FERMAT'S PRINCIPLE TO SURFACE WAVES

Surface waves are waves that travel along surfaces. The surface may be the free boundary of a solid or a fluid, the interface between two solids, or the interface between a fluid and a solid. We distinguish between "true" surface waves, which are waves with their fields confined in the neighborhoods of the surface, and loosely bound or leaky waves, which are waves that lose energy by radiation as they propagate. The surface may be planar or curved.

On a planar surface, different types of acoustic surface waves have been found. Among the most popular ones are the following (see appendix C):

1. The Rayleigh wave traveling at the free boundary of an elastic solid
2. The Stoneley wave traveling at the interface between two elastic solids
3. The Scholte-Stoneley wave and the generalized Rayleigh wave, both traveling at the interface between a fluid and an elastic solid

These waves are all true surface waves, except the generalized Rayleigh wave, which is a loosely bound wave. The true surface waves are not attenuated along the surface, but their amplitude decreases exponentially in the direction perpendicular to the surface.

On a curved surface, similar surface waves have been found. They are classified in different groups that take the same names as their limits when the local radius of curvature tends to infinity (i.e., the planar limit). On the free boundary of an elastic body, a generalized Rayleigh wave has been found that tends to the classical Rayleigh wave at the planar limit. Similarly, at the curved interface between an elastic body and a fluid, a generalized Scholte-Stoneley wave and a generalized Rayleigh wave have been found that behave at the planar limit like those mentioned previously on a planar surface. Another type of surface waves, which have no counterpart on a planar surface, are the creeping waves (also called Franz waves in scientific literature). These waves also exist on the surface of a rigid or soft body. At the planar limit, they tend to a homogeneous plane wave traveling in a direction parallel to the surface.

We now restrict our considerations to the interface between a fluid and an elastic or rigid body, which is of primary importance in diffraction theory. Since our theory is geometrical, it should be valid when the wavelength of the sound is small compared to all other dimensions.

1.3.1. PLANAR INTERFACE

Let us first consider a planar interface. We will restrict our considerations to the two-dimensional case with a line source parallel to the interface and perpendicular to the x-axis. The line source and the observation point P are both located in the fluid. Without restrictions (due to the generality of the problem), we suppose that P lies in the xy plane, with the y-axis perpendicular to the interface. Let Q be the interaction of the line source with the xy plane (fig. 1.8). By means of Fermat's principle, we may define a ray joining the two points Q and P in the fluid and having a path on the interface S. The ray is composed of three segments: a space ray from Q to some point Q_1 on S, a surface ray from Q_1 to another point P_1 on S, and a space ray from P_1 to P. We suppose that both media are homogeneous. Then according to the conditions (1.5), all segments of rays are straight lines. We denote by θ the acute angle of the incident ray QQ_1, with the normal to the interface and denote by θ_1 the acute angle of the diffracted ray P_1P with the same normal (fig. 1.8).

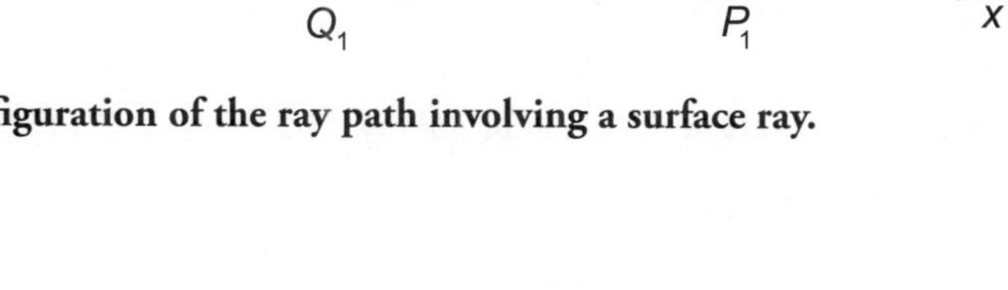

Figure 1.8. Configuration of the ray path involving a surface ray.

Applying the conditions (1.6) in the form of (1.7), since the velocity C_1 of the space waves and the velocity C_S of the surface wave are different, we see that the three segments QQ_1, Q_1P_1, and P_1P are coplanar with the normal and that

$$C_1^{-1} \sin \theta = C_1^{-1} \sin \theta_1 = C_S^{-1}. \tag{1.20}$$

According to (1.20), we have $\theta = \theta_1$. However, if the surface wave is a Scholte-Stoneley wave, we know that $C_S < C_1$ (see appendix C). Hence equation (1.20) cannot be satisfied with a real angle θ. A similar situation is encountered when the surface wave is a generalized Rayleigh wave. In that case, the surface wave is attenuated when it propagates along the surface—which is a consequence of the fact that its wave number $k_S = \omega / C_S$ is complex—and again, (1.20) can only be satisfied with a complex angle θ.

In order to apply the ray theory to the excitation and diffraction of surface rays, two methods are possible. The first one is to extend Fermat's principle to complex ray paths. The second one is founded on a geometrical interpretation of the real part of the phase.

Fermat's principle can easily be extended to complex rays in the case of a planar interface between homogeneous media. In the preceding example, let Q and P be two real points in the fluid with a complex ray joining them and a complex path on the interface. It follows from Fermat's principle extended to complex rays that such a ray consists of a complex space ray from Q to some complex point Q_1 on the complex extension of the interface S, a complex surface ray from Q_1 to another complex point P_1 on S, and a complex space ray from P_1 to the real observation point P. Applying conditions (1.5) and (1.7) extended to complex rays, we see that all complex ray segments are complex straight lines and that at both Q_1 and P_1 the unit tangents $\mathbf{i}$, $\mathbf{r}$, and $\mathbf{t}_S$ to the space and surface rays, respectively (fig. 1.9), must satisfy the following condition:

$$C_1^{-1}\,\mathbf{i} - C_S^{-1}\,\mathbf{t}_S = C_S^{-1}\,\mathbf{t}_S - C_1^{-1}\,\mathbf{r} = \lambda\,\mathbf{n}, \tag{1.21}$$

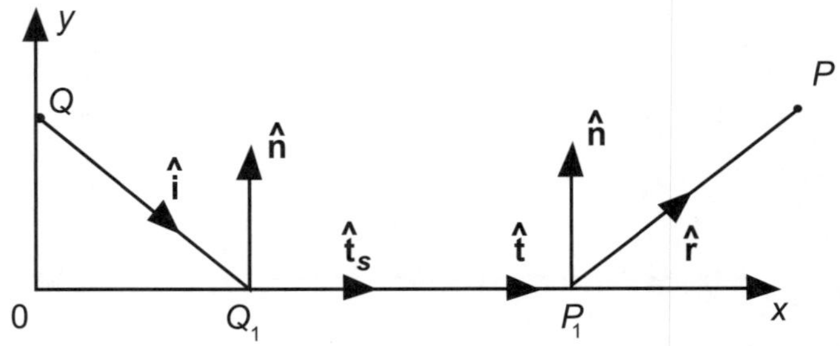

Figure 1.9. Schematic representation of the complex ray path joining Q to P.

where **n** denotes the unit vector normal to S at Q_1 and P_1 and where λ is a complex number.

Equation (1.21) is equivalent to the following geometrical conditions: (1) that **i** (respectively **r**) and $\mathbf{t}_S$ be coplanar with **n** and lie on opposite sides of **n** and (2) that the angle θ between (–**i**) and **n** (between **r** and **n**, respectively) satisfy the Snell-Descartes law,

$$C_S \sin \theta = C_1, \tag{1.22}$$

where C_S and θ are complex. If we suppose that the source Q lies on the y-axis, then its coordinates are $(0, Y_o)$. The complex phase variation along the complex ray QQ_1P_1P is given by

$$\phi = \frac{\omega}{C_1} d\left(Q, Q_1\right) + \frac{\omega}{C_S} d\left(Q_1, P_1\right) + \frac{\omega}{C_1} d\left(P_1, P\right), \tag{1.23}$$

where $d(\mathbf{x}, \mathbf{x}') = \left[\left(\mathbf{x} - \mathbf{x}'\right)^2\right]^{1/2}$ is the distance between two points, which in general is complex. If we let (x, y) be the coordinates of P, then

$$d\left(Q, Q_1\right) = \frac{Y_o}{\cos\theta},\ d\left(P_1, P\right) = \frac{Y}{\cos\theta},\ d\left(Q_1, P_1\right) = x - \left(Y + Y_o\right) tg\,\theta, \tag{1.24}$$

and by using (1.22), we obtain

$$\phi = \frac{\omega}{C_1}\left[x \sin\theta + \left(Y_o + Y\right)\cos\theta\right]. \tag{1.25}$$

Substituting $\theta = \theta' + i\theta''$ in (1.25) leads to

$$\phi = \frac{\omega}{C_1} ch\theta'' \left[x \sin\theta' + \left(Y + Y_o\right)\cos\theta'\right] + \frac{i\omega}{C_1} sh\theta''\left[x\cos\theta' - \left(Y_o + Y\right)\sin\theta'\right]. \tag{1.26}$$

The second term on the right-hand side of (1.26) shows that the wave is exponentially attenuated.

A second method for applying the ray theory to this problem is to give a geometrical interpretation of the real part of ϕ by using a real ray path. Consider a real ray starting at Q and making an angle θ' with the normal to the interface. This ray intercepts the x-axis at the real point Q_1' of coordinates $(Y_o tg\theta', 0)$. Then suppose that this ray follows the interface from Q_1' to a real point P_1' having the coordinates $(x - Y_o tg\theta', 0)$ and, finally, follows a straight line from P_1' to the observation point P (fig. 1.10). Moreover, suppose that the wave propagating along the surface has a propagation constant equal to the real part of k_S and that the space waves from Q to Q_1 and from P_1 to P propagate with the wave number $K' = k_1 ch\theta''$. It is, then, easy to verify by comparison with (1.26) that

$$R_e\phi = K' Q\, Q_1' + R_e\left(k_s\right) Q_1' P_1' + K' P_1' P, \tag{1.27}$$

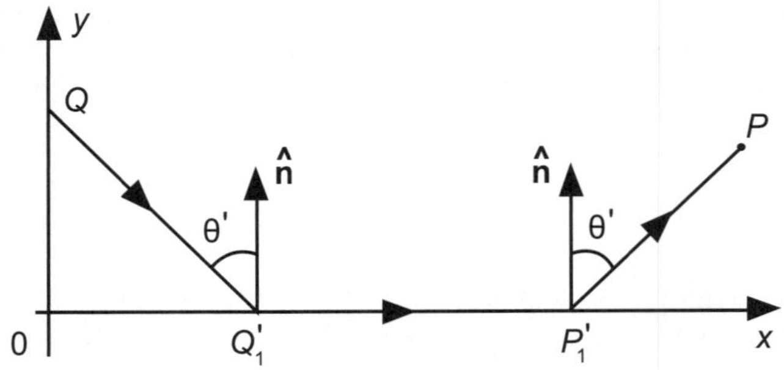

Figure 1.10. Real ray joining Q to P and having a path on the interface.

where, according to equation (1.22),

$$R_e(k_s) = k_1 \sin\theta' \, ch\theta''. \tag{1.28}$$

It is therefore possible to calculate the phase of the field at P by following the real ray $Q\,Q'_1PP'_1$, which is uniquely defined by the characteristic angle θ'. This angle can be determined if we know the solution k_S of the characteristic equation defining the surface wave. If we assume that k_S is known (for its determination in special cases, see appendix C), then θ' is given by the solution of

$$\sin\theta' \, ch\theta'' = R_e(k_s / k_1), \quad \cos\theta' \, sh\theta'' = I_m(k_s / k_1), \tag{1.29}$$

which is given by

$$\theta' = \text{Arc}\cos\left(\frac{1}{\sqrt{2}}\left[\sqrt{(\xi^2-1)^2+4\xi_I^2}-(\xi^2-1)\right]^{1/2}\right), \tag{1.30}$$

where $R_e(k_s / k_1) = \xi_R, \text{Im}(k_s / k_1) = \xi_I, \xi^2 = \xi_R^2 + \xi_I^2$.

By writing equation (1.27) in the form

$$R_e\phi = \mathbf{K}'_i \cdot \mathbf{Q}\,\mathbf{Q}'_1 + \mathbf{K}'_s \cdot \mathbf{Q}'_1\mathbf{P}'_1 + \mathbf{K}'_r \cdot \mathbf{P}'_1\mathbf{P}, \tag{1.31}$$

we define three wave vectors: $\mathbf{K}'_i$, $\mathbf{K}'_r$, and $\mathbf{K}'_S$, directed along $\mathbf{QQ}'_1$, $\mathbf{Q}'_1\mathbf{P}'_1$, and $\mathbf{P}'_1\mathbf{P}$, respectively, with modulus $|\mathbf{K}'_i| = |\mathbf{K}'_r| = K' = k_1 ch\theta''$ and $|\mathbf{K}'_s| = K'\sin\theta'$.

The ray $\mathbf{QQ}_1'$ does not correspond to a local homogeneous plane wave, since such a wave has a real wave number along the x-axis and consequently does not satisfy the Snell-Descartes law (1.22) at $\mathbf{Q}_1'$. We will show that it is an inhomogeneous local plane wave defined by a complex wave number, $\mathbf{K}$, given by

$$\mathbf{K} = \mathbf{K}' + i\mathbf{K}'', \qquad (1.32)$$

with (see equations (C.30) to (C.35) in appendix C)

$$\mathbf{K}' \cdot \mathbf{K}'' = 0, \qquad (1.33)$$

$$K'^2 - K''^2 = k_1^2. \qquad (1.34)$$

According to (1.22), we have for the incident wave $\mathbf{K}_i' \cdot \mathbf{x} = k_s = k_1 \sin\theta$. Hence

$$\mathbf{K}_i' \cdot \mathbf{x} = k_1 \sin\theta' \, ch\theta'' = K_i' \sin\theta', \qquad (1.35)$$

$$\mathbf{K}_i'' \cdot \mathbf{x} = k_1 \cos\theta' \, sh\theta'' = K_i'' \cos\theta', \qquad (1.36)$$

where

$$K_i' = k_1 \, ch\theta'', \quad K_i'' = k_1 \, sh\theta''. \qquad (1.37)$$

It follows from (1.35) and (1.36) that $\mathbf{K}_i'$ is therefore directed along QQ_1', whereas $\mathbf{K}_i''$ is orthogonal to QQ_1'. The conditions (1.33) and (1.34) are therefore satisfied. Hence the incident wave is an inhomogeneous local plane wave propagating along QQ_1' with its phase front (surface of constant phase) orthogonal to QQ_1', decaying exponentially in the direction of the vector $\mathbf{K}_i''$, and oriented according to (1.36) such that its projection on the x-axis is positive. Similarly, the radiated wave is also an inhomogeneous local plane wave propagating along $P_1'P$ and decaying exponentially in the direction of the vector $\mathbf{K}_r''$, orthogonal to $P_1'P$, and oriented such that its projection on the x-axis is positive (fig. 1.11).

Since $\mathbf{K}_i'' \cdot \mathbf{Q}\,\mathbf{Q}_1' = \mathbf{K}_r'' \cdot \mathbf{P}_1'\mathbf{P} = 0$, the phase variation of the incident and radiated inhomogeneous plane waves along QQ_1' and $P_1'P$, respectively, is real. The imaginary part of the phase at P is therefore

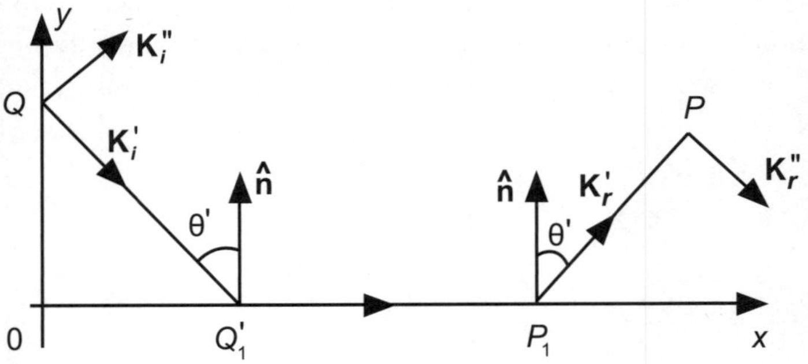

Figure 1.11. Incident and radiated inhomogeneous local plane waves.

$$I_m\phi = I_m(k_s)\ Q_1'P_1' = k_1\ sh\theta''\left[x\cos\theta' - (Y + Y_o)\sin\theta'\right], \tag{1.38}$$

which is in agreement with (1.26).

Applying (1.22) to the radiated wave, we obtain

$$\mathbf{K}_r' \cdot \mathbf{x} = k_1 \sin\theta'\ ch\theta'' = K_r' \sin\theta', \tag{1.39}$$

$$\mathbf{K}_r'' \cdot \mathbf{x} = k_1 \cos\theta'\ sh\theta'' = K_r'' \cos\theta', \tag{1.40}$$

where,

$$K_r' = k_1\ ch\theta'', \quad K_r'' = k_1\ sh\theta''.$$

Hence

$$K_i' = K_r' = K' = k_1\ ch\theta'', \tag{1.41}$$

$$K_i'' = K_r'' = K'' = k_1\ sh\theta'', \tag{1.42}$$

and

$$R_e(k_s) = K'\sin\theta', \tag{1.43}$$

$$I_m(k_s) = K''\cos\theta'. \tag{1.44}$$

Equations (1.35), (1.39), and (1.43) show that the Snell-Descartes law apply at the points Q_1' and P_1' to the real ray joining Q to P. However, the wave number of the space rays QQ_1' and $P_1'P$ is no longer k_1 but K', and the wave number of the surface ray $Q_1'P_1'$ is equal to $R_e(k_S)$.

According to (1.43) and (1.44), the surface wave can also be represented by an inhomogeneous plane wave with a phase front propagating away from the surface in the direction θ' and an amplitude decaying exponentially in the direction orthogonal to the propagation of the wave front. The surface ray shed, therefore, continuously rays in the direction θ'. One of these rays passes through the observation point P (fig. 1.12).

The geometrical theory presented thus far applies directly to loosely bound or leaky waves. For a Scholte-Stoneley wave, the wave number $k_S = \omega / C_S$ is real. Since

$$k_s = k_1 \sin\theta = k_1 \left(\sin\theta' ch\theta'' + i\cos\theta' sh\theta''\right), \tag{1.45}$$

we have, therefore, $\theta' = \dfrac{\pi}{2}$ and

$$k_s = k_1\, ch\theta'' = K'. \tag{1.46}$$

Hence, in order to excite this surface wave, Q must lie on the interface. Since its amplitude decreases exponentially in the Y direction, we have, according to (1.34),

$$K'' = \sqrt{k_s^2 - k_1^2} = k_1\, sh\theta''. \tag{1.47}$$

The sound field corresponding to a Scholte-Stoneley wave excited by a point source can therefore be represented locally by an inhomogeneous plane wave propagating along the surface. (For additional information on the Scholte-Stoneley surface wave, see appendix C.)

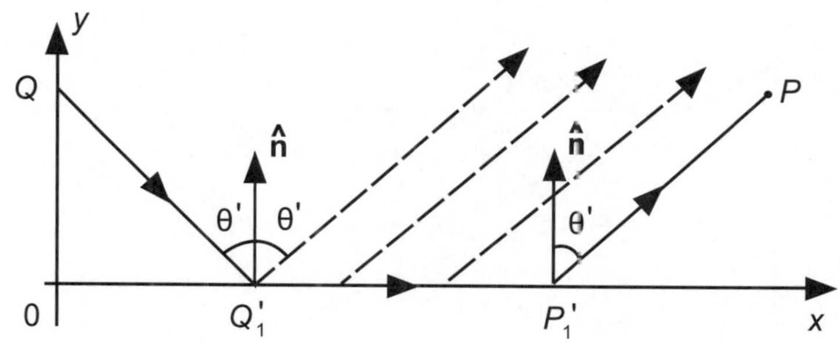

Figure 1.12. Excitation of a surface wave at the characteristic angle θ' and continuous shading of rays.

1.3.2. CURVED INTERFACE

We consider now a curved interface S separating a fluid from a convex elastic body. A source Q and an observation point P are both located in the fluid. By means of the Fermat principle, we may define a ray joining the points P and Q in the fluid and having an arc on the interface S. Such a ray consists of a complex space ray from Q to some point Q_1 on S, a complex surface ray from Q_1 to another point Q_2 on S, and a complex space ray from Q_2 to P. According to (1.7) extended to complex rays, we see that QQ_1 and Q_2P are straight lines and that at both points Q_1 and Q_2, the unit tangents to the surface ray must satisfy the conditions

$$C_1^{-1}\hat{i} - C_s^{-1}(Q_1)\,\hat{t}_1 = \lambda\,\mathbf{n}_1, \tag{1.48}$$

$$C_s^{-1}(Q_2)\,\hat{t}_2 - C_1^{-1}\hat{r} = \lambda'\,\mathbf{n}_2, \tag{1.49}$$

where $\mathbf{t}_1$ and $\mathbf{t}_2$ are unit vectors tangent to the surface ray at Q_1 and Q_2, respectively; $\mathbf{n}_1$ and $\mathbf{n}_2$ are unit vectors normal to S at Q_1 and Q_2, oriented toward the fluid; and λ and λ' are complex numbers. All unit tangents $\mathbf{i}$, $\mathbf{r}$, $\mathbf{t}_1$, and $\mathbf{t}_2$ are oriented in the direction of increasing arc length along the ray. Equations (1.48) and (1.49) are equivalent to the geometrical conditions that $\mathbf{i}$ and $\mathbf{t}_1$ ($\mathbf{r}$ and $\mathbf{t}_2$, respectively) be coplanar with $\mathbf{n}_1$ (respectively $\mathbf{n}_2$) and lie on opposite sides of $\mathbf{n}_1$ (respectively $\mathbf{n}_2$; fig. 1.13). Moreover, the angle θ_1 (respectively θ_2) between $-\mathbf{i}$ and $\mathbf{n}_1$ ($\mathbf{r}$ and $\mathbf{n}_2$, respectively) satisfies the the Snell-Descartes law:

$$C_s(Q_1)\,\sin\theta_1 = C_1\,, \quad C_s(Q_2)\,\sin\theta_2 = C_1. \tag{1.50}$$

It is important to note that figure 1.13 is only a schematic representation and not a geometrical representation of a complex ray path. We now apply the condition (1.5) to complex rays. Since

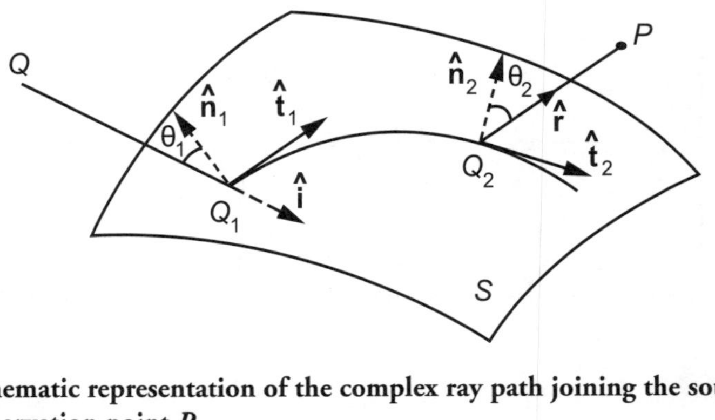

Figure 1.13. Schematic representation of the complex ray path joining the source Q to the observation point P.

the arc Q_1Q_2 lies on S, $\delta\mathbf{M}$ lies in the plane tangential to S, and consequently $\mathbf{n} \cdot \delta\mathbf{M} = 0$, where $\mathbf{n}$ is the unit vector normal to S at an arbitrary point M on the arc Q_1Q_2. Moreover, since $\delta\mathbf{M}$ takes an arbitrary value in the tangent plane to the surface S at M, the condition (1.5) can only be satisfied if $\delta\mathbf{M} \cdot \mathbf{dt} = 0$. Hence $\mathbf{dt}$ is normal to S, and we can write

$$\frac{\mathbf{dt}}{ds} = \lambda\,\mathbf{n}. \tag{1.51}$$

According to Frenet's formula,

$$\frac{\mathbf{dt}}{ds} = \frac{\mathbf{h}}{R(s)}, \tag{1.52}$$

where $\mathbf{h}$ is the principal normal to the surface ray. Equations (1.51) and (1.52) show that $\mathbf{h}$ and $\mathbf{n}$ coincide. This property is characteristic of the geodesics on a surface.

As in the case of a planar interface, it is possible to calculate the complex phase ϕ associated to the ray QQ_1Q_2P and separate the real part corresponding to the acoustical path length from the imaginary part corresponding to the exponential decay of the amplitude of the wave. Although in the case of a two-dimensional problem this can be done for a cylindrical interface whose cross section is a circle, parabola, or arbitrary quadric [1.11–1.12], it becomes difficult to apply this method to curves of a higher-order degree or to a general three-dimensional problem with a path Q_1Q_2 on the interface S represented by a torsional geodesic.

However, we will see in chapter 2 that the generalized Rayleigh waves propagating along the interface between a fluid and a convex elastic body are weakly attenuated so that we can write

$$R_e\, k_s\left(Q_1\right) = k_1\, R_e \sin\theta_1 \cong k_1 \sin\theta_1', \quad R_e k_s\left(Q_2\right) = k_1\, R_e \sin\theta_2 \cong k_1 \sin\theta_2', \tag{1.53}$$

$$I_m\, k_s\left(Q_1\right) = k_1 I_m \sin\theta_1 \cong k_1\, \theta_1'' \cos\theta_1', \quad I_m k_s\left(Q_2\right) = k_1\, I_m \sin\theta_2 \cong k_1\, \theta_2'' \cos\theta_2'. \tag{1.54}$$

In this case, the real part of the phase can be calculated by extending the second method presented for a planar interface to a curved interface. This consists of introducing a real ray path $QQ_1'Q_2'P$, where QQ_1' and $Q_2'P$ are straight lines and where the arc $Q_1'Q_2'$ is a real geodesic of the surface.

According to (1.53), the incident ray makes an angle θ_1' with the normal to the interface S at Q_1', and the radiated ray makes an angle θ_2' with the normal to S at Q_2'. Let $\mathbf{t}_1'$ (respectively $\mathbf{t}_2'$) be the unit vector tangent to the geodesic at Q_1' (respectively Q_2'), $\mathbf{n}_1'$ (respectively $\mathbf{n}_2'$) be the unit vector normal to S at Q_1' (respectively Q_2'), and $\mathbf{i}'$(respectively $\mathbf{r}'$) be the unit vector along the incident (respectively radiated) ray (fig. 1.14). Then according to (1.7), $\mathbf{t}_1'$ lies in the incident plane formed by $\mathbf{i}'$and $\mathbf{n}_1'$, and $\mathbf{t}_2'$ lies in the radiated plane formed by $\mathbf{r}'$and $\mathbf{n}_2'$.

According to (1.53), the phase variation along the real ray path $QQ_1'Q_2'P$ is equal to the principal order of $R_e\phi$ for I_mk_s small compared to R_ek_s, where ϕ is the exact phase variation along the complex ray path QQ_1Q_2P. Moreover, the surface ray continuously sheds energy

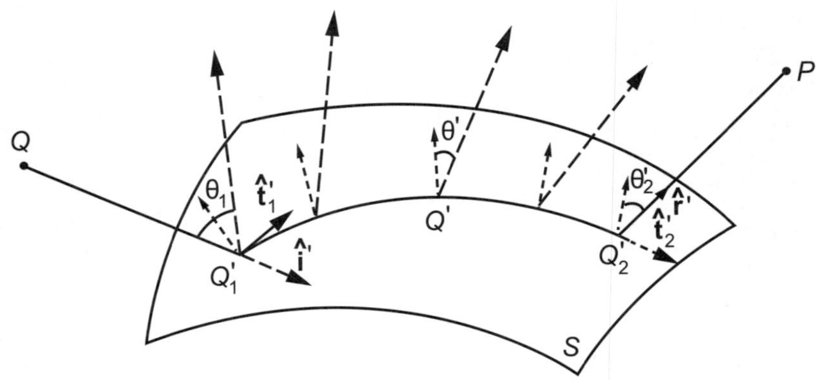

Figure 1.14. Geometrical representation of the real ray path joining the source Q to the observation point P and having an arc on the surface S. The generalized Rayleigh wave is excited on the convex surface S at the critical angle θ'_1 and radiates off the surface into the fluid under the local critical angle along the geodesic $Q'_1Q'_2$.

in the direction $\mathbf{r}(Q')$, defined by $R_e k_s(Q') \cong k_1 \sin\theta'$, where Q' is an arbitrary point along the geodesic $Q'_1Q'_2$ and where $\mathbf{r}(Q')$ is in the plane formed by the unit normal $\mathbf{n}(Q')$ and the unit vector $\mathbf{t}(Q')$ of the tangent to the geodesic at (Q'). In the special case of a circular cylindrical interface, when Q and P lie on a plane perpendicular to its axis, the imaginary part of ϕ that corresponds to the exponential attenuation of the wave can easily be calculated at first order by using (1.54).

It can then be extended to a more general situation by using a local cylindrical approximation of the surface. This technique will be used in chapters 2 and 3.

For a Scholte-Stoneley wave or for a creeping (i.e., Franz) wave, we will show in chapter 2 that in the principal order we have $\theta'_1 = \theta'_2 = \dfrac{\pi}{2}$. In this case, the incident and the radiated rays are both tangent to the geodesic at Q'_1 and Q'_2, respectively, and the surface ray continuously sheds energy when it propagates along the geodesic $Q'_1Q'_2$ in the direction tangent to this geodesic (fig. 1.15).

1.4. FUNDAMENTALS OF ASYMPTOTIC EXPANSIONS

The term *asymptotic expansions* was first introduced by Poincaré in 1892 [1.2]. However, the underlying concept was known much earlier by all those attempting to solve a problem by using a perturbation procedure. Applied to differential equations, this procedure generally leads to a

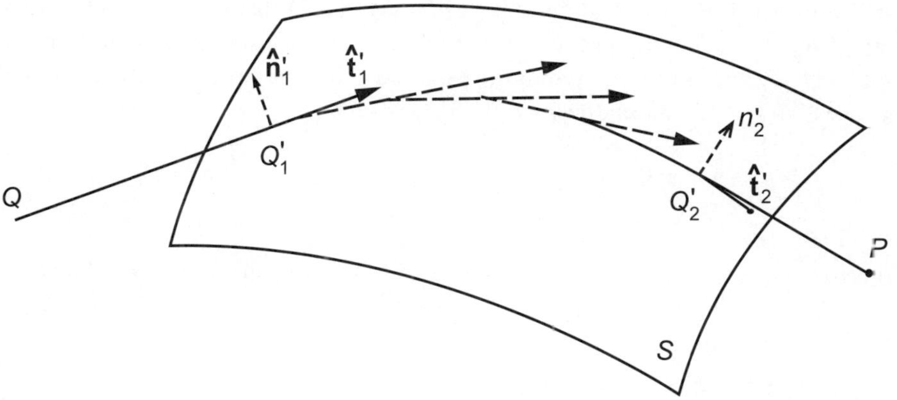

Figure 1.15. Geometrical representation of the real ray path joining the source *Q* to the observation *P* and having an arc on the surface *S*. This ray path corresponds to a Scholte-Stoneley wave or a creeping wave propagating on a convex surface *S*.

solution expressed in a series expansion in terms of integer powers of a small parameter ε satisfying the equation

$$f(\varepsilon) - \sum_{n=0}^{N} a_n \, \varepsilon^n = o\left(\varepsilon^N\right), \tag{1.55}$$

where $f(o)$ is the solution of the unperturbed problem and $o(\,)$ is the Landau symbol small "*o*," having the following definition: we say $f = o(g)$ as ε tends to zero if given any number $\eta > o$, as small as we want, there exists a neighborhood R_o of the origin such that $|f| \leq \eta |g|$ for all $\varepsilon \in R_o$.

Equation (1.55) means that if ε is sufficiently small, then the N-terms expansion represents an approximation of the solution that converges to the exact solution of the problem as ε tends toward o. A representation such as (1.55), which is limited to a finite number of terms, is called an asymptotic expansion.

1.4.1. ASYMPTOTIC SEQUENCE

Consider a sequence $\{v_n(\varepsilon)\}$, $n = 1, 2, \ldots$ of real and positive functions of ε defined and continuous in a neighborhood R_o of the origin. Such a sequence is called an asymptotic sequence if

$$v_{n+1}(\varepsilon) = o\left(v_n(\varepsilon)\right), \text{ as } \varepsilon \to 0. \tag{1.56}$$

If the sequence (1.56) is valid uniformly in n, which means that the choice of η and R_o in the definition of small "o" does not depend on n, then the sequence is said to be uniform in n. It is possible to extend the concept of asymptotic expansions by using the sequence $\nu_N(\varepsilon)$ instead of the sequence ε^n in (1.55). Indeed, owing to (1.56), we have

$$f(\varepsilon)-\sum_{n=0}^{N} a_n\, \nu_n(\varepsilon)=o(\nu_N(\varepsilon)), \text{ as } \eta\to 0. \tag{1.57}$$

Generally $f(\varepsilon)$ depends on other variables (e.g., x) belonging to a given domain D. The expansion

$$F(x,\varepsilon)=\sum_{n=0}^{N} a_n(x)\ \nu_n(\varepsilon) \tag{1.58}$$

is said to be uniform if

$$f(x,\varepsilon)-F(x,\varepsilon)=o(\nu_N(\varepsilon)), \text{ as } \varepsilon\to 0 \tag{1.59}$$

is uniformly valid in all the domain D. This implies that the error in f remains of the order of $\nu_N(\varepsilon)$ not only when ε tends to zero for a fixed x but also for all possible variations of x, provided that x continues to remain in the domain D.

It is possible to construct by recurrence all the terms of an asymptotic expansion. When N tends to infinity, we get an infinite series that we call an asymptotic series. An asymptotic series $F(\varepsilon)$ is therefore an expansion of a given function $f(\varepsilon)$ in an infinite series with respect to an asymptotic sequence $\nu_n(\varepsilon)$. It is a purely formal expansion that we write as

$$F(\varepsilon)=\sum_{n=0}^{\infty} a_n\, \nu_n(\varepsilon). \tag{1.60}$$

This series may be convergent or divergent. When it is divergent, we can still give a meaning to (1.60) by truncating the series after $N+1$ terms and write

$$f(\varepsilon)-F_N(\varepsilon)=o(\nu_N(\varepsilon)), \text{ as } \varepsilon\to 0,$$

where $F_N(\varepsilon)$ is an asymptotic expansion defined by

$$F_N(\varepsilon)=\sum_{n=0}^{N} a_n\, \nu_n(\varepsilon). \tag{1.61}$$

1.4.2. COMPATIBLE ASYMPTOTIC SEQUENCE

It is, in principle, possible to determine the coefficients of an asymptotic expansion by taking the limits:

$$a_o = \lim_{\varepsilon \to 0} \frac{f(\varepsilon)}{v_o(\varepsilon)},$$

$$a_1 = \lim_{\varepsilon \to 0} \frac{f(\varepsilon) - a_o v_o(\varepsilon)}{v_1(\varepsilon)}, \tag{1.62}$$

$$a_K = \lim_{\varepsilon \to 0} \frac{f(\varepsilon) - \sum_{n=o}^{K-1} a_n v_n(\varepsilon)}{v_K(\varepsilon)}.$$

However, it may happen that these limits are all equal to zero or infinity. In such a case, we say that the sequence $v_n(\varepsilon)$ is not compatible with the function $f(\varepsilon)$. For example, we consider the function of the small parameter ε:

$$f(\varepsilon) = \cos\varepsilon.$$

If we choose the asymptotic sequence

$$v_n(\varepsilon) = \varepsilon^n \quad (n = 0,1,2,....),$$

then we get the following asymptotic expansion:

$$F(\varepsilon) = 1 - \frac{\varepsilon^2}{2!} + \frac{\varepsilon^4}{4!} - \frac{\varepsilon^6}{6!} + o(\varepsilon^6).$$

Hence the asymptotic sequence ε^n is compatible with the function cos ε.

However, if we choose the asymptotic sequence

$$v_n(\varepsilon) = \varepsilon^{n+\frac{1}{2}} \quad (n = 0,1,2,....),$$

then all terms of the asymptotic expansion are infinite. Hence the sequence $\varepsilon^{n+\frac{1}{2}}$ is not compatible with the function cos ε. The same conclusion arises if we choose

$$v_n(\varepsilon) = \varepsilon^{2n+1} \quad (n = 0,1,2,....).$$

This example shows that it is necessary to impose certain restrictions to the choice of the asymptotic sequence $\nu_n(\varepsilon)$: We write $f = O[g(\varepsilon)]$ in R_1, where "O" is the Landau symbol large "O" if there exists a constant A independent of ε such that $|f| \le A|g|$ for all $\varepsilon \in R_1$. We say that f is of order g, which we write as $f = Ord(g)$ if we have simultaneously $f = O(g)$ and $g = O(f)$. With these definitions, we can now formulate the restrictions imposed on $\nu_n(\varepsilon)$. An asymptotic sequence $\nu_n(\varepsilon)$ is compatible with the function $f(\varepsilon)$ if it contains a sequence $\mu_n(\varepsilon)$, defined as follows:

$$\begin{aligned} &Ord\left[\mu_0(\varepsilon)\right] = Ord\left[f(\varepsilon)\right], \\ &Ord\left[\mu_1(\varepsilon)\right] = Ord\left[f(\varepsilon) - b_o\mu_o(\varepsilon)\right], \\ &Ord\left[\mu_K(\varepsilon)\right] = Ord\left[f(\varepsilon) - \sum_{n=o}^{K-1} b_n\mu_n(\varepsilon)\right], \end{aligned} \tag{1.63}$$

where $b_o, b_1, \dots, b_n$ denote the nonvanishing coefficients of (1.62):

$$\begin{aligned} b_o &= \lim_{\varepsilon \to 0} \frac{f(\varepsilon)}{\nu_0(\varepsilon)}, \\ b_n &= \lim_{\varepsilon \to 0} \frac{f(\varepsilon) - \sum_{q=o}^{n-1} b_q \mu_q(\varepsilon)}{\mu_n(\varepsilon)}. \end{aligned} \tag{1.64}$$

For example, if we choose

$$\nu_n(\varepsilon) = (\sin\varepsilon)^n \quad (n = 0,1,2,\dots.)$$

to represent the function

$$f(\varepsilon) = \cos\varepsilon,$$

then we get the asymptotic expansion

$$F(\varepsilon)=1-\frac{(\sin\varepsilon)^2}{2}-\frac{(\sin\varepsilon)^4}{4}-\frac{1}{16}(\sin\varepsilon)^6+o\left((\sin\varepsilon)^6\right),$$

which verifies (1.63). The asymptotic sequence (sin $\varepsilon)^n$ is therefore compatible with the function cos ε.

1.4.3. PROPERTIES OF AN ASYMPTOTIC EXPANSION

The properties of an asymptotic expansion are as follows:

1. An asymptotic expansion is unique, given a compatible asymptotic sequence.
2. An asymptotic expansion represents an approximation to the solution when ε is small.

According to (1.57), for a fixed value of N (e.g., $N = N_1$) and for $\eta > 0$ that is arbitrarily small, there exists $\varepsilon = \varepsilon_1$ such that

$$\left|f(\varepsilon_1)-F_{N_1}(\varepsilon_1)\right|<\eta,$$

where $F_{N_1}(\varepsilon)$ is given by (1.61) with $N = N_1$.

Conversely, if $\varepsilon = \varepsilon_1$ is fixed and if we augment N from zero to infinity, then the difference $\left|f(\varepsilon_1)-F_N(\varepsilon_1)\right|$ passes through a value smaller than η when N passes through N_1 and then augments indefinitely if the series $F_N(\varepsilon_1)$ diverges or tends to zero if the series converges. In general, the series $F_N(\varepsilon_1)$ diverges, and a typical shape of the curve giving the variation of the error Δ with N is shown in figure 1.16.

The minimum value Δm is a function of ε and tends to zero as $\varepsilon \to 0$. Hence for a given value of ε there is an optimal value of N for which the difference between the exact solution and its asymptotic expansion is the smallest. Unfortunately, there does not exist a general rule by which we can predict the value of this optimum, and a certain amount of experience is needed to determine the desired number of terms to be retained in the asymptotic expansion. However, this is not too severe of a problem, since in most practical situations it is not possible to construct more than the first two terms. The asymptotic expansion is therefore useful for ε sufficiently small, for which it is generally much more efficient for numerical evaluations of the solutions than expansions that are uniformly and absolutely convergent.

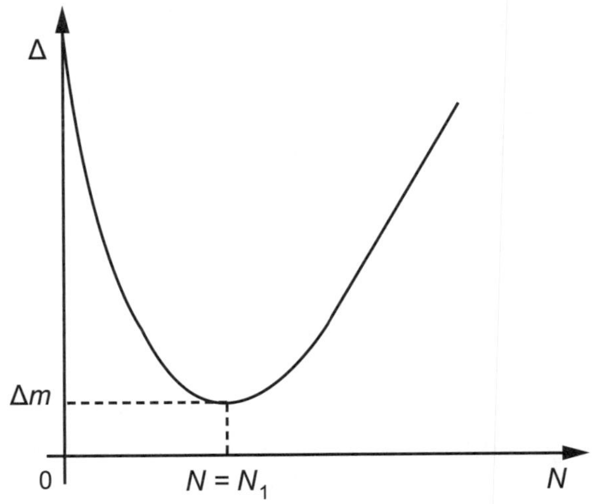

Figure 1.16. Typical curve giving the variation of the error Δ as a function of the number of terms N of an asymptotic expansion.

Asymptotic expansions are widely used to solve the equations of mathematical physics. They are usually encountered when the theory of perturbations is applied to a differential or a partial differential equation, to an integral equation, or more generally, to an operational equation. Other procedures also lead to asymptotic expansions. Among them is the asymptotic evaluation of an integral by the saddle point method. More information concerning asymptotic expansions can be found in references [1.13–1.17].

In the next section we will apply the perturbation theory to the Helmholtz equation and show how the GTD can be justified by the concept of asymptotic expansions.

1.5. ASYMPTOTIC SOLUTION OF THE WAVE EQUATION IN A SOURCE-FREE, UNBOUNDED MEDIUM

We consider successively a source-free (i.e., in the absence of excitation), unbounded, homogeneous fluid and a source-free, unbounded, homogeneous, isotropic elastic medium.

1.5.1. DERIVATION OF THE ASYMPTOTIC EXPANSION OF THE SOLUTION

Perturbation theory starts with the choice of a small parameter. In an unbounded homogeneous fluid, the geometrical parameters are the principal radii of curvature R_1, R_2 of the wave front and the wavelength λ. Hence the parameters that are important in diffraction theory are kR_1 and kR_2, where $k = 2\pi / \lambda$ is the wave number.

In an unbounded homogeneous isotropic elastic medium, longitudinal and transverse waves may propagate, each with a different wavelength. The important parameters are $k_\ell R_{\ell 1}$, $k_\ell R_{\ell 2}$, $k_t R_{t1}$, $k_t R_{t2}$, where $k_\ell = 2\pi / \lambda_\ell$, $k_t = 2\pi / \lambda_t$ are the wave numbers of the longitudinal and transverse waves, respectively, and $(R_{\ell 1}, R_{\ell 2})$, (R_{t1}, R_{t2}) are the principal radii of curvature of the corresponding incident wave fronts.

We suppose that all the characteristic dimensions of the incident wave fronts are large compared to the corresponding wavelengths:

$$k R_1 \gg 1 ,\ k R_2 \gg 1 ,\ k_\ell R_{\ell 1} \gg 1 ,\ k_\ell R_{\ell 2} \gg 1 ,\ k_t R_{t1} \gg 1 ,\ k_t R_{t2} \gg 1.$$

A small parameter that characterizes the diffraction problem in an unbounded homogeneous media is therefore $1 / kL$, where L is the smallest of the characteristic lengths R_1, R_2 for a fluid and $R_{\ell 1}$, $R_{\ell 2}$ R_{t1}, R_{t2} for an elastic medium. We can always take characteristic length L equal to 1. Then the small parameters are $\frac{1}{k}$ for a fluid and $\frac{1}{k_\ell}, \frac{1}{k_t}$ for an elastic medium.

We first consider the scalar Helmholtz equation (see appendix A, section A.4) verified by the pressure $P(\mathbf{r})$ corresponding to a time-periodic wave in an unbounded homogeneous source-free fluid,

$$\left(\Delta + k^2\right)P = 0, \tag{1.65}$$

where k is the wave number.

To derive a solution using the perturbation approach, the small parameter being $1/k$, we divide the Helmholtz equation (1.65) by k^2 to obtain the following:

$$\left(\frac{1}{k^2}\Delta + 1\right)P = 0.$$

The term $k^{-2}\Delta$ could be viewed as a perturbation, but the solution of the unperturbed equation would then simply be a trivial solution, $P = 0$. To avoid this pitfall, we perform the following transformation:

$$P(\mathbf{r}) = e^{ikS(\mathbf{r})}\, p(\mathbf{r}), \tag{1.66}$$

which is known as the quasi-optics ansatz. This statement was introduced by Sommerfeld and Runge in 1911 [1.1], who were the first to derive the laws of GO from the wave equation. If we introduce (1.66) in (1.65) and order the different terms according to the parameter k^{-1}, which is supposed small, the wave equation is converted into a new form as follows:

$$\left(1-|\nabla S|^2\right)+\frac{i}{k}\left(\Delta S+2\nabla S\cdot\frac{\nabla p}{p}\right)+\frac{1}{k^2}\frac{\Delta p}{p}=0. \tag{1.67}$$

If we neglect the last term, we are led to the well-known eikonal and transport equations:

$$|\nabla S|^2=1, \tag{1.68}$$

$$\Delta S+2\mathbf{\nabla}S\cdot\frac{\mathbf{\nabla}p}{p}=0. \tag{1.69}$$

It will be shown in sections 1.5.2 and 1.5.4 that the solutions of these equations contain all the laws of GO.

We now focus our attention on the third term of (1.67), which has been neglected. We see that the wave optics analysis leading to (1.67) allows us to establish the condition for the validity of the GO approximation given by (1.68) and (1.69)—a necessary and sufficient condition being that $k^2\Delta p\,/\,p$ is truly negligible. It fails, obviously, in regions where the derivatives of the amplitude p of the pressure field has a rapid spatial variation within a wavelength. For example, this happens near a focus or a caustic surface or a shadow boundary. Outside these regions, the third term of (1.67) is small and can be treated as a perturbation. Since it involves only the amplitude function p and not the phase function s, the perturbation modifies only the function p. So the function p can be written formally as an asymptotic expansion with respect to powers of the small parameter k^{-1}, which we have replaced by $(ik)^{-1}$ for convenience:

$$p(\mathbf{r})=\sum_{n=o}^{N}\frac{1}{(ik)^n}p_n(\mathbf{r})+o\left(k^{-N}\right). \tag{1.70}$$

If we insert (1.70) into (1.66) and (1.65) and set equal to zero the terms of equal powers of $1/k$, we see that the eikonal equation remains the same as (1.68) and is independent of the order n. However, only the first-order term p_o verifies the transport equation (1.69); all other terms $(n > 0)$ are solutions of a system of coupled modified transport equations,

$$p_n\,\Delta S+2\nabla S\cdot\nabla p_n=-\Delta\,p_{n-1}, \tag{1.71}$$

which can be solved by iteration if the unperturbed solution p_o is known.

Consider now the vector Helmholtz equations verified by the particle displacement vectors $\mathbf{U}_\ell$ and $\mathbf{U}_t$ of the longitudinal and transverse waves, respectively (see appendix B):

$$\Delta \mathbf{U}_\ell + k_\ell^2 \mathbf{U}_\ell = 0, \tag{1.72}$$

$$\Delta \mathbf{U}_t + k_t^2 \mathbf{U}_t = 0, \tag{1.73}$$

where $\mathbf{U}_\ell$ and $\mathbf{U}_t$, furthermore, satisfy

$$\text{curl}\, \mathbf{U}_\ell = 0, \quad \text{div}\, \mathbf{U}_t = 0. \tag{1.74}$$

According to the procedure followed by Sommerfeld and Runge, starting with the quasi-optics ansatz (1.66) with the scalar function $p(\mathbf{r})$ replaced by the vector function $\mathbf{u}$, we get

$$\left(1 - |\nabla S|^2\right) \mathbf{u} + \frac{i}{k}\left(\Delta S + 2\nabla S \cdot \nabla\right) \mathbf{u} + \frac{1}{k^2} \Delta \mathbf{u} = 0, \tag{1.75}$$

where $\mathbf{u}$ and k stand either for $\mathbf{u}_\ell$, k_ℓ or for $\mathbf{u}_t$, k_t.

Equation (1.75) is equivalent to three scalar equations for each component of the displacement vector, which are similar to (1.67) and can therefore be submitted to the same treatment. We see that if we neglect the last term of (1.75), we find the same eikonal equation as for a scalar field. The transport equation is also the same as in (1.69) for each component of the displacement vector.

Inserting the quasi-optics ansatz into equations (1.74) leads to

$$\nabla S \cdot \mathbf{u}_t + \frac{i}{k_t} \nabla \cdot \mathbf{u}_t = 0, \tag{1.76}$$

$$\nabla S \times \mathbf{u}_\ell + \frac{i}{k_\ell} \nabla \times \mathbf{u}_\ell = 0. \tag{1.77}$$

Equations (1.76) and (1.77) give us information concerning the polarization.

At first order, we have

$$\nabla S \cdot \mathbf{u}_t = 0, \quad \nabla S \times \mathbf{u}_\ell = 0. \tag{1.78}$$

Since ∇S is orthogonal to the wave front $S(\mathbf{r})$ = constant, we see that $\mathbf{u}_\ell$ is orthogonal to the wave front and directed along the ray trajectory, whereas $\mathbf{u}_t$ is tangent to the wave front and orthogonal to the ray trajectory. These properties, associated with the solutions of the eikonal and transport equations given in sections 1.5.2 and 1.5.4, generalize the laws of GA to elastic media. Furthermore, since the term of order k^{-2} in (1.75) is small outside the regions where the derivatives of the components of $\mathbf{u}$ have a rapid spatial variation within a wavelength that happens near a focus or a caustic surface or a shadow boundary, it can be treated as a perturbation, and $\mathbf{u}_\ell$

and $\mathbf{u}_t$ can be written formally as an asymptotic expansion with respect to the powers of the small parameters k_ℓ^{-1} and k_t^{-1}, respectively:

$$\mathbf{U}_\ell(\mathbf{r}) = e^{ik_\ell S_\ell(\mathbf{r})} \sum_{n=0}^{N} (ik_\ell)^{-n} \mathbf{u}_n^\ell(\mathbf{r}) + o(k_\ell^{-N}), \tag{1.79}$$

$$\mathbf{U}_t(\mathbf{r}) = e^{ik_t S_t(\mathbf{r})} \sum_{n=0}^{N} (ik_t)^{-n} \mathbf{u}_n^t(\mathbf{r}) + o(k_t^{-N}). \tag{1.80}$$

Inserting (1.79) and (1.80) in (1.72) and (1.73), respectively, and in (1.74), we obtain the following set of equations for the longitudinal displacement:

$$\begin{aligned} &|\nabla S_\ell|^2 = 1, \\ &(\Delta S_\ell + 2\nabla S_\ell \cdot \nabla)\, \mathbf{u}_n^\ell = -\Delta \mathbf{u}_{n-1}^\ell, \\ &\nabla S_\ell \cdot \mathbf{u}_n^\ell = -\nabla \cdot \mathbf{u}_{n-1}^\ell, \end{aligned} \tag{1.81}$$

and we obtain a similar set of equations for the transverse displacement:

$$\begin{aligned} &|\nabla S_t|^2 = 1, \\ &(\Delta S_t + 2\nabla S_t \cdot \nabla)\, \mathbf{u}_n^t = -\Delta \mathbf{u}_{n-1}^t, \\ &\nabla S_t \cdot \mathbf{u}_n^t = -\nabla \cdot \mathbf{u}_{n-1}^t. \end{aligned} \tag{1.82}$$

Since $\mathbf{u}_{-1}^\ell = \mathbf{u}_{-1}^t = 0$, we see that $\mathbf{u}_o^\ell$ and $\mathbf{u}_o^t$ verify the transport equations,

$$(\Delta S_\ell + 2\nabla S_\ell \cdot \nabla)\, \mathbf{u}_o^\ell = 0, \tag{1.83}$$

$$(\Delta S_t + 2\nabla S_t \cdot \nabla)\, \mathbf{u}_o^t = 0. \tag{1.84}$$

In addition, we have

$$\nabla S_t \cdot \mathbf{u}_o^t = 0, \quad \nabla S_\ell \times \mathbf{u}_o^\ell = 0. \tag{1.85}$$

According to (1.83) to (1.85), we see that the GA approximation for the elastic wave propagation in homogeneous isotropic media is the first term of an asymptotic expansion, with respect to the asymptotic sequences $(ik_\ell)^{-n}$ and $(ik_t)^{-n}$ for the longitudinal and transverse waves, respectively.

1.5.2. RESOLUTION OF THE EIKONAL EQUATION

In a Cartesian system of coordinates (x_1, x_2, x_3), a point in space is represented by $\mathbf{r} = \mathbf{r}\,(x_1, x_2, x_3)$, and the eikonal equation in (1.81) or (1.82) may be written

$$\left(\frac{\partial S}{\partial x_1}\right)^2 + \left(\frac{\partial S}{\partial x_2}\right)^2 + \left(\frac{\partial S}{\partial x_3}\right)^2 - 1 = 0. \tag{1.86}$$

The first-order differential equation is also called the Hamilton-Jacobi equation. It has the general form $F(x_i, \xi_i) = 0$, where $\xi_i = \partial S / \partial x_i$. An equation of this type is usually solved by applying the method of characteristics that consists of writing the total derivative of F,

$$dF = \frac{\partial F}{\partial x_i} d x_i + \frac{\partial F}{\partial \xi_i} d\xi_i = 0,$$

leading to

$$\frac{dx_i}{\partial F / \partial \xi_i} = -\frac{d\xi_i}{\partial F / \partial x_i} = ds$$

or equivalently

$$\frac{d\xi_i}{ds} = -\frac{\partial F}{\partial x_i}, \quad \frac{dx_i}{ds} = \frac{\partial F}{\partial \xi_i}. \tag{1.87}$$

The system of equations (1.87) are the parametric equations of the characteristic curves. According to (1.86), we have

$$\frac{dx_i}{ds} = 2\xi_i, \quad \frac{d\xi_i}{ds} = 0. \tag{1.88}$$

The first equation (1.88) shows that the tangent to a characteristic curve is directed along the gradient of the phase:

$$\frac{d\mathbf{r}}{ds} = 2\nabla S. \tag{1.89}$$

The characteristic curves are therefore the rays.

The second equation (1.88) means that the tangent is constant along a characteristic curve:

$$\frac{d(\nabla S)}{ds} = 0. \tag{1.90}$$

The characteristic curves (or rays) are therefore straight lines in a homogeneous medium, directed along the gradient of the phase. If σ is the curvilinear coordinate along the characteristic curve, we have $|d\mathbf{r} / d\sigma| = 1$. Hence according to (1.89) and (1.86) we see that $d\sigma = 2ds$ and that

$$\frac{d\mathbf{r}}{d\sigma} = \boldsymbol{\sigma} = \nabla S, \tag{1.91}$$

where $\boldsymbol{\sigma}$ is the unit vector tangent to the characteristic curve, since $|\nabla S| = 1$.

1.5.3. PROPERTIES OF THE CHARACTERISTIC CURVES

The properties of characteristic curves are as follows:

1. The characteristic curves (or rays) are orthogonal to the surface $S(\mathbf{r})$ = constant, since ∇S is a unit vector orthogonal to S.
2. The characteristic curves form a family of curves depending on two parameters (also called a congruence of curves), since one characteristic curve passes through each point $\mathbf{r}_o$ on $S(\mathbf{r})$ = constant and a point of S is defined by two parameters.

It is a mathematical property that a congruence of curves always has an envelope, which in general is a surface. In some cases, the surface can be degenerated into a line or a point. In GO (or GA), the following vocabulary is used: the surface $S(\mathbf{r})$ = constant is called a wave front, the characteristic curves are called the rays, and the envelope of the characteristic curves is called a caustic (fig. 1.17).

The phase variation from $\mathbf{r}_o$ to $\mathbf{r}_1$ is given by

$$k\left[S(\mathbf{r}_1) - S(\mathbf{r}_o)\right] = k\int_o^\sigma \frac{\partial S}{\partial \sigma} d\sigma = k\int_o^\sigma \nabla S \cdot \frac{d\mathbf{r}}{d\sigma} d\sigma$$

$$= k\int_o^\sigma d\sigma = k\sigma = \mathbf{k}\cdot(\mathbf{r}_1 - \mathbf{r}_o).$$

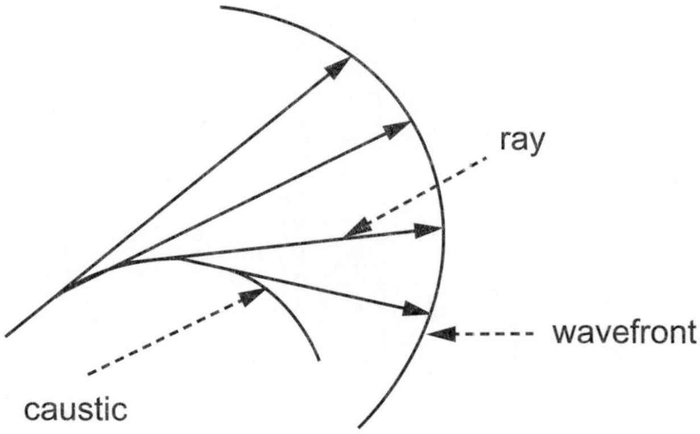

Figure 1.17. Wave front, rays, and caustic.

1.5.4. RESOLUTION OF THE TRANSPORT EQUATION

The transport equations for the zero-order term of the pressure in an ideal fluid and for the zero-order term of the longitudinal and transverse displacements in an elastic medium (see equations (1.69), (1.83), and (1.84)) all have the same form. If u is the amplitude p_o of the pressure or a Cartesian component of the amplitudes $\mathbf{u}_o^\ell$ or $\mathbf{u}_o^t$ of the longitudinal and transverse displacements, we need to solve the following equation:

$$u\,\Delta S + 2\nabla S \cdot \nabla u = 0. \tag{1.92}$$

In scientific literature, we can find different methods of resolution of this equation. The most familiar one consists of multiplying (1.92) by u. We then obtain the following:

$$\nabla \cdot \left(u^2\, \nabla S\right) = 0. \tag{1.93}$$

This shows that $u^2\, \nabla S$ has a vanishing divergence, and consequently its flux in a tube of rays is conserved. The flux through the walls formed by the rays vanishes since $u^2\, \nabla S$ is parallel to the rays (fig. 1.18). The flux through the surfaces $d\Sigma(o)$ and $d\Sigma(\sigma)$, delimited by the ray tube on the wave fronts $\Sigma(o)$ and $\Sigma(\sigma)$, is given by

$$u^2(\sigma)\,d\sum(\sigma) = u^2(o)\,d\sum(o), \tag{1.94}$$

and since (1.93) and (1.94) are verified by the pressure p_o and all three components of $\mathbf{u}_o^\ell$ and $\mathbf{u}_o^t$, we see that the flux of the square of the amplitude of the pressure and the flux of the square of

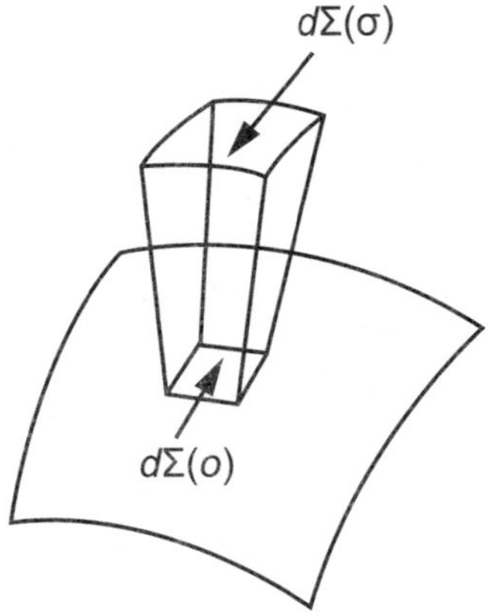

Figure 1.18. Ray tube.

the amplitudes of the longitudinal and transverse displacements, at zero order in the asymptotic expansion, are conserved in a tube of rays:

$$|p_o(\sigma)|^2 d\Sigma(\sigma) = |p_o(o)|^2 d\Sigma(o), \tag{1.95}$$

$$|\mathbf{u}_o^{\ell}(\sigma)|^2 d\Sigma(\sigma) = |\mathbf{u}_o^{\ell}(o)|^2 d\Sigma(o), \tag{1.96}$$

$$|\mathbf{u}_o^{t}(\sigma)|^2 d\Sigma(\sigma) = |\mathbf{u}_o^{t}(o)|^2 d\Sigma(o). \tag{1.97}$$

This result corresponds to the second postulate of GO (or GA). We see that in an elastic medium, (1.96) and (1.97) do not provide any information about the direction of the displacement. The transport equation (1.92) can also be solved in a curvilinear coordinate system (σ, σ_1, σ_2), where σ is along a ray and σ_1, σ_2 are coordinates along the principal directions of the wave front (fig. 1.19).

This coordinate system is orthogonal. The general expression of the Laplacian is

$$\Delta \mathbf{u} = \frac{1}{h_1 h_2 h}\left\{ \frac{\partial}{\partial\sigma}\left(\frac{h_1 h_2}{h}\frac{\partial u}{\partial\sigma}\right) + \frac{\partial}{\partial\sigma_1}\left(\frac{h_2 h}{h_1}\frac{\partial u}{\partial\sigma_1}\right) + \frac{\partial}{\partial\sigma_2}\left(\frac{h h_1}{h_2}\frac{\partial u}{\partial\sigma_2}\right)\right\}. \tag{1.98}$$

Here,

$$h_1 = 1 + \frac{\sigma}{R_1}, \quad h_2 = 1 + \frac{\sigma}{R_2}, \quad h = 1, \tag{1.99}$$

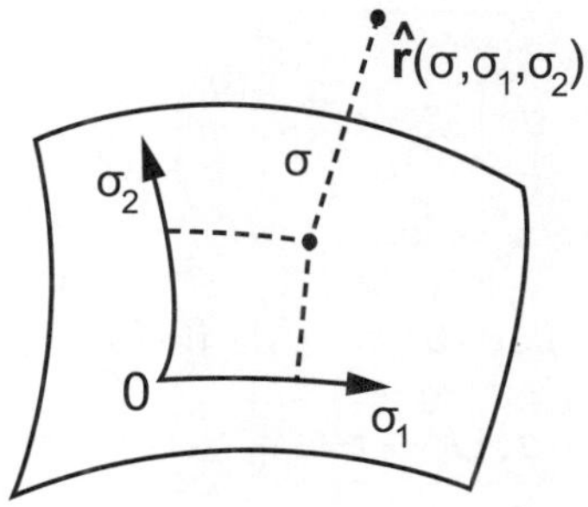

Figure 1.19. Curvilinear (ray) coordinate system.

where R_1 and R_2 are, respectively, the principal radii of curvature along the coordinate curves σ_1 and σ_2. We suppose that R_1 and R_2 are both positive, which means that the ray pencil from o to σ in the direction of wave propagation is divergent. Since $S(\mathbf{r})$ is constant on a wave front, we have

$$\frac{\partial S}{\partial \sigma_1} = \frac{\partial S}{\partial \sigma_2} = 0.$$

Hence, according to (1.98) applied to the function $S(\mathbf{r})$, we obtain

$$\Delta S = \frac{1}{R_1 + \sigma} + \frac{1}{R_2 + \sigma}. \tag{1.100}$$

According to (1.91), along a ray we have

$$\nabla S \cdot \nabla u = \boldsymbol{\sigma} \cdot \nabla u = \frac{\partial u}{\partial \sigma}, \tag{1.101}$$

and since σ_1 and σ_2 are constant along a ray, we can replace $\frac{\partial u}{\partial \sigma}$ with $\frac{du}{d\sigma}$.

Inserting (1.100) and (1.101) in the transport equation (1.92), it follows that

$$2\frac{du}{d\sigma} + \left(\frac{1}{R_1 + \sigma} + \frac{1}{R_2 + \sigma}\right) u = 0. \tag{1.102}$$

Hence

$$\frac{u(\sigma)}{u(o)} = \sqrt{\frac{R_1 R_2}{(R_1 + \sigma)(R_2 + \sigma)}}, \tag{1.103}$$

and

$$\mathbf{u}(\sigma) = \sqrt{\frac{R_1 \, R_2}{(R_1+\sigma)(R_2+\sigma)}} \; \mathbf{u}(o), \tag{1.104}$$

where u is p_o and where $\mathbf{u}$ is either $\mathbf{u}_o^{\ell}$ or $\mathbf{u}_o^{t}$.

Equation (1.104) shows that the direction of the field $\mathbf{u}$ is invariant along a ray. Since we have

$$d\Sigma(\sigma) \cong \frac{(R_1+\sigma)(R_2+\sigma)}{R_1 \, R_2} \, d\Sigma(o), \tag{1.105}$$

one can easily check that (1.104) implies (1.94) and (1.95) to (1.97). This result corresponds to the third postulate of GO (or GA).

Combining all the properties we have found leads us to the following important result: the GO (or GA) applied to the wave propagation in a fluid or an elastic medium is the first term of the asymptotic expansions (1.70) or (1.79) and (1.80), respectively. These asymptotic expansions are also called the Luneburg-Kline expansions.

We will now solve the transport equation for $n \neq 0$, in which case it takes the form

$$p_n \, \Delta S + 2\nabla S \cdot \nabla \, p_n = -\Delta \, p_{n-1} \tag{1.106}$$

for a scalar field p_n in a fluid, and

$$(\Delta S + 2\nabla S \cdot \nabla)\mathbf{u}_n = -\Delta \, \mathbf{u}_{n-1} \tag{1.107}$$

for a vector displacement field in an elastic medium. Proceeding as we did for $n = 0$, we need to solve the equation

$$u_n \, \Delta S + 2\nabla S \cdot \nabla \, u_n = -\Delta \, u_{n-1}, \tag{1.108}$$

where u_n stands for p_n or for each component of $\mathbf{u}_n$. Applying the first method, we see that there is no power conservation in a tube of rays because the right-hand side of (1.108) is different from zero for $n \neq 0$.

In the curvilinear coordinate system (σ, σ_1, σ_2), equation (1.108) reduces to the differential equation along the ray

$$2\frac{d\,u_n}{d\sigma} + \left(\frac{1}{R_1+\sigma} + \frac{1}{R_2+\sigma} \right) u_n = -\Delta \, u_{n-1}. \tag{1.109}$$

In order to solve this equation, we rewrite it in another form by introducing the function

$$\left|\frac{J(\sigma)}{J(o)}\right| = \frac{(R_1+\sigma)(R_2+\sigma)}{R_1\ R_2}, \tag{1.110}$$

which verifies the identity

$$\frac{1}{|J(\sigma)|}\frac{d|J(\sigma)|}{d\sigma} = \frac{1}{R_1+\sigma} + \frac{1}{R_2+\sigma}. \tag{1.111}$$

Inserting (1.111) in the transport equation (1.109), we obtain

$$2\frac{du_n}{d\sigma} + \frac{u_n}{|J|}\frac{d|J|}{d\sigma} = \frac{2}{\sqrt{|J|}}\frac{d}{d\sigma}\left(\sqrt{|J|}\,u_n\right) = -\Delta u_{n-1},$$

which can be integrated immediately to give

$$u_n(\sigma) = \sqrt{\left|\frac{J(o)}{J(\sigma)}\right|}\,u_n(o) - \frac{1}{2}\int_o^\sigma \sqrt{\left|\frac{J(\sigma')}{J(\sigma)}\right|}\,\Delta u_{n-1}(\sigma')d\sigma'. \tag{1.112}$$

This equation is verified by the pressure p_n in a fluid and by each component of the longitudinal and transverse displacements in an elastic medium. Hence we also have

$$\mathbf{u}_n(\sigma) = \sqrt{\left|\frac{J(o)}{J(\sigma)}\right|}\,\mathbf{u}_n(o) - \frac{1}{2}\int_o^\sigma \sqrt{\left|\frac{J(\sigma')}{J(\sigma)}\right|}\,\Delta \mathbf{u}_{n-1}(\sigma')d\sigma', \tag{1.113}$$

where $\mathbf{u}_n$ stands for $\mathbf{u}_n^\ell$ or $\mathbf{u}_n^t$. For $n = 0$, we see that (1.113) gives again (1.104).

It is possible to extend the formula (1.104) to situations where the ray pencil from o to σ is not divergent. In figure 1.20, four rays are shown defining an astigmatic pencil of rays. By convention, the radii of curvature are taken positive if the rays emanating from the corresponding focus are divergent and negative if the rays are convergent. When $R_2 < 0$, the focus line F_2 lies ahead of the wave front. Hence when we progress along a ray in the direction of wave propagation, we cross a focal line. Then a change of $-\pi/2$ has to be introduced in the phase to take into account the crossing of a caustic surface (i.e., the passing through a point of tangency of a ray with a caustic surface). This condition is verified if the following convention for the square root in (1.104) is adopted: $\sqrt{R_{1,2}+\sigma}$ takes positive real, negative imaginary, or zero values. The selection of the correct square root can be justified by the behavior of the field near a caustic surface, which must be analyzed with a different ansatz for the asymptotic expansion of the pressure in a fluid or for the longitudinal and transverse displacements in an elastic medium.

We now come back to (1.112) and (1.113). These formulas enable one to continue p_n or $\mathbf{u}_n$ along a given ray. For the zero-order term p_o (respectively $\mathbf{u}_o$), it is only necessary to know the initial value p_o (respectively $\mathbf{u}_o$) at a reference point in order to carry out this continuation. At a caustic where $\sigma = -R_1$ or $\sigma = -R_2$, p_o (respectively $\mathbf{u}_o$) become infinite, equations (1.112) and (1.113) fail.

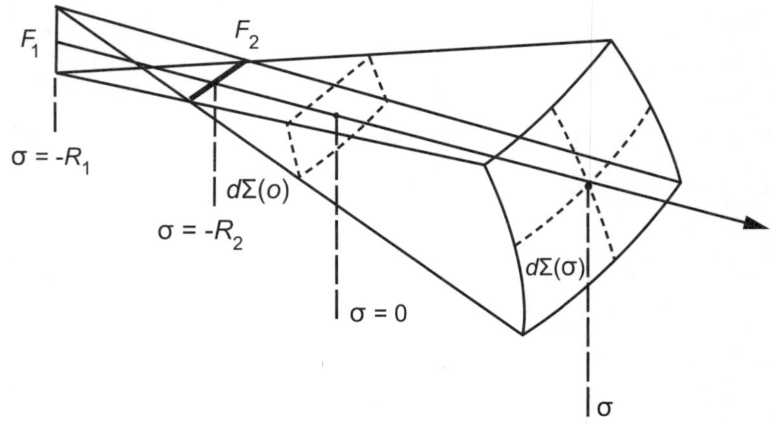

Figure 1.20. Astigmatic pencil of rays.

For a higher-order term ($n > 0$), more information is needed in order to continue p_n (respectively $\mathbf{u}_n$) along a given ray. In addition to the initial value $p_n(o)$ (respectively $\mathbf{u}_n(o)$), the function $\Delta p_{n-1}(\sigma')$ (respectively $\Delta \mathbf{u}_{n-1}(\sigma')$) must be known for all σ' in the range $o < \sigma' < \sigma$. For instance, for $n = 1$, one needs $\Delta p_o(\sigma')$ (respectively $\Delta \mathbf{u}_o(\sigma')$). This implies the knowledge of the first and second derivatives of R_1 and R_2 with respect to the transverse coordinates (σ_1, σ_2) on the initial wave front.

1.6. ACOUSTIC FIELD REFLECTED BY A SMOOTH, RIGID, SOFT, OR IMPEDANCE SURFACE

The simplest model problem for the reflection of a sound wave by a smooth object is the reflection of a sound wave by an infinite plane. Image theory shows that the reflected field has the same asymptotic structure as the incident field. It can therefore be represented by an asymptotic expansion having the same asymptotic sequence.

We assume that the incident acoustic pressure field is represented by the asymptotic expansion

$$p^i(\mathbf{r}) = e^{ikS^i(\mathbf{r})} \sum_{n=o}^{N} (ik)^{-n} p_n^i(\mathbf{r}) + o\left(k^{-N}\right), \tag{1.114}$$

where the phase verifies the eikonal equation

$$\left|\nabla S^i\right|^2 = 1 \tag{1.115}$$

and where the amplitudes p_n^i verify the transport equation

$$p_n^i \, \Delta S^i + 2\nabla S^i \cdot \nabla p_n^i = -\Delta p_{n-1}^i, \tag{1.116}$$

with $\Delta p_{-1}^i \equiv 0$. A field of the form (1.114) to (1.116) is called a ray field. If it is limited to its zero-order term, it is a GA field.

We have seen that in a homogeneous medium, the rays are straight lines in the direction of ∇S. Some of them intercept the surface of the body and divide the space into an illuminated region and a shadow region separated by a surface Σ, which is called the shadow boundary of the incident field (fig. 1.21)

The shadow region is delimited by the rays tangent to the surface S of the body along the curve Γ separating the lit region S_e from the shadow region S_o on S. In the lit region away from the shadow boundary and possible caustics, the incident ray field $p^i(\mathbf{r})$ gives rise to a reflected ray field $p^R(\mathbf{r})$. We represent this field by an asymptotic expansion similar to (1.114):

$$p^R(\mathbf{r}) = e^{ikS^R(\mathbf{r})} \sum_{n=o}^{N} (ik)^{-n} \, p_n^R(\mathbf{r}) + o\left(k^{-N}\right), \tag{1.117}$$

where the phase $S^R(\mathbf{r})$ and the amplitudes p_n^R verify the eikonal equation (1.115) and the transport equation (1.116), respectively.

As for the incident field, the characteristics or rays of the eikonal equation are orthogonal to the wave front $S^R(\mathbf{r})$ = constant and form a congruence in $\Re^3$. They therefore have an envelope or caustic that can be located outside the body (i.e., a real caustic) or inside the body (i.e., a virtual caustic).

The asymptotic expansions (1.114) and (1.117) are valid at every point in the illuminated region of $\Re^3$, with the exception of those points located on or in the vicinity of the shadow boundaries and caustics. Therefore, we can use these expansions at observation points located on the lit side S_e of S (not too close to the shadow boundary Γ) and apply the boundary conditions at those points.

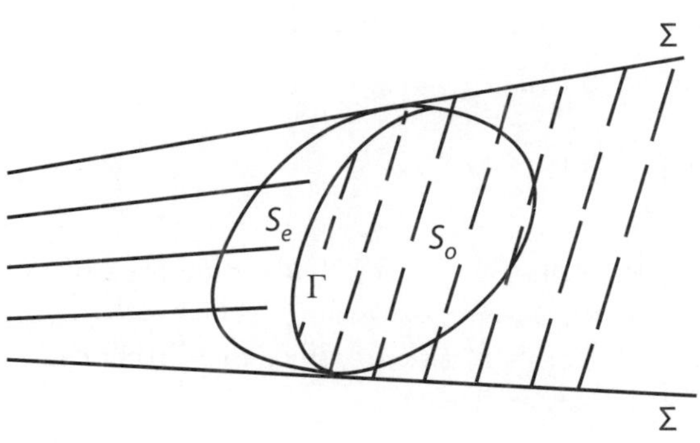

Figure 1.21. Incident ray field intercepted by a smooth body.

For a rigid body, we have (see appendix A):

$$\frac{\partial p}{\partial n} = \mathbf{n} \cdot \boldsymbol{\nabla} p = 0, \quad \mathbf{r} \in S, \tag{1.118}$$

where p is the total pressure field $p = p^i + p^R$ and $\mathbf{n}$ is the unit normal oriented to the outside of the volume delimited by S.

For a soft body, the total pressure verifies

$$p = 0, \quad \mathbf{r} \in S, \tag{1.119}$$

and for an impedance boundary condition characterized by an impedance Z, we have

$$\frac{\partial p}{\partial n} + ik\overline{Z}\, p = 0, \quad \mathbf{r} \in S, \tag{1.120}$$

where $\overline{Z} = \dfrac{Z}{Z_o}$ is the impedance Z of the surface divided by the impedance Z_o of the medium where the waves propagate.

The asymptotic expansions (1.114) and (1.117) do not give the total pressure field on S—they only give an approximation of it. Other diffraction phenomena occur and contribute to the diffracted field, such as creeping waves or waves diffracted by singularities of the tangent plane or of the curvature if S is not regular. However, since the asymptotic expansions that are associated with these diffracted phenomena are defined with respect to different asymptotic sequences, other than k^{-n}, the boundary conditions (1.118) to (1.120) are separately verified by each species of waves. This property holds for multiple reflected rays, since the corresponding phase functions are different. This implies that the boundary conditions must be applied separately for each order of diffraction.

Accordingly, we impose the conditions (1.118) to (1.120) to the asymptotic expansions (1.114) and (1.117). The conditions (1.118) to (1.120) involve the continuity of the phase at each point on S_e:

$$S^i(\mathbf{r}) = S^R(\mathbf{r}), \quad r \in S_e. \tag{1.121}$$

The surface gradient of the eikonal S^i is simply the projection of $\boldsymbol{\nabla} S^i = \hat{s}^i$ of the incident ray on the tangent plane to the surface. According to (1.121), the projection of $\boldsymbol{\nabla} S^i$ must be equal to the projection of $\boldsymbol{\nabla} S^R = \hat{s}^R$. Hence, since the direction $\hat{s}^R$ of the reflected field is pointed outward from the surface, we have

$$\hat{s}^R = \hat{s}^i - 2\hat{n}\left(\hat{n} \cdot \hat{s}^i\right), \tag{1.122}$$

which is the law of reflection.

On the surface, the amplitude p_n of the nth term of the asymptotic expansion verifies

$$\frac{\partial S^i}{\partial n}\, p_n^i + \frac{\partial S^R}{\partial n}\, p_n^R + \frac{\partial p_{n-1}^i}{\partial n} + \frac{\partial p_{n-1}^R}{\partial n} = 0 \tag{1.123}$$

for a rigid body and

$$p_n^i + p_n^R = 0 \tag{1.124}$$

for a soft body. For an impedance boundary condition, we find

$$\frac{\partial S^i}{\partial n}\, p_n^i + \overline{Z}\, p_n^i + \frac{\partial S^R}{\partial n}\, p_n^R + \overline{Z}\, p_n^R + \frac{\partial p_{n-1}^i}{\partial n} + \frac{\partial p_{n-1}^R}{\partial n} = 0. \tag{1.125}$$

For $n = 0$, equation (1.103) holds for $p_o^R(\sigma)$, where σ is the coordinate along the ray. Therefore, we have

$$p_o^R(\sigma) = \sqrt{\frac{\rho_1^R\,\rho_2^R}{\left(\rho_1^R + \sigma\right)\left(\rho_2^R - \sigma\right)}}\; p_o^R(o), \tag{1.126}$$

where ρ_1^R and ρ_2^R are the principal radii of curvature of the reflected wave front.

For a hard body, (1.123) for $n = 0$ can also be written as

$$\left(\mathbf{n}\cdot\hat{s}^i\right)\, p_o^i(o) + \left(\mathbf{n}\cdot\hat{s}^R\right)\, p_o^R(o) = 0.$$

Hence

$$p_o^R(o) = -\frac{\left(\mathbf{n}\cdot\hat{s}^i\right)}{\left(\mathbf{n}\cdot\hat{s}^R\right)}\, p_o^i(o) \cong R_h\, p_o^i(o), \tag{1.127}$$

where R_h is the reflection coefficient on a hard body. Since according to the law of reflection $\mathbf{n}\cdot\hat{s}^i = -\,\mathbf{n}\cdot\hat{s}^R$, we have $R_h = 1$.

Similarly, for a soft boundary, (1.124) gives $R_s = -1$. For an impedance boundary, equation (1.125) leads to

$$R = \frac{\cos\theta - \overline{Z}}{\cos\theta + \overline{Z}}, \tag{1.128}$$

where θ is the angle of reflection (fig. 1.22).

Finally, according to (1.126), the pressure field p^R at an observation point P, due to the leading term of the asymptotic expansion, is given by

$$p^R(P) = p^i(Q)\, R \sqrt{\frac{\rho_1^R \rho_2^R}{(\rho_1^R + \sigma)(\rho_2^R + \sigma)}}\, e^{ik\sigma}, \tag{1.129}$$

where R is the reflection coefficient and where Q is the point of reflection with $\sigma \cong |\mathbf{QP}|$ and

$$p^i(Q) = p_o^i(Q)\, e^{ik\, s^i(Q)}. \tag{1.130}$$

Equation (1.129) is the GO (GA) approximation.

For the calculation of higher-order terms ($n > 0$), (1.126) is replaced by

$$p_n^R(\sigma) = \sqrt{\left|\frac{J(o)}{J(\sigma)}\right|}\, p_n^R(o) - \frac{1}{2}\int_o^\sigma \left|\frac{J(\sigma')}{J(\sigma)}\right| \Delta p_{n-1}^R(\sigma')\, d\sigma'. \tag{1.131}$$

The function $J(\sigma)$ is the Jacobian of the transformation $(x_1, x_2, x_3) \to (\sigma_1, \sigma_2, \sigma)$ from the Cartesian coordinates to the ray coordinates.

According to (1.110), we have

$$\left|\frac{J(\sigma')}{J(\sigma)}\right| = \frac{(\rho_1^R + \sigma')(\rho_2^R + \sigma')}{(\rho_1^R + \sigma)(\rho_2^R + \sigma)}. \tag{1.132}$$

General formulas for ρ_1^R and ρ_2^R are given in appendix D.

When the principal radii of curvature of the surface S and the incident wave front are all large compared to the wavelength, the higher-order terms are small and can be neglected in practical applications. However, the second-order term ($n = 1$) ceases to be negligible when kR_1 and kR_2 are of the order or smaller than 5, where R_1 and R_2 are the principal radii of curvature of the surface S.

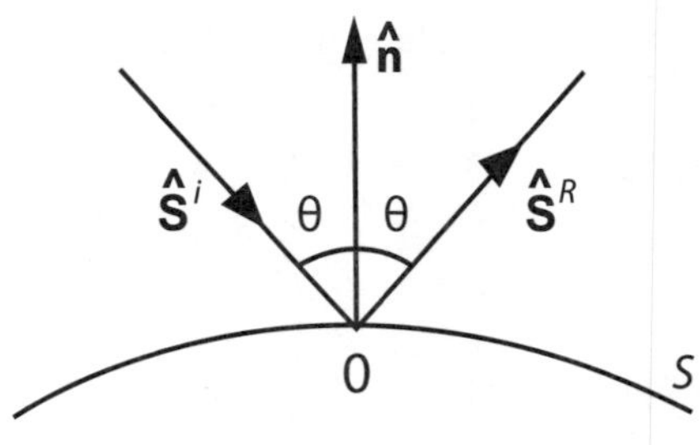

Figure 1.22. Incident and reflected rays.

1.7. REFLECTED AND TRANSMITTED WAVES AT A SMOOTH INTERFACE BETWEEN A FLUID AND AN ELASTIC MEDIUM

The model problem corresponding to the interaction between an incident acoustic wave with a smooth curved boundary separating a fluid from an elastic body is a similar problem in which the incident acoustic wave is a plane wave interacting with a planar interface between a fluid and an elastic half-space. We know from the solution of this problem (given in section B.8 of appendix B) that the incident acoustic wave generates a reflected acoustic wave in the fluid and two refracted elastic waves propagating independently in the elastic medium.

The interface between a fluid and an elastic solid can, in addition, support elastic surface waves and creeping waves. In this section we consider only the generation of reflected and refracted waves. Creeping waves and elastic surface waves will be treated in sections 1.9 and 2.3.4.2, respectively.

As mentioned in the preface, our interest is restricted to waves in homogeneous ideal fluids and in homogeneous, isotropic, linearly elastic solids. In these media, the rays are straight lines that are normal to the surfaces of constant phases (i.e., the wave fronts).

We have seen in section 1.5 that, analogously to plane waves, ray fields of sound waves can propagate in an unbounded fluid. Similarly, ray fields of longitudinal and transverse wave motions can propagate in an unbounded elastic medium.

We assume that the incident acoustic pressure field is given by the asymptotic expansion (1.114), which satisfies the eikonal and transport equations (1.115) and (1.116). Let $\mathbf{u}^i(\mathbf{r})$ be the displacement in the fluid due to the incident wave. According to Euler's equation for small displacements (see appendix A), the displacement $\mathbf{u}$ in the fluid due to an acoustic wave motion is related to the pressure by

$$\mathbf{u} = \frac{1}{\rho_o \omega^2} \mathbf{grad}\; p, \tag{1.133}$$

where ρ_o is the density of the fluid at rest.

Hence, since the Laplacian operator commutes with the gradient operator $(\Delta \nabla = \nabla \Delta)$, we have

$$\Delta \mathbf{u} + k^2 \mathbf{u} = \frac{1}{\rho_o \omega^2}\, grad\left(\Delta p + k^2 p\right) = 0.$$

If the asymptotic expansion of the displacement due to the incident wave is written in the form

$$\mathbf{u}^i(\mathbf{r}) = e^{ik S^i(\mathbf{r})} \sum_{n=o}^{N} (ik)^{-n} u_n^i(\mathbf{r}) + o\left(k^{-N}\right), \tag{1.134}$$

then by substituting (1.134) in the Helmholtz equation verified by $\mathbf{u}^i$, and equating terms of similar order in k^{-1}, we see that S^i and $\mathbf{u}^i_n$ verify the eikonal and transport equations:

$$\left|\nabla S^i\right|^2 = 1, \tag{1.135}$$

$$\left(\Delta S^i + 2\nabla S^i \cdot \nabla\right) \mathbf{u}_n^i = -\Delta \mathbf{u}_{n-1}^i. \tag{1.136}$$

Now knowing the asymptotic expansion of the pressure field, we can derive the corresponding terms of the asymptotic expansion of the displacement by applying (1.133) to (1.114) and identifying the terms of the same order in k^{-1}. The result is

$$\mathbf{u}_n^i = \frac{i}{\rho_o \omega c_o}\left(\nabla S^i p_n^i + \nabla p_{n-1}^i\right), \tag{1.137}$$

which for the first two terms, $n = 0$ and $n = 1$, gives

$$\mathbf{u}_o^i = \frac{i}{\rho_o \omega c_o} \nabla S^i p_o^i, \tag{1.138}$$

$$\mathbf{u}_1^i = \frac{i}{\rho_o \omega c_o}\left(\nabla S^i p_1^i + \nabla p_o^i\right). \tag{1.139}$$

We assume that the displacement of the reflected ray field is given by an asymptotic expansion similar to (1.134):

$$\mathbf{u}^R(\mathbf{r}) = e^{ik S^R(\mathbf{r})} \sum_{n=o}^{N} (ik)^{-n}\, \mathbf{u}_n^R(\mathbf{r}) + o\left(k^{-N}\right), \tag{1.140}$$

where $S^R(\mathbf{r})$ and $\mathbf{u}_n^R(\mathbf{r})$ verify, respectively, the eikonal and transport equations, which are

$$\left|\nabla S^R\right|^2 = 1, \tag{1.141}$$

$$\left(\Delta S^R + 2\nabla S^R \cdot \nabla\right) \mathbf{u}_n^R = -\Delta \mathbf{u}_{n-1}^R. \tag{1.142}$$

Similarly, we assume that the refracted longitudinal and transverse waves in the solid are also ray fields, represented by the asymptotic expansions (1.79) and (1.80), verifying the eikonal and transport equations (1.81) and (1.82), respectively. These ray fields must satisfy the following boundary conditions at the interface:

1. The normal displacement is continuous through the interface.
2. The normal stress is continuous through the interface.
3. The tangential stress vanishes on the interface.

We consider the first boundary condition, which states that

$$\mathbf{n}\cdot\left(\mathbf{u}^i+\mathbf{u}^R\right)=\mathbf{n}\cdot\left(\mathbf{u}_\ell+\mathbf{u}_t\right),\quad \mathbf{r}\in S, \tag{1.143}$$

where $\mathbf{n}$ is the unit vector normal to the interface S at a point $M(\mathbf{r})$, pointing in the fluid.

Applying (1.143) to the asymptotic expansions (1.79), (1.80), (1.134), and (1.140), we see that this first condition involves the continuity of the phase at every point on the interface

$$k\,S^i\left(\mathbf{r}\right)=k\,S^R\left(\mathbf{r}\right)=k_\ell S_\ell\left(\mathbf{r}\right)=k_t S_t\left(\mathbf{r}\right),\quad \mathbf{r}\in S_e, \tag{1.144}$$

where S_e is that part of the interface S, which is directly submitted to the incident pressure field.

In equation (1.144), k is the wave number of the incident and reflected acoustic waves in the fluid (see appendix A), which supports only longitudinal waves, and k_ℓ and k_t are the wave numbers of the longitudinal and transverse elastic waves refracted in the solid. By taking the surface gradient of the eikonals in (1.144) and noting that

$$\boldsymbol{\nabla}S^i=\hat{s}^i\,,\ \boldsymbol{\nabla}S^R=\hat{s}^R\,,\ \boldsymbol{\nabla}S_\ell=\hat{s}^\ell\,,\ \boldsymbol{\nabla}S_t=\hat{s}^t$$

and that the surface gradient is equal to the projection of the gradient on the surface, we obtain

$$k\,\hat{s}^i\cdot\mathbf{t}=k\,\hat{s}^R\cdot\mathbf{t}=k_\ell\hat{s}^\ell\cdot\mathbf{t}=k_t\hat{s}^t\cdot\mathbf{t}, \tag{1.145}$$

where $\mathbf{t}$ is the unit vector in the direction of the projections of the gradients of the eikonals on the surface, since according to (1.144) all these projections are colinear. Equation (1.145) shows that $\hat{s}^i$, $\hat{s}^R$, $\hat{s}^\ell$, and $\hat{s}^t$ are all in the same plane. Hence $\hat{s}^R$, $\hat{s}^\ell$, and $\hat{s}^t$ are in the plane of incidence defined by the direction of propagation of the incident wave and the normal to the surface. If we denote by θ_i, θ_R, θ_ℓ, and θ_t the acute angles of the incident and refracted rays with the normal to the surface pointing in the fluid (see fig. 1.23), equation (1.145) can also be written in the following form:

$$k\sin\theta_i=k\sin\theta_R=k_\ell\sin\theta_\ell=k_t\sin\vartheta_t. \tag{1.146}$$

Thus the reflected wave satisfies the law of reflection $\theta_i=\theta_R=\theta$, and the refracted wave satisfies the Snell-Descartes law of refraction:

$$\sin\theta_i=\frac{k_\ell}{k}\sin\theta_\ell=\frac{k_t}{k}\sin\theta_t. \tag{1.147}$$

Since $k_\ell/k=c_o/c_\ell$, $k_t/k=c_o/c_t$ where c_o is the sound velocity in the fluid and where c_ℓ and c_t are the longitudinal and transverse elastic wave velocities in the solid, we also have

$$\sin\theta_i = \frac{c_o}{c_\ell}\sin\theta_\ell = \frac{c_o}{c_t}\sin\theta_t. \tag{1.148}$$

We know from the theory of elastic waves (see appendix B) that the velocity of longitudinal waves is always greater than that of transverse waves: $c_\ell > c_t$. Moreover, in most cases of interest, the velocity in the fluid is less than that of the longitudinal waves in the solid. It can also be less than the velocity of the transverse waves. For $c_o < c_t < c_\ell$, the configuration of the incident, reflected, and refracted waves are shown in figure 1.23.

We have therefore recovered the results obtained previously in section 1.2, by applying Fermat's principle. The discussion concerning the domain of existence of the refracted rays, given in section 1.2.2, also remains valid here.

In the present case, the incident, reflected, and refracted fields are more general than the GO field. They are ray fields—the amplitudes of which are represented by asymptotic expansions, including higher-order terms than GO. However, like the GO field, ray fields also verify the eikonal equation that governs the phase variation along a ray in accordance with Fermat's principle.

In order to derive formulas for the amplitudes of the different ray fields that intervene in the problem, we express the asymptotic expansions (1.79) and (1.80) giving the displacements generated by the longitudinal and transverse waves in the solid, with respect to the same asymptotic sequence as that of the incident ray field. By using the identity

$$\left(ik_\ell\right)^{-n} = \left(ik_o\right)^{-n}\left(\frac{c_o}{c_\ell}\right)^{-n}, \quad \left(ik_t\right)^{-n} = \left(ik_o\right)^{-n}\left(\frac{c_o}{c_t}\right)^{-n}, \tag{1.149}$$

the asymptotic expansions (1.79) can be written in the following form:

$$\mathbf{u}_\ell(\mathbf{r}) = e^{ik_\ell s^\ell(\mathbf{r})}\sum_{n=o}^{N}\left(ik\right)^{-n}\left(\frac{c_o}{c_\ell}\right)^{-n}\mathbf{u}_n^\ell(\mathbf{r}) + o\left(k_\ell^{-N}\right), \tag{1.150}$$

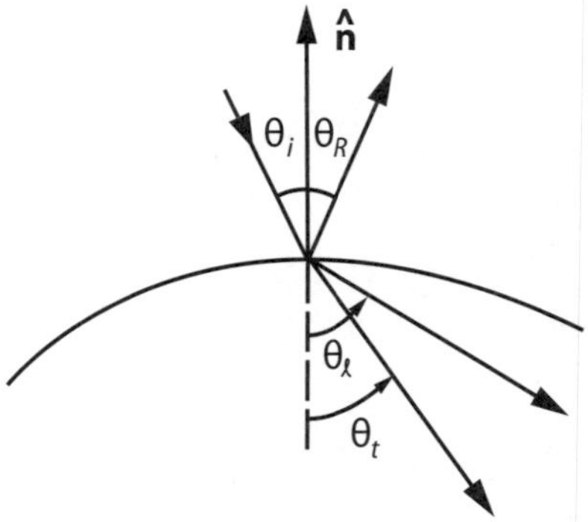

Figure 1.23. Incident, reflected, and refracted rays at a curved interface between a fluid and an elastic solid.

$$\mathbf{u}_t(\mathbf{r}) = e^{ik_t S^t(\mathbf{r})} \sum_{n=o}^{N} (ik)^{-n} \left(\frac{c_o}{c_t}\right)^{-n} \mathbf{u}_n^t(\mathbf{r}) + o\left(k_t^{-N}\right). \tag{1.151}$$

Applying (1.143) to (1.134), (1.140), (1.150), and (1.151) and identifying the terms of the same order, we obtain

$$\mathbf{n}\cdot\mathbf{u}_n^i + \mathbf{n}\cdot\mathbf{u}_n^R = \left(\frac{c_o}{c_\ell}\right)^{-n} \mathbf{n}\cdot\mathbf{u}_n^\ell + \left(\frac{c_o}{c_t}\right)^{-n} \mathbf{n}\cdot\mathbf{u}_n^t, \quad \mathbf{r}\in S. \tag{1.152}$$

For $n = 0$, (1.152) reduces to

$$\mathbf{n}\cdot\mathbf{u}_o^i + \mathbf{n}\cdot\mathbf{u}_o^R = \mathbf{n}\cdot\mathbf{u}_o^\ell + \mathbf{n}\cdot\mathbf{u}_o^t.$$

Moreover, $\mathbf{u}_o^i$ verifies (1.138), and a similar equation exists for $\mathbf{u}_o^R$. Hence $\mathbf{u}_o^i$ and $\mathbf{u}_o^R$ are oriented along ∇S^i and ∇S^R, respectively.

Similarly, according to (1.85), $\mathbf{u}_o^\ell$ is oriented along ∇S_ℓ while $\mathbf{u}_o^t$ is orthogonal to ∇S_t. Taking $\mathbf{u}_o^t$, as shown in figure 1.24, we obtain

$$\left(\mathbf{u}_o^R - \mathbf{u}_o^i\right)\cos\theta = -\mathbf{u}_o^\ell \cos\theta_\ell + \mathbf{u}_o^t \sin\theta_t. \tag{1.153}$$

For $n = 1$, according to (1.137) and the third equation of (1.81) and (1.82), we have

$$\mathbf{u}_1^i = \frac{i}{\rho_o \omega c_o}\left(\nabla S^i p_1^i + \nabla p_o^i\right), \; \mathbf{u}_1^R = \frac{i}{\rho_o \omega c_o}\left(\nabla S^R p_1^R + \nabla p_o^R\right),$$
$$\nabla S_\ell \times \mathbf{u}_1^\ell = -\nabla \times \mathbf{u}_o^\ell, \tag{1.154}$$

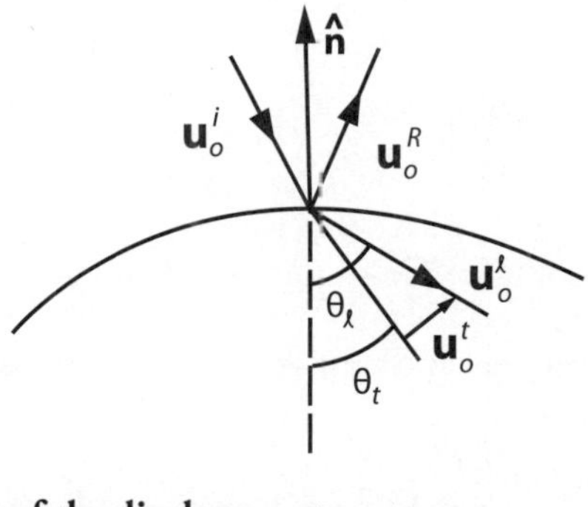

Figure 1.24. Orientation of the displacement vectors.

$$\nabla S_t \cdot \mathbf{u}_1^t = -\nabla \cdot \mathbf{u}_o^t.$$

These equations show that $\mathbf{u}^i_1$, $\mathbf{u}^R_1$, and $\mathbf{u}_1^\ell$ are no longer oriented along the corresponding rays and that $\mathbf{u}^t_1$ is not perpendicular to the associated refracted ray.

We consider now the second and third boundary conditions. In order to express the stress tensor, we need to introduce a coordinate system. Let x_1, x_2, x_3 be a local rectangular coordinate system, and let the $x_1\, x_2$ plane coincide with the incident plane. Let $\mathbf{i}_1$, $\mathbf{i}_2$, $\mathbf{i}_3$ be the unit vectors of the coordinate system with $\mathbf{i}_2$ along the normal to the interface pointing toward the fluid. If we denote by σ_{ij} the stress tensor (see appendix B), the second and third boundary conditions imply that

$$\sigma_{22}^i + \sigma_{22}^R = \sigma_{22}^\ell + \sigma_{22}^t \text{ on } s, \tag{1.155}$$

$$\sigma_{21}^\ell + \sigma_{21}^t = 0 \text{ on } s. \tag{1.156}$$

The other tangential components of the stress tensor are identically zero since in a fluid the stress corresponds to the pressure, which is normal to the interface, and since $\mathbf{u}_\ell$ and $\mathbf{u}_t$ are in the $x_1\, x_2$ plane. Thus

$$\sigma_{23}^i = \sigma_{23}^R = \sigma_{23}^\ell = \sigma_{23}^t = 0. \tag{1.157}$$

Taking into account that the total pressure in the fluid is normal to the interface and in opposite direction to the normal, the forces acting on the interface by the wave motion in the fluid are given by

$$\sigma_{22}^i + \sigma_{22}^R = p^i + p^R. \tag{1.158}$$

The components of the stress tensor in the elastic medium are given by

$$\sigma_{ij} = \lambda \delta_{ij} u_{k,k} + \mu\left(u_{i,j} + u_{j,i}\right), \tag{1.159}$$

where $u_{i,j} = \dfrac{\partial u_i}{\partial x_j}$ and where λ and μ are the Lamé coefficients (see appendix B).

Applying (1.159) to the longitudinal and transverse wave motions, we obtain the following:

$$\sigma_{22}^\ell = \lambda \nabla \cdot \mathbf{u}_\ell + 2\mu \frac{\partial u_2^\ell}{\partial x_2}, \tag{1.160}$$

$$\sigma_{22}^{t} = \lambda \nabla \cdot \mathbf{u}_{t} + 2\mu \frac{\partial u_2^t}{\partial x_2}, \tag{1.161}$$

$$\sigma_{21}^{\ell} = \mu \left(\frac{\partial u_2^\ell}{\partial x_1} + \frac{\partial u_1^\ell}{\partial x_2} \right), \tag{1.162}$$

$$\sigma_{21}^{t} = \mu \left(\frac{\partial u_2^t}{\partial x_1} + \frac{\partial u_1^t}{\partial x_2} \right). \tag{1.163}$$

By substituting to $\mathbf{u}^i$, $\mathbf{u}^R$, $\mathbf{u}_\ell$, and $\mathbf{u}_t$ their asymptotic expansions, the boundary conditions (1.155) and (1.156) can be written as follows:

$$\begin{aligned} p_n^i + p_n^R = ik_\ell & \left\{ \lambda \left(\mathbf{\nabla} S_\ell \cdot \mathbf{u}_n^\ell + \mathbf{\nabla} \cdot \mathbf{u}_{n-1}^\ell \right) + 2\mu \left[\left(\mathbf{i}_2 \cdot \mathbf{\nabla} S_\ell \right) u_{2,n}^\ell + \frac{\partial u_{2,n-1}^\ell}{\partial x_2} \right] \right\} \\ + ik_t & \left\{ \lambda \left(\mathbf{\nabla} S_t \cdot \mathbf{u}_n^t + \mathbf{\nabla} \cdot \mathbf{u}_{n-1}^t \right) + 2\mu \left[\left(\mathbf{i}_2 \cdot \mathbf{\nabla} S_t \right) u_{2,n}^t + \frac{\partial u_{2,n-1}^t}{\partial x_2} \right] \right\}, \end{aligned} \tag{1.164}$$

$$\begin{aligned} & ik_\ell \left(\mathbf{i}_1 \cdot \mathbf{\nabla} S_\ell \right) u_{2,n}^\ell + \frac{\partial u_{1,n-1}^\ell}{\partial x_2} + ik_t \left(\mathbf{i}_1 \cdot \mathbf{\nabla} S_t \right) u_{2,n}^t + \frac{\partial u_{1,n-1}^t}{\partial x_2} \\ & + ik_\ell \left(\mathbf{i}_2 \cdot \mathbf{\nabla} S_\ell \right) u_{1,n}^\ell + \frac{\partial u_{2,n-1}^\ell}{\partial x_2} + ik_t \left(\mathbf{i}_2 \cdot \mathbf{\nabla} S_t \right) u_{1,n}^t + \frac{\partial u_{2,n-1}^t}{\partial x_2} = 0. \end{aligned} \tag{1.165}$$

For $n = 0$, these equations reduce to

$$\begin{aligned} & p_o^i + p_o^R = ik_\ell \left[\lambda \left(\mathbf{\nabla} S_\ell \cdot \mathbf{u}_o^\ell \right) + 2\mu \left(\mathbf{i}_2 \cdot \mathbf{\nabla} S_\ell \right) u_{2,o}^\ell \right] + \\ & ik_t \left[\lambda \left(\mathbf{\nabla} S_t \cdot \mathbf{u}_o^t \right) + 2\mu \left(\mathbf{i}_2 \cdot \mathbf{\nabla} S_t \right) u_{2,o}^t \right], \end{aligned} \tag{1.166}$$

$$k_\ell \left(\mathbf{i}_1 \cdot \mathbf{\nabla} S_\ell \right) u_{2,o}^\ell + k_t \left(\mathbf{i}_1 \cdot \mathbf{\nabla} S_t \right) u_{2,o}^t + k_\ell \left(\mathbf{i}_2 \cdot \mathbf{\nabla} S_\ell \right) u_{1,o}^\ell + k_t \left(\mathbf{i}_2 \cdot \mathbf{\nabla} S_t \right) u_{1,o}^t = 0. \tag{1.167}$$

Since $\mathbf{u}_o^\ell$ is directed along the ray and $\mathbf{u}_o^t$ is orthogonal to the ray, we have

$$\begin{aligned} & \mathbf{\nabla} S_\ell \cdot \mathbf{u}_o^\ell = u_o^\ell, \\ & \mathbf{\nabla} S_\ell \cdot \mathbf{u}_o^t = 0, \end{aligned} \tag{1.168}$$

and consequently

$$u_{1,o}^{\ell} = u_o^{\ell} \sin \theta_{\ell}, \quad u_{2,o}^{\ell} = -u_o^{\ell} \cos \theta_{\ell},$$
$$u_{1,o}^{t} = u_o^{t} \cos \theta_{t}, \quad u_{2,o}^{t} = u_o^{t} \sin \theta_{t}. \tag{1.169}$$

Hence equations (1.166) and (1.167) reduce, respectively, to equations (1.170) and (1.171), given by

$$p_o^i + p_o^R = ik_{\ell}\, u_o^{\ell}\left(\lambda + 2\mu \cos^2 \theta_{\ell}\right) - ik_t\, u_o^t \sin 2\theta_t, \tag{1.170}$$

$$k_{\ell} \sin 2\theta_{\ell}\, u_o^{\ell} - k_t \cos 2\theta_t\, u_o^t = 0. \tag{1.171}$$

According to (1.138) and (1.135) and their homologues for p_o^R, the first member of (1.170) can also be expressed in terms of the amplitudes u_o^i and u_o^R of the displacements:

$$p_o^i + p_o^R = -i\,\rho_o \omega c_o \left(u_o^i + u_o^R\right). \tag{1.172}$$

Substituting (1.172) in (1.170), we obtain with (1.153) and (1.171) a system of three linear equations for the unknowns u_o^R, u_o^{ℓ}, u_o^t. Since these equations are identical to those obtained in section B.8 of appendix B for a planar interface, their solutions are the same.

For the reflection coefficient R in the fluid and the transmission coefficients T^{ℓ} and T^{t} of the longitudinal and transverse waves in the elastic solid, we have (see section B.8 of appendix B)

$$R = \frac{u_o^R}{u_o^i} = \frac{Z_{\ell} \cos^2 2\theta_t + Z_t \sin^2 2\theta_t - Z_o}{Z_{\ell} \cos^2 2\theta_t + Z_t \sin^2 2\theta_t + Z_o}, \tag{1.173}$$

$$T_{\ell} = \frac{u_o^{\ell}}{u_o^i} = \frac{2c_o}{c_{\ell}} \rho_1^{-1} \rho_o Z_{\ell} \frac{\cos 2\theta_t}{Z_{\ell} \cos^2 2\theta_t + Z_t \sin^2 2\theta_t + Z_o}, \tag{1.174}$$

$$T_t = \frac{u_o^t}{u_o^i} = -\frac{2c_o}{c_t} \rho_1^{-1} \rho_o Z_t \frac{\sin 2\theta_t}{Z_{\ell} \cos^2 2\theta_t + Z_t \sin^2 2\theta_t + Z_o}, \tag{1.175}$$

where Z_o, Z_{ℓ}, Z_t are given by (see equation (B.70) of appendix B)

$$Z_o = \frac{\rho_o c_o}{\cos \theta}, \quad Z_{\ell} = \frac{\rho_1 c_{\ell}}{\cos \theta_{\ell}}, \quad Z_t = \frac{\rho_1 c_t}{\cos \theta_t}, \tag{1.176}$$

where ρ_1 is the mass density of the solid.

For $n > 0$, equations (1.168) need to be replaced by the following expressions, given in (1.81) and (1.82):

$$\nabla S_\ell \times \mathbf{u}_n^\ell = -\nabla \times \mathbf{u}_{n-1}^\ell, \tag{1.177}$$

$$\nabla S_t \cdot \mathbf{u}_n^t = -\nabla \cdot \mathbf{u}_{n-1}^t. \tag{1.178}$$

Moreover, according to (1.137), equation (1.172) needs to be replaced by

$$\mathbf{u}_n^i + \mathbf{u}_n^R = \frac{i}{\rho_o \omega c_o}\left(\nabla S^i\, p_n^i + \nabla S^R\, p_n^R + \nabla\, p_{n-1}^i + \nabla\, p_{n-1}^R\right). \tag{1.179}$$

By inserting (1.179) into (1.152), we obtain

$$\begin{aligned}\mathbf{n}\cdot\left(\mathbf{u}_n^i + \mathbf{u}_n^R\right) &= \frac{i}{\rho_o \omega c_o}\left[\left(p_n^R - p_n^i\right)\cos\theta + \frac{\partial p_{n-1}^i}{\partial x_2} + \frac{\partial p_{n-1}^R}{\partial x_2}\right] \\ &= \left(\frac{c_o}{c_\ell}\right)^{-n} u_{2,n}^\ell + \left(\frac{c_o}{c_t}\right)^{-n} u_{2,n}^t.\end{aligned} \tag{1.180}$$

The system of five linear equations (1.164), (1.165), (1.177), (1.178), and (1.180) can, in principle, be solved recursively for the five unknowns constituted by the components $u_{1,n}^\ell$, $u_{2,n}^\ell$, $u_{1,n}^t$, and $u_{2,n}^t$ of the displacements $\mathbf{u}_n^\ell$ and $\mathbf{u}_n^t$ and by the pressure p_n^R. However, the solution of this system of equations is cumbersome, even for $n = 1$. The reason is that at each order n, we have to calculate the derivatives of the solution obtained at order $n - 1$. For $n = 1$, for instance, we need the values of the derivatives $\partial p_o^R / \partial x_2$, $\partial u_{1,o}^\ell / \partial x_2$, $\partial u_{2,o}^\ell / \partial x_2$, $\partial u_{1,o}^t / \partial x_2$, and $\partial u_{2,o}^t / \partial x_2$ on the interface S. These derivatives can be calculated by using (1.112) and (1.113). They imply the knowledge of the first and second derivatives of the principal radii of curvature of the reflected and refracted wave fronts.

1.8. ACOUSTIC FIELD DIFFRACTED BY THE EDGE OF AN IMPENETRABLE WEDGE

We consider now an object, the surface of which has a discontinuity line in the tangent plane forming a sharp edge that may be straight or curved (fig. 1.25). Away from this discontinuity line, the surface is assumed to be smooth and its principal radii of curvature are large in terms of the wavelength. The object is embedded in a fluid.

An incident ray field of acoustic pressure $p^i(\mathbf{r})$, defined by the asymptotic expansion (1.114), propagates in the fluid and hits the boundary surface of the object, giving rise at an arbitrary point $M(\mathbf{r})$ in the fluid to a scattered field $p^s(\mathbf{r})$, which can be written as the sum of a general reflected field $p^R(\mathbf{r})$ given by (1.117) and an additional field $p^d(\mathbf{r})$, due to the presence of the edge, called the diffracted field:

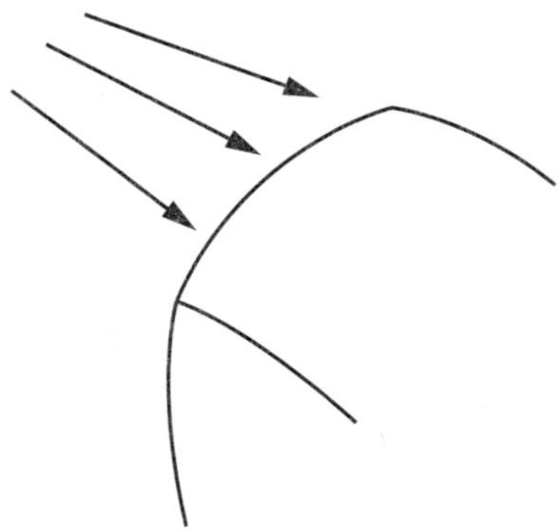

Figure 1.25. General shape of a curved wedge.

$$p^s(\mathbf{r}) = p^R(\mathbf{r}) + p^d(\mathbf{r}). \tag{1.181}$$

We suppose that p^d has the general form

$$p^d(\mathbf{r}) = \frac{1}{\sqrt{k}} e^{ik S^d(\mathbf{r})} \sum_{n=o}^{N} (ik)^{-n} p_n^d(\mathbf{r}) + o(k^{-N}). \tag{1.182}$$

The ansatz (1.182) together with the decomposition (1.181) are suggested by the solution of the diffraction problem of an incident plane wave by a straight wedge, which is the corresponding model or canonical problem whose solution is given in chapter 2 for different kinds of boundary conditions.

The form of the asymptotic solution given by (1.182) is not valid close to the edge and to the shadow boundaries of the incident and reflected fields where the field of rays is singular. If we insert (1.182) into the Helmholtz equation (1.65) and equate the terms of equal powers in k^{-1}, we see that $S^d(\mathbf{r})$ and $p^d(\mathbf{r})$ verify the eikonal and transport equations,

$$|\nabla S^d|^2 = 1, \tag{1.183}$$

$$p_n^d \Delta S^d + 2\nabla S \cdot \nabla p_n^d = -\Delta p_{n-1}^d, \tag{1.184}$$

the solutions of which have been established in section 1.5.

The diffracted field is therefore a ray field. Moreover, for an impenetrable object, the diffracted rays emanate from the edge since those emanating from a regular point of the surface are already accounted for in the expression of the general reflected field. Figure 1.26 shows an astigmatic pencil of diffracted rays. Since all diffracted rays pass through the edge, the edge is one of the caustic surfaces (here degenerated into a line) of the family of edge-diffracted rays.

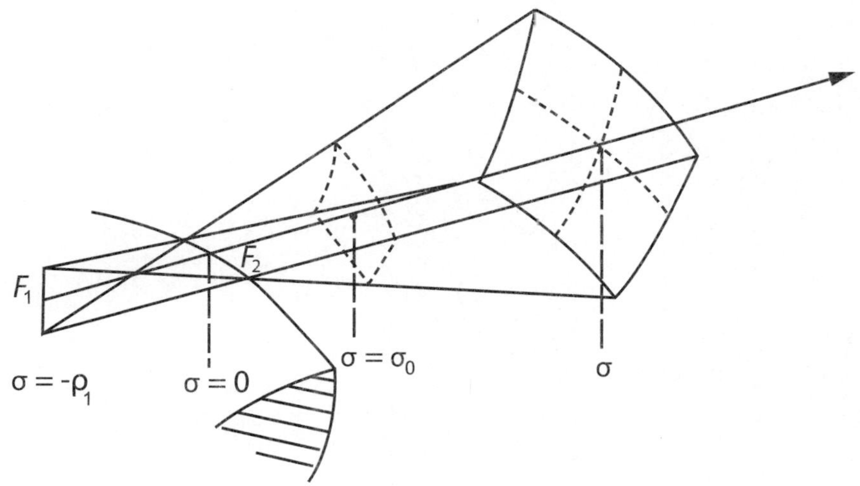

Figure 1.26. Astigmatic pencil of diffracted rays.

The eikonal equation is a first-order partial differential equation, the solution of which is completely determined if we know the value of the phase function on an initial wave front. On a regular surface, the phase function of the reflected field is related to the phase function of the incident field by the boundary conditions. This procedure cannot be applied to the present problem, since the boundary conditions are not defined on an edge. However, since the boundary conditions on a regular surface impose the continuity of phase, we can extend this property by imposing the continuity of phase on the edge:

$$S^i(\mathbf{r}) = S^d(\mathbf{r}), \quad \mathbf{r} \in C, \tag{1.185}$$

where C is the edge. This condition completely defines the solution of the eikonal equation.

According to (1.185), the projection of the gradient ∇S^i of the eikonal of the incident field, on the tangent $\mathbf{t}$ to the edge, must be equal to the projection of the gradient ∇S^d of the eikonal of the diffracted field on the same tangent. Hence, since $\nabla S^i = \mathbf{s}^i$, $\nabla S^d = \mathbf{s}^d$, we have

$$\mathbf{s}^i \cdot \mathbf{t} = \mathbf{s}^d \cdot \mathbf{t} = \cos\beta, \tag{1.186}$$

which is the law of edge diffraction, already established in section 1.2.2 by applying the generalized Fermat principle: the diffracted rays at a given point on the edge make the same angle β with the tangent to the edge as the incident ray. The diffracted rays therefore form a cone that is known as the Keller cone (fig. 1.27).

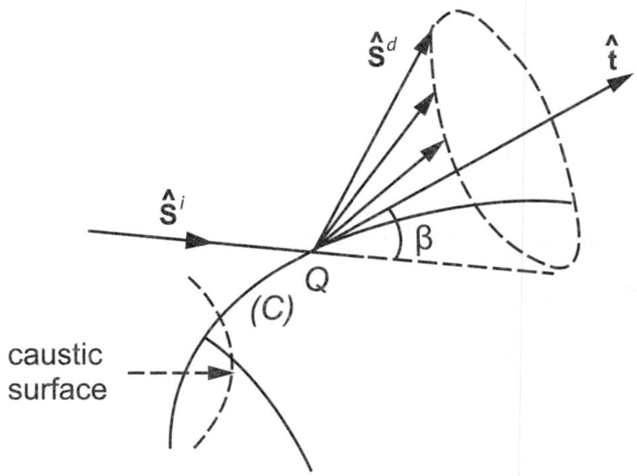

Figure 1.27. Keller's cone and unit vectors s^i, s^d, t.

In order to completely determine the diffracted field, it is necessary to find a way to relate the amplitudes p_n^d of the diffracted field to the amplitudes p_n^i of the incident field, which are known. Since the edge C is a caustic of the diffracted field, the field predicted by the solution (1.112) of the transport equation is infinite on C, and it is therefore not possible to match the amplitude of the diffracted field with that of the incident field on C. However, if we relate the diffracted field at a point σ on the diffracted ray to the diffracted field at a reference point σ_o (not situated on the edge ($\sigma_o \neq 0$)) by using (1.112), we obtain

$$p_n^d(\sigma) = \sqrt{\left|\frac{J(\sigma_o)}{J(\sigma)}\right|}\, p_n^d(\sigma_o) - \frac{1}{2}\int_{\sigma_o}^{\sigma} \sqrt{\left|\frac{J(\sigma')}{J(\sigma)}\right|}\, \Delta p_{n-1}^d(\sigma')d\sigma'. \tag{1.187}$$

By writing the integral along the ray in the form

$$\int_{\sigma_o}^{\sigma} = \int_{o}^{\sigma} - \int_{o}^{\sigma_o}, \tag{1.188}$$

where the finite parts of the divergent integrals are taken on the right-hand side of (1.188), we get an alternative form for (1.187):

$$\begin{aligned}\sqrt{|J(\sigma)|}\, p_n^d(\sigma) + \frac{1}{2}\int_{o}^{\sigma}\sqrt{|J(\sigma')|}\, \Delta p_{n-1}^d(\sigma')d\sigma', \\ = \sqrt{|J(\sigma_o)|}\, p_n^d(\sigma_o) + \frac{1}{2}\int_{o}^{\sigma_o}\sqrt{|J(\sigma')|}\, \Delta p_{n-1}^d(\sigma')d\sigma'.\end{aligned} \tag{1.189}$$

Since σ_o is arbitrary, we see that the quantity

$$\delta_n = \sqrt{|J(\sigma)|}\, p_n^d(\sigma) + \frac{1}{2}\int_o^{\sigma} \sqrt{|J(\sigma')|}\, \Delta p_{n-1}^d(\sigma')\,d\sigma' \tag{1.190}$$

is independent of σ. If δ_n is known, we can calculate the pressure along the diffracted ray by applying the recurrence relation

$$p_n^d(\sigma) = \frac{\delta_n}{\sqrt{|J(\sigma)|}} - \frac{1}{2}\int_{\sigma_o}^{\sigma} \sqrt{\left|\frac{J(\sigma')}{J(\sigma)}\right|}\, \Delta p_{n-1}^d(\sigma')\,d\sigma'. \tag{1.191}$$

Since δ_n is constant along a diffracted ray, it can be determined by taking the limit of the right-hand side of (1.190) when σ tends to zero:

$$\delta_n = \lim_{\sigma \to o} \left(p_n^d(\sigma)\sqrt{|J(\sigma)|} \right). \tag{1.192}$$

According to (1.110) and the fact that the caustic line F_2 lies on the edge, we have

$$\sqrt{|J(\sigma)|} = a\sqrt{(\rho_1 + \sigma)\sigma}, \tag{1.193}$$

where a is a constant and ρ_1 is the radius of curvature of the wave front at the diffraction point on the edge. Inserting (1.193) into (1.192) yields

$$\delta_n = a\sqrt{\rho_1}\, \lim_{\sigma \to o} \left(\sqrt{\sigma}\, p_n^d(\sigma) \right). \tag{1.194}$$

Since δ_n is a constant, the limit on the right-hand side of (1.194) exists. If we denote this limit by d_n and insert (1.194) into (1.191), we obtain, by taking into account (1.193),

$$p_n^d(\sigma) = d_n\sqrt{\frac{\rho_1}{\sigma(\rho_1 + \sigma)}} - \frac{1}{2}\int_{\sigma_o}^{\sigma} \sqrt{\frac{\sigma'(\rho_1 + \sigma')}{\sigma(\rho_1 + \sigma)}}\, \Delta p_{n-1}^d(\sigma')\,d\sigma'. \tag{1.195}$$

For $n = 0$, (1.195) reduces to

$$p_o^d(\sigma) = d_o\sqrt{\frac{\rho_1}{\sigma(\rho_1 + \sigma)}}. \tag{1.196}$$

In order to determine d_o, we compare (1.196) to the zero-order term of the asymptotic solution of the corresponding canonical problem—namely, the diffraction of an incident plane wave by the straight wedge tangent to the original curved wedge at the point of diffraction Q and submitted to the same boundary conditions as the curved wedge (fig. 1.28).

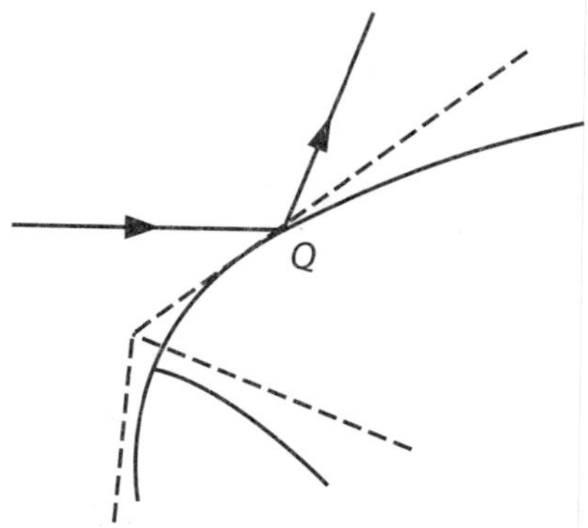

Figure 1.28. Straight wedge tangent to the curved wedge at the diffraction point *Q*.

Solutions of the canonical problem of a straight, impenetrable wedge for different boundary conditions are given in chapter 2. The dominant term of their asymptotic expansion has the general form

$$p_o^d(\sigma) = p_o^i(Q) D \frac{1}{\sqrt{\sigma}}, \tag{1.197}$$

where D is the diffraction coefficient and $p_o^i(Q)$ is the zero-order term of the amplitude of the incident field at the point of diffraction Q.

Comparing (1.197) to (1.196) for σ tending to zero yields

$$d_o = p_o^i(Q) D \tag{1.198}$$

and finally

$$p_o^d(\sigma) = p_o^i(Q) D \sqrt{\frac{\rho_1}{\sigma(\rho_1 + \sigma)}}. \tag{1.199}$$

In order to derive the relation (1.199), we have made the assumption that close to the edge, the zero-order term of the field diffracted by a curved wedge is identical to the zero-order term of the same field diffracted by a straight wedge tangent to the curved wedge at the diffraction point Q. We present hereafter a way to justify this statement.

Consider a rectangular coordinate system $OXYZ$ with its origin O at the diffraction point Q, the OX axis tangent to the edge at Q, the OX axis tangent to the face S_1 of the curved wedge at Q, and the OZ axis oriented such that the system $OXYZ$ is direct (fig. 1.29).

Let C_1 and C_2 be the intersections of the plane perpendicular to the edge at O with the faces S_1 and S_2, respectively. Since the edge is a line caustic, the amplitude of the diffracted field varies rapidly in a direction perpendicular to the edge, in a domain around O of radius λ, where λ is the

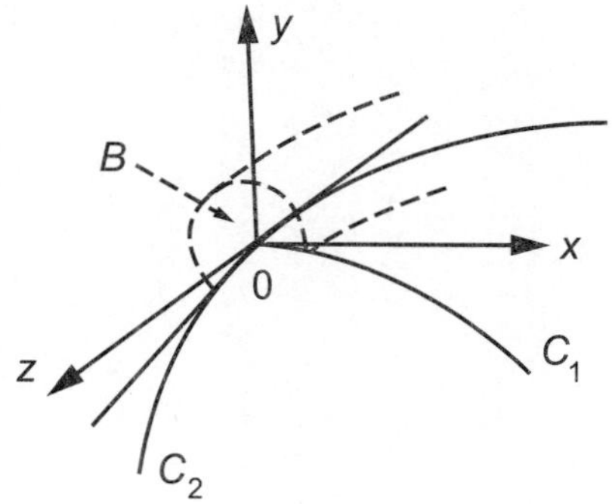

Figure 1.29. Curved wedge: Local coordinate system and caustic boundary layer *B*.

wavelength (see fig. 1.29). On the other hand, it varies very slowly in the same domain, in a direction parallel to the edge. We can therefore replace the original three-dimensional curved wedge with a two-dimensional curved wedge with its cross-section delimited by the curves C_1 and C_2.

Let the face C_1 be defined by the equation $y = f(x)$. Close to $x = o$, $f(x)$ can be expanded in a Taylor series. Since OX is tangent to C_1 at O, $f(o) = f'(x) = 0$. Thus the Taylor series gives

$$y = f(x) = \frac{x^2}{2} f''(o) + 0(x^3). \tag{1.200}$$

In the caustic region close to the edge, the pressure field is the solution of the Helmholtz equation,

$$(\Delta + k^2) p(x, y) = 0, \tag{1.201}$$

which satisfies the boundary conditions on C_1. Introducing the stretched variables

$$x' = kx, \quad y' = ky, \tag{1.202}$$

where k is the wave number in the fluid, the Helmholtz equation transforms in

$$(\Delta' + 1) p(k^{-1}x'; k^{-1}y') = 0, \tag{1.203}$$

where $\Delta' = \frac{\partial^2}{\partial x'^2} + \frac{\partial^2}{\partial y'^2}$. Inserting (1.202) into (1.200) gives

$$y' = \frac{1}{2} k^{-1} f''(o) x'^2. \tag{1.204}$$

Hence y' is of a lower order in k^{-1} than x' and can therefore be put equal to zero in the boundary conditions. For a soft body or a hard body, for instance, the corresponding boundary conditions are

$$\begin{aligned} &\text{Soft: } p(x,y) = 0 \text{ on } C_1, \\ &\text{Hard: } \frac{\partial p}{\partial n}(x,y) = 0 \text{ on } C_1, \end{aligned} \tag{1.205}$$

where n is the distance along the outward normal to C_1. These transform into

$$\begin{aligned} &p\left(k^{-1}x',o\right) = 0, \\ &\frac{\partial p}{\partial y'}\left(k^{-1}x',o\right) = 0, \end{aligned} \tag{1.206}$$

which corresponds to the boundary conditions on a planar surface.

Similar equations can be written for the boundary conditions on C_2 by taking the OX axis tangent to C_2 at O. Hence at zero order in k^{-1} the solution of the original problem in the caustic region for a soft or hard wedge is identical to that of the tangent plane approximation. The generalization of this result to a mixed boundary condition is straightforward.

Another comment concerning the derivation of the relation (1.199) is of importance. The constant δ_o has been obtained as the limit of a quantity (see equation (1.194)) involving the zero-order term of the diffracted pressure field when σ tends to zero. But as we mentioned before, the asymptotic expansion (1.182) of the diffracted field is not valid in the caustic region close to the edge. Similarly the same restriction applies when we take the limit, when σ tends to zero, of the zero-order term of the field diffracted by a straight wedge.

However, since in the caustic region the zero-order term of the exact asymptotic expansion is the same for a curved wedge or a straight wedge tangent to the curved wedge at the point of diffraction, the zero-order terms of the asymptotic expansions (1.182) and (1.197), valid outside the caustic region, must be the same at the peripheries of this region in order to match to the common solution inside the caustic region. Therefore they have the same limit when σ tends to zero.

The same procedure cannot be extended to higher-order terms, since the tangent plane approximation is not valid in the caustic region for these terms.

For $n = 1$, the constant d_1 depends on the radius of curvature of the edge and on the radii of curvature of the faces, at the diffraction point. For $n > 1$, higher-order derivatives of these geometrical parameters enter in the constants d_n.

In order to solve the problem for $n > 0$, another technique called the *boundary layer method* [1.18] can be used. It constructs a different asymptotic expansion in the caustic region, written in stretched coordinates. This expression is substituted into the Helmholtz equation and into the boundary, radiation, and edge conditions. Usually a recurrent sequence of simpler problems appear by equating terms of similar order in k^{-1}. If the ansatz is correct, these problems can be solved step-by-step up to any chosen order. In the caustic region, the expansion is then matched with the outer expansion. However, even for $n = 1$, this problem has not been solved in the most general case of a curved three-dimensional wedge. Some results can be found in scientific literature [1.19] for the first-order term of the asymptotic expansion, in the case of a two-dimensional curved wedge.

For our purposes, we do not present the techniques that permit us to have the first-order term of the asymptotic expansion of the diffracted pressure field. Since this term is of order $k^{-3/2}$, its influence at high frequencies is of the same order as the diffraction by a discontinuity in curvature and is smaller than a double first-order diffraction by two wedges, which is of order k^{-1}. It can therefore be neglected in practical applications of the theory.

Coming back to the fundamental formula (1.199) for curved wedge diffraction, we see that in order to calculate the diffracted field, we need the diffraction coefficient D of a straight wedge and the radius of curvature ρ_2 of the diffracted wave front, which is also the distance from Q to the second nappe of the caustic (the first nappe being degenerated into the line formed by the edge of the curved wedge). Formulas for the diffraction coefficient are derived in chapters 2 and 3 from the asymptotic solution of the canonical problem of a plane wave diffracted by a straight wedge, with different boundary conditions. These formulas can be directly inserted into (1.199) for the calculation of the diffracted pressure field.

The calculation of the caustic distance ρ_2 in (1.199) is not a trivial matter for curved wedge diffraction. We will derive hereafter a formula for ρ_2, for a curved torsionless edge, by employing differential geometry.

We start with the law of edge diffraction (see (1.186), here $\boldsymbol{\sigma}^i = \mathbf{s}^i, \boldsymbol{\sigma} = \mathbf{s}^d$),

$$\mathbf{t} \cdot \boldsymbol{\sigma}^i = \mathbf{t} \cdot \boldsymbol{\sigma} = \cos\beta, \tag{1.207}$$

where $\mathbf{t}$ is the unit vector tangent to the edge at the point of diffraction Q, $\boldsymbol{\sigma}^i$ is the unit vector along the incident ray, and $\boldsymbol{\sigma}$ is the unit vector along the diffracted ray (fig. 1.30[a]).

Consider a point Q' close to Q. If we denote by η the curvilinear abscissa along the edge and by $d\eta$ the small displacement QQ', the law of diffraction applied at Q' gives

$$\left(\mathbf{t} + \frac{d\mathbf{t}}{d\eta}d\eta\right) \cdot \left(\boldsymbol{\sigma}^i + \frac{d\boldsymbol{\sigma}^i}{d\eta}d\eta\right) = \left(\mathbf{t} + \frac{d\mathbf{t}}{d\eta}d\eta\right) \cdot \left(\boldsymbol{\sigma} + \frac{d\boldsymbol{\sigma}}{d\eta}d\eta\right). \tag{1.208}$$

Taking (1.207) into account and retaining only the first-order terms in (1.208) yields

$$\frac{d\mathbf{t}}{d\eta} \cdot \left(\boldsymbol{\sigma}^i - \boldsymbol{\sigma}\right) = \mathbf{t} \cdot \left(\frac{d\boldsymbol{\sigma}}{d\eta} - \frac{d\boldsymbol{\sigma}^i}{d\eta}\right). \tag{1.209}$$

Since the diffracted ray pencil converges at the caustic point O, we have $\rho_2 = OQ$ and $d\boldsymbol{\sigma} = \mathbf{s}\rho_2^{-1}ds$, where s is the curvilinear abscissa along the circle of center O situated in the plane $\mathbf{t}$, $\boldsymbol{\sigma}$ and passing through Q and where $\mathbf{s}$ is the unit vector tangent at Q to this circle (see fig. 1.30[b]). Hence $ds = d\eta \sin\beta$, $\mathbf{t} \cdot \mathbf{s} = \sin\beta$, and

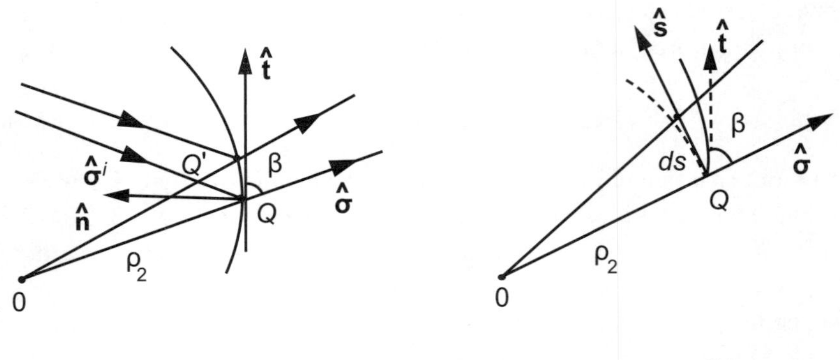

Figure 1.30. Diffracted rays emanating from the virtual caustic point *O* in the plane formed by *O* and the tangent t.

$$\mathbf{t}\cdot\frac{d\boldsymbol{\sigma}}{d\eta}=\frac{1}{\rho_2}\sin^2\beta. \tag{1.210}$$

Similarly,

$$\mathbf{t}\cdot\frac{d\boldsymbol{\sigma}^i}{d\eta}=\frac{1}{\rho^i}\sin^2\beta, \tag{1.211}$$

where ρ^i is the radius of curvature of the incident wave front in the plane formed by $\boldsymbol{\sigma}^i$ and $\mathbf{t}$. Inserting (1.210) and (1.211) into (1.209) and noting (the Frenet-Serret formula)

$$\frac{d\mathbf{t}}{d\eta}=\frac{\mathbf{n}}{a},$$

where a is the principal radius of curvature of the edge at Q and $\mathbf{n}$ is the unit vector of the principal normal to the edge at Q directed toward the center of curvature, we obtain the final result

$$\frac{1}{\rho_2}=\frac{1}{\rho^i}+\frac{\mathbf{n}\cdot\left(\boldsymbol{\sigma}^i-\boldsymbol{\sigma}\right)}{a\sin^2\beta}. \tag{1.212}$$

In conclusion, we see that formulas (1.199) and (1.212) allow us to calculate the field diffracted by an impenetrable curved wedge in the principal order, whatever the boundary conditions are, once we know the solution of the canonical problem.

1.9. ACOUSTIC FIELD IN THE SHADOW ZONE OF A SMOOTH CONVEX OBJECT

We suppose that the object is impenetrable, its boundary conditions corresponding to a soft, hard, or impedance surface or to a convex elastic shell separating a fluid from vacuum. In section 1.6, we saw that GA predicts a vanishing field in the shadow zone of a smooth convex object. The reason is that the asymptotic expansion of the Luneburg-Kline type given by (1.114) is not adapted to this zone. In reality, the solution of the circular cylinder problem, which is the associated canonical problem derived in chapter 2, shows that the field decreases as $\exp\left(-\alpha k^{1/3}\right)$, $\alpha > 0$. In other words, the decay is faster than algebraic. Locally, the field has the character of an inhomogeneous plane wave. To represent such a wave, we have two possibilities: The first one is to extend the Luneburg-Kline expansion so that it represents a complex wave with phase and amplitude functions that are both complex. The second one is to retain a real representation and to augment the phase factor $\exp[ik\, S^d(\mathbf{r})]$ by an exponential attenuation factor.

When adopting the second representation, the solution of the canonical problem in the deep shadow zone (outside the penumbra region) and far from the surface suggests the following form for the total pressure field, known as the Friedlander Keller expansion [1.20]:

$$p(\mathbf{r}) = e^{ikS(\mathbf{r}) + ik^{1/3}\varphi(\mathbf{r})} \sum_{n=o}^{N} (ik)^{-\frac{n}{3}} p_n(\mathbf{r}) + O\left(k^{-\frac{N}{3}}\right). \tag{1.213}$$

Inserting this expansion into the Helmholtz equation and ordering the terms according to increasing powers of $k^{1/3}$, we obtain to $O(k^2)$ the following result:

$$|\nabla S|^2 = 1.$$

Hence far from the object, the eikonal $\mathbf{S}(\mathbf{r})$ satisfies the eikonal equation of GA.

We have seen in section 1.2.2 that the Fermat principle predicts creeping rays that are excited at the shadow boundary Γ, follow geodesics on the object, and detach tangentially from it, as shown in figure 1.31.

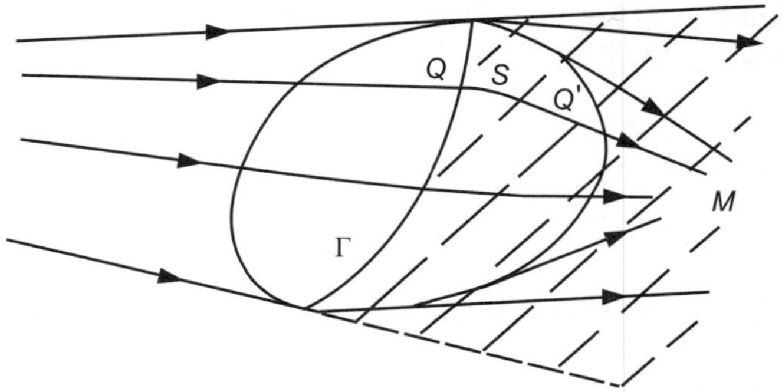

Figure 1.31. Creeping rays.

Let us now introduce the ray coordinates. We denote by s the curvilinear abscissa on the geodesic followed by a ray at the launching point Q' of the space ray and denote by ℓ the total length of the ray equal to $s + \sigma$, where σ is the length of the diffracted ray.

A point M on the ray is then defined by

$$\mathbf{OM} = \mathbf{OQ}' + (\ell - s)\hat{t}, \tag{1.214}$$

where $\hat{t}$ is the vector tangent to the geodesic at Q'. From (2.214), we obtain

$$\frac{\partial \mathbf{OM}}{\partial s} = \frac{\partial \mathbf{OQ}'}{ds} - \hat{t} + (\ell - s)\frac{\partial \hat{t}}{\partial s},$$

and since $\dfrac{\partial \mathbf{OQ}'}{ds} = \hat{t}$ and $\dfrac{\partial \hat{t}}{\partial s} = -\dfrac{\hat{n}}{\rho}$, we have

$$\frac{\partial \mathbf{OM}}{\partial s} = -(\ell - s)\frac{\hat{n}}{\rho}, \tag{1.215}$$

where ρ is the radius of curvature of the geodesic at Q' and where $\hat{n}$ is the unit vector normal to the surface and directed outside of it. We also have

$$\frac{\partial \mathbf{OM}}{\partial \ell} = \hat{t}. \tag{1.216}$$

Equations (1.215) and (1.216) imply that the ray coordinates s and ℓ are orthogonal. Using these coordinates, we can calculate the gradient to find

$$|\boldsymbol{\nabla}\ell|^2 = 1. \tag{1.217}$$

The total length of the creeping ray is therefore a solution of the eikonal equation. In order to complete the determination of the eikonal, we use the phase matching principle and apply it at the point Q on the shadow boundary Γ. The phase at that point is equal to the phase of the incident field. Hence $S(Q) = S^i(Q)$.

We consider now the next order term in the expression, obtained by inserting the asymptotic expansion (1.213) into the Helmholtz equation. For $O\left(k^{5/3}\right)$, we obtain the result

$$\nabla S \cdot \nabla \varphi = 0, \tag{1.218}$$

which upon replacing φ by $R_e\varphi + iI_m\varphi$, shows that the equiphase surfaces of (1.213) are orthogonal to the surfaces of equal amplitude defined by $I_m\varphi(\mathbf{r})$ = constant on each ray.

It can also be shown that the amplitude p_n verifies linear, ordinary, differential equations along the rays, similar to (1.71), and that p_0 verifies the transport equation of GA. However, the right-hand side of the equations verified by the higher-order terms are different and depend on the derivatives of the function $\varphi(\mathbf{r})$.

Since the phase function $S(\mathbf{r})$ is known, the arbitrary elements in the construction of the solution are the value $p_n(\mathbf{r})$ on some surface and the value of $\varphi(\mathbf{r})$ on each ray. These quantities may be adjusted so that the expansion corresponds to the solution of a particular boundary value problem.

The diffracted rays shed from the surface are tangent to the surface, which is therefore a caustic surface of these rays. Hence the boundary conditions on the surface cannot be applied to the diffracted field given by (1.213), which is infinite there. However, a procedure can be used that is similar to the one utilized for the edge-diffracted field.

From the zero-order transport equation, it follows that the pressure field along a diffracted ray is given by

$$p_0(\sigma) = \sqrt{\left|\frac{J(\sigma_0)}{J(\sigma)}\right|}\, p_0(\sigma_0), \tag{1.219}$$

where σ_0 is an arbitrary point on the ray. Hence the quantity

$$\delta_0 = \sqrt{|J(\sigma)|}\, p_0(\sigma) = \sqrt{|J(\sigma_0)|}\, p_0(\sigma_0) \tag{1.220}$$

is independent of σ. Consequently, if δ_0 is known we can calculate the pressure along the ray by applying the relation

$$p_0(\sigma) = \frac{\delta_0}{\sqrt{|J(\sigma)|}}. \tag{1.121}$$

Now δ_0 can be determined by taking the following limit:

$$\delta_0 = \lim_{\sigma \to 0} \left(p_0(\sigma)\sqrt{|J(\sigma)|}\right).$$

Since the rays leaving the surface form an astigmatic pencil of rays with a caustic line element F_2 located on the surface (see fig. 1.20), we have

$$J(\sigma) = a\sqrt{(\rho+\sigma)\,\sigma}, \tag{1.222}$$

where a is a constant and where ρ is the radius of curvature of the wave front at the point on the surface where the creeping ray launches a space ray. Hence, similar to the edge-diffraction problem, we have

$$\delta_0 = a\sqrt{\rho}\; d_0, \tag{1.223}$$

where $d_0 = \lim_{\sigma\to 0}\left(\sqrt{\sigma}\, p_0(\sigma)\right)$, and by inserting (1.223) and (1.222) into (1.221), we obtain

$$p_0(\sigma) = d_0\sqrt{\frac{\rho}{\sigma(\rho+\sigma)}}. \tag{1.224}$$

It is apparent that when $\sigma = -\rho$, the pressure $p_0(\sigma)$ becomes infinite and this asymptotic approximation is no longer valid. In this case, the observation point is on a caustic. As the observation point passes through a caustic in the direction of observation, $\rho + \sigma$ changes sign and a phase shift of $-\frac{\pi}{2}$ appears in the formula (1.224). The selection of the correct square root $e^{-i\pi/2}$ can be justified by the behavior of the field near a regular caustic analyzed in chapter 4.

By taking (1.224) into account in (1.213), we obtain for the zero-order term of the diffracted field in the deep shadow

$$p(M) = p^i(Q)\, e^{iks+ik^{1/3}\varphi(s)}\, d_0\, e^{ikQ'M}\sqrt{\frac{\rho}{Q'M(\rho+Q'M)}}, \tag{1.225}$$

where the factor $\exp[ik(s + Q'M)]$ comes from (1.217), in which $\ell = s + Q'M$. The relation (1.225) may be considered the variation of the field along the ray $QQ'M$. As a matter of fact, only the incident field at Q, the radiated field along $Q'M$, and the phase variation along the arc QQ' are known. The factor $\exp\left[i\left(ks + k^{1/3}\varphi(s)\right)\right] d_0$ multiplied by $p^i(Q)$ does not give the pressure field on the surface along the creeping ray. It merely serves as a transfer function between the incident field at Q and the diffracted field away from Q', which we call the surface ray field. This terminology was introduced by Kouyoumjian [1.8]. A way to determine the unknown terms $\varphi(s)$ and d_0 consists of comparing (1.225) to the solution of a canonical problem.

From the canonical problems treated in chapter 2 (sections 2.3.1 and 2.3.3), it is found that the surface ray field is composed of an infinite amount of modes that propagate independently of each other along the creeping ray path.

Let the field associated with one of these modes be

$$a(s') = A(s')\, e^{iks'},$$

where s' is the distance along the surface ray measured from Q, which is chosen as the phase reference point. The surface ray sheds rays tangentially as it propagates along the geodesic on the curved surface. This energy is lost from the surface ray field, and the field of each mode is attenuated. Assuming that the energy flux between adjacent surface rays is conserved, we have

$$\frac{d}{ds'}\left(\left|A(s')\right|^2 d\eta(s')\right) = -2\alpha\left(\left|A(s')\right|^2 d\eta(s')\right), \tag{1.226}$$

where α is the attenuation constant of the surface ray mode considered and where $d\eta(s')$ is the distance measured along the surface wave front between two adjacent rays (fig. 1.32).

Integrating (1.226) between o and s, we obtain

$$\left|A(s)\right| = \left|A(o)\right| \sqrt{\frac{d\eta(o)}{d\eta(s)}}\, e^{-\int_o^s \alpha(s')\,ds'}. \tag{1.227}$$

Then assuming that (1.225) is verified by each creeping wave mode, we obtain for the contribution of the mode of order q to the diffracted field at M

$$p^{(q)}(M) = p^i(Q)e^{iks}\sqrt{\frac{d\eta(o)}{d\eta(s)}}\; e^{ik^{1/3}\varphi_q(s)}\; d_0^{(q)}\; e^{ikQ'M}\sqrt{\frac{\rho}{Q'M(\rho + Q'M)}} \tag{1.228}$$

with

$$k^{1/3} I_m\, \varphi_q(s) = \int_0^s \alpha_q(s')\, ds'. \tag{1.229}$$

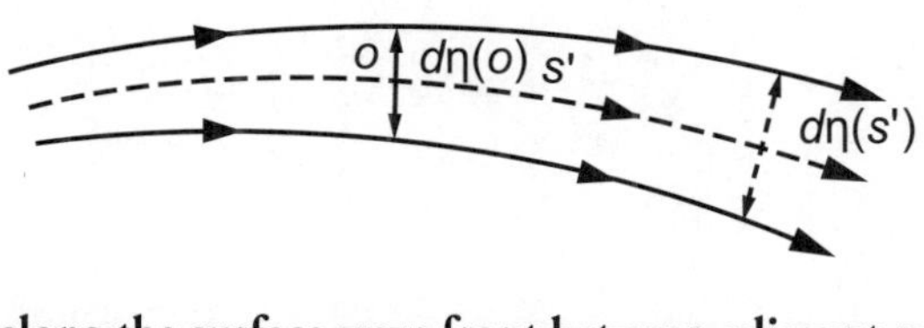

Figure 1.32. Distance along the surface wave front between adjacent rays.

Adding the modes in (1.228) gives the general expression of the zero-order term of the diffracted field in the deep shadow:

$$p(M) = p^i(Q)e^{iks}\sqrt{\frac{d\eta(o)}{d\eta(s)}}\left(\sum_{q=1}^{\infty} e^{ik^{1/3}\varphi_q(s)}\, d_0^{(q)}\right)e^{ikQ'M}\sqrt{\frac{\rho}{Q'M(\rho+Q'M)}}. \tag{1.230}$$

The term in parentheses remains to be determined.

For a circular cylinder submitted to the pressure field of a line source parallel to its axis, (1.230) yields

$$p(M) = p^i(Q)e^{iks}\left(\sum_{q=1}^{\infty} e^{ik^{1/3}\varphi_q(s)}\, d_0^{(q)}\right)\frac{e^{ikQ'M}}{\sqrt{Q'M}}. \tag{1.231}$$

By comparing (1.231) to the asymptotic form of the exact solution of this canonical problem (given in chapter 2, sections 2.3.1 and 2.3.3), we obtain formulas for φ_q and $d_0^{(q)}$ that depend on the boundary conditions. For $d_0^{(q)}$, see equations (2.321), (2.325), and (2.330). However, these formulas are only valid for a circular cylinder. In order to extend them to more general convex surfaces, we need to attach a physical interpretation to these coefficients. Since the geodesic followed by the creeping ray on the circular cylinder has a constant radius of curvature, (1.229) gives

$$k^{1/3} I_m\, \varphi_q(s) = \alpha_q s. \tag{1.232}$$

Recalling that the field behaves locally on the surface of the cylinder as an inhomogeneous plane wave, we deduce from this relation that the real part of $\varphi_q(s)$ must also be proportional to s. Hence we can write

$$k^{1/3}\varphi_q(s) = b_q s,\quad I_m b_q = \alpha_q. \tag{1.233}$$

Next, observing that the cylinder surface is a caustic for the diffracted rays, the diffracted field at M behaves like the field radiated by a line source parallel to the cylinder axis and passing through Q'. Now the field of a line source of unit strength, at large distance from the source, is given by

$$\frac{i}{4}H_0^{(1)}(kr) \approx D_s\frac{e^{ikr}}{\sqrt{r}}, \tag{1.234}$$

where

$$D_s = \frac{e^{i\frac{\pi}{4}}}{\sqrt{8\pi k}}. \tag{1.235}$$

We can therefore separate this coefficient from $d_0^{(q)}$ and write

$$d_0^{(q)} = D^{(q)}\ D_s. \tag{1.236}$$

Moreover, we make the hypothesis that the diffraction coefficient $D^{(q)}$ is a product of two coefficients,

$$D^{(q)} = D^{(q)}_{sr,\,cr}(Q)\ D^{(q)}_{cr,\,sr}(Q'), \tag{1.237}$$

where $D^{(q)}_{sr,cr}$ is the excitation coefficient of the creeping wave mode q by a space wave and $D_s D^{(q)}_{cr,sr}$ is the radiation coefficient of a space wave by the creeping mode q.

Applying the reciprocity theorem, we obtain

$$D^{(q)}_{sr,\,cr}(Q)\ D^{(q)}_{cr,\,sr}(Q') = D^{(q)}_{sr,\,cr}(Q')\ D^{(q)}_{cr,\,sr}(Q) \tag{1.238}$$

or equivalently

$$\frac{D^{(q)}_{cr,sr}(Q)}{D^{(q)}_{sr,cr}(Q)} = \frac{D^{(q)}_{cr,sr}(Q')}{D^{(q)}_{sr,cr}(Q')} = C^{(q)}, \tag{1.239}$$

where $C(q)$ is a constant, which may depend on the mode.

If we assume that the diffraction processes at Q and Q' are symmetric (which means that $D_{sr,cr}(Q)$ has the same functional behavior with respect to the local radii of curvature of the geodesic as $D_s D_{cr,sr}(Q')$), then $C(q) = D_s^{-1}$. Adopting this result, the general expression of the diffracted field in the shadow region of a three-dimensional convex surface submitted to an incident pressure field $p^i(Q)$ is given by

$$p(M) = p_i(Q)\, e^{iks} \sqrt{\frac{d\eta(o)}{d\eta(s)}} \sum_{q=s}^{\infty} e^{ik^{1/3}\varphi_q(s)}\ D^{(q)}_{sr,cr}(Q)\, D^{(q)}_{cr,sr}(Q')\, D_s\, e^{ikQ'M} \sqrt{\frac{\rho}{Q'M(\rho + Q'M)}}, \tag{1.240}$$

where ρ is the distance of Q' to the caustic of the diffracted rays in the plane of diffraction, defined by the direction of observation and the binormal to the geodesic at Q', and where the coefficients $D^{(q)}_{sr,cr}$ and $D^{(q)}_{cr,sr}$ verify (1.239) with $C(q) = D_s^{-1}$.

Expressions for these coefficients for a circular cylinder surface and a cylindrical convex surface at normal and oblique incidence for hard, soft, impedance, and elastic surface boundary conditions are derived in chapter 2, together with some heuristic extensions of these coefficients for a general, three-dimensional convex surface.

REFERENCES

[1.1] Sommerfeld, A., and J. Runge. 1911. "Anwendung der Vektorrechnung auf die Grundlagen der Geometrischen Optik." *Annalen der Physik* 35: 277–98.

[1.2] Poincaré, H. (1892) 1957. *Les Méthodes Nouvelles de la Mécanique*. Vol. 2. Reprint, New York: Dover.

[1.3] Kline, M. 1951. "An Asymptotic Solution of Maxwell's Equations." *Communications on Pure and Applied Mathematics* 4: 225–62.

[1.4] Keller, J. B. 1958. "A Geometrical Theory of Diffraction." In *Proceedings of Symposia in Applied Mathematics*. Vol. 3, *Calculus of Variations and its Applications*, edited by L. M. Graves, 27–52. New York: McGraw-Hill.

[1.5] ———. 1956. "Diffraction by a Convex Cylinder." *IRE Transactions on Antennas and Propagation* 24: 312–21.

[1.6] ———. 1957. "Diffraction by an Aperture." *Journal of Applied Physics* 28: 426–44.

[1.7] ———. 1962. "Geometrical Theory of Diffraction." *Journal of Optical Society of America* 52: 116–30.

[1.8] Kouyoumjian, R. G. 1975. "The Geometrical Theory of Diffraction and Its Applications." In *Numerical and Asymptotic Techniques in Electromagnetics*, edited by R. Mittra. Berlin: Springer-Verlag.

[1.9] Babich, V. M., and V. S. Buldyrev. 1991. *Short Wavelength Diffraction Theory: Asymptotic Methods*. Berlin: Springer-Verlag.

[1.10] Franz, W. 1957. *Theorie der Beugung Elektromagnetischer Wellen*. Berlin: Springer-Verlag.

[1.11] Keller, J. B., and F. C. Karal. 1969. "Surface Wave Excitation and Propagation." *Journal of Applied Physics* 31 (6): 1039–46.

[1.12] Wang, W. D., and G. Deschamps. 1974. "Application of Complex Ray Tracing to Scattering Problems." *Proceedings of the IEEE* 62 (11): 1541–51.

[1.13] Erdelyi, A. 1957. *Asymptotic Expansions*. New York: Dover.

[1.14] Van Dyke, M. 1975. *Perturbation Methods in Fluid Mechanics*. Stanford, CA: Parabolic Press.

[1.15] Kevorkian, J., and J. D. Cole. 1981. *Perturbation Methods in Applied Mathematics*. New York: Springer-Verlag.

[1.16] Olver, F. W. J. 1974. *Asymptotics and Special Functions*. San Diego: Academic Press.

[1.17] François, C. 1981. *Les Méthodes de Perturbation en Mécanique*. Paris: Ecole Nationale Supérieure de Techniques Avancées (ENSTA).

[1.18] Zauderer, E. 1970. "Boundary Layer and Uniform Asymptotic Expansions for Diffraction Problems." *SIAM Journal on Applied Mathematics* 19: 575–600.

[1.19] Borovikov, V. A. 1979. "Diffraction by a Wedge with Curved Faces." *Soviet Physics—Acoustics* 25 (6): 825–35.

[1.20] Friedlander, F. C., and J. B. Keller. 1955. "Asymptotic Expansions of Solutions of $(\Delta + k^2)U = 0$." *Communications on Pure and Applied Mathematics* 8: 387–94.

Chapter 2

CANONICAL PROBLEMS AND NONUNIFORM ASYMPTOTIC THEORY OF ACOUSTIC WAVE DIFFRACTION

2.1. INTRODUCTION

As mentioned in the introduction to the geometrical theory of diffraction (GTD) in chapter 1, a canonical problem is the simplest boundary value problem having the same local geometry as the original problem near the point of diffraction, which can be solved exactly, usually by the method of separation of variables. This definition of a canonical problem has many difficulties: The exact solution of a canonical problem is generally given in the form of a series (e.g., a series of Bessel and Hankel functions); it is not a simple matter to obtain an asymptotic expansion of this. Moreover, asymptotic solutions depend on the domain of observation of the scattered field. In order to obtain similar asymptotic formulas between the original problem and the canonical problem, it is necessary that both problems have a similar ray geometry in the angular domain of interest. In particular, they must have the same singularities in the ray geometry (e.g., shadow boundaries or caustics). There are additional restrictions, especially those concerning the local geometry retained in the canonical problem, which depend on the order of the terms in the asymptotic expansion of the diffracted field that we want to determine. We have seen in chapter 1, section 1.8 that, in the case of a curved wedge, the diffraction coefficient of the zero-order term of the asymptotic diffracted field is given by the tangent plane approximation of the wedge at the point of diffraction and that, for higher-order terms, the local geometry must be refined. Higher-order diffraction phenomena (e.g., the diffraction by a line discontinuity in the curvature of the surface) also need to consider refined local geometry, taking into account the radii of curvature of the adjacent surfaces close to the edge. This is the case even for the dominant term in the asymptotic expansion of the diffracted field.

In this chapter, we focus our investigations on three canonical problems: (1) the straight wedge, (2) the circular cylinder, and (3) the concave circular cylindrical surface. These canonical problems cover the most important diffraction phenomena encountered on a body embedded in a fluid (e.g., air or sea water)—namely, edge diffraction, convex surface diffraction (e.g., creeping waves and elastic surface waves), and concave surface diffraction (e.g., whispering gallery modes). For each canonical problem, we derive first an exact solution in terms of a series of entire or fractional Bessel and Hankel functions. Then we transform this solution in an integral form, generally in a complex plane. Finally, by applying the saddle point method, we extract the dominant term of the asymptotic expansion of the solution from this integral.

2.2. THE WEDGE

2.2.1. HARD AND SOFT STRAIGHT WEDGE AT NORMAL AND OBLIQUE INCIDENCE

The diffraction by a wedge has received a great deal of attention in the last century. In 1902, MacDonald published an integral representation for the eigenfunction expansions of the total field (see pages 186–98 in [2.1]). Plane, cylindrical, and spherical wave illuminations of the wedge by scalar waves were treated. The first asymptotic solution to the wedge problem is due to Sommerfeld [2.2]. He treated the case of scalar plane wave illumination and obtained the diffraction coefficient valid outside the transition regions at the shadow and reflection boundaries. At the end of section 2.2.1.1.3, this is denoted by Keller's "nonuniform diffraction coefficient" of the wedge.

This section deals with the scalar wave diffraction by an impenetrable wedge. Both hard (Neumann) and soft (Dirichlet) boundary conditions are treated.

2.2.1.1. Line Source Excitation

We start with an incident cylindrical wave excited by a z-directed, uniform line source of unit strength that is parallel to the edge of an impenetrable infinite wedge and located at a distance ρ' from its edge in the surrounding fluid. The result for the plane wave case may then be obtained by letting ρ' tend to infinity. The coordinate system and the notations are defined in figure 2.1.

The ox axis is chosen along one of the faces of the wedge. It is denoted by the reference face. The location of the source S and the location of the observation point P are defined by the circular cylindrical coordinates (ρ', ϕ') and (ρ, ϕ), respectively. The exterior wedge angle is $n\pi$, where n is an arbitrary positive scalar with $0 \leq n \leq 2$. The total field, which consists of the incident and scattered components of the field outside the wedge, is given by a two-dimensional scalar Green function, $G(\boldsymbol{\rho}, \boldsymbol{\rho}')$, verifying the scalar Helmholtz equation,

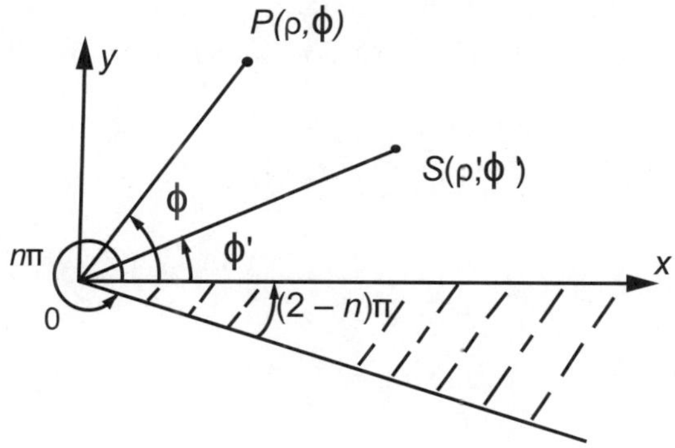

Figure 2.1. The two-dimensional wedge and line source in the circular cylindrical coordinate system.

$$\left(\Delta + k^2\right) G(\boldsymbol{\rho}, \boldsymbol{\rho}') = -\delta\left(\left|\boldsymbol{\rho} - \boldsymbol{\rho}'\right|\right), \tag{2.1}$$

where Δ is the two-dimensional Laplace operator in cylindrical coordinates, $\delta\left(\left|\boldsymbol{\rho} - \boldsymbol{\rho}'\right|\right)$ is the Dirac delta function, and k is the wave number in the fluid.

The Green function $G(\boldsymbol{\rho}, \boldsymbol{\rho}')$ satisfies the following boundary conditions at the wedge surface:

$$G_s\left(\boldsymbol{\rho}, \boldsymbol{\rho}'\right) = 0, \text{ at } \phi = 0, n\pi \tag{2.2}$$

for the soft wedge and

$$\frac{\partial G_h}{\partial \phi}\left(\boldsymbol{\rho}, \boldsymbol{\rho}'\right) = 0, \text{ at } \phi = 0, n\pi \tag{2.3}$$

for the hard wedge.

Furthermore, both Green functions G_s and G_h satisfy the Sommerfeld radiation condition [2.3],

$$\lim_{\rho \to \infty} \sqrt{\rho} \left(\frac{\partial G}{\partial \rho} - ikG \right) = 0, \tag{2.4}$$

and the Meixner edge condition [2.4],

$$G = O(1), \ \rho \to 0. \tag{2.5}$$

The solution of (2.1), which verifies the conditions presented in (2.2) through (2.5), is unique and given by a convergent eigenfunction expansion, which for a time dependence $\exp(-i\omega t)$, is given by

$$G_{s_b}(\boldsymbol{\rho},\boldsymbol{\rho}') = G_{s_b}(\rho,\phi;\rho',\phi') = \frac{i}{4n}\sum_{m=0}^{\infty} \varepsilon_m J_{\frac{m}{n}}(k\rho_<) H^{(1)}_{\frac{m}{n}}(k\rho_>) \left[\cos\frac{m}{n}(\phi-\phi') \mp \cos\frac{m}{n}(\phi+\phi')\right], \tag{2.6}$$

where $\rho_< = min\ \rho,\ \rho',\ \rho_> = max\ \rho,\ \rho'$ and where $O \le \phi,\phi' \le n\pi$ and $\varepsilon_m = 1$ for $m = 0$, $\varepsilon_m = 2$ for $m \neq 0$. $J_{\frac{m}{n}}(k\rho)$ and $H^{(1)}_{\frac{m}{n}}(k\rho)$ represent the fractional cylindrical Bessel function and the fractional cylindrical Hankel function of the first kind, respectively.

We give hereafter some details concerning the construction of the solutions (2.6) by the method of separation of variables.

2.2.1.1.1. Solution by the Method of Separation of Variables

With the Laplacian expressed in cylindrical coordinates, (2.1) yields

$$\frac{1}{\rho}\frac{\partial}{\partial\rho}\left(\rho\frac{\partial G}{\partial\rho}\right) + \frac{1}{\rho^2}\frac{\partial^2 G}{\partial\phi^2} + k^2 G = -\frac{\delta(|\boldsymbol{\rho}-\boldsymbol{\rho}'|)\,\delta(|\phi-\phi'|)}{\rho}. \tag{2.7}$$

For $\rho \neq \rho'$, the second member of (2.7) equals zero. The Helmholtz equation is then a homogeneous equation. Following the method of separation of variables, we seek to find solutions of the homogeneous Helmholtz equation of the form

$$G = R(\rho)\Phi(\phi). \tag{2.8}$$

Substitution of (2.8) into (2.7) and division by G yields, after multiplication by ρ^2,

$$\frac{\rho}{R}\frac{d}{d\rho}\left(\rho\frac{dR}{d\rho}\right) + \frac{1}{\Phi}\frac{d^2\Phi}{d\phi^2} + k^2\rho^2 = 0\,,\ \text{for } \boldsymbol{\rho} \neq \boldsymbol{\rho}'. \tag{2.9}$$

The second term is independent of ρ, and the other terms are independent of ϕ. Hence

$$\frac{1}{\Phi}\frac{d^2\Phi}{d\phi^2} = -q^2, \tag{2.10}$$

where q is a constant. Equation (2.9) then becomes

$$\frac{\rho}{R}\frac{d}{d\rho}\left(\rho\frac{dR}{d\rho}\right)-q^2+k^2\rho^2=0, \tag{2.11}$$

which is an equation in ρ only.

The wave equation is now separated. The Φ equation is a harmonic equation, giving rise to harmonic functions. The R equation is a Bessel equation of order q.

A general solution of (2.10) is

$$\Phi = Ae^{iq\phi}+Be^{-iq\phi}, \tag{2.12}$$

where A and B are constants.

Consider first the soft case. Application of the boundary conditions (2.2) yields $A = -B$ and $2iA \sin nq\pi = 0$. A nontrivial solution of these equations is $nq = m$, where m is an integer taking the values $m = 0, \pm1, \pm2, \ldots$.

Hence to satisfy the boundary conditions on the wedge faces at $\phi = 0$ and $\phi = n\pi$, the solution of (2.10) must be of the form

$$\Phi_m(\phi,\phi') = a_m \sin\frac{m}{n}\phi\, g(\phi'). \tag{2.13}$$

Let us now consider $a_m \sin\frac{m}{n}\phi$ to be a term in the Fourier sine series of a function $f(\phi)$ over a period $n\pi$ so that

$$\begin{aligned} f(\phi) &= \sum_{m=o}^{\infty} a_m \sin\frac{m}{n}\phi \\ a_m &= \frac{2}{n\pi}\int_o^{n\pi} f(\phi')\sin\frac{m}{n}\phi'\, d\phi'. \end{aligned} \tag{2.14}$$

If we rewrite (2.14) as

$$f(\phi)=\int_o^{n\pi} f(\phi')\left[\frac{2}{n\pi}\sum_{m=o}^{\infty}\sin\frac{m}{n}\phi'\sin\frac{m}{n}\phi\right]d\phi',$$

then the term inside the brackets must be the delta function $\delta(\phi-\phi')$:

$$\delta(\phi-\phi')=\frac{2}{n\pi}\sum_{m=o}^{\infty}\sin\frac{m}{n}\phi'\sin\frac{m}{n}\phi. \tag{2.15}$$

Equation (2.15) is the completeness relation. It means that the functions $\sqrt{\frac{2}{n\pi}}\sin\left(\frac{m}{n}\phi\right)$ form a closed set of angular eigenfunctions.

Now we consider the various solutions of the Bessel equation (2.11) with $q = \frac{m}{n}$. We note that only the $J_m(k\rho)$ functions are not singular at $\rho = 0$ and that $H_q^{(1)}(k\rho)$, with the time domain dependence $\exp(-i\omega t)$ chosen in this book, represents outward traveling waves. Such waves verify the Sommerfeld radiation condition. Hence for the elementary solutions $R_q(\rho, \rho')$ with $q = \frac{m}{n}$ we must choose

$$R_q(\rho, \rho') = \begin{cases} b_m J_q(k\rho) H_q^{(1)}(k\rho'), & \rho < \rho' \\ b_m J_q(k\rho') H_q^{(1)}(k\rho), & \rho > \rho' \end{cases} \tag{2.16}$$

to ensure finiteness of $R_q(\rho, \rho')$ at $\rho = 0, \infty$ and its continuity at $\rho = \rho'$.

Choosing equation (2.15) for the angular eigenfunctions, possible solutions of the Helmholtz equation for the soft case are

$$G_s(\rho, \phi; \rho', \phi') = \frac{2}{n\pi} \sum_{m=0}^{\infty} b_m J_q(k\rho_<) H_q^{(1)}(k\rho_>) \sin q\phi \sin q\phi', \tag{2.17}$$

where $\rho_< = \min \rho, \rho'$; $\rho_> = \max \rho, \rho'$; and $q = \frac{m}{n}$.

Substituting (2.17) with (2.15) into the homogeneous Helmholtz equation (2.7) gives

$$\left(\frac{1}{\rho} \frac{\partial}{\partial \rho} \rho \frac{\partial}{\partial \rho} - \frac{q^2}{\rho^2} + k^2 \right) \frac{2}{n\pi} \sum_{m=0}^{\infty} R_q(\rho, \rho') \sin q\phi \sin q\phi'$$

$$= -\frac{\delta(\rho - \rho')}{\rho} \frac{2}{n\pi} \sum_{m=0}^{\infty} \sin q\phi \sin q\phi'.$$

Since $\sqrt{\frac{2}{n\pi}} \sin q\phi$ form a complete basis of angular eigenfunctions, we must have

$$\left(\frac{1}{\rho} \frac{\partial}{\partial \rho} \rho \frac{\partial}{\partial \rho} - \frac{q^2}{\rho^2} + k^2 \right) R_q(\rho, \rho') = -\frac{\delta(\rho - \rho')}{\rho}. \tag{2.18}$$

If we multiply both sides of (2.18) by $\rho d\rho$ and integrate from $\rho = \rho' + \Delta$ to $\rho = \rho' - \Delta$, with $\Delta \to 0$, then

$$\frac{\partial}{\partial \rho} R_q(\rho^+, \rho') - \frac{\partial}{\partial \rho} R_q(\rho^-, \rho') = -\frac{1}{\rho'}, \tag{2.19}$$

which yields

$$b_m\left[J_q(k\rho')H_q^{(1)'}(k\rho')-J_q'(k\rho')H_q^{(1)}(k\rho')\right]=-\frac{1}{k\rho'}. \tag{2.20}$$

The expression in the brackets is a Wronskian of the Bessel equation, equal to $2i/\pi k\rho'$ so that $b_m=\pi i/2$. Our solution of the Helmholtz equation for cylindrical wave incidence and soft boundary conditions is now completely defined and given by

$$G_s(\rho,\phi;\rho',\phi')=\frac{i}{n}\sum_{m=0}^{\infty}J_q(k\rho_<)H_q^{(1)}(k\rho_>)\sin q\phi\sin q\phi', \tag{2.21}$$

where $q=\dfrac{m}{n}$.

Consider now the hard case. Application of the boundary conditions (2.3) to (2.12) yields $A=B$ and $-2Aq\sin nq\pi=0$. Again, a nontrivial solution is given by $nq=m$, where m is an integer with

$$\Phi_m(\phi,\phi')=a_m\cos\frac{m}{n}\phi\, g(\phi'). \tag{2.22}$$

Proceeding as in the soft case, we find the completeness relation

$$\delta(\phi-\phi')=\frac{1}{n\pi}\sum_{m=0}^{\infty}\varepsilon_m\cos\frac{m}{n}\phi\cos\frac{m}{n}\phi', \tag{2.23}$$

where $\varepsilon_m=1$ for $m=0$ and $\varepsilon_m=2$ for $m\geq 1$. Hence the functions $\sqrt{\dfrac{\varepsilon_m}{n\pi}}\cos\dfrac{m}{n}\phi$ form a closed set of angular eigenfunctions for the hard boundary conditions.

The equation for $R_q(\rho,\rho')$ remains the same as in the soft case. Hence (2.20) is still verified and the final result is given by

$$G_h(\rho,\phi;\rho',\phi')=\frac{i}{2n}\sum_{m=0}^{\infty}\varepsilon_m J_q(k\rho_<)H_q^{(1)}(k\rho_>)\cos q\phi\cos q\phi', \tag{2.24}$$

where $q=\dfrac{m}{n}$. It is possible to write both Green functions in a symmetric form by introducing $\varepsilon_m/2$ in (2.21), which does not change the formula since the second member is equal to zero for $m = 0$ and for $m \geq 1$, $\varepsilon_m / 2 = 1$. We can also replace the product of sine functions in (2.21) and the product of cosine functions in (2.24), respectively, by

$$\frac{1}{2}\left[\cos q(\phi-\phi') - \cos q(\phi+\phi')\right] \text{ and } \frac{1}{2}\left[\cos q(\phi-\phi') + \cos q(\phi+\phi')\right].$$

Then we obtain the solutions of the Helmholtz equation in the form given by equation (2.6).

2.2.1.1 2. Integral Representation of the Solution

The series (2.6) converges slowly and becomes rapidly impracticable for numerical calculations when $k\rho$ or $k\rho'$ becomes large (typically > 100). It is therefore important to transform this series into an integral form adapted for its asymptotic evaluation.

An integral representation for the product $J_q(k\rho)H_q^{(1)}(k\rho')$, where $q = \frac{m}{n}$, is

$$J_q(k\rho)H_q^{(1)}(k\rho') = \frac{1}{\pi i}\int_{o}^{c+i\infty} e^{\frac{1}{2}\left[t - k^2(\rho^2+\rho'^2)t^{-1}\right]} I_q\left(\frac{k^2\rho\rho'}{t}\right)\frac{dt}{t}, \tag{2.25}$$

where $c > 0$ and $I_q\left(\frac{k^2\rho\rho'}{t}\right)$ is the modified cylindrical Bessel function of the first kind. A useful integral representation for $I_q(x)$ is ([2.1] and [2.5])

$$I_q(x) = -\frac{1}{2\pi}\int_{\gamma-i\infty}^{\gamma'-i\infty} e^{x\cos\xi + iq\xi}\, d\xi, \tag{2.26}$$

where

$$\pi < \gamma < 2\pi,$$
$$-\pi < \gamma' < 0.$$

Thus

$$I_q\left(\frac{k^2\rho\rho'}{t}\right) = -\frac{1}{2\pi}\int_L e^{\frac{k^2\rho\rho'}{t}\cos\xi - iq\xi}\, d\xi, \tag{2.27}$$

or

$$I_q\left(\frac{k^2\rho\rho'}{t}\right) = -\frac{1}{2\pi}\int_{L'} e^{\frac{k^2\rho\rho'}{t}\cos\xi + iq\xi}\, d\xi, \tag{2.28}$$

where the contours L and L' are two contours in the complex plane ξ, symmetric with respect to the origin (fig. 2.2). Replacing the cosine terms in (2.6) by exponentials; utilizing (2.25), (2.27), and (2.28) in (2.6); and interchanging the orders of summation and integration allows us to write (2.6) as

$$G_{s_b}(\boldsymbol{\rho},\boldsymbol{\rho}') = G\left(\boldsymbol{\rho},\boldsymbol{\rho}';\overset{-}{\beta}\right) \mp G\left(\boldsymbol{\rho},\boldsymbol{\rho}';\overset{+}{\beta}\right), \tag{2.29}$$

where $\overset{\mp}{\beta} = \phi \mp \phi'$ and

$$G(\boldsymbol{\rho},\boldsymbol{\rho}';\beta) = -\frac{1}{8n\pi^2}\int_0^{c+i\infty} e^{\frac{1}{2}\left[t-k^2\left(\rho^2+\rho'^2\right)t^{-1}\right]} \left\{\int_L e^{\frac{k^2\rho\rho'}{t}\cos\xi}\sum_{m=1}^{\infty} e^{-iq(\xi+\beta)}\,d\xi + \int_{L'} e^{\frac{k^2\rho\rho'}{t}\cos\xi}\sum_{m=0}^{\infty} e^{+iq(\xi+\beta)}\,d\xi\right\}\frac{dt}{t}, \tag{2.30}$$

where β stands for either $\overset{-}{\beta}$ or $\overset{+}{\beta}$.

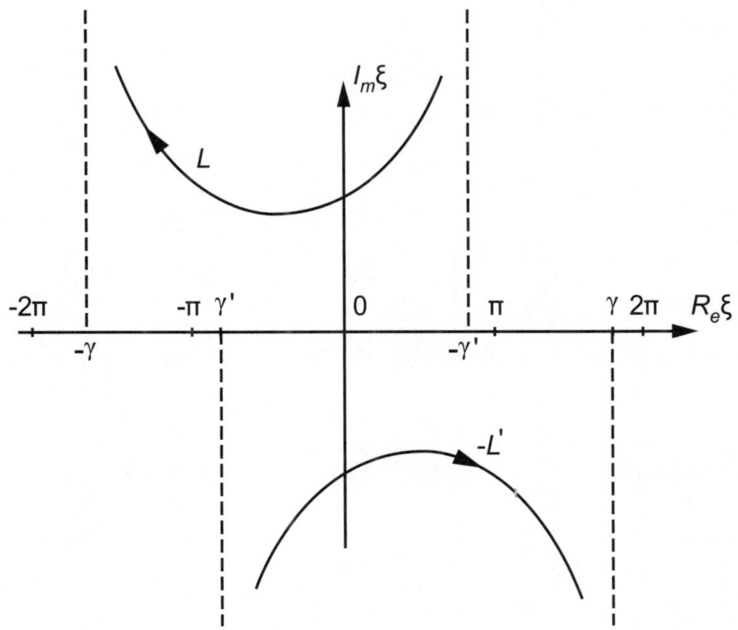

Figure 2.2. The contours L and L' for the integral representation of the Bessel function of the first kind $I_q(x)$.

The $m = 0$ term in (2.6) has been incorporated in the L' contour representation of $I_q(x)$. Equation (2.30) can be further simplified by noting that the terms depending on $q = m/n$ are geometrical series that can be replaced by their sums. We have

$$\begin{aligned} \sum_{m=1}^{\infty} e^{-i\frac{m}{n}(\xi+\beta)} &= \sum_{m=o}^{\infty} e^{-i\frac{m}{n}(\xi+\beta)} - 1 = \frac{1}{2i}\cot\left(\frac{\xi+\beta}{2n}\right) - \frac{1}{2}, \\ \sum_{m=o}^{\infty} e^{+i\frac{m}{n}(\xi+\beta)} &= -\frac{1}{2i}\cot\left(\frac{\xi+\beta}{2n}\right) + \frac{1}{2}. \end{aligned} \tag{2.31}$$

Substituting (2.31) into (2.30) and interchanging the orders of integration gives

$$G(\mathbf{\rho},\mathbf{\rho}';\beta) = -\frac{1}{8n\pi^2}\int_{L-L'}\left[-\frac{1}{2i}\cot\left(\frac{\xi+\beta}{2n}\right)+\frac{1}{2}\right]\left(\int_0^{c+i\infty} e^{\frac{t}{2}-\frac{z^2(\xi)}{2t}}\frac{dt}{t}\right)d\xi, \tag{2.32}$$

where $z(\xi) = k\left[\rho^2 + \rho'^2 - 2\rho\rho'\cos\xi\right]^{1/2}$.

We recognize in (2.32) the modified cylindrical Bessel function of the second kind $K_o(iz)$ defined by (see [2.5])

$$K_o(-iz) = \frac{1}{2}\int_o^{c+i\infty} e^{\frac{t}{2}-\frac{z^2(\xi)}{2t}}\frac{dt}{t}. \tag{2.33}$$

Furthermore, since $z(\xi) = z(-\xi)$, we have

$$\int_{L-L'} K_o(-iz)\,d\xi = 0.$$

Finally, (2.32) can be written as

$$G(\mathbf{\rho},\mathbf{\rho}';\beta) = \frac{1}{8n\pi^2 i}\int_{L-L'}\cot\left(\frac{\xi+\beta}{2n}\right)K_o\left(-ik\sqrt{\rho^2+\rho'^2-2\rho\rho'\cos\xi}\right)d\xi. \tag{2.34}$$

Equation (2.34) inserted in (2.29) directly leads to a complex contour representation for $G_{s_h}(\mathbf{\rho},\mathbf{\rho}')$:

$$G_{s_h}(\mathbf{\rho},\mathbf{\rho}') = \frac{1}{8\pi^2 ni}\int_{L-L'} K_o\left(-ik\sqrt{\rho^2+\rho'^2-2\rho\rho'\cos\xi}\right)\Bigg|\, \mathrm{c}$$

$$\left[\cot\left(\frac{\xi+\phi-\phi'}{2n}\right) \mp \cot\left(\frac{\xi+\phi+\phi'}{2n}\right)\right]d\xi. \tag{2.35}$$

It is important to note that the integral representation (2.35) is the exact form of the solution of the Helmholtz equation for a line source excitation of a soft or hard wedge.

The contours L and L' in figure 2.2 can be shifted in a direction parallel to the real axis, since γ and γ' in (2.26) are allowed to take values in the domains $(\pi, 2\pi)$ and $(-\pi, 0)$, respectively. By choosing $\gamma = \frac{3\pi}{2}$ and $\gamma' = -\frac{\pi}{2}$, one obtains the Sommerfeld contours shown in figure 2.3. These contours will be chosen for the asymptotic evaluation of the integral (2.35).

If the argument of the modified Bessel function $K_o\left(-iz(\xi)\right)$ is large with respect to the order (i.e., $\left|z(\xi)\right| >> 1$), then one may use the large argument approximation (see [2.5]),

$$K_o\left(-i z(\xi)\right) \approx \sqrt{\frac{\pi}{-2i\, z(\xi)}}\, e^{i z(\xi)}, \quad -\pi \le \arg\left(-i z\right) \le \pi. \tag{2.36}$$

Inserting (2.36) into (2.35) gives

$$G_{s_b}(\boldsymbol{\rho}, \boldsymbol{\rho}') \cong G_1\left(\boldsymbol{\rho}, \boldsymbol{\rho}'; \bar{\beta}\right) \mp G_1\left(\boldsymbol{\rho}, \boldsymbol{\rho}'; \overset{+}{\beta}\right), \tag{2.37}$$

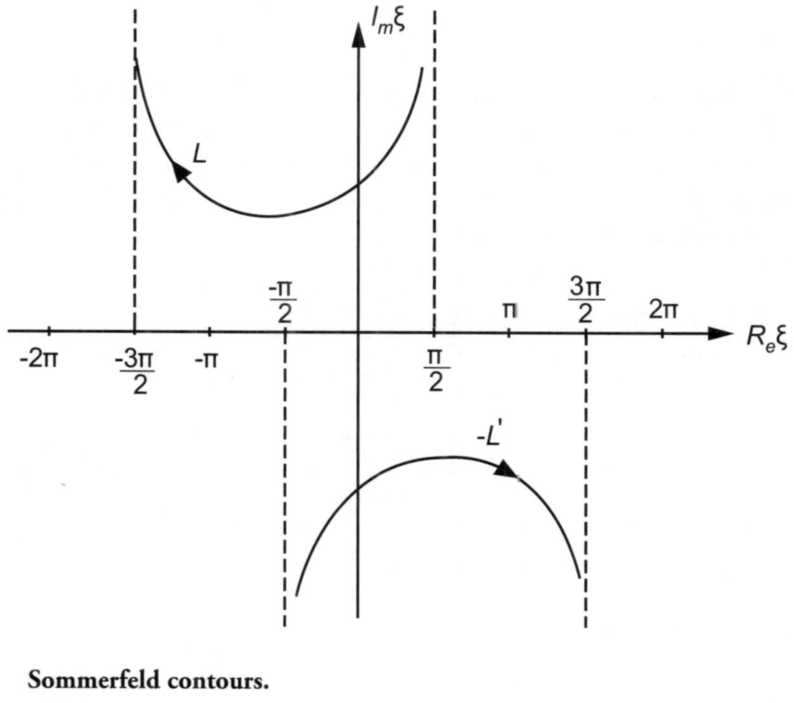

Figure 2.3. Sommerfeld contours.

where $\overset{\mp}{\beta} = \varphi \mp \varphi'$ and

$$G_1(\mathbf{\rho},\mathbf{\rho}';\beta) = \frac{1}{4\pi in}\frac{e^{i\frac{\pi}{4}}}{\sqrt{8\pi k}} \int_{L-L'} \frac{e^{ik\sqrt{\rho^2+\rho'^2-2\rho\rho'\cos\xi}}}{\left(\rho^2+\rho'^2-2\rho\rho'\cos\xi\right)^{1/4}} \cot\left(\frac{\xi+\beta}{2n}\right) d\xi, \tag{2.38}$$

which can be written

$$G_1(\mathbf{\rho},\mathbf{\rho}';\beta) = \int_{L-L'} F(\xi,\beta)\, d\xi, \tag{2.39}$$

with

$$F(\xi,\beta) = \frac{1}{4\pi in}\frac{e^{i\frac{\pi}{4}}}{\sqrt{8\pi k}} \frac{e^{ik\sqrt{\rho^2+\rho'^2-2\rho\rho'\cos\xi}}}{\left(\rho^2+\rho'^2-2\rho\rho'\cos\xi\right)^{1/4}} \cot\left(\frac{\xi+\beta}{2n}\right). \tag{2.40}$$

When $k\rho$ and $k\rho'$ are both large but finite, the exponential in (2.40) may be approximated by

$$\begin{aligned} e^{ik\sqrt{\rho\ +\rho'^2-2\rho\rho'\cos\xi}} &= e^{ik\sqrt{(\rho+\rho')^2-2\rho\rho'(1+\cos\xi)}} \\ &\cong e^{ik(\rho+\rho')}\, e^{-ik\frac{\rho\rho'}{(\rho+\rho')^2}(1+\cos\xi)} . \end{aligned} \tag{2.41}$$

This approximation is justified when ξ is close to $\pm\pi$. As we will see later, this condition is satisfied at the saddle points of the integrand. In the frame of this approximation, which corresponds to a line source and an observation point at finite distance from the edge under the high-frequency conditions $k\rho >> 1, k\rho' >> 1$, the integral representation of the infinite series solution (2.6) of the Helmholtz equation is given by

$$\begin{aligned} G_{s_b}(\mathbf{\rho},\mathbf{\rho}') = \frac{1}{4\pi in}\frac{e^{i\frac{\pi}{4}}}{\sqrt{8\pi k}} e^{ik(\rho+\rho')} \int_{L-L'} & \frac{e^{-ik\frac{\rho\rho'}{\rho+\rho'}(1+\cos\xi)}}{\left(\rho^2+\rho'^2-2\rho\rho'\cos\xi\right)^{1/4}} \\ & \left[\cot\left(\frac{\xi+\phi-\phi'}{2n}\right) \mp \cot\left(\frac{\xi+\phi+\phi'}{2n}\right)\right] d\xi. \end{aligned} \tag{2.42}$$

We will use (2.42) when ξ is close to $\pm\pi$ and (2.37) through (2.40) elsewhere.

2.2.1.1.3. Asymptotic Form of the Solution

Let

$$F_1(\xi,\beta) = \frac{1}{4\pi in}\frac{e^{i\frac{\pi}{4}}}{\sqrt{8\pi k}} e^{ik(\rho+\rho')}\left(\rho^2+\rho'^2-2\rho\rho'\cos\xi\right)^{-1/4}\cot\left(\frac{\xi+\beta}{2n}\right) \tag{2.43}$$

and

$$f(\xi) = -i(1+\cos\xi), \tag{2.44}$$

$$\Omega = \frac{k\rho\rho'}{\rho+\rho'}, \tag{2.45}$$

where Ω is assumed to be large.

Consider the integral over an arbitrary contour Λ in the complex ξ-plane:

$$I(\beta) = \int_\Lambda F_1(\xi,\beta)e^{\Omega f(\xi)d\xi}. \tag{2.46}$$

This integral is in the proper form for an asymptotic evaluation via the method of steepest descent. The saddle points of $f(\xi)$ occur at

$$\frac{df(\xi)}{d\xi} = i\sin\xi = 0 \rightarrow \xi_s = \pm m\pi \text{ where } m = 0, 1, 2, 3, \ldots.$$

However, $\xi_s = \pm\pi$ are the only ones chosen because the steepest descent paths (SDP) through $\xi_s = \pm\pi$ allow one to close the $L - L'$ contour, thereby facilitating the use of the Cauchy residue theorem.

The steepest descent equations are found from the condition

$$I_m f(\xi) = I_m f(\xi_s) \tag{2.47}$$

or $\cos u \cosh v = -1$, where $\xi = u + iv$. In order to ensure convergence of the integral in (2.46), one must have $R_e f(\xi) < 0$ as $|I_m(\xi)| \rightarrow \infty$. Since $R_e f(\xi) = -\sin u \sinh v$, the portions of the complex ξ-plane for which the convergence condition is satisfied are shown as shaded vertical strips in figure 2.4. The paths chosen through $\pm\pi$ lie in the proper domain, and because they

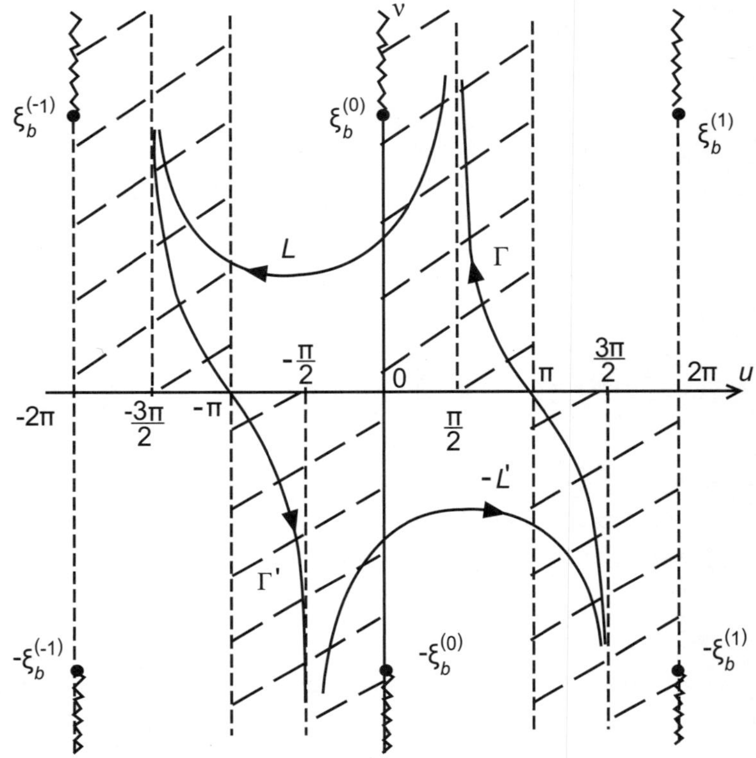

Figure 2.4. SDP Γ and Γ' in the complex ξ-plane ($\xi = u + iv$).

blend in with the $L - L'$ contours as $\left|I_m(\xi)\right| \to \infty$, they allow one to obtain a closed path of integration, $C = L - L' + \Gamma + \Gamma'$.

For sufficiently large Ω, the major contribution to the integrals evaluated over the SDP Γ and Γ' occurs only from the vicinity of the saddle points $\xi_s = \pm\pi$. Hence it follows that the approximation (2.41) justified as $\left|1 + \cos\xi\right|$ is small for $\xi \approx \xi_s$.

We study now the singularities of $G_{s_b}(\mathbf{\rho}, \mathbf{\rho}')$ in the complex ξ-plane. According to (2.40), $F(\xi, \beta)$ has branch point singularities at $\xi = \xi_b$ that are roots of

$$\rho^2 + \rho'^2 - 2\rho\rho' \cos\xi_b = 0$$

and pole singularities that occur at $\xi = \xi_p$, where $\frac{\xi+\beta}{2n} = N\pi, N = 0, \pm 1, \pm 2, \ldots$. For the case where $\rho \neq \rho'$, it follows that $\rho^2 + \rho'^2 > 2\rho\rho'$, and therefore

$$\xi_b = \pm i \cosh^{-1}\left(\frac{\rho^2 + \rho'^2}{2\rho\rho'}\right) + 2m\pi \text{ where } m = 0, \pm 1, \pm 2, \ldots . \tag{2.48}$$

According to (2.48), the branch points are not located in the vicinity of the saddle points at $\xi_s = \pm\pi$. These locations, together with the branch cuts, are indicated in figure 2.4. The contours L and L' can always be chosen so that the branch cuts lie outside the closed path formed by $L - L'$ and Γ, Γ'. Hence the integrand is single valued on and within the closed path, and the branch cuts do not contribute to the integral.

The residues corresponding to ξ_p are evaluated for $\left[\xi_p\right] \leq \pi$, as only these poles lie within the closed path of integration, $C = L - L' + \Gamma + \Gamma'$.

Application of the residue theorem gives

$$\int_C F(\xi, \beta) d\xi = 2\pi i \sum_p R_p, \tag{2.49}$$

where R_p are the residues of the poles situated inside the closed contour C given by

$$R_p = \frac{P(\xi_p, \beta)}{Q'(\xi_p, \beta)} \tag{2.50}$$

with

$$P(\xi, \beta) = \frac{1}{4\pi i n} \frac{e^{i\frac{\pi}{4}}}{\sqrt{8\pi k}} \frac{e^{ik\sqrt{\rho^2+\rho'^2-2\rho\rho'\cos\xi}}}{\left(\rho^2 + \rho'^2 - 2\rho\rho'\cos\xi\right)^{1/4}} \cos\left(\frac{\xi+\beta}{2n}\right)$$

and

$$Q(\xi, \beta) = \sin\left(\frac{\xi+\beta}{2n}\right)$$

with

$$Q'(\xi, \beta) = \frac{1}{2n}\cos\left(\frac{\xi+\beta}{2n}\right).$$

The residue at $\xi_p = -\beta + 2nN\pi$ is thus given by

$$R_p = \frac{1}{2\pi i} \frac{e^{i\frac{\pi}{4}}}{\sqrt{8\pi k}} \frac{e^{ik\sqrt{\rho^2+\rho'^2-2\rho\rho'\cos(-\beta+2nN\pi)}}}{\left(\rho^2+\rho'^2-2\rho\rho'\cos\left(-\beta+2nN\pi\right)\right)^{1/4}} U\left(\pi-\left|-\beta+2nN\pi\right|\right), \tag{2.51}$$

where $U(t)$ is the Heaviside function: $U(t) = 0$ for $t < 0$, $U(t) = 1$ for $t > 0$, and $U(t) = 1/2$ for $t = 0$. Equation (2.51) may be recognized as the asymptotic form of

$$\frac{1}{2\pi i}\left[\frac{i}{4} H_o^{(1)} k\sqrt{\rho^2+\rho'^2-2\rho\rho'\cos\left(-\beta+2nN\pi\right)}\right] U\left(\pi-\left|-\beta+2nN\pi\right|\right), \tag{2.52}$$

and therefore it represents the field, due to the line source or its image. It corresponds to the geometrical optics (GO) or geometrical acoustics (GA) contribution, which is a superposition of the incident and reflected fields.

The incident field is described by $N = 0$ and $\beta = \bar{\beta} = \phi - \phi'$. The reflected field from the surface $\phi = 0$, resulting from the incident field, is described by $N = 0$ and $\beta = \overset{+}{\beta} = \phi + \phi'$. Other values of N describe fields that may be reflected from either the surface $\phi = 0$ or the surface $\phi = n\pi$.

Since the total field is composed of the GO field and the diffracted field, it follows that the saddle point contributions must yield the diffracted ray field. The first-order saddle point approximation of the integral $I(\beta)$ given by (2.46) for a pole not close to the saddle point is obtained by the following procedure.

The integral $I(\beta)$ is transformed into a "canonical" form, wherein the function $f(\xi)$ is replaced by a polynomial $\tau(s)$ that describes, in the simplest fashion, the saddle point arrangement at ξ_s. Since $f'(\xi)$ has a first-order zero at ξ_s and no other zeros nearby, we may select the polynomial $\tau(s)$ as

$$f(\xi) = \tau(s) = f(\xi_s) - s^2, \tag{2.53}$$

the point ξ_s in the complex ξ-plane being chosen to correspond to $s = 0$ in the complex s-plane. The SDP P in the s-plane, along which $I_m\tau(s)$ = constant, is clearly the real s-axis. Thus the integral $I(\beta)$ over the SDP Γ or Γ' becomes

$$I(\beta) = e^{\Omega f(\xi_s)} \int_{-\infty}^{+\infty} G(s)\, e^{-\Omega s^2}\, ds, \tag{2.54}$$

with $G(s)$ given as

$$G(s) = F_1(\xi)\frac{d\xi}{ds}, \quad \frac{d\xi}{ds} = -\frac{2s}{f'(\xi)}. \tag{2.55}$$

Since $G(s)$ is assumed regular near $s = 0$, it can be expanded into a Taylor series

$$G(s) = G(o) + G'(o)s + G''(o)\frac{s^2}{2!} + \cdots + G^{(n)}\frac{s^n}{n!} + \cdots, \tag{2.56}$$

which converges uniformly inside a circle with finite radius r at $s = 0$ (r being the distance to the nearest singularity of $G(s)$).

In order to determine the first coefficient of the expansion (2.56), we develop $f(\xi)$ near ξ_s in the Taylor series and identify the result to (2.53),

$$f(\xi) \cong f(\xi_s) + \frac{(\xi - \xi_s)^2}{2} f''(\xi_s) + \cdots \cong f(\xi_s) - s^2,$$

which gives near ξ_s,

$$\xi - \xi_s \approx s\sqrt{-\frac{2}{f''(\xi_s)}},$$

and

$$G(o) = F_1(\xi_s)\left(\frac{d\xi}{ds}\right)_{s=o} = F_1(\xi_s)\sqrt{-\frac{2}{f''(\xi_s)}}, \tag{2.57}$$

where $f''(\xi_s) \neq 0$ at the first-order saddle point. Since ds is positive along the path of integration and in particular at $s = 0$, arg $(d\xi / ds)$ at $s = 0$ must be chosen equal to arg $d\xi$ at ξ_s along the SDP. This requirement specifies the square root function in (2.57). If $G(s)$ is approximated by $G(o)$, one obtains as the first-order asymptotic approximation for $I(\beta)$,

$$I(\beta) \approx G(o)\, e^{\Omega f(\xi_s)} \int_{-\infty}^{+\infty} e^{-\Omega s^2}, \tag{2.58}$$

and noting that the integral in (2.58) equals to $\sqrt{\pi / \Omega}$,

$$I(\beta) \approx \sqrt{-\frac{2\pi}{\Omega f''(\xi_s)}} F_1(\xi_s)\, e^{\Omega f(\xi_s)}, \quad \Omega \to \infty. \tag{2.59}$$

For the SDP Γ and Γ', we have $\arg(d\xi) = \frac{3\pi}{4}$ and $-\frac{\pi}{4}$, respectively, at the corresponding saddle points. Hence it follows that

$$I(\beta, \xi_s = \pm\pi) = -\frac{e^{ik(\rho+\rho')}}{8\pi n k\sqrt{\rho\rho'}}\, e^{\mp\frac{i\pi}{2}} \cot\left(\frac{\beta \pm \pi}{2n}\right) \tag{2.60}$$

and

$$-G^d_{s_b} = I\left(\bar{\beta}, \xi_s = \pi\right) + I\left(\bar{\beta}, \xi_s = -\pi\right) \pm \left[I\left(\overset{+}{\beta}, \xi_s = \pi\right) + I\left(\overset{+}{\beta}, \xi_s = -\pi\right)\right]$$

or

$$G^d_{s_b} = \frac{e^{ik(\rho+\rho')}}{8\pi n k\sqrt{\rho\rho'}}\left\{ e^{-\frac{i\pi}{2}} \cot\left(\frac{\bar{\beta}+\pi}{2n}\right) + e^{\frac{i\pi}{2}} \cot\left(\frac{\bar{\beta}-\pi}{2n}\right) \mp \left[e^{-\frac{i\pi}{2}} \cot\left(\frac{\overset{+}{\beta}+\pi}{2n}\right) + e^{\frac{i\pi}{2}} \cot\left(\frac{\overset{+}{\beta}-\pi}{2n}\right)\right]\right\} \tag{2.61}$$

$$= \frac{ie^{ik(\rho+\rho')}}{4\pi n k\sqrt{\rho\rho'}}\left[\frac{\frac{1}{n}\sin\frac{\pi}{n}}{\cos\frac{\pi}{n} - \cos\frac{\bar{\beta}}{n}} \mp \frac{\frac{1}{n}\sin\frac{\pi}{n}}{\cos\frac{\pi}{n} - \cos\frac{\overset{+}{\beta}}{n}}\right].$$

For a line source, the incident field at the edge is given by

$$p^i = \frac{i}{4} H_o^{(1)}\left(k\rho'\right) \approx \frac{e^{\frac{i\pi}{4}}}{\sqrt{8\pi k\rho'}} e^{ik\rho'} \tag{2.62}$$

for large ρ'.

According to (1.199), the dominant term of the diffracted field due to an incident cylindrical wave propagating in a direction normal to a straight wedge takes the form

$$p_o^d(\sigma) = p_o^i(Q)\frac{D}{\sqrt{\sigma}}, \tag{2.63}$$

where $\sigma = \rho$ and $p_o^i(Q)$ is the incident field on the edge. Hence by comparing (2.63) to (2.61) it follows the expression of the diffraction coefficient:

$$D_{s_b}\left(\phi,\phi',k\right) = \frac{e^{\frac{i\pi}{4}}}{\sqrt{2\pi k}} \frac{\sin\frac{\pi}{n}}{n}\left[\frac{1}{\cos\frac{\pi}{n} - \cos\left(\frac{\phi-\phi'}{n}\right)} \mp \frac{1}{\cos\frac{\pi}{n} - \cos\left(\frac{\phi+\phi'}{n}\right)}\right]. \tag{2.64}$$

The result indicated in (2.64) is identical to that deduced by Sommerfeld [2.2]. The total field surrounding the wedge is given asymptotically by superposing the results of (2.51) and (2.61). The diffraction coefficient (2.64) becomes singular at incident and diffraction angles that verify $\phi \pm \phi' = \pi + 2nN\pi$, which correspond to the shadow and reflection boundaries.

In the transition regions of the shadow and reflection boundaries, a pole is close to the saddle point. When this happens, the ordinary steepest descent method is not applicable since $G(s)$, given by (2.55), is singular at $s = 0$ and cannot be expended into the Taylor series (2.56). The diffraction coefficient given by (2.64) is usually called Keller's nonuniform diffraction coefficient for the wedge. A uniform diffraction coefficient for the wedge that is valid inside the transition regions and tends to Keller's diffraction coefficient away from the transition regions will be derived in chapter 3.

2.2.1.2. Plane Wave Incidence

The result for the plane wave case may be obtained by letting $\rho' \to \infty$ in the line source case. When $\rho' \to \infty$, (2.36) reduces to

$$K_o\left(-iz(\xi)\right) \approx \sqrt{\frac{\pi}{2}}\, e^{\frac{i\pi}{4}} \frac{e^{ik\rho'}}{\sqrt{k\rho'}} e^{-ik\rho\cos\xi} \tag{2.65}$$

and accordingly

$$G(\boldsymbol{\rho},\boldsymbol{\rho}';\beta) \approx \frac{1}{4\pi i n}\frac{e^{i\frac{\pi}{4}}}{\sqrt{8\pi k}}\frac{e^{ik\rho'}}{\sqrt{\rho'}}\int_{L-L'} e^{-ik\rho\cos\xi}\cot\left(\frac{\xi+\beta}{2n}\right)d\xi. \tag{2.66}$$

The sound field of a line source is given by (2.62). If one suppresses the "line source factor" appearing in (2.66), then the resulting expressions, which we denote by $g(\rho, \beta)$, corresponds to a plane wave normally incident on the edge of the wedge and having a unit amplitude and zero phase at the edge. Hence it follows that for a plane wave we have

$$g(\rho,\beta) = \approx \frac{1}{4\pi i n}\int_{L-L'} e^{-ik\rho\cos\xi}\cot\left(\frac{\xi+\beta}{2n}\right)d\xi. \tag{2.67}$$

The total field is therefore given by

$$g_{s_h}(\rho,\phi,\phi') = g\left(\rho,\overset{-}{\beta}\right) \mp g\left(\rho,\overset{+}{\beta}\right). \tag{2.68}$$

The integral (2.67) over $L - L'$ is evaluated, as done earlier for a line source excitation. The saddle points of the integrand of (2.67) are again $\xi_s = \pm m\pi$, where $m = 0, 1, 2, \ldots$, and the pole singularities occur at $\xi_p = -\beta + 2nN\pi$, where $N = 0, \pm 1, \pm 2, \ldots$. Hence it follows that the SDP Γ and Γ' passing, respectively, through $\xi_s = \pi$ and $\xi_s = -\pi$, close the $L - L'$ contour thereby allowing the use of Cauchy's residue theorem.

The complex ξ-plane topology is the same as in figure 2.4 with $L - L'$ and $\Gamma + \Gamma'$ contours unchanged. The integrand of (2.67), however, has no branch points.

Applying the steepest descent method to (2.67), one obtains, for poles not close to the saddle points, the following expression for the diffracted field corresponding to a plane wave that is normally incident on the edge of the wedge and has a unit amplitude and zero phase at the edge:

$$g_{s_b}^{d} \approx D_{s_b} \frac{e^{ik\rho}}{\sqrt{\rho}},$$

where the diffraction coefficient D_{s_b} is unchanged, as expected, and given by (2.64).

If the field of the incident plane wave at the edge is given by $g^{i}(o)$, then the diffracted field is given by

$$g_{s_b}^{d} = g^{i}(o)\, D_{s_b} \frac{e^{ik\rho}}{\sqrt{\rho}}. \tag{2.69}$$

2.2.1.2.1. Oblique Incidence

Let $G_{s_b}(\mathbf{r})$ represent the total field at an observation $\mathbf{r}$, due to a plane sound wave obliquely incident on an impenetrable wedge. It satisfies the Helmholtz equation,

$$\left(\Delta + k^2\right) G_{s_b}(\mathbf{r}) = 0. \tag{2.70}$$

In cylindrical coordinates (ρ, ϕ, z) with the z-axis along the edge and the x-axis along the reference face $\phi = 0$, the direction of propagation of the incident wave is defined by the angle ϕ' of its projection on the xy plane and by the oblicity angle β_o (fig. 2.5).

A traveling wave dependence in z allows one to express the total field as

$$G_{s_b}(\mathbf{r}) = g_{s_b}(\boldsymbol{\rho}; \phi', k_t) e^{ik_z z}, \tag{2.71}$$

where $k_z = k \cos \beta$. Inserting (2.71) into (2.70) yields

$$\left(\Delta_t + k_t^2\right) g_{s_b}(\rho, \phi; \phi', k_t) = 0, \tag{2.72}$$

where Δ_t is the transverse Laplacian operator.

The function g satisfies the following boundary conditions at the wedge surface:

$$g_s(\rho, \phi; \phi', k_t) = 0, \text{ at } \phi = 0,\ n\pi \tag{2.73}$$

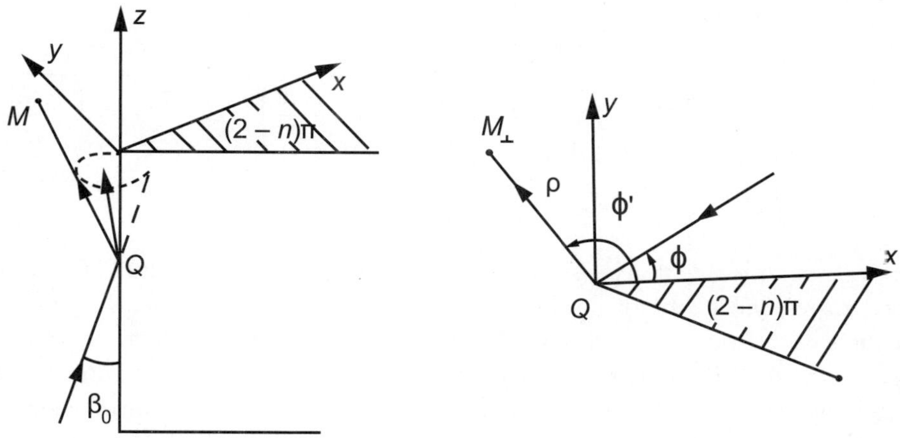

Figure 2.5. Geometry for the ray coordinate system for oblique incidence.

for the soft wedge and

$$\frac{\partial g_h}{\partial \phi}\left(\rho,\phi;\phi',k_t\right)=0, \text{ at } \varphi=0\,,\; n\pi \tag{2.74}$$

for the hard wedge.

Both functions g_s and g_h also satisfy the Sommerfeld radiation condition and the Meixner edge condition given by (2.4) and (2.5), respectively, with G replaced by g.

The equation (2.72) verified by g and the boundary and edge conditions are the same as for normal incidence. The total field is therefore given by (2.68), in which $g(\boldsymbol{\rho},\beta)$ is identical to (2.67), with k replaced by k_t.

Moreover, if the field of the incident ray at the diffraction point Q is given by $g^i(Q)$, then the expression of the diffracted field at an observation point $M(\rho, \phi, z)$ on a diffracted ray is given by

$$G_{s_h}\left(M\right)=g^i\left(Q\right)D_{s_h}\left(\phi,\phi';k,\sin\beta_0\right)\frac{e^{ik\left(\rho\sin\beta_0+z\cos\beta_0\right)}}{\sqrt{\rho/\sin\beta_0}}, \tag{2.75}$$

where $D_{s_h}\left(\phi,\phi';k,\sin\beta_0\right)$ is the diffraction coefficient.

Equation (2.75) has the general form (1.199) since, according to (1.212), $\rho_1=\rho_i=\infty$, $\rho=QM\sin\beta_0$, and $\rho\sin\beta_0+z\cos\beta_0=QM=\sigma$. Hence it follows that the diffraction coefficient for oblique incidence is given by

$$D_{s_b}(\phi,\phi';k,\sin\beta_0)=\frac{e^{\frac{i\pi}{4}}}{\sin\beta_0\sqrt{2k}}\frac{\sin\frac{\pi}{n}}{n}\left[\frac{1}{\cos\frac{\pi}{n}-\cos\left(\frac{\phi-\phi'}{n}\right)}\mp\frac{1}{\cos\frac{\pi}{n}-\cos\left(\frac{\phi+\phi'}{n}\right)}\right].\tag{2.76}$$

It is important to note that the singularities of the diffraction coefficient (2.76), for oblique incidence, correspond to the shadow and reflection boundaries of the incident field. This is because those boundaries are completely defined by their projections on a plane perpendicular to the edge and the angle β_0. Moreover, since the diffraction coefficient in (1.199) does not depend on the radii of curvature of the incident and diffracted wave fronts, it remains the same for an incident spherical wave and, more generally, for an incident wave having different principal radii of curvature at the diffraction point.

2.2.2. IMPEDANCE WEDGE AT NORMAL AND OBLIQUE INCIDENCE: MALIUZHINETS'S SOLUTION

Consider an impenetrable wedge with planar faces that make the angle $\pm\Phi$ with the *y*-axis. Let the wedge be impinged by an incident plane wave with propagation vector $\mathbf{k}^i=(-k\sin\varphi_o,-k\cos\varphi_o)$, normal to the edge of the wedge and pressure:

$$p^i(x,y)=\exp\left[-ik\left(x\sin\varphi_o+y\cos\varphi_o\right)\right].\tag{2.77}$$

The coordinate system and the notations are defined in figure 2.6.

The surface impedances of the upper and lower faces of the wedge are supposed constant and are denoted by Z^+ and Z^-, respectively. Let $p(x, y)$ be the total pressure field at an observation point $M(x, y)$ in the fluid outside the wedge. The function $p(x, y)$ must verify the Helmholtz equation in the domain D occupied by the fluid:

$$\left(\Delta+k^2\right)p=0,\quad M\in D.\tag{2.78}$$

It also must verify the boundary conditions on the faces. In polar coordinates, $x=r\sin\varphi$, $y=r\cos\varphi$, $p(r,\varphi)$ must verify

$$-\frac{1}{r}\frac{\partial p}{\partial\varphi}\pm ikZ^{\pm}p=0,\quad\varphi=\pm\Phi.\tag{2.79}$$

In addition, the energy associated with the pressure field solution must be bounded at the edge.

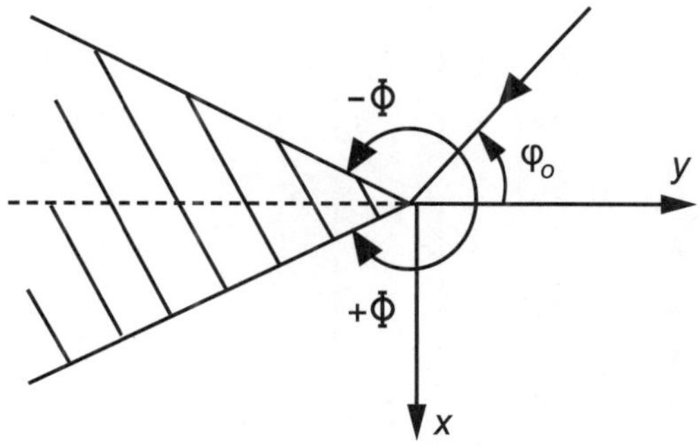

Figure 2.6. Geometry of the diffraction by a wedge.

An exact solution to this problem was first established by Maliuzhinets [2.6]. The total field outside the wedge is written in the form of a spectrum of homogeneous and inhomogeneous plane waves along a contour γ in the complex α-plane:

$$P(r,\varphi) = \frac{1}{2\pi i} \int_{\gamma} e^{-ik\, r \cos\alpha} S(\alpha + \varphi)\, d\alpha. \tag{2.80}$$

The contour γ is known as the Sommerfeld contour (fig. 2.7). It comprises two branches, L and $-L'$, which are symmetric with respect to the origin. We must realize at this point that the representation (2.80) is strictly formal in nature, and we must verify the convergence of the integral, as well as assure that the result is physically meaningful.

It can be directly verified that along the infinite branches of γ, the imaginary part of cos α is negative, and hence the exponential in (2.80) tends to zero. We can therefore assure that the integral is convergent for a large class of functions, $S(\alpha)$.

A justification of the spectral representation (2.80) due to Maliuzhinets [2.7] will be given hereafter, but we can also consider this representation as an ansatz, the validity of which is justified a posteriori by showing that it leads to a unique solution after having imposed the boundary conditions on the faces and the Meixner condition on the edge. The origin of this ansatz is the general integral form of the solution in the particular cases where the impedance Z is either zero (hard surface) or infinite (soft surface).

2.2.2.1. Analytical Details for the Justification of the General Form of the Solution

In order to justify (2.80), Maliuzhinets started with the Fourier integral:

$$P(x,y) = \int_{-\infty}^{+\infty} e^{i\left(x w + y\sqrt{k^2 - w^2}\right)} S_o(w)\, dw. \tag{2.81}$$

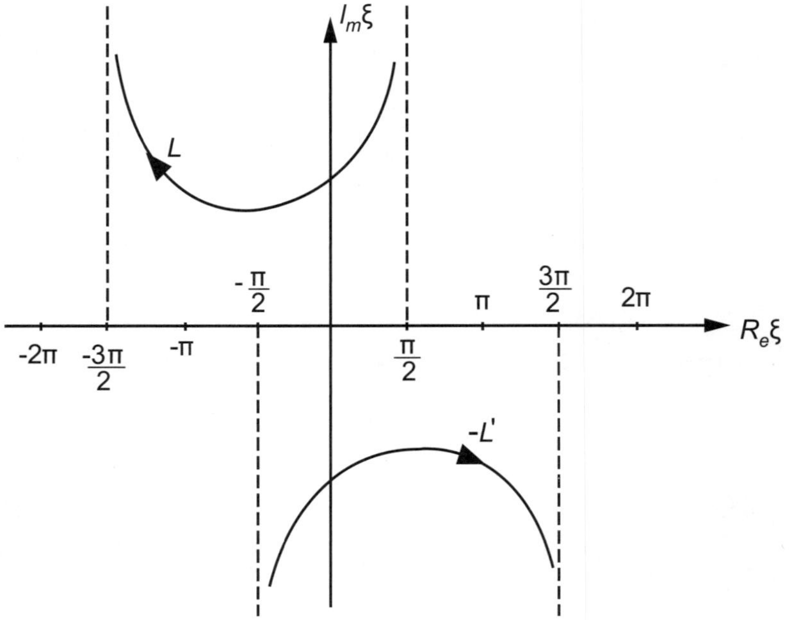

Figure 2.7. Sommerfeld contour $\gamma = L - L'$.

If we choose a positive value for the imaginary part of the square root when $|w| > k$ and if $S_o(w)$ is finite in the whole integration interval, then (2.81) is a general solution of the Helmholtz equation in the half-space $y > 0$ (fig. 2.8).

If we replace the half-space by a wedge-shaped domain, as shown in figure 2.6, the representation (2.81) may still be used for $y > 0$. However, for $y < 0$, the Fourier integral does not converge.

The problem does not change when we go over to cylindrical coordinates $x = r \cos \varphi$, $y = r \sin \varphi$ and replace the integration variable with $w = k \sin \alpha$. Now if we choose the following determination of the square root in (2.81),

$$\sqrt{k^2 - w^2} = k \cos \alpha,$$

then the Fourier integral (2.81) will take the form

$$P(r,\varphi) = \int_C e^{ikr\cos(\varphi-\alpha)} S_o(k \sin \alpha) k \cos \alpha \, d\alpha, \tag{2.82}$$

where C is the contour in the complex α-plane defined by $I_m(k \sin \alpha) = 0$.

If the wave number k is complex with $k = k' + ik''$, then we have the following for $\alpha = \alpha' + i\alpha''$:

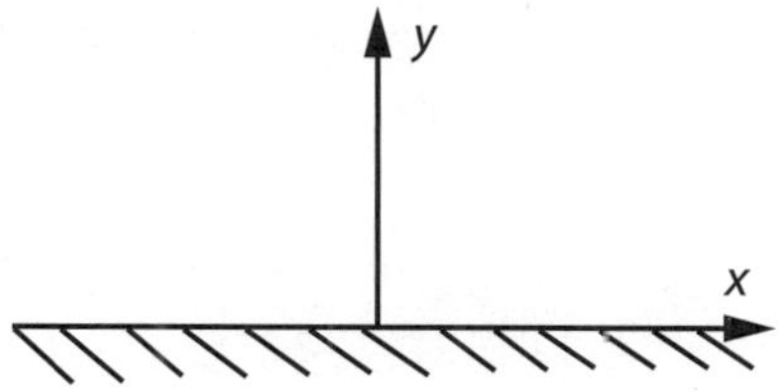

Figure 2.8. Half-space $y > 0$.

$$I_m\left(k\sin\alpha\right) = k'\cos\alpha'\,sh\alpha'' + k''\sin\alpha' ch\alpha''.$$

Hence the contour C is defined by

$$th\,\alpha'' = -\frac{k''}{k'}\tan\alpha'. \tag{2.83}$$

For a passive system, we have $k'' > 0$, and since k' is supposed positive, the contour C has the form represented in figure 2.9.

In a similar way, it can be shown that the contours for which $I_m(k\cos\alpha) > 0$ are defined by

$$th\,\alpha'' = -\frac{k''}{k'}\cot\alpha', \tag{2.84}$$

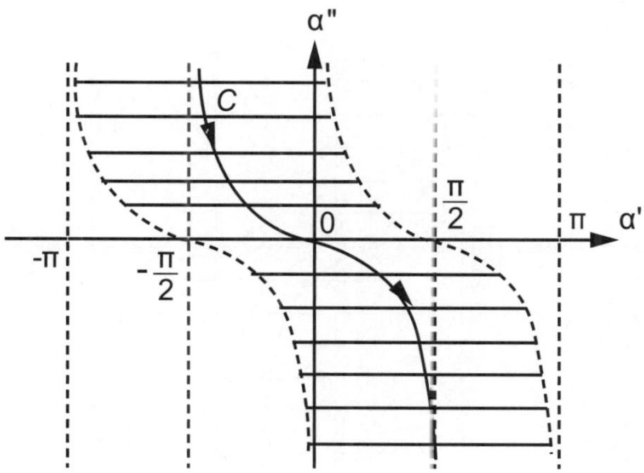

Figure 2.9. Integration contour in the complex α-plane and the domain in which $I_m(k\cos\alpha) > 0$.

and one verifies readily that $I_m(k\cos\alpha) > 0$ in the horizontally shaded region of figure 2.9.

If we change the variable α in (2.82) into $\alpha + \varphi$, we obtain

$$P(r,\varphi) = \int_{C_\varphi} e^{ikr\cos\alpha} S_o\left[k\sin(\alpha+\varphi)\right] k\cos(\alpha+\varphi)\, d\alpha, \tag{2.85}$$

where C_φ is the contour obtained by translating the contour C along the axis α' by $-\varphi$ (fig. 2.10).

When $\varphi \in \left[-\pi/2, \pi/2\right]$, the contour C_φ is located in the horizontally shaded region of figure 2.9, which corresponds precisely to the region where $I_m(\cos(\alpha)) \geq 0$. Since $S_o(\alpha)$ is supposed bounded in the same region, it is possible to shift the contour C_φ along α' and apply it on C without modifying the integral (2.85), except for the contribution of possible poles located in the region between C_φ and C.

Therefore, we have

$$P(r,\varphi) = \int_{C} e^{ikr\cos\alpha} S_o\left[k\sin(\alpha+\varphi)\right] k\cos(\alpha+\varphi)\, d\alpha + 2\pi i R, \tag{2.86}$$

where R are the residues at the poles.

The representation (2.86) is valid for $|\varphi| < \pi/2$. In order to extend it to the domain $\pi/2 < |\varphi| < \Phi$, we consider the region in the complex α-plane where $I_m(k\cos\alpha) < 0$. Applying

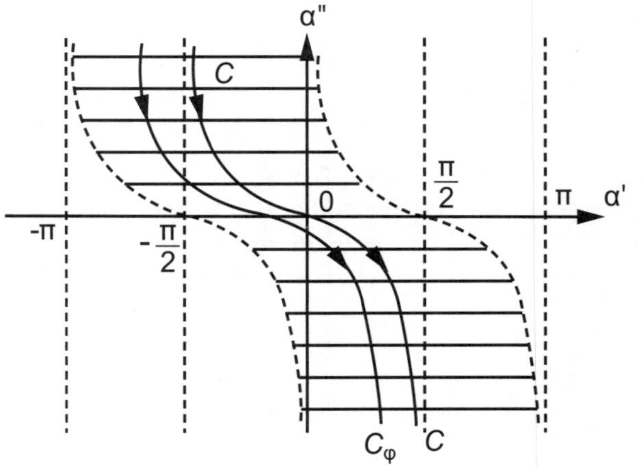

Figure 2.10. Contour C_φ.

(2.84), we see that $I_m(k\cos\alpha) = 0$ along the contours passing through $\alpha' = \frac{\pi}{2} + n\pi$ and $\alpha' = -\frac{\pi}{2} - n\pi$ (n = 1, 2, …). For n = 1, one verifies readily that $I_m(k\cos\alpha) < 0$ in the horizontally shaded region of figure 2.11.

Changing the determination of cos α in (2.82) and writing

$$P(r,\varphi) = \int_C e^{-ikr\cos(\alpha-\varphi)} S_1(\alpha)\,d\alpha, \quad \varphi > 0 \tag{2.87}$$

and

$$P(r,\varphi) = \int_C e^{-ikr\cos(\alpha-\varphi)} S_2(\alpha)\,d\alpha, \quad \varphi < 0, \tag{2.88}$$

where $S_1(\alpha)$ and $S_2(\alpha)$ are two new unknown functions, we obtain the following after replacing α by $\alpha + \varphi$:

$$P(r,\varphi) = \int_{C_\varphi^+} e^{-ikr\cos\alpha} S_1(\alpha+\varphi)\,d\alpha, \quad \varphi > 0, \tag{2.89}$$

$$P(r,\varphi) = \int_{C_\varphi^-} e^{-ikr\cos\alpha} S_2(\alpha+\varphi)\,d\alpha, \quad \varphi < 0. \tag{2.90}$$

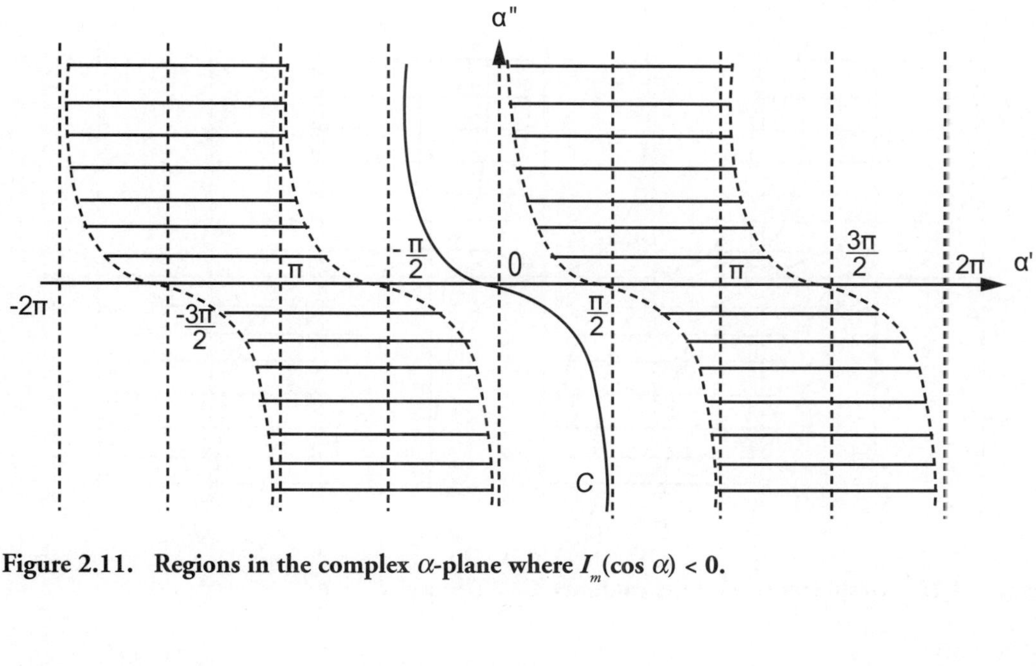

Figure 2.11. Regions in the complex α-plane where $I_m(\cos\,\alpha) < 0$.

The contours C_φ^+ and C_φ^- are shown in figure 2.12 for $|\varphi| > \pi/2$. They are respectively lying in the right and left horizontally shaded regions of figure 2.11.

When φ takes positive values between $\pi/2$ and π, C_φ^+ is located in the cross-hatched region on the side $\alpha' > 0$. Similarly, when φ takes negative values between $-\pi/2$ and $-\pi$, C_φ^- is located in the cross-hatched region on the side $\alpha' < 0$. Hence the representations (2.89) and (2.90) are valid for $\pi/2 < |\varphi| < \pi$ and can therefore be used to represent the solutions in the domain $\pi/2 < |\varphi| < \Phi$ corresponding to $y < 0$.

Again, we suppose that $S_1(\alpha)$ and $S_1(\alpha)$ are bounded in the domain where $I_m(k \cos \alpha) < 0$ and may have pole singularities. Thus by shifting the contours C_φ^+ on C^+ and C_φ^- on C^-, where C^+ and C^- are contours passing through π and $-\pi$, the representations (2.89) and (2.90) can be written

$$P(r,\varphi) = \int_{C^+} e^{-ikr\cos\alpha} S_1(\alpha+\varphi)\, d\alpha + 2\pi i A, \tag{2.91}$$

$$P(r,\varphi) = \int_{C^-} e^{-ikr\cos\alpha} S_2(\alpha+\varphi)\, d\alpha + 2\pi i B, \tag{2.92}$$

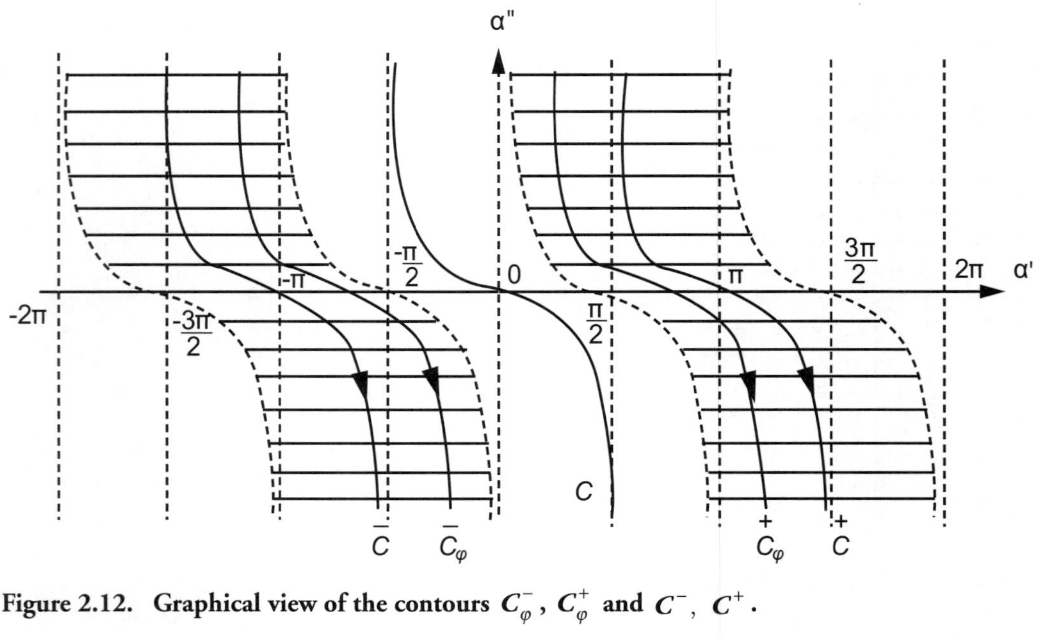

Figure 2.12. Graphical view of the contours C_φ^-, C_φ^+ and C^-, C^+.

where A and B are the residues at the poles located respectively in the domain between C_φ^+ and C^+ and in the domain between C_φ^- and C^-.

It is desirable to reduce the number of unknown functions while keeping the representation sufficiently general. By changing the integration variable α into $\alpha + \pi$ in (2.91) and $\alpha - \pi$ in (2.92), we obtain

$$P(r,\varphi) = \int_C e^{ikr\cos\alpha}\, S_1(\alpha + \varphi + \pi) + 2\pi i A, \tag{2.93}$$

$$P(r,\varphi) = \int_C e^{ikr\cos\alpha}\, S_2(\alpha + \varphi - \pi) + 2\pi i B. \tag{2.94}$$

If we impose the condition

$$S_2(\alpha + \varphi - \pi) = -S_1(\alpha + \varphi - \pi), \tag{2.95}$$

then we obtain $B = A$, and hence, by adding (2.93) and (2.94), we get

$$P(r,\varphi) = \frac{1}{2}\int_C e^{ikr\cos\alpha}\left[S_1(\alpha + \varphi + \pi) - S_1(\alpha + \varphi - \pi)\right] d\alpha + 2\pi i A. \tag{2.96}$$

The representation (2.96) is valid for $\pi/2 < |\varphi| < \Phi$. If we compare it to (2.86), then we see that we can extend its validity to $0 < |\varphi| < \Phi$ by searching a function $S_1(\alpha)$, verifying

$$S_1(\alpha + \pi) - S_1(\alpha - \pi) = 2S_o(k\sin\alpha)k\cos\alpha. \tag{2.97}$$

In addition, if $S_o(\alpha)$ has poles at α_p in the domain $-\pi/2 < R_e\alpha_p < \pi/2$, then it follows from (2.97) that $S_1(\alpha + \pi)$ and $S_1(\alpha - \pi)$ must have poles at $\alpha_p + \pi$ and $\alpha_p - \pi$, and one verifies readily that $A = R$.

By once again separating (2.96) into two integrals like (2.93) and (2.94), changing the integration variable α into $\alpha + \pi$ in the first integral and $\alpha - \pi$ in the second integral, and summing both integrals, one is led to the expression

$$P(r,\varphi) = \frac{1}{2}\left(\int_{C^+} - \int_{C^-}\right) e^{-ikr\cos\alpha}\, S_1(\alpha + \varphi)\, d\alpha + 2\pi i R, \tag{2.99}$$

where the contours C^+ and C^- are shown in figure 2.11. By deforming these contours along two paths that are symmetric with respect to the origin as shown in figure 2.13, we obtain

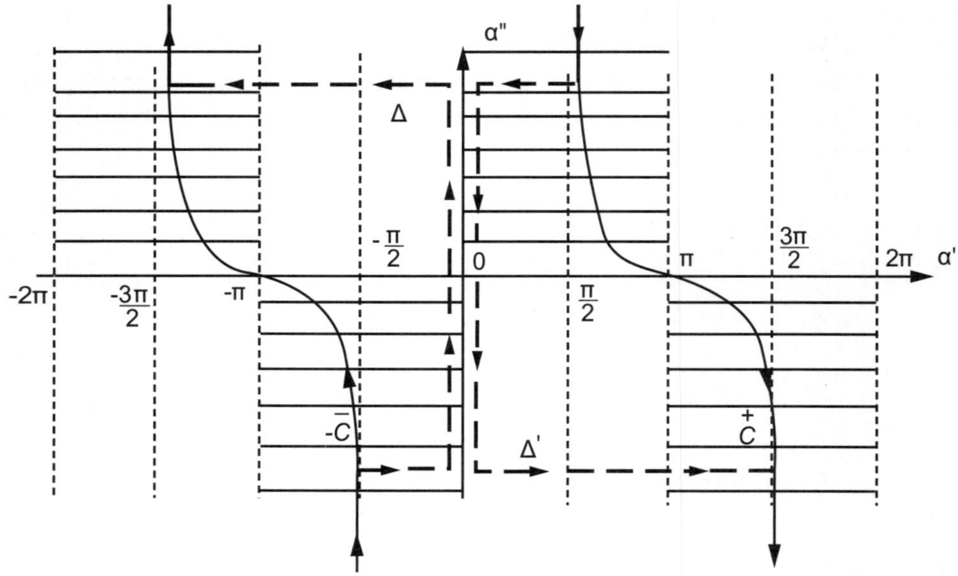

Figure 2.13. Deformation of the contours C^+ and C^-.

$$\frac{1}{2}\left(\int_{C^+} - \int_{C^-}\right) e^{-ikr\cos\alpha} S_1(\alpha+\varphi)\,d\alpha = \frac{1}{2}\left(\int_{\Delta} + \int_{\Delta'}\right) e^{-ikr\cos\alpha} S_1(\alpha+\varphi)\,d\alpha - 2\pi i R, \quad (2.100)$$

where R are the residues at the pole $\alpha_p - \pi$ or $\alpha_p + \pi$. It can be easily verified that only one of these two groups of poles may be intercepted by the contour deformation since $-\pi < R_e\alpha_p < \pi/2$ (see fig. 2.10).

The contours Δ and Δ' are shown in figure 2.13. Inserting (2.100) into (2.99) leads to the following expression for $P(r,\varphi)$:

$$P(r,\varphi) = \frac{1}{2}\left(\int_{\Delta} + \int_{\Delta'}\right) e^{-ikr\cos\alpha} S_1(\alpha+\varphi)\,d\alpha. \quad (2.101)$$

Since the integrals on the portions of Δ and Δ' along the imaginary axis cancel each other, the contours Δ and Δ' can be replaced by the Sommerfeld contour γ, represented in figure 2.14. The final result is therefore

$$P(r,\varphi) = \frac{1}{2\pi i}\int_{\gamma} e^{-ikr\cos\alpha}\, S(\alpha+\varphi)\,d\alpha, \quad (2.102)$$

where we have stated $S = \pi i S_1$. This result is identical to (2.80) and thus gives a justification of this formula.

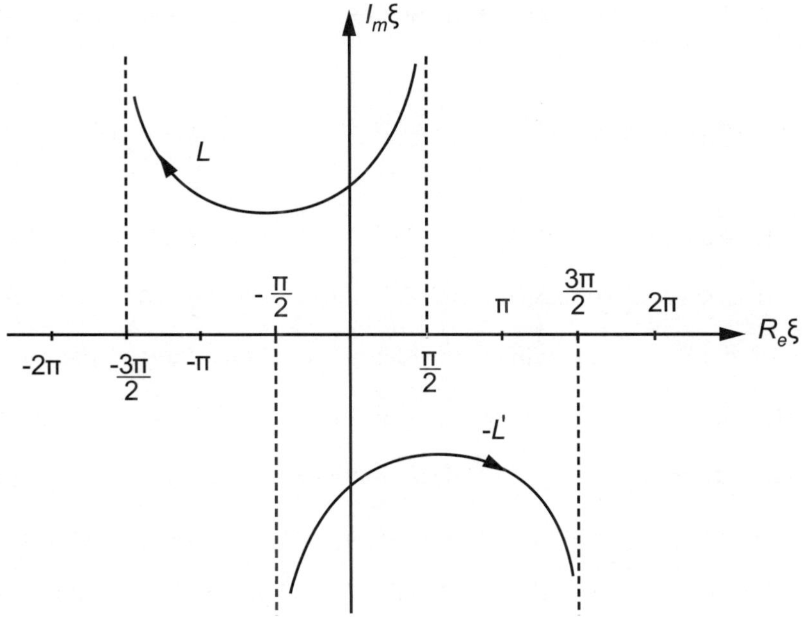

Figure 2.14. Sommerfeld contour $\gamma = \Delta + \Delta' = \boldsymbol{L} - \boldsymbol{L}'$ **.**

2.2.2.2. Derivation of the Solution

We start with the Fourier representation (2.102) of the field, which obviously verifies the Helmholtz equation (2.78). Equation (2.102) is a plane wave spectrum of components:

$$\Pi(\alpha,\varphi) = e^{-ik\,r\cos\alpha} S(\alpha+\varphi). \tag{2.103}$$

On the faces $\varphi = \Phi$ and $\varphi = -\Phi$ of the wedge, each elementary plane wave verifies the boundary condition (2.79). Hence

$$-\frac{1}{r}\frac{\partial \Pi}{\partial \varphi} \pm ik\,Z^{\pm}\Pi = 0, \quad \varphi = \pm\Phi. \tag{2.104}$$

Owing to (2.103), we obtain

$$\begin{aligned}\frac{\partial \Pi}{\partial \varphi}(\alpha,\varphi) &= e^{-ik\,r\cos\alpha}\frac{\partial S}{\partial \alpha}(\alpha+\varphi) \\ &= \frac{\partial}{\partial \alpha}\left[e^{-ik\,r\cos\alpha}\, S(\alpha+\varphi)\right] - ikr\sin\alpha\, S(\alpha+\varphi)e^{-ik\,r\cos\alpha}.\end{aligned} \tag{2.105}$$

Substituting (2.105) into (2.104) and integrating over α along the Sommerfeld contour γ, one obtains

$$\int_{\gamma} ikr\, S(\alpha \pm \Phi)\left(\sin\alpha \pm Z^{\pm}\right)e^{-ikr\cos\alpha}d\alpha + \left[e^{-ikr\cos\alpha}\, S(\alpha \pm \Phi)\right]_{|I_m(\alpha)|\to\infty} = 0, \qquad (2.106)$$

where the integrated term in (2.106) has to be taken at the limits $|I_m\alpha| \to \infty$ of the integration path. This term is equal to zero if $S(\alpha)$ increases less than $\exp\left[kr\left|\sin\alpha' \, sh\alpha''\right|\right]$ for all r. This requirement is fulfilled since, according to the Meixner condition, $P\left(r,\varphi\right)$ must be finite when r tends to zero (on the edge). Therefore the integral (2.102), which for $r \to 0$ reduces to the integral of $S(\alpha + \varphi)$, must vanish for all values of φ. The fact that the integration path is symmetric with respect to the origin implies that for $|I_m\alpha| \to \infty$, $S(\alpha + \varphi)$ tends to an even function of α independent of φ; thus it tends to a constant. Eliminating the integrated term in (2.106), we obtain

$$\int_{\gamma} S(\alpha \pm \Phi)\left(\sin\alpha \pm Z^{\pm}\right)e^{-ikr\cos\alpha}d\alpha = 0. \qquad (2.107)$$

Since γ is symmetric with respect to the origin, a necessary and sufficient condition that equation (2.107) holds is that its integrand be even. Hence the unknown function $S(\alpha)$ must verify

$$\left(\sin\alpha + Z^{+}\right)S(\alpha + \Phi) = \left(Z^{+} - \sin\alpha\right) S(\Phi - \alpha), \qquad (2.108)$$

$$\left(\sin\alpha - Z^{-}\right)S(\alpha - \Phi) = -\left(Z^{-} + \sin\alpha\right)S(-\alpha - \Phi). \qquad (2.109)$$

These equations can also be written in the form

$$S\left[\pm(\Phi + \alpha)\right] = -R_{\pm}(\alpha)S\left[\pm(\Phi - \alpha)\right], \qquad (2.110)$$

where $R_{\pm}$ is the reflection coefficient on the faces $\varphi = \pm\Phi$, given by

$$R_{\pm}(\alpha) = \frac{\sin\alpha - Z^{\pm}}{\sin\alpha + Z^{\pm}}. \qquad (2.111)$$

To solve the equations (2.110), we observe that if $\psi(\alpha)$ is a particular solution of (2.110), then we obtain a class of solutions by stating

$$S(\alpha) = \sigma(\alpha)\,\psi(\alpha) \tag{2.112}$$

with

$$\sigma\left[\pm(\Phi+\alpha)\right] = \sigma\left[\pm(\Phi-\alpha)\right]. \tag{2.113}$$

According to (2.112), the poles of $S(\alpha)$ can be distributed over the functions $\sigma(\alpha)$ and $S(\alpha)$. Since the integral (2.102) represents the total field, $S(\alpha)$ must have a pole at $\alpha = \varphi_0$ in order to recover the incident field (2.77), which in polar coordinates reads

$$p_i(r,\varphi) = p_0^i \exp\left[-ik\cos(\varphi-\varphi_0)\right] \tag{2.114}$$

This pole is not dependent on the boundary conditions and will be attributed to $\sigma(\alpha)$. On the other hand, the poles of $\psi(\alpha)$ will depend on the boundary conditions. Particularly, if the reflection coefficient $R_\pm(\alpha)$ has poles at $\alpha = \alpha_p^\pm$, then $\psi(\alpha)$ has poles at $\alpha = \pm(\Phi + \alpha_p^\pm)$.

Now for a passive system, we know that $R_e Z^\pm < 0$, which implies $R_e\,\alpha_p^\pm > 0$. Hence the poles of $\psi(\alpha)$ are located outside the strip $|R_e(\alpha)| \le \Phi$. The same result is obtained for the zeros of $\psi(\alpha)$, given by $\alpha = \pm(\Phi + \alpha_s^\pm)$ where $\alpha_s^\pm$ are the zeros of $R_\pm(\alpha)$. This led us to make the following hypotheses:

1. $\psi(\alpha)$ has no zeros nor poles in the strip $|R_e(\alpha)| \le \Phi$.
2. $\sigma(\alpha)$ contains all the poles of $S(\alpha)$ other than those of $R_\pm(\alpha)$. In the strip $|R_e(\alpha)| \le \Phi$, the function $\sigma(\alpha)$ has a simple pole at $\alpha = \varphi_o$.

Now consider the equation verified by the functions $\psi(\alpha)$, obtained by inserting (2.113) into (2.110), which reads

$$\psi\left[\pm(\Phi+\alpha)\right] = -R_\pm(\alpha)\,\psi\left[\pm(\Phi-\alpha)\right]. \tag{2.115}$$

According to hypothesis 1, it is possible to solve the equation (2.115) separately by Fourier transforming it in the strip $\left|R_e(\alpha)\right| \le \Phi$. For deriving the solution, we follow the procedure established by Vaccaro [2.8].

In the first step, equation (2.115) is transformed in a sum of terms by taking the logarithm of each member:

$$\frac{\psi'\left[\pm(\alpha+\Phi)\right]}{\psi\left[\pm(\alpha+\Phi)\right]} = \pm\frac{R'_\pm(\alpha)}{R_\pm(\alpha)} - \frac{\psi'\left[\pm(\Phi-\alpha)\right]}{\psi\left[\pm(\Phi-\alpha)\right]}. \tag{2.116}$$

In order to simplify the notation, we set

$$g(\alpha) = \frac{\psi'(\alpha)}{\psi(\alpha)}, \quad r_\pm(\alpha) = \frac{R'_\pm(\alpha)}{R_\pm(\alpha)}. \tag{2.117}$$

Then equation (2.116) reads

$$g\left[\pm(\alpha+\Phi)\right] + g\left[\pm(\Phi-\alpha)\right] = \pm r_\pm(\alpha). \tag{2.118}$$

In the second step, we apply the Fourier transform to (2.118). We set

$$H(\omega) = TF\left[g(\alpha)\right] = \int_{-i\infty}^{+i\infty} g(\alpha)e^{-\omega\alpha}d\alpha, \tag{2.119}$$

$$S_\pm(\omega) = TF\left[r_\pm(\alpha)\right] = \int_{-i\infty}^{+i\infty} r_\pm(\alpha)e^{-\omega\alpha}d\alpha, \tag{2.120}$$

where the Fourier transform is taken along the imaginary axis of the complex α-plane, since $\psi(\alpha)$ is assumed to be regular in the strip $\left|R_e(\alpha)\right| \le \Phi$, and where $-\infty < \omega < +\infty$ is real.

Applying the translation theorem, we obtain

$$H(\omega) = e^{-\omega\Phi}\int_{-i\infty}^{+i\infty} g(\Phi+\alpha)e^{-\omega\alpha}\,d\alpha = e^{\omega\Phi}\int_{-i\infty}^{+i\infty} g\left[-(\Phi-\alpha)\right]e^{-\omega\alpha}\,d\alpha, \tag{2.121}$$

$$H(-\omega) = e^{-\omega\Phi}\int_{-i\infty}^{+i\infty} g\left[-(\Phi+\alpha)\right]e^{-\omega\alpha}\,d\alpha = e^{\omega\Phi}\int_{-i\infty}^{+i\infty} g(\Phi-\alpha)e^{-\omega\alpha}\,d\alpha. \tag{2.122}$$

According to (2.121) and (2.122), the Fourier transform of (2.118) reads

$$e^{\omega\Phi}H(\omega)+e^{-\omega\Phi}H(-\omega)=+S_{+}(\omega),$$
$$e^{\omega\Phi}H(-\omega)+e^{-\omega\Phi}H(\omega)=-S_{-}(\omega). \tag{2.123}$$

The solution of the system (1.123) is given by

$$H(\omega)=\frac{S_{+}(\omega)e^{\omega\Phi}+S_{-}(\omega)e^{-\omega\Phi}}{2sh(2\omega\Phi)}. \tag{2.124}$$

The third step is to take the inverse Fourier transform of (2.114). To do this, we note that

$$r_{\pm}(\alpha)=\frac{\cos\alpha}{\sin\alpha-Z^{\pm}}-\frac{\cos\alpha}{\sin\alpha+Z^{\pm}}. \tag{2.125}$$

Setting $Z^{\pm}=\sin\,\theta^{\pm}$, it can be shown that

$$\frac{\cos\alpha}{\sin\alpha\pm\sin\theta^{\pm}}-\frac{1}{\sin\left(\alpha\pm\theta^{\pm}\right)}=\frac{-\sin\alpha}{\cos\alpha+\cos\theta^{\pm}}. \tag{2.126}$$

Equation (1.126) can be readily verified by calculating

$$\frac{\cos\alpha}{\sin\alpha\pm\sin\theta^{\pm}}+\frac{\sin\alpha}{\cos\alpha+\cos\theta^{\pm}}=\frac{1+\cos\left(\alpha\pm\theta^{\pm}\right)}{\left(\sin\alpha\pm\sin\theta^{\pm}\right)\left(\cos\alpha+\cos\theta^{\pm}\right)} \tag{2.127}$$

and by expressing the right-hand side member of (2.127) with the half angle $\frac{1}{2}\left(\alpha\pm\theta^{\pm}\right)$. According to (2.126), equation (2.125) reads

$$r_{\pm}(\alpha)=\frac{1}{\sin\left(\alpha-\theta^{\pm}\right)}-\frac{1}{\sin\left(\alpha+\theta^{\pm}\right)}. \tag{2.128}$$

Inserting (2.128) into (2.118) and (2.120) yields

$$S_{+}(\omega)=-TF\left\{\left[\cos\left(\alpha+\theta^{+}-\frac{\pi}{2}\right)\right]^{-1}+\left[\cos\left(\alpha+\frac{\pi}{2}-\theta^{+}\right)\right]^{-1}\right\},$$
$$S_{-}(\omega)=-TF\left\{\left[\cos\left(\alpha+\theta^{-}-\frac{\pi}{2}\right)\right]^{-1}+\left[\cos\left(\alpha+\frac{\pi}{2}-\theta^{-}\right)\right]^{-1}\right\}, \tag{2.129}$$

and by applying the translation theorem to (2.129), we obtain

$$S_+(\omega) = -\left[e^{-\omega\left(\frac{\pi}{2}-\theta^+\right)} + e^{\omega\left(\frac{\pi}{2}-\theta^+\right)}\right] TF\left(\frac{1}{\cos\alpha}\right),$$
$$S_-(\omega) = -\left[e^{-\omega\left(\frac{\pi}{2}-\theta^-\right)} + e^{\omega\left(\frac{\pi}{2}-\theta^-\right)}\right] TF\left(\frac{1}{\cos\alpha}\right). \tag{2.130}$$

Finally, by inserting (2.130) into (2.124), we obtain

$$H(\omega) = -\left[e^{\omega\left(\Phi+\frac{\pi}{2}-\theta^+\right)} + e^{\omega\left(\Phi+\theta^+-\frac{\pi}{2}\right)} + e^{\omega\left(-\Phi+\frac{\pi}{2}-\theta^-\right)} + e^{\omega\left(-\Phi+\theta^--\frac{\pi}{2}\right)}\right] L(\omega), \tag{2.131}$$

where

$$L(\omega) = \frac{TF\left(\frac{1}{\cos\alpha}\right)}{2sh(2\omega\Phi)}.$$

The inverse Fourier transform of $L(\omega)$ reads

$$\ell(\alpha) = TF^{-1}(L(\omega)) = \left[TF^{-1}\left(\frac{1}{2sh(2\omega\Phi)}\right)\right] * \frac{1}{\cos\alpha}, \tag{2.132}$$

where the sign $(*)$ is the convolution product. Noting that (see Gradshteyn and Ryshik [2.9])

$$TF^{-1}\left(\frac{1}{2sh(2\omega\Phi)}\right) = \frac{1}{2\pi}\int_{-\infty}^{+\infty} \frac{e^{i\omega x}}{2sh(2\omega\Phi)} d\omega = \frac{i}{2\pi}\int_0^{\infty} \frac{\sin\omega x}{sh(2\omega\Phi)} d\omega$$
$$= \frac{i}{8\Phi} th\left(\frac{\pi x}{4\Phi}\right) = \frac{1}{8\Phi}\tan\left(\frac{\pi i x}{4\Phi}\right), \tag{2.133}$$

we obtain

$$\ell(\alpha) = \frac{1}{8\Phi}\int_{-\infty}^{+\infty} \tan\left(\frac{\pi i x}{4\Phi}\right) \frac{dx}{\cos(\alpha - ix)} = -\frac{i}{8\Phi}\int_{-i\infty}^{+i\infty} \tan\left(\frac{\pi x}{4\Phi}\right)\frac{dx}{\cos(\alpha - x)}. \tag{2.134}$$

According to (2.119), we have

$$g(\alpha) = TF^{-1} H(\alpha),$$

where $H(\alpha)$ is given by (2.131). Applying the translation theorem to the inverse Fourier transform of $H(\alpha)$ and taking into account (2.132) to (2.134), we obtain the following expression for $g(\alpha)$:

$$g(\alpha)=\ell\left(\alpha+\Phi+\frac{\pi}{2}-\theta^{+}\right)+\ell\left(\alpha+\Phi-\frac{\pi}{2}+\theta^{+}\right)+\ell\left(\alpha-\Phi+\frac{\pi}{2}-\theta^{-}\right)+\ell\left(\alpha-\Phi-\frac{\pi}{2}+\theta^{-}\right). \tag{2.135}$$

According to (2.117), we have

$$\psi(\alpha)=\psi(0)\exp\left(\int_{0}^{\alpha}g(\mu)d\mu\right), \tag{2.136}$$

where $\psi(0)$ is a constant that we make equal to 1.

Inserting (2.135) into (2.136) and taking (2.134) into account, we finally obtain

$$\psi(\alpha)=\psi_{\Phi}\left(\alpha+\Phi+\frac{\pi}{2}-\theta^{+}\right)\psi_{\Phi}\left(\alpha+\Phi-\frac{\pi}{2}+\theta^{+}\right)\psi_{\Phi}\left(\alpha-\Phi-\frac{\pi}{2}+\theta^{-}\right)\psi_{\Phi}\left(\alpha-\Phi+\frac{\pi}{2}-\theta^{-}\right), \tag{2.137}$$

where

$$\psi_{\Phi}(\alpha)=\exp\left[\frac{i}{8\Phi}\int_{0}^{\alpha}d\mu\int_{-i\infty}^{+i\infty}\tan\left(\frac{\pi v}{4\Phi}\right)\frac{dx}{\cos(\mu-x)}\right]. \tag{2.138}$$

The function $\psi_{\Phi}(\alpha)$ is known as the Maliuzhinets function. By integrating (2.138) over μ, it is possible to get

$$\psi_{\Phi}(\alpha)=\sqrt{\cos\left(\frac{\pi\alpha}{4\Phi}\right)}\exp\left\{\frac{1}{2\pi}\int_{-\infty}^{+\infty}\mathrm{Log}\left[1-i\tan\left(\frac{\pi\alpha}{4\Phi}\right)th\left(\frac{\pi v}{4\Phi}\right)\right]\right\}\frac{dv}{chv},$$

and by performing the change of variable $w=th\left(\frac{\pi v}{4\Phi}\right)$, we finally obtain

$$\psi_{\Phi}(\alpha)=\sqrt{\cos\left(\frac{\pi\alpha}{4\Phi}\right)}\exp\left[\frac{2\Phi}{\pi^{2}}\int_{0}^{1}\frac{\mathrm{Log}\left(1+w^{2}\tan^{2}\frac{\pi\alpha}{4\Phi}\right)}{\left(1-w^{2}\right)ch\left(\frac{4\Phi}{\pi}\arg th\,w\right)}dw\right]. \tag{2.139}$$

The function $\psi_{\Phi}(\alpha)$ has no singularity in its domain of definition $-\Phi\le\alpha\le\Phi$ and takes the value 1 for $\alpha=0$. It is a meromorphic function taking values outside the domain $[-\Phi,\Phi]$ that can be calculated by transferring its argument in the fundamental domain with the help of one of the following relations established by Maliuzhinets [2.6]:

$$\frac{\psi_\Phi(\alpha+2\Phi)}{\psi_\Phi(\alpha-2\Phi)} = \cot\left[\frac{1}{2}\left(\alpha+\frac{\pi}{2}\right)\right],$$

$$\psi_\Phi\left(\alpha+\frac{\pi}{2}\right)\psi_\Phi\left(\alpha-\frac{\pi}{2}\right) = \psi_\Phi^2\left(\frac{\pi}{2}\right)\cos\left(\frac{\pi\alpha}{4\Phi}\right), \tag{2.140}$$

$$\psi_\Phi(\alpha+\Phi)\psi_\Phi(\alpha-\Phi) = \psi_\Phi^2(\Phi)\psi_{\frac{\Phi}{2}}(\alpha).$$

In general, the formula (2.139) cannot be directly applied to the calculation of $\psi(\alpha)$ via (2.137), since one or several arguments of the functions $\psi_\Phi(\alpha)$ may have values falling outside the fundamental domain $[-\Phi,\Phi]$. In this case, the formulas (2.140) need to be used in order to transfer these arguments into the fundamental domain.

We now complete the solution $S(\alpha)$, given by (2.112), by determining the function σ. A particular solution of (2.113) is given by

$$\sigma_0(\alpha) = \sin\left(\frac{\pi\alpha}{2\Phi}\right). \tag{2.141}$$

Any analytical function of $\sigma_0(\alpha)$ is also a solution of (2.113). We can therefore construct a general solution of (2.113) from a function $f(z)$ with $z = \sigma_0(\alpha)$, analytic in the strip $|R_e(\alpha)| \le \Phi$, outside a simple pole $z_0 = \sigma_0(\varphi_0)$ corresponding to $\alpha = \varphi_0$, where φ_0 is the angle defining the propagation of the incident field. If we expand $f(z)$ in a Laurent series in the vicinity of z_0, then by analytical continuation $f(z)$ can be extended to the entire strip $|R_e(\alpha)| \le \Phi$. Hence

$$f(z) = \frac{A_{-1}}{z-z_0} + \sum_{n=0}^{\infty} A_n(z-z_0)^n. \tag{2.142}$$

Coming back to (2.139), we see that when $|I_m\alpha| \to \infty$, the argument of the logarithmic function remains finite since $I_m\left(tg\frac{\pi\alpha}{4\Phi}\right) \to 1$. Therefore the function $\psi_\Phi(\alpha)$ behaves like $\sqrt{\cos\left(\frac{\pi\alpha}{4\Phi}\right)}$. Hence

$$\lim_{|I_m\alpha|\to\infty} \psi_\Phi(\alpha) = O\left[\exp\left(\frac{\pi|I_m\alpha|}{8\Phi}\right)\right], \tag{2.143}$$

and consequently

$$\lim_{|I_m\alpha|\to\infty} \psi(\alpha) = O\left[\exp\left(\frac{\pi I_m(\alpha)}{2\Phi}\right)\right], \tag{2.144}$$

where $O(\cdot)$ is the Landau symbol ("large Oh").

Moreover, we have shown after equation (2.106) that, owing to the Meixner condition, $S(\alpha)$ tends to a constant when $|I_m\alpha| \to \infty$ or, equivalently,

$$\lim_{|I_m\alpha|\to\infty} S(\alpha) = O(1). \tag{2.145}$$

Equations (2.145) and (2.144) lead to

$$\lim_{|I_m\alpha|\to\infty} f\left[\sigma_0(\alpha)\right] = O\left[\exp\left(\frac{-\pi|I_m\alpha|}{2\Phi}\right)\right], \tag{2.146}$$

which implies that the constant term ($n = 0$) and the terms $n > 0$ in (2.142) are zero. It remains

$$f(z) = \frac{A_{-1}}{z - z_0},$$

where A_{-1} is the residue at the pole z_0. Coming back to the variable α, we obtain

$$S(\alpha) = \frac{A_{-1}}{\sin\left(\frac{\pi\alpha}{2\Phi}\right) - \sin\left(\frac{\pi\varphi_0}{2\Phi}\right)}\psi(\alpha), \tag{2.147}$$

and since the residue at the pole $\alpha = \varphi_0$ must be equal to the incident field given by (2.114), it follows that

$$P_0^i = \frac{A_{-1}\psi(\varphi_0)}{\frac{\pi}{2\Phi}\cos\left(\frac{\pi\varphi_0}{2\Phi}\right)} \tag{2.148}$$

or

$$A_{-1} = \frac{\pi}{2\Phi}\,\frac{\cos\left(\frac{\pi\varphi_0}{2\Phi}\right)}{\psi(\varphi_0)}P_0^i \tag{2.149}$$

and

$$\sigma(\alpha) = \frac{\frac{\pi}{2\Phi}\cos\left(\frac{\pi\varphi_0}{2\Phi}\right)}{\sin\left(\frac{\pi\alpha}{2\Phi}\right) - \sin\left(\frac{\pi\varphi_0}{2\Phi}\right)}\,\frac{P_0^i}{\psi(\varphi_0)}. \tag{2.150}$$

This formula can be simplified by expressing the numerator and the denominator with respect to the half angle $\pi x / 4\,\Phi$, where x stands either for α or φ_0, and by noting that

$$\cos 2y = \cos^2 y - \sin^2 y = \cos(x-y)\cos(x+y) + \sin(x-y)\sin(x+y).$$

The final expression reads

$$\sigma(\alpha) = \frac{\pi}{4\Phi}\left[\cot\frac{\pi}{4\Phi}(\alpha-\varphi_0) + \tan\frac{\pi}{4\Phi}(\alpha+\varphi_0)\right]\frac{P_0^i}{\psi(\varphi_0)}. \tag{2.151}$$

Inserting (2.151) into (2.112), the integral form (2.102) of the solution reads

$$P(r,\varphi) = \frac{P_0^i}{8\Phi i}\int_\gamma e^{-ikr\cos\alpha}\frac{\psi(\alpha+\varphi)}{\psi(\varphi_0)}\left[\cot\frac{\pi}{4\Phi}(\alpha+\varphi+\varphi_0)\right.$$
$$\left. -\tan\frac{\pi}{4\Phi}(\alpha+\varphi+\varphi_0)\right]d\alpha. \tag{2.152}$$

This form of the solution has been first obtained by Maliuzhinets [2.7].

2.2.2.3. Asymptotic Evaluation of the Solution

When the observation point is far away from the edge of the wedge ($kr >> 1$), the integral (2.152) can be evaluated asymptotically. The contour γ in (2.152) is a double loop that can be closed by the steepest descent path (SDP) Γ and Γ' passing through the stationary points $\alpha = \pi$ and $\alpha = -\pi$, respectively (fig. 2.15).

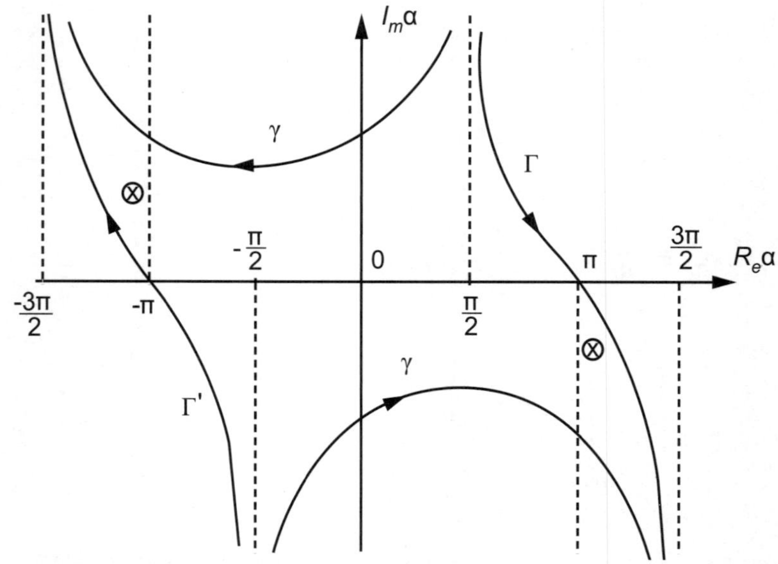

Figure 2.15. The double loop path γ and the SDP Γ and Γ'.

This contour is the same as the one introduced earlier in section 2.2.1, figure 2.4. The procedure that will be used for asymptotically evaluating the integral (2.152) is also similar to that employed in section 2.2.1. Recall that the phase of the integrand in (2.152) is stationary when sin $\alpha = 0$ or $\alpha = N\pi$ with $N = 0, \pm 1, \pm 2, \ldots$ and that the SDP are defined by $I_m(-i \cos \alpha)$ constant and $R_e(-i \cos \alpha) < 0$, or, equivalently, for $\alpha = \alpha' + i\alpha''$:

$$\cos \alpha' ch\alpha'' = \text{constant}, \quad \sin \alpha' sh\alpha'' > 0.$$

The SDP passing through $\alpha = \pm\pi$ are therefore defined by

$$\cos \alpha' ch\alpha'' = -1, \ \sin \alpha' sh\alpha'' > 0,$$

and the asymptotes are $\alpha' = -3\pi/2$ and $\alpha' = -\pi/2$ for the SDP passing through $\alpha = -\pi$ and $\alpha' = \pi/2$, $\alpha' = 3\pi/2$ for the SDP passing through $\alpha = \pi$. These asymptotes are the same as those of the contour γ. Owing to the orientation of the SDP Γ and Γ', it can be easily verified that Γ and Γ' make the angles $-\pi/4$ and $3\pi/4$, respectively, with the real axis at $\alpha = \pi$ and $\alpha = -\pi$.

Applying the residue theorem to the closed contour $\gamma - \Gamma - \Gamma'$, we obtain

$$\int_{\gamma} F(\alpha) d\alpha = \int_{\Gamma + \Gamma'} F(\alpha) d\alpha + 2\pi i \sum_{p=1}^{N} R(\alpha_p), \tag{2.153}$$

where $R(\alpha_p)$ are the residues at the poles α_p of $F(\alpha)$, located inside the closed contour. In order to obtain an asymptotic evaluation of the integral along the contour γ, we must localize the poles inside the closed contour $\gamma - \Gamma - \Gamma'$ and asymptotically evaluate the integrals along the SDP Γ and Γ'.

We will now apply this procedure to the integral (2.102), where $S(\alpha)$ stands for $\sigma(\alpha)\psi(\alpha)$, with $\psi(\alpha)$ and $\sigma(\alpha)$ given by (2.137) and (2.150), respectively.

The poles of $S(\alpha + \varphi)$ are those of $\sigma(\alpha + \varphi)$ and $\psi(\alpha + \varphi)$. According to (2.150), the poles of $\sigma(\alpha)$ are real and given by

$$\alpha = \varphi_0 + 4N\Phi,$$
$$\alpha = 2\Phi - \varphi_0 + 4N\Phi \qquad N = 0, \pm 1, \pm 2, \ \ldots .$$

Consequently, for a sharp wedge $\left(\Phi > \dfrac{\pi}{2}\right)$, the poles that may be located in the interval $(-\pi, \pi)$ are defined by

$$\begin{aligned} \alpha_1 &= \varphi_0 - \varphi, \\ \alpha_2 &= -(\varphi + \varphi_0) + 2\Phi, \\ \alpha_3 &= -(\varphi + \varphi_0) - 2\Phi. \end{aligned} \tag{2.154}$$

Substituting (2.150) into (2.102), we obtain

$$P\left(r,\varphi\right)=\frac{P_0^i}{4\Phi i}\frac{\cos\left(\frac{\pi\varphi_0}{2\Phi}\right)}{\psi\left(\varphi_0\right)}\int_\gamma e^{-ik\,r\cos\alpha}\frac{\psi\left(\alpha+\varphi\right)}{\sin\frac{\pi\left(\alpha+\varphi\right)}{2\Phi}-\sin\frac{\pi\varphi_0}{2\Phi}}\,d\alpha. \tag{2.155}$$

The contribution of the pole α_1 is obtained readily and is given by

$$P_1\left(r,\varphi\right)=p_0^i\exp\left[ikr\cos\left(\varphi-\varphi_0\right)\right], \tag{2.156}$$

which is equal to the pressure of the incident field, as expected.

The calculation of the residues at the poles α_2 and α_3 is slightly more involved. At the pole α_2, we find directly

$$P_2\left(r,\varphi\right)=P_0^i\cos\frac{\pi\varphi_0}{2\Phi}\frac{\psi\left(\alpha_2+\varphi\right)e^{-ik\,r\cos\alpha_2}}{\psi\left(\varphi_0\right)\cos\left[\frac{\pi}{2\Phi}\left(\alpha_2+\varphi\right)\right]}=-P_0^i\frac{\psi\left(2\Phi-\varphi_0\right)}{\psi\left(\varphi_0\right)}e^{-ik\,r\cos\left[2\Phi-\left(\varphi+\varphi_0\right)\right]}.$$

According to (2.115), we have

$$\psi\left(2\Phi-\varphi_0\right)=\psi\left(\Phi+\Phi-\varphi_0\right)=-R_+\left(\Phi-\varphi_0\right)\psi\left(\varphi_0\right).$$

Hence

$$P_2\left(r,\varphi\right)=P_0^i\,R_+\left(\Phi-\varphi_0\right)e^{-ik\,r\cos\left[2\Phi-\left(\varphi+\varphi_0\right)\right]}. \tag{2.157}$$

Applying the same procedure for the derivation of the contribution of the pole α_3, one finds readily

$$P_3\left(r,\varphi\right)=P_0^i\,R_-\left(\Phi+\varphi_0\right)e^{-ik\,r\cos\left[2\Phi+\left(\varphi+\varphi_0\right)\right]}. \tag{2.158}$$

It is easy to verify that $P_2(r,\varphi)$ and $P_3(r,\varphi)$ are the pressures at the observation point $M(r,\varphi)$ of the fields reflected by the faces $\varphi=+\Phi$ and $\varphi=-\Phi$, respectively. We observe also that when $\alpha_1=\pm\pi$, the direction of observation lies on the shadow boundary of the incident field, whereas when $\alpha_2=\pi$ and $\alpha_3=-\pi$, it lies, respectively, on the shadow boundaries of the field reflected by the faces $\varphi=+\Phi$ and $\varphi=-\Phi$.

We consider now the poles of $\psi(\alpha)$. According to (2.115), they are given by

$$\alpha = \pm\left(\Phi + \alpha_p^{\pm}\right), \tag{2.159}$$

where $\alpha_p^{\pm}$ are the poles of $R_{\pm}(\alpha)$. Consequently, the poles of $\psi(\alpha+\varphi)$ are

$$\begin{aligned} \varphi_p^{+} &= +\left(\Phi + \alpha_p^{+}\right) - \varphi, \\ \varphi_p^{-} &= -\left(\Phi + \alpha_p^{-}\right) - \varphi. \end{aligned} \tag{2.160}$$

The poles of the reflection coefficient are given by

$$\sin\alpha_p^{\pm} = -Z^{\pm} = -\sin\theta^{\pm}. \tag{2.161}$$

Therefore, $\alpha_p^{\pm} = \pi + \theta^{\pm} + 2N\pi$ and $\alpha_p^{\pm} = -\theta^{\pm} + 2N\pi$, where $N = 0, \pm 1, \pm 2, \ldots$.

Since (2.111) gives the reflection coefficient of an elementary plane wave of the spectrum (2.103), making an angle α with a planar impedance surface, we find that the poles of the reflection coefficient in the complex α-plane are located outside the strip (0, π) when $R_e\,\theta^{\pm} > 0$ (fig. 2.16). This condition is fulfilled when $R_e\,Z^{\pm} > 0$ (passive surface). If in addition $I_m\theta^{\pm} < 0$, then the corresponding waves are known to be surface waves.

The position of the poles of $R(\alpha)$ are symmetric with respect to $\alpha = \frac{\pi}{2}$. The one that has a positive real part corresponds to a wave with a positive phase velocity, whereas the one with a negative real part corresponds to a wave with a negative phase velocity. Since these waves emanate from the edge of the wedge, the trace of which is the origin of the coordinate system, only the waves with a positive phase velocity need to be retained. We have, therefore, $\alpha_p^{+} = \pi + \theta^{+}$ and $\alpha_p^{-} = \pi + \theta^{-}$.

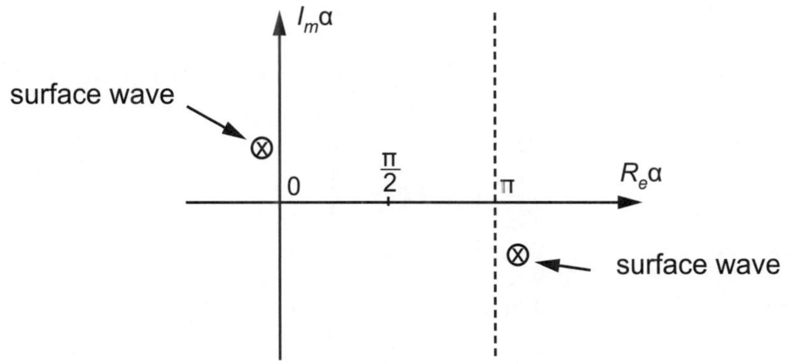

Figure 2.16. Location of the poles of $R(\alpha)$ in the complex α-plane for a passive surface.

Coming back to the poles of $\psi(\alpha + \varphi)$ given by (2.160), we find that for $\varphi = \pm\Phi$ these poles are given by $\varphi_p^{\pm} = \pm\alpha_p^{\pm}$. They are located outside the strip $(-\pi, \pi)$, but because we always have $\left|I_m\,\theta^{\mp}\right| > R_e\,\theta^{\mp}$ for a passive surface, these poles are located inside the closed contour $\gamma - \Gamma - \Gamma'$, as shown in figure 2.15.

When $0 < \varphi < \Phi$ and $\varphi \to \Phi$, the pole φ_p^+ given by (2.160) moves toward the SDP Γ passing through $\alpha = \pi$. On the other hand, when $-\Phi < \varphi < 0$ and $\varphi \to -\Phi$, the pole φ_p^- moves toward the SDP Γ' passing through $\alpha = -\pi$. When $|\varphi|$ decreases, each pole crosses the SDP for some value of φ beyond which the corresponding surface wave no longer exists.

The pole φ_p^+ is located on the contour Γ when $\cos\left(R_e\,\varphi_p^+ - \pi\right) = \left[ch\left(I_m\,\varphi_p^+\right)\right]^{-1}$. It lies inside the closed contour $\gamma - \Gamma - \Gamma'$ when

$$\cos\left(R_e\,\varphi_p^+ - \pi\right) < \left[ch\left(I_m\,\varphi_p^+\right)\right]^{-1}$$

or

$$R_e\,\varphi_p^+ < \pi + \operatorname{Arc\,cos}\left[\frac{1}{ch\left(I_m\,\varphi_p^+\right)}\right]. \tag{2.162}$$

Alternatively, the pole φ_p^- is located inside the closed contour $\gamma - \Gamma - \Gamma'$ when

$$R_e\,\varphi_p^- > -\pi - Arc\cos\left[\frac{1}{ch\left(I_m\,\phi_p^-\right)}\right], \tag{2.163}$$

and since $I_m\,\varphi_p^+ = I_m\,\alpha_p^+$ and $I_m\,\varphi_p^- = -I_m\,\alpha_p^-$, the poles are located inside the closed contour $\gamma - \Gamma - \Gamma'$ when

$$\left|R_e\,\varphi_p^{\pm}\right| < \pi + \operatorname{Arc\,cos}\left[\frac{1}{ch\left(I_m\,\alpha_p^{\pm}\right)}\right]. \tag{2.164}$$

The condition (2.162) can also be written

$$R_e\,\varphi_p^+ + GD\left(I_m\,\varphi_p^+\right) < \pi, \tag{2.165}$$

where

$$GD(x) = \sin(x)\,\mathrm{Arc}\cos\left[\frac{1}{ch\,x}\right] \tag{2.166}$$

is the Gudermann function.

According to (2.160), the condition (2.165) reads

$$\Phi_p^+ - \varphi < \pi,$$

where

$$\Phi_p^+ = \Phi + R_e\,\alpha_p^+ + GD\left(I_m\,\alpha_p^+\right). \tag{2.167}$$

Alternatively, the condition (2.163) can be written as

$$\Phi_p^- - \varphi > -\pi,$$

where, by noting that $GD\left(-I_m\,\alpha_p^-\right) = -GD\left(I_m\,\alpha_p^+\right)$,

$$\Phi_p^- = -\Phi - R_e\,\varphi_p^- - GD\left(I_m\,\alpha_p^+\right). \tag{2.168}$$

With these notations, the condition for the existence of the surface waves is

$$\left|\Phi_p^\pm - \varphi\right| < \pi, \tag{2.169}$$

where

$$\Phi_p^\pm = \pm\left[\Phi + R\alpha_p^\pm + GD\left(I_m\,\alpha_p^+\right)\right]. \tag{2.170}$$

In order to calculate the residues at the poles of $\psi(\alpha + \varphi)$, we observe that the residue at the pole does not change by a translation of the coordinates. In other words, if A is the residue of $\psi(\alpha + \varphi)$ at the pole φ_p^+, which by definition is given by

$$A = \lim_{\alpha \to \varphi_p^+} \left(\alpha - \varphi_p^+\right)\psi\left(\alpha + \varphi\right), \tag{2.171}$$

and if we make the change of variable $\alpha = \alpha' + \Phi - \varphi$, then we have also

$$A = \lim_{\alpha' \to \alpha_p^+} \left(\alpha' - \alpha_p^+\right) \psi\left(\alpha' + \Phi\right). \tag{2.172}$$

According to (2.172), the residue of $\psi(\alpha + \varphi)$ at the pole φ_p^+ is the same as the residue of $\psi(\alpha + \Phi)$ at the pole α_p^+. Coming back to (2.115), we obtain for the residue of $\psi(\alpha + \Phi)$ at the pole α_p^+, the expression

$$\text{Residue } \psi\left(\alpha + \Phi\right) = -\frac{2\sin\alpha_p^+}{\cos\alpha_p^+} \psi\left(\Phi - \alpha_p^+\right). \tag{2.173}$$

Similarly, the change of variable $\alpha = -\left(\alpha' + \Phi\right) - \varphi$ in (2.171), with φ_p^+ replaced by the pole φ_p^- given by (2.160), yields

$$A = \lim_{\alpha' \to \alpha_p^-} \left(\alpha' - \alpha_p^-\right) \psi\left[-\left(\Phi + \alpha'\right)\right] = \text{ Residue } \psi\left[-\left(\Phi + \alpha\right)\right] \text{ at } \alpha_p^-. \tag{2.174}$$

Hence according to (2.174) the residue of $\psi(\alpha + \varphi)$ at the pole φ_p^- is equal to the residue of $\psi\left[-\left(\Phi + \alpha\right)\right]$ at the pole α_p^-. According to (2.115), the residue of $\psi\left[-\left(\Phi + \alpha\right)\right]$ at the pole α_p^- is given by

$$\text{Residue } \psi\left[-\left(\Phi + \alpha\right)\right] = -\frac{2\sin\alpha_p^-}{\cos\alpha_p^-} \psi\left[-\left(\Phi - \alpha_p^-\right)\right].$$

Finally the evaluation of the integral (2.155) at the poles of $\psi(\alpha + \varphi)$, which gives the contribution of the surface waves to the total field, yields

$$P^s\left(r,\varphi\right) = +P_0^i \, U\left(\pi - \left|\Phi_p^{\pm} - \varphi\right|\right) F_1\left(\alpha_p^{\pm}, \varphi_0\right) \psi\left[\pm\left(\Phi - \alpha_p^{\pm}\right)\right] e^{-ikr\cos\left[\pm\left(\Phi + \alpha_p^{\pm}\right) - \varphi\right]}, \tag{2.175}$$

with

$$F_1\left(\alpha_p^{\pm}, \varphi_0\right) = \frac{\pi}{\Phi} \frac{\cos\left(\dfrac{\pi\varphi_0}{2\Phi}\right)}{\psi\left(\varphi_0\right)} \frac{\tan\alpha_p^{\pm}}{\pm\cos\dfrac{\pi}{2\Phi}\alpha_p^{\pm} - \sin\dfrac{\pi}{2\Phi}\varphi_0}. \tag{2.176}$$

When $\varphi = \pm\Phi$, the surface waves propagate, respectively, along the faces (+) and (–) of the wedge with a phase given by

$$e^{ikr\cos\theta^{\pm}} = e^{ikr\cos\left(\theta' - i\theta''\right)} = e^{ikr\left(\cos\theta' ch\theta'' + i\sin\theta' sh\theta''\right)}, \tag{2.177}$$

where $\theta^{\pm} = \pm(\theta' - i\theta'')$, with $\theta' \geq 0, \theta'' > 0$.

The surface wave along each face of the wedge is therefore an inhomogeneous plane wave. For a purely imaginary impedance ($\theta' = 0$), it propagates along the faces $\varphi = \pm\Phi$ with the wave number $K' = kch\theta''$, which is real. In this case, the surface wave is not attenuated in the direction $\varphi = \pm\Phi$. However, in the direction orthogonal to each face, it decreases exponentially. Such a wave is called a true surface wave. Using the bivector representation of the inhomogeneous plane wave introduced in chapter 1, section 1.3 and in appendix C, we write for $\theta' = 0$ and $|\varphi| \leq \Phi$,

$$e^{ikr\cos\left[-\varphi\pm(\Phi-i\theta'')\right]} = e^{ikr\left[ch\theta''\cos(\Phi\pm\varphi)+ish\theta''\sin(\Phi\pm\varphi)\right]} = e^{i\mathbf{K}\cdot\mathbf{r}}, \tag{2.178}$$

where

$$\mathbf{K} = \mathbf{K}' + i\mathbf{K}'', \quad \mathbf{K}'\cdot\mathbf{K}'' = 0, \tag{2.179}$$

and

$$\mathbf{K}' = k\,ch\theta'', \quad \mathbf{K}'' = k\,sh\theta''. \tag{2.180}$$

Since the wave number $\mathbf{K}$ is in the direction $\varphi = \pm\Phi$, the wave number $\mathbf{K}''$ is orthogonal to the face $\varphi = +\Phi$ or $\varphi = -\Phi$. Hence the lines of constant amplitude are parallel to each face, whereas the lines of constant phase are orthogonal to each face. These lines are limited by the angles $\pm\left[\Phi - GD(\theta'')\right]$, as shown in figure 2.17.

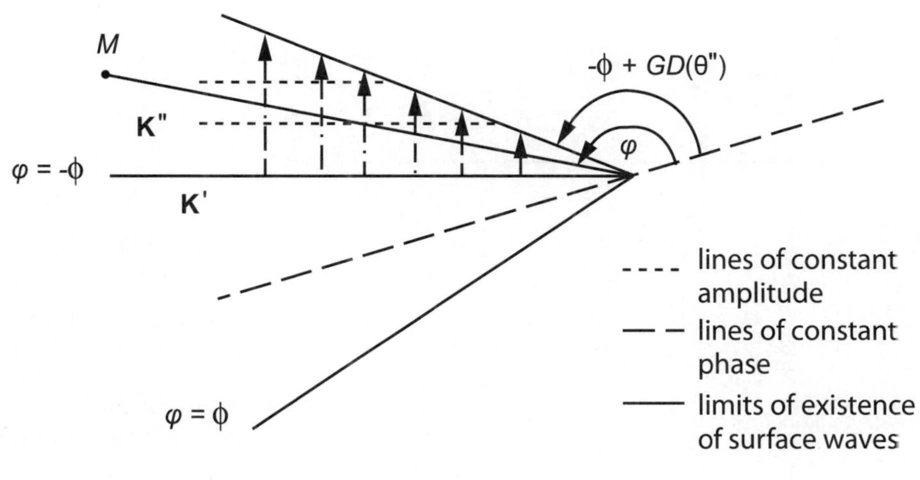

Figure 2.17. Surface waves for a pure imaginary impedance.

When $\theta' > 0$, (2.177) is an inhomogeneous plane wave propagating in the direction $\mathbf{K}'$, making an angle $-\theta'$ with the face, and decaying exponentially in the direction $\mathbf{K}''$ orthogonal to $\mathbf{K}'$. Hence the lines of constant amplitude are parallel to $\mathbf{K}'$, and the lines of constant phase are parallel to $\mathbf{K}''$. These lines are limited by the angles $\varphi_\ell^{\pm} = \pm\left[\Phi + \theta' - GD\left(\theta''\right)\right]$, as shown on the side $\varphi < 0$ in figure 2.18. When $|\varphi| < \Phi + \theta' - GD\left(\theta''\right)$, the surface wave does not exist. In this case, the poles $\varphi_p^{\pm}$ are located outside the closed contour $\gamma - L - L'$.

In the direction normal to the surface, the amplitude of the wave decreases when the observation point moves apart from the surface. However, as the observation point leaves the surface, the wave's amplitude falls suddenly to zero when $\varphi = \varphi_\ell^{\pm}$. This limit constitutes a shadow boundary for the surface wave. It will be shown in chapter 3 that the diffracted field interacts strongly with the surface wave field in a region close to this shadow boundary, known as the transition region, and that the total field is continuous there.

So far, we have calculated the contributions of the poles of the integrand in (2.155). According to (2.153), we now need to evaluate the integrals along the SDP Γ and Γ' in order to obtain the asymptotic evaluation of the integral (2.155).

We suppose that all poles are far from these paths. Especially, the poles corresponding to the incident and reflected fields are supposed as sufficiently away from the saddle points $\alpha = \pm\pi$ such that they have a negligibly small influence on the evaluation, by the method of steepest descent, of the integrals along the contours Γ and Γ'. The case where these poles are close to the saddle points $\pm\pi$ needs a special treatment. It corresponds to the transition regions at the shadow boundaries of the direct and reflected fields. The derivation of a uniform solution valid everywhere is much more involved. Such a solution will be presented in chapter 3.

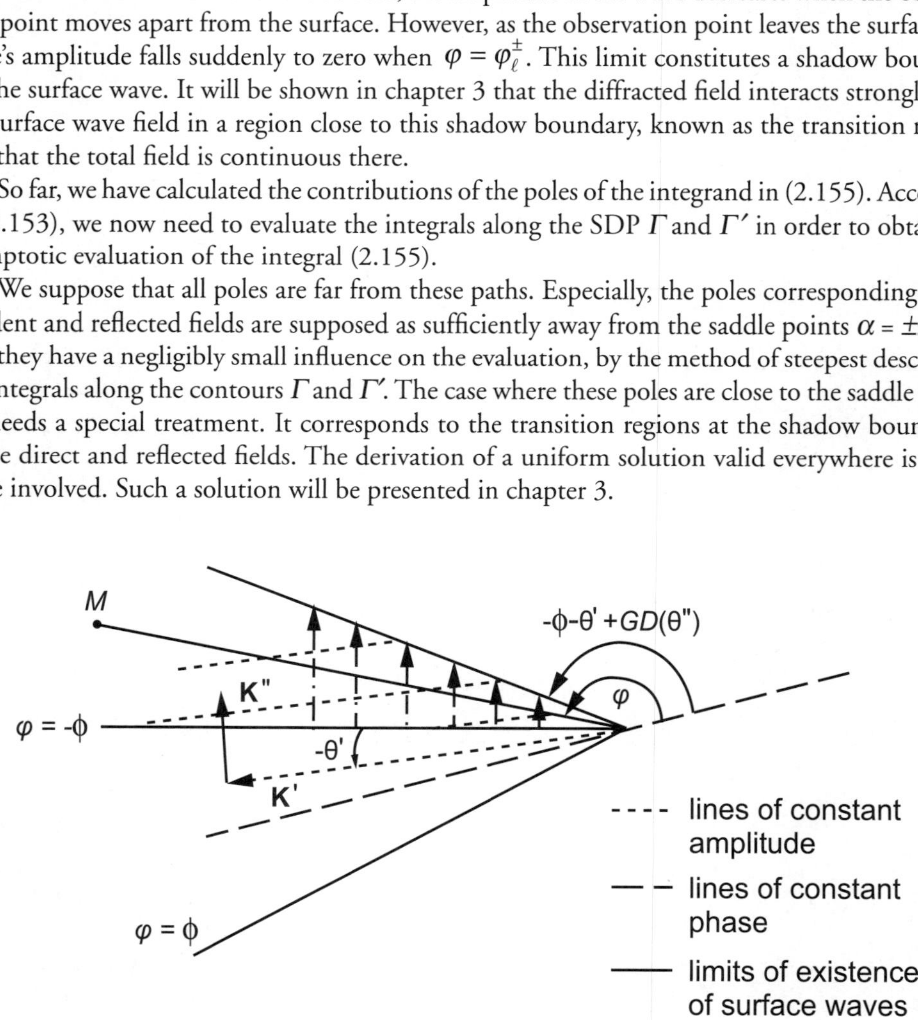

Figure 2.18. Surface waves for $R_e Z^{\pm} > 0$.

A general technique for obtaining the asymptotic evaluation of an integral by the steepest descent method has been presented in section 2.2, equations (2.53) to (2.59). We present here a slightly different method that has the advantage of being less abstract.

Consider the integral

$$I = \int_{\Gamma} e^{-ikr\cos\alpha} F(\alpha)\, d\alpha, \tag{2.181}$$

where, according to (2.155),

$$F(\alpha) = \frac{P_0^i}{4\Phi i} \frac{\cos\frac{\pi\varphi_0}{2\Phi}}{\psi(\varphi_0)} \frac{\psi(\alpha+\varphi)}{\sin\frac{\pi}{2\Phi}(\alpha+\varphi) - \sin\frac{\pi\varphi_0}{2\Phi}}. \tag{2.182}$$

Since the pole of $F(\alpha)$ is far from the saddle point $\alpha = \pi$, $F(\alpha)$ is regular near π and can therefore be expended in a Taylor series that converges uniformly inside a circle with finite radius ρ centered at $\alpha = \pi$ (ρ being the distance to the pole):

$$F(\alpha) = F(\pi) + (\alpha-\pi)F'(\pi) + \frac{(\alpha-\pi)^2}{2} F''(\pi) + \cdots. \tag{2.183}$$

Similarly, close to the saddle point, cos α can also be expended in a Taylor series:

$$\cos\alpha = -1 + \frac{(\alpha-\pi)^2}{2} \cdots. \tag{2.184}$$

Since the tangent to Γ at the saddle point makes the angle $-\pi/4$ with the real α-axis and since the integrand (2.181) decreases very rapidly away from the saddle point when kr is large, we can replace the integration path Γ by the straight line $\left(\infty e^{\frac{3i\pi}{4}}, \infty e^{-\frac{i\pi}{4}}\right)$ passing through $\alpha = \pi$. In order to obtain the first-order term, we associate, as usual, the first-order term of the amplitude with the first- and second-order terms of the phase. Then according to (2.183) and (2.184), we obtain the following for the first-order term:

$$I = F(\pi) e^{ikr} \int_{\infty e^{\frac{3i\pi}{4}}}^{\infty e^{-\frac{i\pi}{4}}} e^{-\frac{ikr\alpha'^2}{2}} d\alpha', \tag{2.185}$$

where $\alpha' = \pi - \alpha$. Performing the change of variable $\alpha' = e^{-\frac{i\pi}{4}} u$, (2.185) yields

$$I = F(\pi)e^{ikr}e^{-\frac{i\pi}{4}}\int_{-\infty}^{+\infty} e^{-\frac{kr u^2}{2}}du = e^{-\frac{i\pi}{4}}\sqrt{\frac{2\pi}{kr}}F(\pi)e^{ikr}. \tag{2.186}$$

Similarly, if we apply the same transformations to the integral

$$I' = \int_{\Gamma'} e^{-ikr\cos\alpha}F(\alpha)d\alpha,$$

then we find

$$I' = e^{\frac{3i\pi}{4}}\sqrt{\frac{2\pi}{kr}}F(-\pi)e^{ikr}, \tag{2.187}$$

and finally, the diffracted field $P^d(r,\varphi) = I + I'$ is given by

$$P^d(r,\varphi) = \frac{\pi P_0^i}{2\Phi}\frac{\cos\frac{\pi\varphi_0}{2\Phi}}{\psi(\varphi_0)}\left[\frac{\psi(-\pi+\varphi)}{\sin\frac{\pi}{2\Phi}(-\pi+\varphi) - \sin\frac{\pi\varphi_0}{2\Phi}} - \frac{\psi(\pi+\varphi)}{\sin\frac{\pi}{2\Phi}(\pi+\varphi) - \sin\frac{\pi\varphi_0}{2\Phi}}\right]\frac{e^{i\left(kr+\frac{\pi}{4}\right)}}{\sqrt{2\pi kr}}. \tag{2.188}$$

Formula (2.188) gives the first or dominant term of the asymptotic expansion of the diffracted field with respect to the small parameter $(kr)^{-1/2}$.

Higher-order terms cannot be directly calculated by keeping higher-order terms in (2.183) and (2.184), since the error introduced by the deformation of the SDP Γ and Γ' in straight lines that are tangent to Γ and Γ', respectively, has not been evaluated. This error can be corrected by keeping only the two first terms in (2.184) and writing

$$\cos\alpha = -1 + \frac{1}{2}(\beta - \pi)^2, \tag{2.189}$$

where β is a new variable. Now by inserting (2.189) into (2.181), we see that the SDP Γ is transformed in another SDP, which coincides with the tangent to Γ, passing through $\alpha = \pi$. Hence by replacing $F(\alpha)$ with $G(\beta) = F[\alpha(\beta)]$ in (2.183) and by using the same technique as in (2.185) through (2.187), we obtain the same result as before for the first-order term of the asymptotic evaluation of the integral (2.181), and higher-order terms can be calculated by keeping higher-order terms in (2.183). It can be easily shown that the next nonzero term is of order $(kr)^{-3/2}$ and, by continuing the same procedure, that the asymptotic sequence (for its definition, see chapter 1) for the diffracted field is given by $(kr)^{-n-1/2}$.

We observe also that formula (2.188) is singular at the poles corresponding to the direct and reflected fields. For example, when the pole $\alpha_1 = \varphi_0 - \varphi$ corresponding to the incident field is equal to π, we have $\varphi = -\pi + \varphi_0$. Consequently, the denominator of the second term in the brackets of (2.188) is equal to zero. Hence when the pole α_1 passes through the SDP, the diffracted field has an infinite jump discontinuity. A similar result holds for the pole corresponding to the reflected field on each face of the wedge. When the poles $\varphi_p^{\pm}$ corresponding to the surface waves cross the SDP Γ or Γ', the term in the brackets (2.188) passes through the value zero and has a finite jump discontinuity. This nonphysical behavior of the diffracted field can be removed by taking into account the proximity of a pole near the SDP Γ and Γ'. Such a procedure, which leads to a uniform asymptotic expansion of the total field, is presented in chapter 3.

By writing (2.188) in the form

$$P^d(M) = P^i(Q)D\frac{e^{ikQM}}{\sqrt{QM}}, \tag{2.190}$$

where $M(r, \varphi)$ is the observation point, Q is the diffraction point on the edge, and D is the diffraction coefficient and by comparing (2.190) to (2.188), we obtain the explicit expression of D given by

$$D(\varphi,\varphi_0;k) = \frac{e^{i\frac{\pi}{4}}}{\sqrt{2\pi k}}\frac{\pi}{2\Phi}\frac{\cos\frac{\pi\varphi_0}{2\Phi}}{\psi(\varphi_0)}\left[\frac{\psi(-\pi+\varphi)}{\sin\frac{\pi}{2\Phi}(-\pi+\varphi)-\sin\frac{\pi\varphi_0}{2\Phi}} - \frac{\psi(\pi+\varphi)}{\sin\frac{\pi}{2\Phi}(\pi+\varphi)-\sin\frac{\pi\varphi_0}{2\Phi}}\right]. \tag{2.191}$$

Equation (2.190) with D given by (2.191) has the general form (1.199) since $QM = \sigma$ and $\rho_1 = \rho_i = \infty$ for an incident plane wave striking a straight wedge. Since the diffraction coefficient in (2.190) does not depend on the radii of curvature of the incident and diffracted wave fronts, it remains the same for a general incident space wave having different principal radii of curvature at the diffraction point. In practical applications of the GTD, it is often suited to refer the incidence and reflection directions to one face of the wedge, as it has been done in section 2.2 (see fig. 2.2). By denoting the exterior wedge angle by $n\pi$, $0 < n \leq 2$ and the incidence and reflection angles referring to the face $n = 0$ by ϕ' and ϕ, respectively (the angular position of the other face being $n\pi$, where all angles are oriented in the trigonometric direction), we have the following relations between the angles:

$$\Phi = n\frac{\pi}{2},$$
$$\varphi = \Phi - \phi = n\frac{\pi}{2} - \phi, \qquad (2.192)$$
$$\varphi_0 = \Phi - \phi' = n\frac{\pi}{2} - \phi'.$$

With these notations, the diffraction coefficient (2.191) reads

$$D(\phi,\phi';k) = \frac{e^{i\frac{\pi}{4}}}{\sqrt{2\pi k}}\frac{1}{n}\frac{\sin\frac{\phi'}{n}}{\psi\left(\frac{n\pi}{2}-\phi'\right)}\left[\frac{\psi\left(-\pi+\frac{n\pi}{2}-\phi\right)}{\cos\left(\frac{\pi+\phi}{n}\right)-\cos\frac{\varphi'}{n}} - \frac{\psi\left(\pi+\frac{n\pi}{2}-\phi\right)}{\cos\left(\frac{\pi-\phi}{n}\right)-\cos\frac{\phi'}{n}}\right]. \qquad (2.193)$$

2.2.2.4. Generalization to Oblique Incidence

The translation invariance of the solution in a direction parallel to the edge allows one to state

$$p\left(\rho,\varphi,z\right) = {}^{ik\,z\cos\beta_0}P\left(\rho,\varphi\right) \qquad (2.194)$$

with

$$P\left(\rho,\varphi\right) = \frac{1}{2\pi i}\int_{\gamma} e^{-ik_{\perp}\rho\cos\alpha}s\left(\alpha+\varphi\right)d\alpha, \qquad (2.195)$$

where β_0 is the angle between the incident direction and the edge.

Also, $k_{\perp} = k\sin\beta_0$ and φ is the angle between the OY axis and the projection of the observation direction on the OXY plane (see fig. 2.19).

By derivation under the integral sign, we see that $P(\rho,\varphi)$ satisfies the Helmholtz equation:

$$\Delta P + k_{\perp}^2 P = 0, \quad k_{\perp} = k\sin\beta_0.$$

On the faces, we impose

$$-\frac{1}{\rho}\frac{\partial\Pi}{\partial\varphi} \pm ik\,Z^{\pm}\Pi = 0, \quad \varphi = \pm\Phi, \qquad (2.196)$$

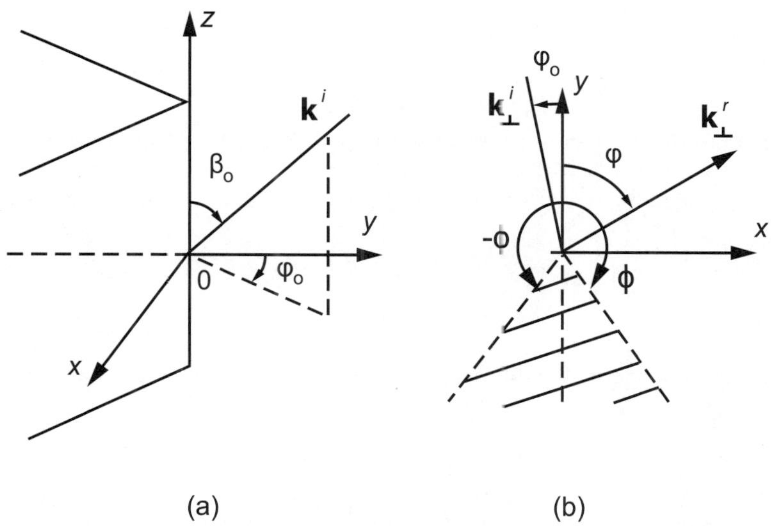

Figure 2.19. Oblique incidence on a straight wedge: (a) coordinate system and angle β_0 and (b) projections of k^i and k^r and angles φ_0 and φ.

where $\Pi = \Pi(\alpha, \varphi; \beta_0)$ is the integrand of (2.195), which can be considered as a plane wave spectrum of components:

$$\Pi(\alpha,\varphi;\beta_0) = e^{-ik\rho \sin\beta_0 \cos\alpha}\, S(\alpha,\varphi)\, d\alpha. \tag{2.197}$$

Proceeding as in the case of normal incidence, we obtain

$$\int_\gamma S(\alpha \pm \Phi)\left(\sin\beta_0 \sin\alpha \pm Z^{\pm}\right) e^{-ik\rho \sin\beta_0 \cos\alpha}\, d\alpha = 0, \tag{2.198}$$

and since γ is symmetric with respect to the origin, (2.198) implies that its integrand be an even function of α. Consequently, instead of (2.108) and (2.109), we now have

$$\left(\sin\beta_0 \sin\alpha + Z^{+}\right) S(\alpha + \Phi) = \left(Z^{+} - \sin\beta_0 \sin\alpha\right) S(\Phi - \alpha), \tag{2.199}$$

$$\left(\sin\beta_0 \sin\alpha - Z^{-}\right) S(\alpha - \Phi) = -\left(Z^{-} + \sin\beta_0 \sin\alpha\right) S(-\alpha - \Phi). \tag{2.200}$$

By dividing both terms of (2.199) and (2.200) by $\sin\beta_0$, we obtain a system of functional equations that is identical to the one obtained for normal incidence, but with modified impedances:

$$Z_{\pm}^{(m)} = \frac{Z^{\pm}}{\sin\beta_0}. \tag{2.201}$$

The solution to the problem is therefore given by the same formulas that are used for normal incidence, provided that $Z^{\pm}$ in (2.111) is replaced by $Z_{\pm}^{(m)}$ and that $\theta^{\pm}$ is now defined by

$$\sin\theta^{\pm} = Z_{\pm}^{(m)}. \tag{2.202}$$

With these modifications, the diffracted field for oblique incidence is given by (2.190) with D given by

$$D(\varphi,\varphi_0;k) = \frac{e^{i\frac{\pi}{4}}}{\sqrt{2\pi k}\sin\beta_0}\frac{\pi}{2\Phi}\frac{\cos\frac{\pi\varphi_0}{2\Phi}}{\psi(\varphi_0)}\left[\frac{\psi(-\pi+\varphi)}{\sin\frac{\pi}{2\Phi}(-\pi+\varphi)-\sin\frac{\pi\varphi_0}{2\Phi}} - \frac{\psi(\pi+\varphi)}{\sin\frac{\pi}{2\Phi}(\pi+\varphi)-\sin\frac{\pi\varphi_0}{2\Phi}}\right], \tag{2.203}$$

where φ_0 and φ are now the projected angles defined in figure 2.19(b) and where the functions $\psi(\alpha)$ are given by (2.137), in which the angles $\theta^{\pm}$ are defined by (2.202).

Again, for practical purposes, the diffraction coefficient for oblique incidence is often written in the form

$$D(\phi,\phi';k) = \frac{e^{i\frac{\pi}{4}}}{\sqrt{2\pi k}\sin\beta_0}\frac{1}{n}\frac{\sin\frac{\phi'}{n}}{\psi\left(\frac{n\pi}{2}-\phi'\right)}\left[\frac{\psi\left(-\pi+n\frac{\pi}{2}-\phi\right)}{\cos\left(\frac{\pi+\phi}{n}\right)-\cos\frac{\phi'}{n}} - \frac{\psi\left(\pi+n\frac{\pi}{2}-\phi\right)}{\cos\left(\frac{\pi-\phi}{n}\right)-\cos\frac{\phi'}{n}}\right], \tag{2.204}$$

where ϕ' and ϕ are the projected angles on a plane perpendicular to the edge, related to the projected angles φ_0 and φ on the same plane by (2.192).

The diffraction coefficients (2.203) or (2.204) verify the reciprocity theorem, which imposes that $D(\varphi, \varphi_0; k) = D(\varphi_0, \varphi; k)$ or, equivalently, $D(\phi, \phi'; k) = D(\phi', \phi; k)$. This property is not evident in the formulas (2.203) or (2.204). Its demonstration is given in appendix E. As shown in chapter 1, section 1.8, the diffraction coefficient (2.204) also applies to a curved wedge, outside the transition regions. These are located at the shadow boundaries of the incident and reflected fields and at the limits of existence of the surface waves. According to equation (1.199), the general formula for the diffracted field by a curved impedance wedge at an observation point M is given by

$$P^d(M) = P^i(Q)D(\varphi,\varphi_0;k\sin\beta)\sqrt{\frac{\rho_1}{s(\rho_1+s)}}e^{iks}, \qquad (2.205)$$

where D is the diffraction coefficient given by (2.203), Q is the diffraction point, $s = QM$, and ρ_1 is the principal radii of curvature of the diffracted wave front, in the plane formed by the tangent to the edge at the diffraction point Q and the direction of observation. A general formula for ρ_1 is given by equation (1.212).

2.2.3. *ELASTIC WEDGE-SHAPED SHELL*

In section 2.2.2, we presented the solution, first established by Maliuzhinets [2.6], for an impenetrable straight wedge with constant impedances on both faces, independent of the direction of propagation of the incident plane wave. No exact solution exists for a coated impenetrable wedge or an elastic hollow wedge separating a fluid from vacuum. However, these problems can be treated with some limitations by a generalization of the Maliuzhinets procedure.

Two types of generalizations have been proposed in scientific literature. The first one, founded on the concept of generalized impedance boundary conditions [2.10], has been applied to a coated wedge by Bernard [2.11]. The second one, founded on the concept of spectral impedance, has also been applied to a coated wedge by Michaeli [2.12]. We have transposed this last procedure to the acoustic diffraction by an elastic wedge-shaped shell.

2.2.3.1. Derivation of the Solution

Consider a hollow elastic wedge embedded in a fluid. The wedge has elastic walls of constant thickness d, and the medium occupying the interior domain is vacuum. The system of coordinates is identical to the one chosen in section 2.2.2. It is defined in figure 2.20, with the z-axis along the edge and the y-axis along the bissectrix of the wedge angle.

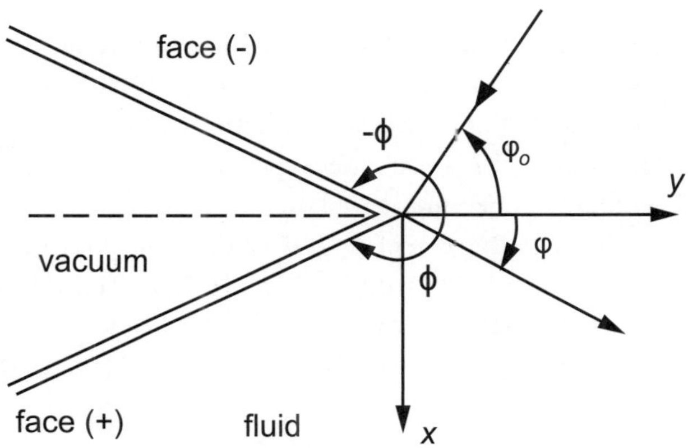

Figure 2.20. Wedge-shaped shell and coordinate system.

A plane wave is incident on the wedge in a direction normal to the edge. Let φ_0 and φ be the angles of incidence and reflection, respectively. They are negative if the rotation from the *y*-axis is in the trigonometric direction.

We suppose that each face can be characterized by an impedance that is identical to the impedance of an infinite elastic plate—composed of the same material and having the same thickness *d* as the faces of the wedge, separating a fluid from vacuum. This assumption is obviously not verified close to the edge where the field on both faces is coupled through the elastic material. This coupling depends on the local geometry of the wedge close to the edge (especially on the wedge angle), the thickness of the walls, and the distance to the edge. However, according to the Meixner condition, the acoustic impedance remains finite near the edge of a penetrable wedge. Hence if the domain close to the edge on which its impedance deviates strongly from that of an infinite elastic plate is small compared to the wavelength λ in the fluid, then the influence of that deviation on the scattered field is also small and can be neglected. Now this domain depends on the thickness of the walls, and therefore its length diminishes with *d*. The physical parameter characterizing the domain of validity of the solution is therefore d / λ. We can also choose the parameter *kd*, where *k* is the wave number in the fluid.

Numerical results on penetrable objects having sharp edges have shown that a good convergence is obtained when the sampling of the surface near the edge is of the order $\lambda / 10$. We can therefore assume that our solution will be accurate for $kd \leq 1$. However, we will see later on, by comparing the results obtained with our high-frequency solution to experimental results, that this limit is rather pessimistic since a good agreement between both results is obtained for $kd \cong 10$.

The impedance of an infinite elastic plate separating a fluid from vacuum has been derived in appendix B (see equation (B.95)). It is a quotient of two determinants 4×4 depending on the incidence angle. The reflection coefficient therefore has the same expression as in (2.111). However, now $Z^{\pm}$ is not a constant but is a function of α:

$$R_{\pm}(\alpha) = \frac{\sin\alpha - Z^{\pm}(\alpha)}{\sin\alpha + Z^{\pm}(\alpha)}. \tag{2.206}$$

Moreover, it is shown in appendix B that $Z(\alpha)$ is an even function of α; hence $Z^{\pm}(\alpha) = Z^{\pm}(-\alpha)$.

As in the case of constant impedance, the solution to the diffraction problem illustrated in figure 2.20 is written in the form (2.80). Its justification is identical to what is given in section 2.2.2.1 for a constant impedance. It is a plane wave spectrum verifying the Helmholtz equation. The derivation of the unknown function $S(\alpha)$ follows the same steps as in section 2.2.2.2. The boundary conditions on the faces $\varphi = \Phi$ and $\varphi = -\Phi$ of the wedge, verified by each elementary plane wave, remain the same as in (2.104) with $Z^{\pm} = Z^{\pm}(\alpha)$, and all equations from (2.105) to (2.111) are unchanged provided that $Z^{\pm}$ be considered as an even function of α. The technique for solving the problem also remains the same. The unknown function is written in the form of a product of two functions

$$S(\alpha) = \sigma(\alpha)\,\psi(\alpha), \tag{2.207}$$

where $\sigma(\alpha)$ verifies (2.113) and where the poles of $S(\alpha)$ are distributed over the functions $\sigma(\alpha)$ and $\psi(\alpha)$, such that

1. $\psi(\alpha)$ has no zeros or poles in the strip $\left|R_e(\alpha)\right| \leq \Phi$.
2. $\sigma(\alpha)$ contains all the poles of $S(\alpha)$ other than those of $R^{\pm}(\alpha)$.

As in the case of a constant impedance, the function $\psi(\alpha)$ verifies the equation (2.115), which can be solved separately by Fourier transforming it in the strip $\left|R_e(\alpha)\right| \leq \Phi$, where, according to the condition 1, it has no singularity.

Proceeding as in the case of a constant impedance, we see that all equations from (2.116) to (2.124) remain unchanged. From that point, however, the equations change since (2.125) is no longer valid and has to be replaced by

$$r_{\pm}(\alpha) = \frac{\cos\alpha - \dfrac{\partial Z^{\pm}(\alpha)}{\partial\alpha}}{\sin\alpha - Z^{\pm}(\alpha)} - \frac{\cos\alpha + \dfrac{\partial Z^{\pm}(\alpha)}{\partial\alpha}}{\sin\alpha + Z^{\pm}(\alpha)} = \frac{2\left(Z^{\pm}\cos\alpha - \sin\alpha\dfrac{\partial Z^{\pm}}{\partial\alpha}\right)}{\sin^2\alpha - \left(Z^{\pm}(\alpha)\right)^2}. \tag{2.208}$$

If we set

$$L_{\pm}(\omega) = \frac{S_{\pm}(\omega)}{2sh(2\omega\Phi)}, \tag{2.209}$$

then (2.124) reads

$$H(\omega) = L_{+}(\omega)e^{\omega\Phi} + L_{-}(\omega)e^{-\omega\Phi}, \tag{2.210}$$

and by denoting $\ell_{\pm}(\alpha)$, the inverse Fourier transform of $L_{\pm}(\omega)$, we have

$$\ell_{\pm}(\alpha) = TF^{-1}\left[L_{\pm}(\omega)\right] = TF^{-1}\left[S_{\pm}(\omega)\right] * TF^{-1}\left[\frac{1}{2sh(2\omega\Phi)}\right],$$

where the sign $(*)$ is the convolution product. Hence

$$\ell_{\pm}(\alpha) = \int_{-\infty}^{+\infty} r_{\pm}(\alpha - ix)\left(\frac{1}{2\pi}\int_{-\infty}^{+\infty}\frac{e^{i\omega x}}{2sh(2\omega\Phi)}d\omega\right)dx. \tag{2.211}$$

Now by using the symmetry properties of the integrand, we have

$$TF^{-1}\left[\frac{1}{2sh(2\omega\Phi)}\right] = \frac{1}{2\pi}\int_{-\infty}^{+\infty}\frac{e^{i\omega x}}{2sh(2\omega\Phi)}d\omega = \frac{i}{2\pi}\int_{0}^{\infty}\frac{\sin\omega x}{sh(2\omega\Phi)}d\omega. \tag{2.212}$$

Inserting (2.212) into (2.211) and performing the change of variable $ix = \nu$, we obtain

$$\ell_{\pm}(\alpha) = -\frac{i}{2\pi}\int_{-i\infty}^{+i\infty} r_{\pm}(\alpha-\nu)\left(\int_{0}^{\infty}\frac{sh\omega\nu}{sh(2\omega\Phi)}d\omega\right)d\nu .$$

According to Gradshteyn and Ryzhik (page 344 in reference [2.9]), we have

$$\int_{0}^{\infty}\frac{sh\,a\,x}{sh\,b\,x}d\,x = \frac{\pi}{2b}\tan\frac{a\pi}{2b}, \quad |a| < b. \tag{2.213}$$

Hence by setting $\alpha - \nu = u$,

$$\ell_{\pm}(\alpha) = -\frac{i}{8\Phi}\int_{-i\infty}^{+i\infty} r_{\pm}(u)\tan\left[\frac{\pi(\alpha-u)}{4\Phi}\right]du. \tag{2.214}$$

Applying the inverse Fourier transform to (2.210) gives

$$g(\alpha) = \ell_{+}(\alpha+\Phi) + \ell_{-}(\alpha-\Phi), \tag{2.215}$$

and since $g(\alpha)$ is the logarithmic derivative of $\psi(\alpha)$, we obtain the following by integrating (2.215):

$$\psi(\alpha) = \psi(o)\exp\left[\int_{0}^{\alpha+\Phi}\ell_{+}(\mu)d\mu + \int_{0}^{\alpha-\Phi}\ell_{-}(\mu)d\mu\right], \tag{2.216}$$

where $\psi(o)$ is a constant that we make equal to 1. We will see later that the solution $S(\alpha)$ depends on the quotient $\psi(\alpha)/\psi(\varphi_0)$. Hence the constant $\psi(o)$ disappears in the final formula.

If we set

$$\psi_{\pm}(\alpha) = \exp\left[\int_{0}^{\alpha}\ell_{\pm}(\mu)d\mu\right], \tag{2.217}$$

then, according to (2.216) and (2.214), we obtain

$$\psi(\alpha) = \psi_{+}(\alpha+\Phi)\,\psi_{-}(\alpha-\Phi) \tag{2.218}$$

with

$$\psi_{\pm}(\alpha) = \exp\left[-\frac{i}{8\Phi}\int_{-i\infty}^{+i\infty} r_{\pm}(\nu)\left(\int_{0}^{\alpha}\tan\left[\frac{\pi(\mu-\nu)}{4\Phi}\right]d\mu\right)d\nu\right]. \tag{2.219}$$

Observing that

$$\int_0^\alpha \tan\left[\frac{\pi(\mu-\nu)}{4\Phi}\right] d\mu = -\frac{4\Phi}{\pi}\int_0^\alpha \frac{d}{d\mu}\text{Log}\left[\cos\frac{\pi}{4\Phi}(\mu-\nu)\right] d\mu = -\frac{4\Phi}{\pi}\text{Log}\left[\frac{\cos\frac{\pi}{4\Phi}(\alpha-\nu)}{\cos\frac{\pi}{4\Phi}\nu}\right], \tag{2.220}$$

we obtain, by stating $\nu = ix$,

$$\psi_\pm(\alpha) = \exp\left\{-\frac{1}{2\pi}\int_{-\infty}^{+\infty} r_\pm(ix)\text{Log}\left[\frac{\cos\frac{\pi}{4\Phi}(\alpha-ix)}{\cos\frac{\pi ix}{4\Phi}\nu}\right] dx\right\}. \tag{2.221}$$

We can further simplify by noting that $Z^\pm(\alpha)$ is an even function of α (see appendix B). Hence $\partial Z^\pm(\alpha)/\partial\alpha$ is an odd function of α, and according to (2.208), $r_\pm(\alpha)$ is an even function of α. By taking this property into account, we finally obtain

$$\psi_\pm(\alpha) = \exp\left[-\frac{1}{2\pi}\int_0^\infty r_\pm(ix)\text{Log}\left[\frac{\cos\frac{\pi}{2\Phi}\alpha + ch\frac{\pi x}{2\Phi}}{1+ch\frac{\pi x}{2\Phi}}\right] dx\right]. \tag{2.222}$$

According to (2.119), the elementary functions $\psi_\pm(\alpha)$ are defined in the strip $|R_e\alpha| < 2\Phi$. When $|R_e\alpha| > \Phi$, $\psi(\alpha)$ cannot be calculated by using (2.218) together with (2.222). However, it is possible to shift the argument of $\psi(\alpha)$ in its fundamental domain by using (2.115). For instance, if $R_e\alpha > \Phi$, we obtain the following by applying (2.115):

$$\psi(\alpha) = -R_+(\alpha-\Phi)\,\psi(2\Phi-\alpha).$$

And since $\psi(2\Phi - \alpha)$ is in its fundamental domain, we have the following according to (2.218):

$$\psi(2\Phi-\alpha) = \psi_+(3\Phi-\alpha)\psi_-(\Phi-\alpha),$$

where for $|R_e\alpha| > \Phi$, $\psi+(3\Phi-\alpha)$ and $\psi-(\Phi-\alpha)$ can be calculated by (2.222).

We will now establish some other properties of the functions $\psi_\pm(\alpha)$, which are necessary for the determination of the function $\sigma(\alpha)$. Especially, it is important to know the behavior of the functions $\psi_\pm(\alpha)$ on the infinite branches of the contour γ. When $|I_m(\alpha)| \to \infty$, we have

$$\mathrm{Log}\left(\frac{\cos\frac{\pi}{2\Phi}\alpha + ch\frac{\pi x}{2\Phi}}{1 + ch\frac{\pi x}{2\Phi}}\right) \approx \frac{\pi}{2\Phi}\left|I_m(\alpha)\right|. \tag{2.223}$$

Consequently

$$\psi_{\pm}(\alpha) \approx \exp\left[-\frac{\left|I_m(\alpha)\right|}{4\Phi}\int_0^{\infty}\frac{R'_{\pm}(ix)}{R_{\pm}(ix)}dx\right]. \tag{2.224}$$

Now

$$I = \int_0^{\infty}\frac{R'_{\pm}(ix)}{R_{\pm}(ix)}dx = -i\,\mathrm{Log}\left[R_{\pm}(ix)\right]\Big|_0^{\infty} \tag{2.225}$$

and

$$R_{\pm}(ix) = \frac{i\,sh\,x - Z^{\pm}(ix)}{i\,sh\,x + Z^{\pm}(ix)}. \tag{2.226}$$

According to equation (B.96) of appendix B, we have

$$\lim_{|x|\to\infty}\left[Z^{\pm}(ix)\right] = 0. \tag{2.227}$$

Hence

$$\lim_{|x|\to\infty}\left[R_{\pm}(ix)\right] = 1, \tag{2.228}$$

and for $x = 0$, $R_{\pm}(0) = -1$. It is possible to show that the determination of Log(–1) must be choosen equal to $+\pi i$. For this, we observe first that in the lossless case, $Z^{\pm} = i\eta^{\pm}$ with $\eta^{\pm} > 0$. Then according to (2.208) we have

$$\frac{R'_{\pm}(ix)}{R_{\pm}(ix)} = -2i\frac{\eta^{\pm}ch\,x - \eta'^{\pm}sh\,x}{sh^2x - \left(\eta^{\pm}\right)^2}. \tag{2.229}$$

In this case, the integration path in (2.225) passes through a pole at $sh\,x = \eta^{\pm}$. If there are some small losses, this pole moves in the fourth quadrant of the complex x-plane, as shown in figure 2.21.

The integration path must therefore pass round the pole in the sense indicated in figure 2.21. Consequently, the integral I in (2.225) may also be written in the form

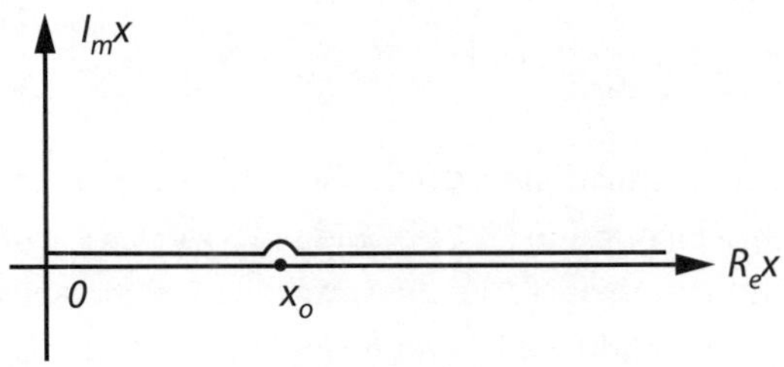

Figure 2.21. Position of the pole and integration path.

$$I = PV\int_0^\infty \frac{R'_\pm(ix)}{R_\pm(ix)}dx - \pi i \times \text{residue}, \tag{2.230}$$

where the first term on the right-hand side of (2.230) is Cauchy's principal value defined by

$$PV\int_0^\infty = \lim_{\varepsilon\to 0}\left(\int_0^{x_0-\varepsilon} + \int_{x_0+\varepsilon}^\infty\right),$$

where $\varepsilon > 0$ and $x_0 = \operatorname{Arg} sh\eta^\pm$. Since (2.225) is also valid for the indefinite integral,

$$\int_0^x \frac{R'_\pm(ix)}{R_\pm(ix)}dx = -i\operatorname{Log} R_\pm(ix),$$

it can be readily shown that

$$PV\int_0^\infty \frac{R'_\pm(ix)}{R_\pm(ix)}dx = 0.$$

Hence by taking the residue of (2.229) we obtain

$$I = -\pi i \times \text{residue} = -\pi. \tag{2.231}$$

This result is in accordance with our choice of the determination of Log (–1) in (2.225). Inserting (2.231) into (2.224) yields

$$\psi_\pm(\alpha) \approx \exp\left(\frac{\pi}{4\Phi}|I_m\alpha|\right), \tag{2.232}$$

when $|I_m(\alpha)| \to \infty$. According to (2.218), it follows the behavior of $\psi(\alpha)$ on the infinite branches of γ:

$$\psi(\alpha) \approx \exp\left(\frac{\pi}{2\Phi}\left|I_m\alpha\right|\right), \quad \left|I_m\alpha\right| \to \infty. \tag{2.233}$$

We now consider the determination of $\sigma(\alpha)$. Since the behavior of $\psi(\alpha)$ when $\left|I_m\alpha\right|$ tends to infinity given by (2.233) is identical to (2.144), and since Meixner's condition at the edge is still given by (2.145), the determination of $\sigma(\alpha)$ follows exactly the same steps as the case of a constant impedance. The result is therefore the same as in formula (2.150).

Finally, the integral form of the solution is given by

$$P(r,\varphi) = \frac{P_0^i}{4\Phi i}\frac{\cos\frac{\pi}{2\Phi}\varphi_0}{\psi(\varphi_0)}\int_\gamma e^{-ikr\cos\alpha}\frac{\psi(\alpha+\varphi)}{\sin\frac{\pi}{2\Phi}(\alpha+\varphi)-\sin\frac{\pi}{2\Phi}\varphi_0}d\alpha, \tag{2.234}$$

where $\psi(\alpha)$ is given by (2.218) and (2.222).

2.2.3.2. Asymptotic Evaluation of the Solution

The technique for asymptotically evaluating the solution (2.234) is the same technique that is used in section 2.2.2.3. It comprises two steps: In a first step, the contour γ is closed by the SDP Γ and Γ', and the solution (2.234) is written in the form

$$\int_\gamma F(\alpha)d\alpha = \int_{\Gamma+\Gamma'} F(\alpha)d\alpha + 2\pi i\sum_{p=1}^{N} R(\alpha_p), \tag{2.235}$$

where $F(\alpha)$ is given by

$$\begin{aligned} F(\alpha) &= \frac{P_0^i}{4\Phi i}\frac{\cos\frac{\pi}{2\Phi}\varphi_0}{\psi(\varphi_0)}e^{-ikr\cos\alpha}\frac{\psi(\alpha+\varphi)}{\sin\frac{\pi}{2\Phi}(\alpha+\varphi)-\sin\frac{\pi}{2\Phi}\varphi_0} \\ &= e^{-ikr\cos\alpha}\sigma(\alpha+\varphi)\psi(\alpha+\varphi) \end{aligned} \tag{2.236}$$

and where $R(\alpha_p)$ are the residues at the poles α_p of $F(\alpha)$ located inside the closed contour $\gamma - \Gamma - \Gamma'$.

In a second step, the integral along the SDP Γ and Γ' are asymptotically evaluated by the method of steepest descent. Since the expression of $F(\alpha)$ given by (2.236) is formally identical to (2.182), the asymptotic evaluation of the integrals along Γ and Γ' leads to the same formal result given by (2.188). However, the function $\psi(\alpha)$ is different and is now given by (2.218) and (2.222). As in section 2.2.2.3, the sum of the contributions of the integrals along Γ and Γ' gives the diffracted field, whereas the contributions of the poles of $F(\alpha)$ give the incident, reflected, and surface waves.

The poles of $F(\alpha)$ are those of $\sigma(\alpha + \varphi)$ and $\psi(\alpha - \varphi)$. Since $\sigma(\alpha)$ is given by the same formula as in section 2.2.2.3, the poles of $\sigma(\alpha + \varphi)$ are the same as those given by (2.154) for a sharp wedge. These poles are real and their contribution gives the incident wave and the waves reflected by each face of the wedge. Moreover, just as for an impedance wedge, $\psi(\alpha)$ also verifies the equation (2.115), and its poles are also those of the reflection coefficient. However, since the impedance is now a function of α, the reflection coefficient has more types of poles. It has been shown in appendix C that for an elastic flat plate separating a fluid from vacuum, the following waves may exist in addition to the reflected wave:

- Surface waves of the Scholte-Stoneley type are found at the interface between a fluid and an elastic half-space. They are similar to those analyzed in section 2.2.2.3.
- Generalized Lamb waves are found when two interfaces are close together.

According to the equation (2.115), the poles of $\psi(\alpha)$ are given by

$$\alpha = \pm\left(\Phi + \alpha_p^{\pm}\right),$$

where $\alpha_p^{\pm}$ are the poles of $R_{\pm}(\alpha)$. Hence the poles of $\psi(\alpha + \varphi)$ are

$$\varphi_p^{\pm} = \pm\left(\Phi + \alpha_p^{\pm}\right) - \varphi. \qquad (2.237)$$

The poles of the reflection coefficient are the roots of the equation

$$\sin\alpha = -Z_{\pm}(\alpha). \qquad (2.238)$$

In the case of an elastic flat plate separating a fluid from vacuum, the roots corresponding to the waves of the Scholte-Stoneley type are characterized by $R_e\left[Z_{\pm}(\alpha)\right] > 0$ and $I_m\left[Z_{\pm}(\alpha)\right] < 0$. The poles corresponding to these waves are therefore located outside the strip $(0, \pi)$, as shown in figure 2.16.

In the wedge problem, for $\varphi = \pm\Phi$, equation (2.237) shows that the poles of $\psi(\alpha + \varphi)$ are given by

$$\phi_p^{\pm} = \pm\alpha_p^{\pm} = \pm\left(\pi + \operatorname{Arc}\sin Z^{\pm}\left(\alpha_p^{\pm}\right)\right), \qquad (2.239)$$

where $R_e\left[Z^{\pm}\left(\alpha_p^{\pm}\right)\right] > 0$ and $I_m\left[Z^{\pm}\left(\alpha_p^{\pm}\right)\right] < 0$. The field of the surface waves corresponding to these poles is given by equation (2.175), where $F_1\left(\alpha_p^{\pm}, \varphi_0\right)$ has to be replaced by $\overline{F}_1\left(\alpha_p^{\pm}, \varphi_0\right)$, given by

$$\overline{F}_1\left(\alpha_p^{\pm},\varphi_0\right)=\frac{\pi}{\Phi}\frac{\cos\left(\frac{\pi\varphi_0}{2\Phi}\right)}{\psi\left(\varphi_0\right)}\frac{\tan\alpha_p^{\pm}}{\pm\cos\frac{\pi}{2\Phi}\alpha_p^{\pm}-\sin\left(\frac{\pi}{2\Phi}\varphi_0\right)}\frac{1}{1+\left(\cos\alpha_p^{\pm}\right)^{-1}\left(\frac{\partial Z^{\pm}(\alpha)}{\partial\alpha}\right)_{\alpha=\alpha_p^{\pm}}}. \tag{2.240}$$

Moreover, $\Phi_p^{\pm}$ is determined by (2.170) with $\alpha_p^{\pm}$, given by (2.239). The lines of constant amplitude and the lines of constant phase of these waves are represented in figures 2.17 and 2.18, where θ' and θ'' have to be replaced by $R_e\left(\text{Arc}\sin Z^{+}\right)$ and $I_m\left(\text{Arc}\sin Z^{+}\right)$, respectively.

The roots of (2.238) corresponding to the generalized Lamb waves are located inside the strip $(0, \pi)$. If we consider the expression

$$R_{\pm}\left(\theta\right)=\frac{\cos\theta-Z^{\pm}\left(\theta\right)}{\cos\theta+Z^{\pm}\left(\theta\right)} \tag{2.241}$$

of the reflection coefficient where the angles are defined with respect to the normal to the flat plate, then we observe that $R_{\pm}\left(\theta\right)$ is an even function of θ since, according to equation (B.93) of appendix B, the equivalent impedances $Z^{\pm}\left(\theta\right)$ are even functions of θ. Hence the positions of the poles θ_p of $R(\theta)$ in the complex θ-plane are symmetric with respect to the origin. Moreover, since the generalized Lamb waves are defined by $R_e\theta_p > 0$ and $I_m\theta_p > 0$ in the strip $(0,\frac{\pi}{2})$, they are located in the first and third quadrants of the complex θ-plane, as shown in figure 2.22.

The transformation from the θ-plane to the α-plane is defined by

$$\alpha=\frac{\pi}{2}-\theta.$$

Since the generalized Lamb waves are inhomogeneous plane waves propagating along the interface with exponentially decreasing amplitude, they must verify $I_m\alpha_p^{+}>0$, $\frac{\pi}{2}<R_e\alpha_p^{+}<\pi$, and $I_m\alpha_p^{-}<0$, $-\pi<R_e\alpha_p^{-}<-\frac{\pi}{2}$.

Hence

$$\alpha_p^{\pm}=\pm\left(\frac{\pi}{2}+\theta_p^{\pm}\right), \tag{2.242}$$

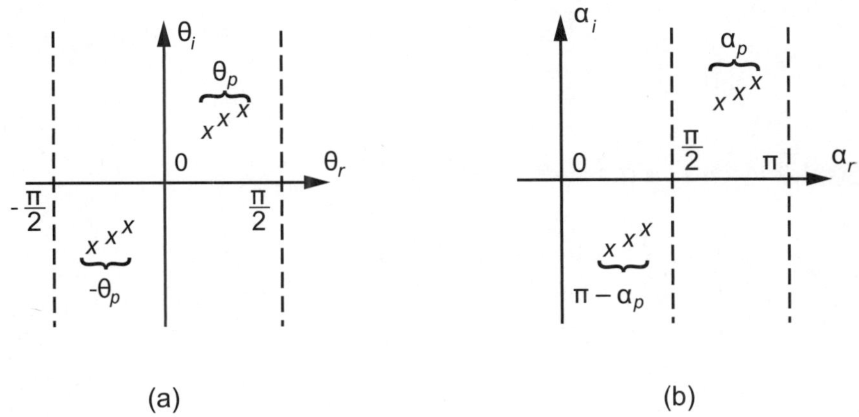

Figure 2.22. Poles of the reflection coefficient (a) in the complex θ-plane and (b) in the complex α-plane.

where $\theta_p^{\pm}$ are the roots of $R_{\pm}(\theta)$ located in the strip $\left(0,\dfrac{\pi}{2}\right)$, with $I_m\theta^{\pm} > 0$.

In order to determine the physical properties of these waves, we insert (2.242) into (2.175) and (2.176). The exponential function in (2.175) yields

$$e^{-ikr\cos\alpha_p^{\pm}} = \exp\left[-ikr\left\{\cos\left[\pm\left(\Phi + R_e\alpha_p^{\pm}\right)-\varphi\right]ch\,I_m\alpha_p^{\pm} \mp i\sin\left[\pm\left(\Phi + R_e\alpha_p^{\pm}\right)-\varphi\right]sh\,I_m\alpha_p^{\pm}\right\}\right].$$

The lines of constant amplitude are defined by

$$\sin\left[\pm\left(\Phi + R_e\alpha_p^{\pm}\right)-\varphi\right] = 0. \qquad (2.243)$$

On the face (+), (2.243) implies $\Phi + \dfrac{\pi}{2} + R_e\theta_p^{+} - \varphi = \pi$, and on the face (–), we must have $\Phi - \dfrac{\pi}{2} - R_e\theta_p^{-} + \varphi = 0.$

If we denote by $\varphi_p^{a\pm}$ the lines of constant amplitude on the faces (+) and (–), then we have

$$\varphi_p^{a+} = \Phi - \frac{\pi}{2} + R_e\theta_p^{+}, \qquad (2.244)$$

$$\varphi_p^{a-} = -\Phi + \frac{\pi}{2} + R_e\theta_p^{-}. \qquad (2.245)$$

Now the generalized Lamb waves exist if

$$\left|\Phi_p^{\pm} - \varphi\right| < \pi. \tag{2.246}$$

On the faces (+) and (–), the condition (2.246) implies

$$\varphi > \Phi_p^+ - \pi = \Phi - \frac{\pi}{2} + R_e\,\theta_p^+ + GD\left(I_m\,\theta_p^+\right) = \varphi_p^{\ell+}, \tag{2.247}$$

$$\varphi < \Phi_p^- + \pi = -\Phi + \frac{\pi}{2} - R_e\,\theta_p^- - GD\left(I_m\,\theta_p^-\right) = \varphi_p^{\ell-}, \tag{2.248}$$

respectively, where $GD(x)$ is the Gudermann function defined by (2.166).

The lines of constant amplitude and the limits of existence of the generalized Lamb wave corresponding to the pole α^-, excited at the edge of a wedge-shaped shell and propagating on the face (–), are represented in figure 2.23. We observe that, as expected, the amplitude of the wave decreases along the face. However, it increases in the direction normal to the face. Therefore, the generalized Lamb wave behaves as a leaky wave.

We also observe that, for an observation point turning around the wedge at constant distance from the edge, the amplitude of the wave decreases exponentially away from the direction defining the limit of existence of the wave.

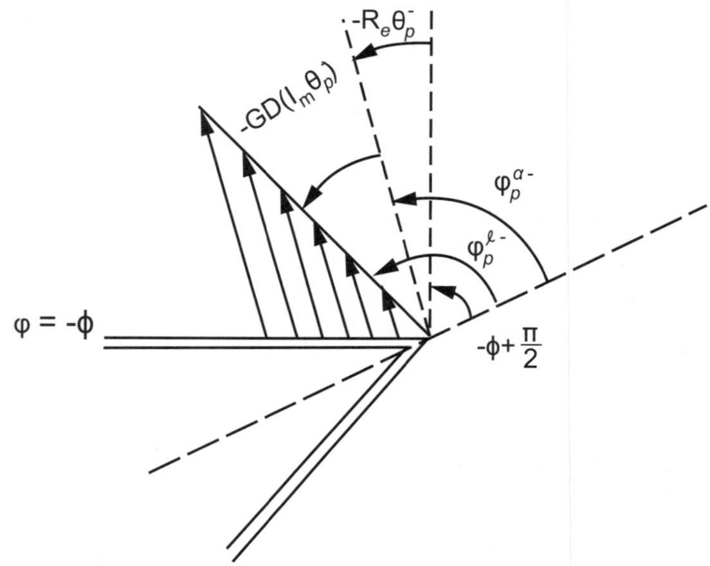

Figure 2.23. Lines of constant amplitude and the limits of existence of the generalized Lamb wave.

As shown in a typical example in section 2.2.3.4, the imaginary part of the pole θ_p of a generalized Lamb wave is very small compared to its real part. Consequently, $GD(I_m\theta_p)$ is also very small, and the line φ_p^ℓ is very close to the line of constant amplitude $R_e\theta_p$. The constant amplitude lines are therefore almost parallel to the direction φ_p^ℓ defining their limit of existence. In this case, the amplitude of the wave remains almost constant in the direction φ_p^ℓ but decreases very rapidly away from this direction, in its domain of existence. In a radiation diagram, it therefore has the characteristic shape shown in figure 2.24.

2.2.3.3. Generalization of the Elastic Wedge-Shaped Shell Solution to Oblique Incidence

As in the case of a constant impedance, the translation invariance of the solution of a plane wave scattered by a straight wedge, in a direction parallel to the edge, leads to

$$p(\rho,\varphi,z) = e^{ikz\cos\beta_0}\, P(\rho,\varphi), \tag{2.249}$$

where

$$P(\rho,\varphi) = \frac{1}{2\pi i}\int_\gamma e^{-ik_\perp \rho\cos\alpha} S(\alpha+\varphi)\, d\alpha \tag{2.250}$$

and where β_0 is the angle between the direction of propagation of the incident wave and the edge of the wedge. The definitions of $k_\perp$ and φ are those given in section 2.2.2.4 (see fig. 2.19).

The derivation of the solution for oblique incidence is identical to that given in section 2.2.2.4. Equations (2.196) to (2.198) still hold, and since $Z^\pm(\alpha)$ is an even function of α, the equations (2.199) and (2.200) also remain unchanged. Hence we obtain a system of functional

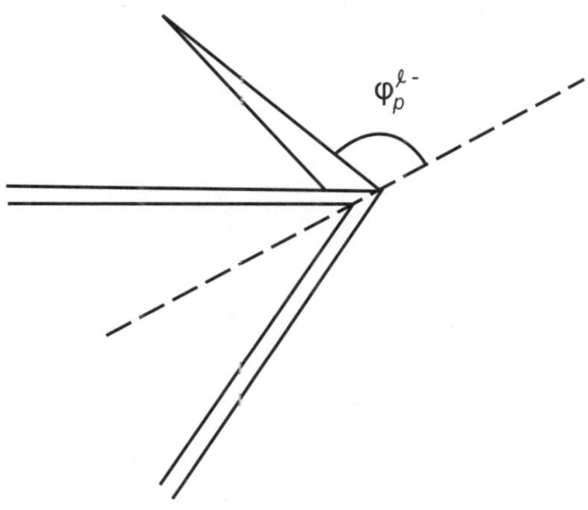

Figure 2.24. Radiation diagram of a generalized Lamb wave.

equations identical to the one obtained for normal incidence, provided that we modify the impedances by replacing them with

$$Z_{\pm}^{(m)}(\alpha) = \frac{Z^{\pm}(\alpha)}{\sin \beta_0}. \tag{2.251}$$

With these modifications, the integral form of the solution is given by

$$p(r,\varphi,z) = e^{ikz\cos\beta_0} \frac{P_0^i}{4\Phi i} \frac{\cos\frac{\pi\varphi_0}{2\Phi}}{\psi(\varphi_0)} \int_{\gamma} e^{-ik\rho\sin\beta_0\cos\alpha} \frac{\psi(\alpha+\varphi)}{\sin\frac{\pi}{2\Phi}(\alpha+\varphi) - \sin\frac{\pi}{2\Phi}\varphi_0} d\alpha, \tag{2.252}$$

where φ_0, φ are now the projected angles defined in figure 2.19(b) and where the functions $\psi(\alpha)$ are given by (2.222), in which $r_{\pm}(\alpha)$ is still given by (2.208), but the impedances $Z^{\pm}(\alpha)$ have to be replaced by $Z_{\pm}^{(m)}(\alpha)$. With these modifications, the diffraction coefficient is still given by (2.203). It can also be introduced in the general formula (2.205) for the field diffracted by a curved, wedge-shaped elastic shell.

2.2.3.4. Typical Examples

A typical example, for which experimental results are available, is given by a right-angle wedge–shaped shell. The faces of the shell are formed with metal plates of brass separating water from vacuum. The physical parameters of brass and water are given in table 2.1. For a brass plate of thickness $d = 0.8$ mm and a frequency of the incident wave equal to 3.4 MHz, the real and imaginary parts of the poles of the reflection coefficient are given in table 2.2.

We see that $I_m\theta_p$ is very small compared to $R_e\theta_p$. It can be shown that the number of poles depends on the characteristic product $F \times d$, where F is the frequency [2.13]. In the previous

Table 2.1. Physical parameters

BRASS

Mass density: $\rho = 8.6$ kg/dm^3

Sound velocity of longitudinal waves: $C_\ell = 4410$ m/s

Sound velocity of transverse waves: $C_t = 2150$ m/s

WATER

Mass density: $\rho_0 = 1$ Kg/dm^3

Sound velocity: $C_0 = 1475$ m/s

Table 2.2. θ_p **in degrees**

$R_e\theta_p$	$I_m\theta_p$
9.23	4.11×10^{-3}
20.42	1.56×10^{-4}
25.80	5.00×10^{-3}
44.99	1.23×10^{-2}
48.32	8.39×10^{-3}

example, $F \times d = 2.72$ (MHz mm). Experimental results giving the generalized Lamb waves excited by a wave incident on a right-angle wedge made of brass have been obtained by de Billy [2.14]. The experimental setup is depicted in figure 2.25.

For $F \times d = 2.2$ MHz $\times 0.9$ mm $= 2$(MHz mm), only four Lamb waves are excited. The measurement diagram is shown in figure 2.26.

The corresponding theoretical results obtained with the formulas established in section 2.2.3.2 for an incident plane wave interacting with an unlimited right-angle wedge–shaped shell separating water from vacuum are drawn in figure 2.27, where the generalized Lamb waves excited on both faces are shown. We see that the number and the position of the lobes agree with those obtained experimentally, but their relative amplitudes are different. It seems to be a consequence of the fact that in the experimental results an integration of the values over about 4° is performed. Hence the amplitudes of thin high lobes appear reduced. Compared to the field diffracted by the edge of the wedge, also shown in figure 2.27, we see that in the angular domain where the Lamb waves exist, their amplitudes are of the same order as the amplitude of the diffracted field.

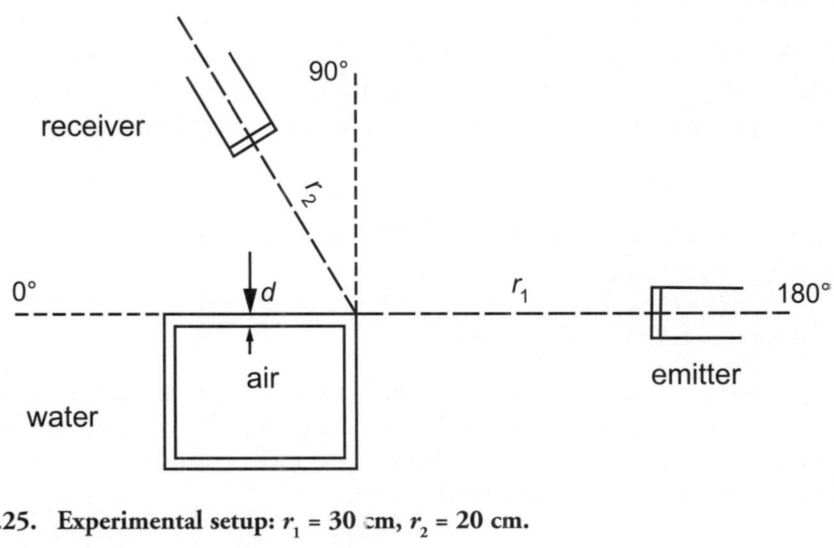

Figure 2.25. Experimental setup: $r_1 = 30$ **cm,** $r_2 = 20$ **cm.**

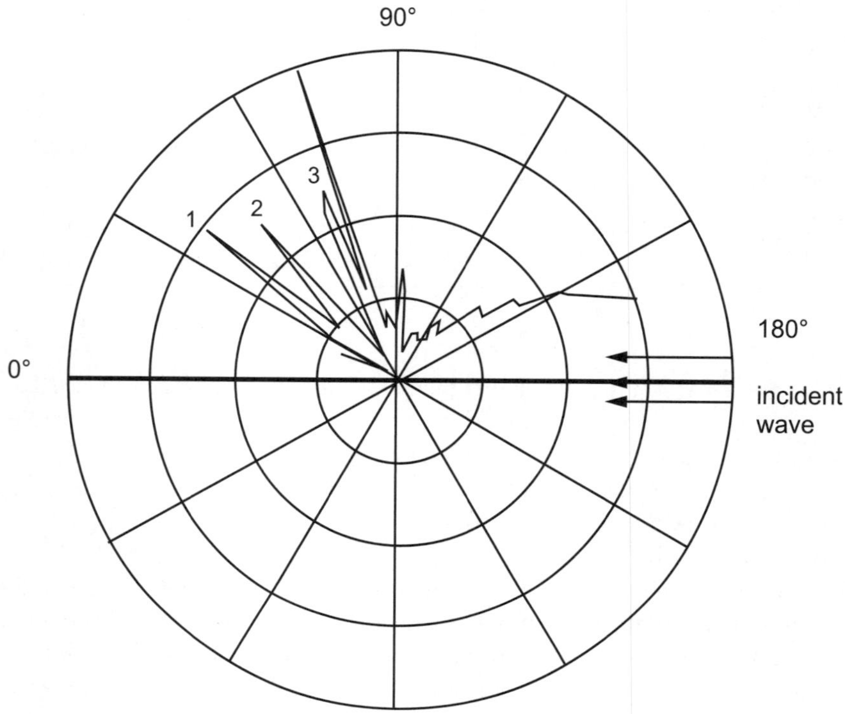

Figure 2.26. Generalized Lamb waves excited at the edge of a shell of brass shaped like a right-angle wedge, surrounded by water: Experimental results.

2.3. THE CIRCULAR CYLINDER

Much of the literature on the subject of scattering from a circular cylinder is devoted to hard or soft boundary conditions. In this section, we will extend our discussion to a circular cylinder whose surface impedance is constant and to a hollow elastic circular cylindrical shell.

2.3.1. CIRCULAR CYLINDER WITH HARD, SOFT, OR IMPEDANCE BOUNDARY CONDITIONS

2.3.1.1. General Solution to the Problem

Consider a line source whose intensity is unity and whose direction is parallel to the generatrix of an infinite cylinder with radius a. We suppose that the cylinder has a constant impedance Z. The problem that we wish to solve is the calculation of the total field in the space exterior to the cylinder, which we suppose to be an ideal fluid.

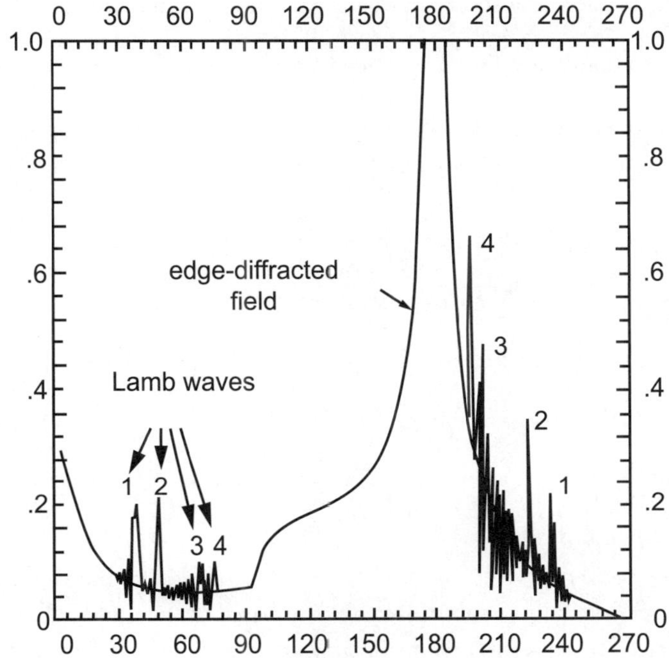

Figure 2.27. Generalized Lamb waves excited at the edge of a shell of brass shaped like a right angle, surrounded by water: Theoretical results.

Let *OXYZ* be a rectangular coordinate system with the *OZ* axis as the axis of revolution of the cylinder. Let us denote ρ and $\theta^i = 0$ as the polar coordinates of the intersection point of the linear source and the plane *xy*. If (r, θ) are the polar coordinates of the observation point, then the total pressure field in the fluid outside the cylinder must satisfy the Helmholtz equation

$$\left(\Delta + k^2\right) p(r,\theta) = -\frac{1}{r}\delta(r-\rho)\delta(\theta), \quad r \geq a; \tag{2.253}$$

the boundary conditions

$$\frac{\partial p}{\partial r} + ikZp = 0, \quad r = a; \tag{2.254}$$

and the radiation condition

$$\lim_{r\to\infty} r^{1/2}\left(\frac{\partial p}{\partial r} - ikp\right) = 0. \tag{2.255}$$

Equation (2.254) includes the special cases of hard and soft boundary conditions, since for $Z = 0$ we have $\dfrac{\partial p}{\partial r} = 0$ (hard) and for $Z = \infty$ we have $p = 0$ (soft).

The solution to (2.253) through (2.255) is well known (see for example [2.15]). By following the method of separation of variables described in section 2.2.1.1, its elaboration is straightforward and leads to

$$p(r,\theta) = \frac{i}{4} \sum_{m=-\infty}^{+\infty} e^{im\theta} \left[J_m(kr_<) - \frac{\Omega J_m(ka)}{\Omega H_m^{(1)}(ka)} H_m^{(1)}(kr_<) \right] H_m^{(1)}(kr_>), \tag{2.256}$$

where we have set

$$\begin{aligned} \Omega &= \frac{\partial}{\partial r} + ikZ, \\ r_< &= \mathrm{Inf}(r,\rho), \quad r_> = \sup(r,\rho). \end{aligned} \tag{2.257}$$

The series (2.256) converges very slowly when $ka >> 1$ and also for points far away from the cylinder. It is a series of normal modes in the terminology used in the theory of open waveguides. The series can be replaced by an alternative representation in a series of angular modes [2.16] that is rapidly convergent when ka becomes large. The method to convert the series of normal modes into angular modes is known as the Watson transformation [2.17].

Let S denote the series

$$S = \sum_{n=-\infty}^{+\infty} f(n). \tag{2.258}$$

Each term of this series can be viewed as the residue of an integral along a closed contour in the complex plane. If we denote by $g(\nu)$ an analytical function such that $g(n) = 0$ for $n = 0$, $\pm 1, \pm 2, \ldots$, then by using the residue theorem we can write

$$S = -\frac{1}{2\pi i} \int_C f(\nu) \frac{dg(\nu)}{d\nu} \frac{d\nu}{g(\nu)}, \tag{2.259}$$

where C is a contour in the complex ν-plane that surrounds the $R_e\nu$ axis, as shown in figure 2.28.

In order that (2.259) be equal to (2.258), $g(\nu)$ must satisfy

$$\frac{1}{g(\nu)} \frac{dg(\nu)}{d\nu} \cong \frac{1}{\nu - n}$$

in the neighborhood of the pole n. If we choose the function $g(\nu) = \sin \nu\pi$, then its behavior near the pole n is given by

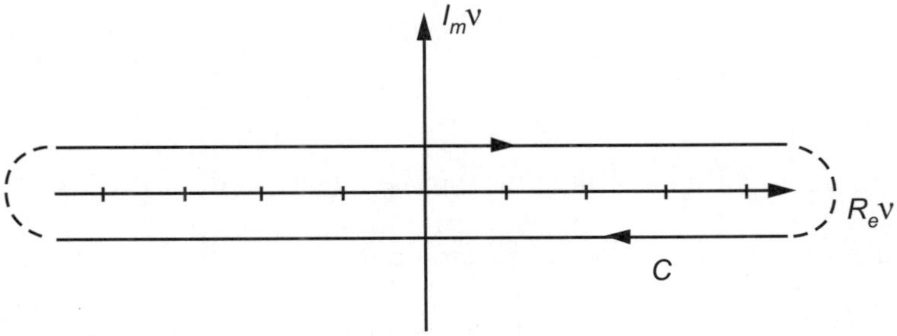

Figure 2.28. Contour C for the Watson transformation.

$$\sin \nu\pi = \pi(\nu - n)\cos n\pi = (-1)^n \pi(\nu - n)$$

and

$$\frac{dg}{d\nu} = \pi \cos n\nu \cong \pi \cos n\pi = \pi(-1)^n .$$

Hence

$$\frac{1}{g(\nu)}\frac{dg(\nu)}{d\nu} = \frac{\pi \cos n\nu}{\sin \nu\pi} \cong \frac{1}{\nu - n} \cong \frac{\pi(-1)^\nu}{\sin \nu\pi} \cong \frac{\pi}{(-1)^\nu \sin \nu\pi} = \frac{\pi}{e^{i\pi\nu}\sin \nu\pi}.$$

The right-hand side of (2.256) can therefore be written in the form

$$p(r,\theta) = -\frac{1}{8}\int_C \frac{e^{i\nu(\theta-\pi)}}{\sin \nu\pi}\left[J_\nu(kr_<) - \frac{\Omega J_\nu(ka)}{\Omega H_\nu^{(1)}(ka)} H_\nu^{(1)}(kr_<)\right] H_\nu^{(1)}(kr_>)d\nu. \tag{2.260}$$

Using the relations

$$\begin{aligned} H_{-\nu}^{(1)}(z) &= e^{i\nu\pi} H_\nu^{(1)}(z), \\ H_{-\nu}^{(2)}(z) &= e^{-i\nu\pi} H_\nu^{(2)}(z), \end{aligned} \tag{2.261}$$

$$J_\nu(z) = \frac{1}{2}\left[H_\nu^{(1)}(z) + H_\nu^{(2)}(z)\right], \tag{2.262}$$

it can be readily shown that the integrand (2.260), without the term $\exp[i\nu(\theta - \pi)]$, is an odd function of ν. Hence by replacing ν by $-\nu$ in the part of the contour C for which $I_m\nu < 0$, we obtain

$$p\left(r,\theta\right)=-\frac{1}{4}\int_{D}\frac{\cos\nu\left(\theta-\pi\right)}{\sin\nu\pi}\left[J_{\nu}\left(kr_{<}\right)-\frac{\Omega J_{\nu}\left(ka\right)}{\Omega H_{\nu}^{(1)}\left(ka\right)}H_{\nu}^{(1)}\left(kr_{<}\right)\right]H_{\nu}^{(1)}\left(kr_{>}\right)d\nu,\quad(2.263)$$

where D is a contour in the upper-half of the ν-plane, parallel to the real ν-axis.

Apart from the poles due to sin $\nu\pi$, which are located on the $R_e\nu$ axis, the integrand of (2.263) has no other poles than those corresponding to the values of ν, for which $\Omega H_{\nu}^{(1)}\left(ka\right)=0$. For a soft or hard cylinder, the poles are, respectively, the zeros of $H_{\nu}^{(1)}\left(ka\right)$ and $H_{\nu}^{(1)'}\left(ka\right)$. These zeros are located on the Stokes lines h_1 for $I_m\nu > 0$ and h_{-1} for $I_m\nu < 0$ of the Hankel function $H_{\nu}^{(1)}\left(x\right)$, shown in figure 2.29, and the absolute value of their imaginary part augments with n. Note that these lines have been introduced in the asymptotic expansion of the Hankel function $H_{\nu}^{(1)}\left(x\right)$ and correspond to an abrupt modification of the asymptotic behavior of this function (see reference [2.18]). In the upper half of the ν-plane, the poles for a soft or hard cylinder are therefore located on the Stokes line h_1. Analytical and numerical results, which are commented on at the end of this section, show that for an impedance cylinder the poles ν_n are generally located in a neighborhood of order $\left(ka\right)^{1/3}$ of the Stokes line h_1 and that their imaginary part augments with n, the more ka is large.

The contour D can be closed by the half circle C_∞ at infinity in the upper-half plane separated in two parts, C_1 and C_2, by the contour C_0 corresponding to the poles of the integrand (fig. 2.29). The integral (2.263) can be evaluated by using the calculus of residues, leading to the following representation for $p(r, \theta)$:

$$p\left(r,\theta\right)=\frac{\pi i}{4}\sum_{n=1}^{\infty}\frac{\cos\nu_n\left(\theta-\pi\right)}{\sin\nu_n\pi}\frac{\Omega H_{\nu_n}^{(2)}\left(ka\right)}{\left[\frac{\partial}{\partial\nu}\Omega H_{\nu}^{(1)}\left(ka\right)\right]_{\nu=\nu_n}}H_{\nu_n}^{(1)}\left(k\rho\right)H_{\nu_n}^{(1)}\left(kr\right),\qquad(2.264)$$

where ν_n are the zeros of $\Omega H_{\nu}^{(1)}\left(ka\right)$. Since the imaginary part of ν_n augments with n, the more ka is large, the more rapidly convergent the series (2.264) is at high frequencies.

Expression (2.264) implies that the integral along C_∞ vanishes. This result can readily be established by replacing the Bessel and Hankel functions with the dominant term of their asymptotic expansion for $|\nu| >> |z|$, given by (see reference [2.16])

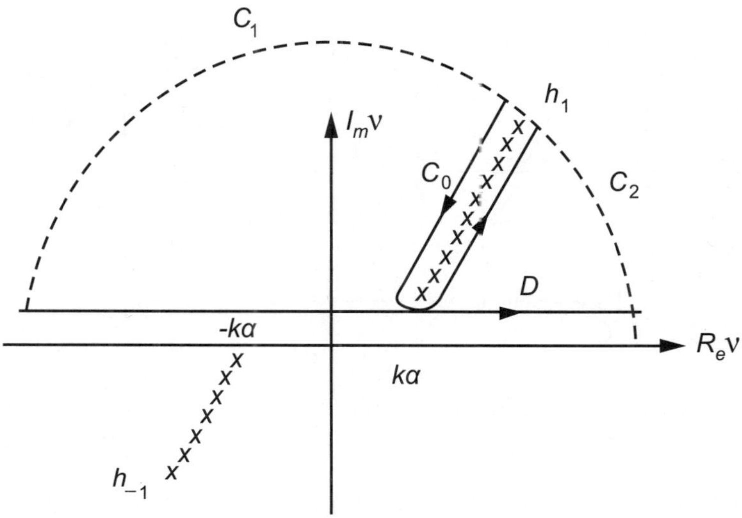

Figure 2.29. Contour of integration D and the Stokes lines.

$$J_\nu(z) \approx \frac{1}{\sqrt{2\pi\nu}}\left(\frac{2\nu}{e\,z}\right)^{-\nu}, \quad |\arg\nu| < \pi.$$

$$H_\nu^{(1)}(z) \approx \begin{cases} e^{i(\pm 1-1)\frac{\pi}{4}}\sqrt{\dfrac{2}{\pi\nu}}\left(\dfrac{2\nu}{e\,z}\right)^{\mp\nu}, & \arg\dfrac{\nu}{z}\begin{vmatrix}>0\\<0\end{vmatrix} \quad \text{in region I} \\ -i\sqrt{\dfrac{2}{\pi\nu}}\left(\dfrac{2\nu}{e\,z}\right)^{\nu}, & \text{in region II}, \end{cases} \tag{2.265}$$

$$H_\nu^{(2)}(z) \approx i\sqrt{\frac{2}{\pi\nu}}\left(\frac{2\nu}{e\,z}\right)^{\nu}, \qquad \text{in region IV},$$

where the different regions in the complex ν-plane for z real are delimited by the Stokes lines of $H_\nu^{(1)}(z)$ and $H_\nu^{(2)}(z)$ and the real ν-axis, as shown in figure 2.30.

We see that on the contour C_2 in region II, the term of the integrand of (2.263) given by

$$T = \left[J_\nu(kr_<) - \frac{\Omega J_\nu(ka)}{\Omega H_\nu^{(1)}(ka)} H_\nu^{(1)}(kr_<)\right] H_\nu^{(1)}(kr_>) \tag{2.266}$$

behaves for $|\nu| \to \infty$, like

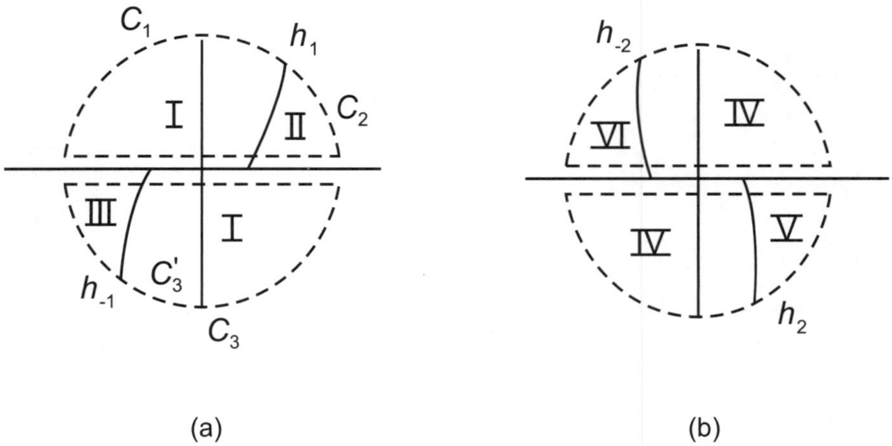

Figure 2.30. Regions of validity of the asymptotic approximations of (a) $H_\nu^{(1)}(kz)$ and (b) $H_\nu^{(2)}(kz)$, in the complex ν-plane.

$$\mathrm{T} \approx -\frac{i}{\pi\nu}\left[\left(\frac{r_<}{r_>}\right)^\nu + A\left(\frac{a^2}{\rho r}\right)^\nu\right], \tag{2.267}$$

where A is a constant depending on the impedance Z. Since both $r_< / r_>$ and $a^2 / r\rho$ are less than unity and $R_e\nu > 0$ in region II, the term T tends to zero as $|\nu| \to \infty$. Moreover, by noting that

$$\frac{\cos\nu(\theta-\pi)}{\sin\nu\pi} = -i\frac{e^{i\nu\theta} + e^{i\nu(2\pi-\theta)}}{1-e^{2\pi i\nu}}, \tag{2.268}$$

we see that when $\theta \neq 0$, since $I_m\nu > 0$, the term (2.268) tends exponentially to zero as $|\nu| \to \infty$. Hence there is no contribution to the integral of the contour C_2 unless $\rho = r$, $\theta = 0$. The contribution C_1 can be analyzed by performing the change of variable $\nu' = -\nu$, which transforms C_1 into C_3. By using the asymptotic formulas (2.265), it can be readily shown that on C_3 the term T of (2.266) behaves again like (2.267) with ν replaced by $-\nu$. Hence on the part of the contour C_3 where $R_e\nu' > 0$, the term T tends to zero as $|\nu| \to \infty$.

We now examine the contribution from the segment C_3' located between the negative imaginary ν-axis and the Stokes line h_{-1}. In a first step, we analyze the behavior of the Hankel function $H_\nu^{(1)}(z)$ in the neighborhood of the Stokes line h_1. To assess the magnitude of the factor $(2\nu / e\,z)^{-\nu}$, we let

$$\nu = \zeta e^{i\psi}, \qquad \zeta = |\nu| \to \infty \tag{2.269}$$

to obtain for z real positive

$$\left(\frac{2\nu}{e\,z}\right)^{-\nu} = \exp\left\{-\nu\left[Log\left(\frac{2\zeta}{e\,z}\right)+i\psi\right]\right\}$$

$$= \exp\left\{-\zeta\left(\left[\cos\psi\,Log\left(\frac{2\zeta}{e\,z}\right)-\psi\sin\psi\right]+i\left[\sin\psi\,Log\left(\frac{2\zeta}{e\,z}\right)+\psi\cos\psi\right]\right)\right\} = e^{-\zeta s(\zeta,\psi)}.$$

The Stokes line h_1 corresponds to $R_e S = 0$, which gives

$$\cos\psi\,\text{Log}\frac{2\zeta}{e\,z} - \psi\sin\psi = 0. \tag{2.270}$$

As $\zeta \to \infty$, equation (2.270) can only be satisfied if $\cos\psi \to 0$ or $\psi \to \frac{\pi}{2}$. If $\psi = \frac{\pi}{2} - \varepsilon$, $0 < \varepsilon << 1$, then for ζ large, (2.270) is satisfied when

$$\varepsilon = \frac{\pi}{2}\left(\text{Log}\frac{2\zeta}{e\,z}\right)^{-1}. \tag{2.271}$$

Equation $\psi = \frac{\pi}{2} - \varepsilon$, where ε is given by (2.271), defines the Stokes line h_1 for ζ large. Similarly, the Stokes line h_{-1}, which is the symmetric of h_1 with respect to the origin, is defined by $\psi = -\frac{\pi}{2} - \varepsilon$, where ε is given by (2.271). According to (2.267), each term in (2.266) has a magnitude characterized by

$$\left|Q^{\nu}\right| = \left|e^{\nu\,\text{Log}\,Q}\right| = \left|e^{\zeta\cos\psi\,\text{Log}\,Q}\right| = e^{\zeta\,\varepsilon\left|\text{Log}\,Q\right|}, \tag{2.272}$$

where $0 < Q < 1$. Consequently, each term in the integrand of (2.263) behaves like $\exp\left[\zeta\left(\varepsilon\left|\text{Log}\,Q\right| - \theta\right)\right]$. Hence for $\theta \neq 0$ this integrand decays as $|\nu| \to \infty$, and there is no contribution to the integral from the path segment C_3', provided that $\theta \neq 0$. It results from the previous discussion that for $\theta \neq 0$, the contour deformation can be carried out and the resulting integral can be evaluated in terms of the residues at the complex zeros of $\Omega\,H_\nu^{(1)}(ka)$ in the first quadrant of the ν-plane.

2.3.1.2. Asymptotic Expansion and Physical Interpretation of the Solution

We now proceed to derive an asymptotic expansion of $p(r, \theta)$ from the expression in (2.264). To this end, we note that

$$\frac{\cos \nu_n (\theta - \pi)}{\sin \nu_n \pi} = -i \frac{e^{i\nu_n \theta} + e^{i\nu_n(2\pi - \theta)}}{1 - e^{2\pi i \nu_n}} \tag{2.273}$$

and that for large values of $k\rho$ and kr, the Hankel functions $H^{(1)}_{\nu_n}(k\rho)$ and $H^{(1)}_{\nu_n}(kr)$ can be replaced by their Debeye asymptotic expansions:

$$H^{(1)}_{\nu}(x) = \sqrt{\frac{2}{\pi x \sin\gamma}}\, e^{ix(\sin\gamma - \gamma\cos\gamma)} e^{-i\frac{\pi}{4}}, \quad \cos\gamma = \frac{\nu}{x}, \quad |\nu - x| > 0\left(|\nu|^{1/3}\right). \tag{2.274}$$

Inserting (2.273) and (2.274) into (2.264), we obtain

$$p(r,\theta) = -i \frac{e^{ik\left[\left(r^2 - a^2\right)^{1/2} + \left(\rho^2 - a^2\right)^{1/2}\right]}}{2k\left(r^2 - a^2\right)^{1/4}\left(\rho^2 - a^2\right)^{1/4}}$$
$$\sum_{n=1}^{\infty} \frac{e^{i\nu_n\theta} + e^{i\nu_n(2\pi-\theta)}}{1 - e^{2\pi i\nu_n}}\, e^{-i\nu_n\left[\operatorname{Arc}\cos\left(\frac{\nu_n}{k\rho}\right) + \operatorname{Arc}\cos\left(\frac{\nu_n}{kr}\right)\right]} \frac{\Omega H^{(2)}_{\nu}(ka)}{\left[\frac{\partial}{\partial\nu}\Omega H^{(1)}_{\nu}(ka)\right]_{\nu=\nu_n}}. \tag{2.275}$$

Since $I_m \nu_n > 0$, we observe that the terms

$$e^{i\nu_n\left[\theta - \operatorname{Arc}\cos\left(\frac{\nu_n}{k\rho}\right) - \operatorname{Arc}\cos\left(\frac{\nu_n}{kr}\right)\right]}, \quad e^{i\nu_n\left[2\pi - \theta - \operatorname{Arc}\cos\left(\frac{\nu_n}{k\rho}\right) - \operatorname{Arc}\cos\left(\frac{\nu_n}{kr}\right)\right]}$$

are evanescent only if θ falls in the range

$$\operatorname{Arc}\cos\left(\frac{\nu_n}{k\rho}\right) + \operatorname{Arc}\cos\left(\frac{\nu_n}{kr}\right) < \theta < 2\pi - \operatorname{Arc}\cos\left(\frac{\nu_n}{k\rho}\right) - \operatorname{Arc}\cos\left(\frac{\nu_n}{kr}\right). \tag{2.276}$$

We will see later that for $ka >> 1$, we have $\nu_n \approx ka$. Hence the condition (2.276) reads

$$\phi_1 + \phi_2 < \theta < 2\pi - \left(\phi_1 + \phi_2\right), \tag{2.277}$$

where the angles ϕ_1 and ϕ_2 are defined by

$$\phi_1 = \operatorname{Arc}\cos\left(\frac{a}{\rho}\right), \quad \phi_2 = \operatorname{Arc}\cos\left(\frac{a}{r}\right) \tag{2.278}$$

and are represented in the figure 2.31.

We will now present a physical interpretation of the representation (2.275) for $p(r, \theta)$. Referring to figure 2.31 and to equation (2.275), we see that for each value of ν_n, the field is the sum of two terms. Looking at the phase of each term, we observe that the first term corresponds to the ray SP_1P_2Q, composed of a space ray from the source S to P_1, a creeping ray from P_1 to P_2, and again a space ray from P_2 to the observation point Q. The path length of the creeping ray from P_1 to P_2 is equal to $\ell = a\left[\theta - (\phi_1 + \phi_2)\right]$. The first inequality of (2.277) indicates that ℓ must be positive. The second term of (2.275) corresponds to the ray $SP_1'P_2'Q$. It is also composed of three parts: a space ray SP_1', a creeping ray $P_1'P_2'$, and a space ray $P_2'Q$. The path length of the creeping ray from P_1' to P_2' is equal to $\ell' = a\left[2\pi - \theta - (\phi_1 + \phi_2)\right]$, and according to the second inequality (2.277), ℓ' must be positive. It is easy to observe in figure 2.31 that when Q crosses the shadow boundary SB_1, ℓ' remains positive whereas ℓ becomes negative. Likewise, when Q crosses the shadow boundary SB_1', ℓ remains positive whereas ℓ' becomes negative. In order for both creeping wave paths to be positive, Q must be located in the shadow region of the source. Consequently, the expressions (2.264) and (2.275) are valid only in the shadow region.

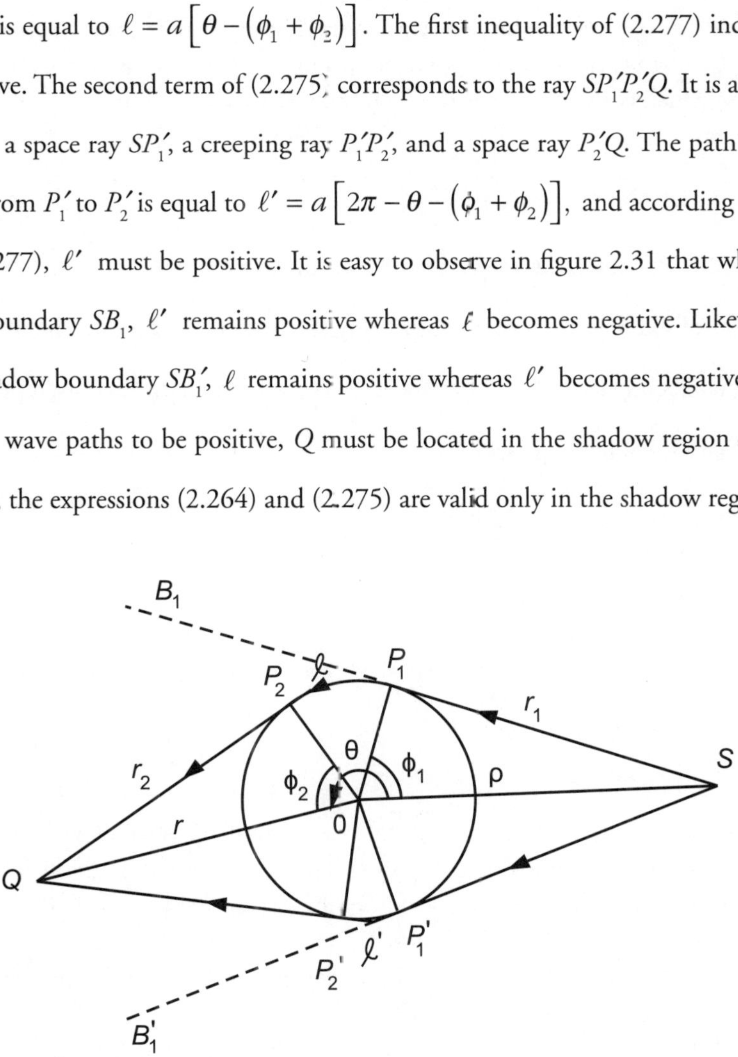

Figure 2.31. Representation of the angles ϕ_1 and ϕ_2.

However, this result does not mean that the residue series (2.264) does not converge when $\theta < \phi_1 + \phi_2$. When $|\nu_n|$ augments and becomes of the same order or large compared to $k\rho$ or kr, the Debye expansion (2.274) is no longer valid and must be replaced by expansions adapted to these situations. As a matter of fact, the series (2.264) converges, but the initial terms increase exponentially and the series is therefore not suitable for numerical evaluation. On the other hand, when $\theta > \phi_1 + \phi_2$, the initial terms of the series converge rapidly.

The reason the residue series (2.264) is not valid in a large domain of the observation angle θ can be explained by a simple physical argument. As soon as the observer is in a position where he can capture the reflected field, then, in addition to the diffracted field, he can also capture the reflected field. Hence if we separate the direct and the reflected fields from the integral (2.263) when these fields exist, then the initial terms of the residue series of the resultant integral will converge rapidly. The mathematical trick for performing this separation consists of using the following identity, which is also known as the Imai transformation [2.19]:

$$\cos\nu(\theta-\pi) = e^{i\nu\pi}\cos\nu\theta - ie^{i\nu\theta}\sin\pi\nu. \tag{2.279}$$

Inserting (2.279) into (2.263), we obtain two integrals $p_1(r,\theta)$ and $p_2(r,\theta)$, corresponding to the first and second terms of (2.279). Applying the residue theorem to the first integral and replacing the Hankel functions $H^{(1)}_{\nu_n}(kr)$ and $H^{(1)}_{\nu_n}(k\rho)$ with their Debye asymptotic expansion (2.274), we obtain the following expression:

$$p_1(r,\theta) = -i\frac{e^{ik\left[\left(r^2-a^2\right)^{1/2}+\left(\rho^2-a^2\right)^{1/2}\right]}}{2k\left(r^2-a^2\right)^{1/4}\left(\rho^2-a^2\right)^{1/4}}$$

$$\sum_{n=1}^{\infty}\frac{e^{i\nu_n(2\pi+\theta)}+e^{i\nu_n(2\pi-\theta)}}{1-e^{2\pi i\nu_n}}\,e^{-i\nu_n\left[\mathrm{Arc}\cos\left(\frac{\nu_n}{k\rho}\right)+\mathrm{Arc}\cos\left(\frac{\nu_n}{kr}\right)\right]}\frac{\Omega H^{(2)}_{\nu}(ka)}{\left[\frac{\partial}{\partial\nu}\Omega H^{(1)}_{\nu}(ka)\right]_{\nu=\nu_n}}. \tag{2.280}$$

We observe now that, contrary to (2.275), the second member of (2.280) is evanescent when n augments for all values of $0 \le \theta \le 2\pi$. If we consider the physical interpretation of (2.280), then we see that when the observation point Q is in the lit region, the path lengths of the two creeping rays reaching Q are both positive and given by $\ell / a = 2\pi + \theta - (\phi_1 + \phi_2)$ and $\ell' / a = 2\pi - \theta - (\phi_1 + \phi_2)$. More precisely, for each value of ν_n, the first term represents a creeping ray that is excited at P_1 by the space ray SP_1. This ray reaches the point P_2 on the cylinder after having encircled the cylinder along the arc $P_1 P_2$ of length ℓ and launches a space ray P_2Q to the observation point Q (fig. 2.32). Likewise, the second term represents a creeping ray that is excited

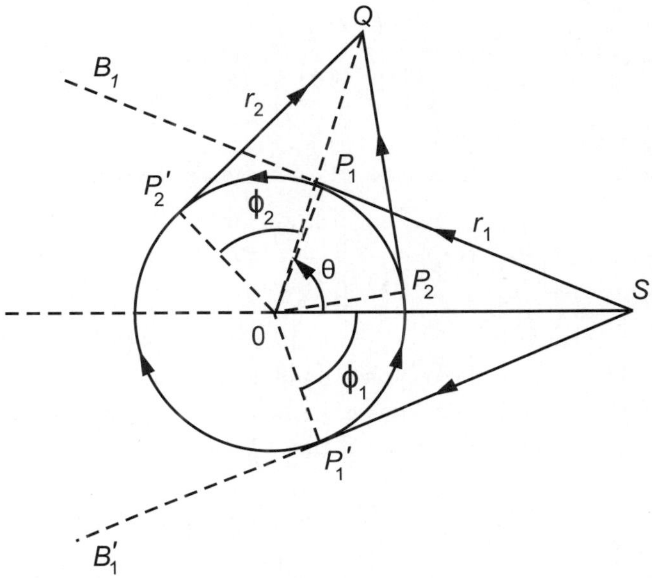

Figure 2.32. Representation of the dominant creeping rays.

at P_1' by the space ray SP', reaches the point P_2' on the cylinder by following the arc $P_1'P_2'$ of length ℓ', and launches a space ray $P_2'Q$ to the observation point Q.

When the observation point Q is in the shadow region above the line SO, the shortest path lengths of the two creeping rays reaching Q are $\ell / a = \theta - (\phi_1 + \phi_2)$ and $\ell' / a = 2\pi - \theta - (\phi_1 + \phi_2)$, respectively. Hence the formula (2.280) contains only the second creeping ray traveling a path length ℓ'. The first creping ray is absent and replaced by the creeping ray, having encircled the cylinder one time before reaching the observation point.

Moreover, since $I_m \nu > 0$,

$$\left(1 - e^{2\pi i \nu_n}\right)^{-1} = \sum_{m=0}^{\infty} e^{2\pi i m \nu_n}. \tag{2.281}$$

By inserting (2.281) into (2.280), we see that for each value of ν_n the result can be interpreted as a sum of two series of creeping rays. The first series representing creeping rays excited at P_1 and reaching Q after having encircled the cylinder m times, whereas the second series represents creeping rays excited at P_1' and reaching Q after having encircled the cylinder m times in the opposite direction. The term $m = 0$ in (2.281) gives the two dominant rays (described previously) when the observation point lies outside the shadow region. When the observation point Q is located outside the shadow region, m takes the values $m = 0, 1, 2, \dots, \infty$. But when Q is in the shadow region, the term $m = 0$ is absent for the shortest path length from S to Q. We will see that this term is contained in the second integral obtained by inserting (2.279) into (2.263), given by

$$p_2(r,\theta)=\frac{i}{4}\int_D e^{i\nu\theta}\left[J_\nu(kr_<)-\frac{\Omega J_\nu(ka)}{\Omega H_\nu^{(1)}(ka)}H_\nu^{(1)}(kr_<)\right]H_\nu^{(1)}(kr_>)\,d\nu, \tag{2.282}$$

where $r_< = \mathrm{Inf}(r,\rho), r_> = \mathrm{Sup}(r,\rho)$.

By using the relation (2.262), equation (2.282) can also be written in the following form:

$$p_2(r,\theta)=\frac{i}{8}\int_D e^{i\nu\theta}\left[H_\nu^{(2)}(kr_<)-\frac{\Omega H_\nu^{(2)}(ka)}{\Omega H_\nu^{(1)}(ka)}H_\nu^{(1)}(kr_<)\right]H_\nu^{(1)}(kr_>)\,d\nu. \tag{2.283}$$

Because the integrand of (2.283) without the term $\exp(i\nu\theta)$ is an even function of ν, the replacement of ν by $-\nu$ yields the same integrand, except for the occurrence of $-\theta$. Hence the replacement of θ in (2.283) by the absolute value $|\theta|$ is of no consequence, and by deforming the contour D as shown in figure 2.29, we obtain

$$p_2(r,\theta)=\frac{i}{8}\int_{C_0} e^{i\nu|\theta|}\left[H_\nu^{(2)}(kr_<)-\frac{\Omega H_\nu^{(2)}(ka)}{\Omega H_\nu^{(1)}(ka)}H_\nu^{(1)}(kr_<)\right]H_\nu^{(1)}(kr_>)\,d\nu. \tag{2.284}$$

Since the integrand in (2.284) has no poles on the real ν-axis, it is possible to deform the contour C_O of figure 2.29 into a contour C_S crossing the real ν-axis as shown in figure 2.33 without changing the integral. We will now show that each term of the integrand in (2.284) has one or two saddle points on the real axis located on one or both sides of ka and that by deforming the contour C_S into the SDP passing through these saddle points and by evaluating the integral by the method of steepest descent, we obtain the incident and reflected field.

Replacing the Hankel functions $H_\nu^{(1)}(x)$ and $H_\nu^{(2)}(x)$, where x equals $kr_>$ or $kr_<$, with their Debye asymptotic expansions, valid for $|\nu - x| > \left(O|\nu|^{1/3}\right)$ where ν is in region I or IV (see fig. 2.30) given by

$$\begin{aligned} H_\nu^{(1)}(x) &= \sqrt{\frac{2}{\pi x \sin\gamma}}\, e^{-i\frac{\pi}{4}}\, e^{ix(\sin\gamma-\gamma\cos\gamma)}, \\ H_\nu^{(2)}(x) &= \sqrt{\frac{2}{\pi x \sin\gamma}}\, e^{i\frac{\pi}{4}}\, e^{-ix(\sin\gamma-\gamma\cos\gamma)}, \end{aligned} \tag{2.285}$$

where $\cos\gamma = \dfrac{\nu}{x}$ and where γ is restricted to $O < R_e\gamma < \pi$, we obtain for the integral I_1 corresponding to the first term of the integrand of (2.284)

$$I_1 = \frac{i}{8}\int_{C_s} e^{i\nu|\theta|} H_\nu^{(2)}(kr_<)H_\nu^{(1)}(kr_>)\,d\nu \approx \frac{i}{4\pi}\int_{C_s} F(\nu)e^{iq(\nu)}\,d\nu, \tag{2.286}$$

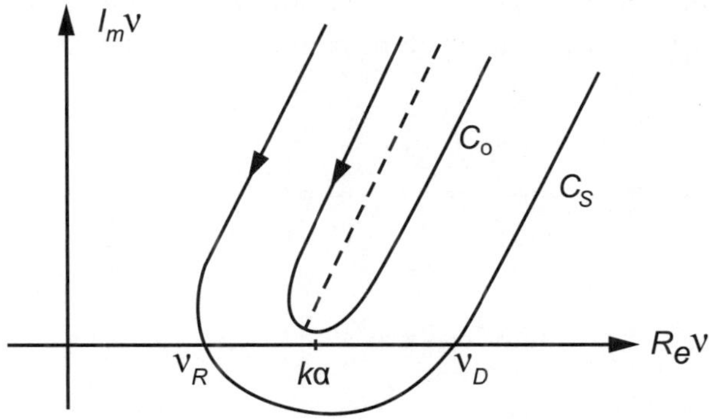

Figure 2.33. Deformation of the contour C_O into a contour C_S passing through the saddle points ν_R and ν_D.

where

$$F(\nu) = \left(k^2\rho^2 - \nu^2\right)^{-1/4}\left(k^2 r^2 - \nu^2\right)^{-1/4}, \tag{2.287}$$

$$q(\nu) = \nu\left(|\theta| + \operatorname{Arc}\cos\frac{\nu}{kr_<} - \operatorname{Arc}\cos\frac{\nu}{kr_>}\right) + \sqrt{\left(kr_>\right)^2 - \nu^2} - \sqrt{\left(kr_<\right)^2 - \nu^2}. \tag{2.288}$$

We observe that I_1 has an oscillating integrand with a stationary phase point ν_s, given by the solution of the equation

$$q'(\nu) = |\theta| + \operatorname{Arc}\cos\frac{\nu}{kr_<} - \operatorname{Arc}\cos\frac{\nu}{kr_>} = 0. \tag{2.289}$$

By denoting

$$\alpha_1 = \operatorname{Arc}\cos\frac{\nu}{kr_<}, \quad \alpha_2 = \operatorname{Arc}\cos\frac{\nu}{kr_>},$$

the saddle point condition (2.289) may also be written in the form

$$\alpha_1 - \alpha_2 + |\theta| = 0,$$

and the solution ν_s verifies $\nu_s = kr_< \cos\alpha_{1s} = kr_> \cos\alpha_{2s}$, where α_{1s} and α_{2s} denote the values of α_1 and α_2 corresponding to ν_s. When ν_s is real, the saddle point condition admits the graphical interpretation shown in figure 2.34. One observes that a real solution is possible only when $SH > SQ$, where H is the projection of the origin O in the direction of observation.

Moreover, since

$$q''(\nu_s) = -\frac{1}{\sqrt{(kr_<)^2 - \nu_s^2}} + \frac{1}{\sqrt{(kr_>)^2 - \nu_s^2}} < 0, \tag{2.290}$$

the SDP makes an angle $-\frac{\pi}{4}$ with the real ν-axis.

Near the stationary point ν_s, we deform the contour C_s into the SDP and evaluate the integral I_1 by the method of steepest descent.

According to the general formula (2.59), giving the leading term of the asymptotic expansion of an integral derived in chapter 2 by the steepest descent method, we obtain the following:

$$\begin{aligned} I_1 &\approx \frac{i}{4\pi} F(\nu_s)\, e^{iq(\nu_s)} \sqrt{\frac{2\pi}{|q''(\nu_s)|}}\; e^{i\frac{\pi}{4}} \\ &= \frac{e^{i\frac{\pi}{4}}}{\sqrt{8\pi k}} \frac{\exp\left[i\sqrt{r_>^2 - \left(\frac{\nu_s}{k}\right)^2} - \sqrt{r_<^2 - \left(\frac{\nu_s}{k}\right)^2}\right]}{\left(\sqrt{r_>^2 - \left(\frac{\nu_s}{k}\right)^2} - \sqrt{r_<^2 - \left(\frac{\nu_s}{k}\right)^2}\right)^{1/2}}. \end{aligned} \tag{2.291}$$

According to (2.289), we have

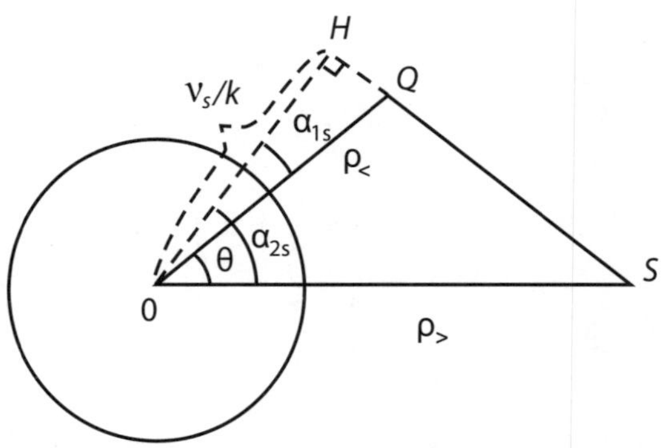

Figure 2.34. Graphical interpretation of the condition represented by equation (2.289).

$$\cos|\theta| = \frac{\nu_s^2}{k^2 r\rho} + \sqrt{1-\left(\frac{\nu_s}{k\rho}\right)^2}\sqrt{1-\left(\frac{\nu_s}{kr}\right)^2},$$

and since

$$\begin{aligned}|\overrightarrow{SQ}|^2 &= |\overrightarrow{SO}|^2 + |\overrightarrow{OQ}|^2 - 2|\overrightarrow{SO}||\overrightarrow{OQ}|\cos\theta \\ &= \rho^2 + r^2 - 2r\rho\cos\theta,\end{aligned}$$

we obtain

$$\begin{aligned}|\overrightarrow{SQ}|^2 &= \rho^2 - \left(\frac{\nu_s}{k}\right)^2 + r^2 - \left(\frac{\nu_s}{k}\right)^2 - 2\sqrt{\rho^2 - \left(\frac{\nu_s}{k}\right)^2}\sqrt{r^2 - \left(\frac{\nu_s}{k}\right)^2} \\ &= \left(\sqrt{r_>^2 - \left(\frac{\nu_s}{k}\right)^2} - \sqrt{r_<^2 - \left(\frac{\nu_s}{k}\right)^2}\right)^2.\end{aligned}$$

This expression can also be obtained directly from figure 2.34. Thus we obtain

$$I_1 = \frac{e^{i\frac{\pi}{4}}}{\sqrt{8\pi k}}\frac{e^{iSQ}}{\sqrt{SQ}}, \tag{2.292}$$

which is the field of a line source at a distance SQ or the incident or direct field at the observation point Q.

Similarly, the integral I_2 corresponding to the second term of the integrand (2.284) verifies

$$\begin{aligned}I_2 &= -\frac{i}{8}\int_{C_s} e^{i\nu|\theta|}\frac{\Omega H_\nu^{(2)}(ka)}{\Omega H_\nu^{(1)}(ka)} H_\nu^{(1)}(k\rho)H_\nu^{(1)}(kr)\,d\nu \\ &\approx -\frac{i}{4\pi}\int_{C_s} e^{i\nu|\theta|}\frac{\Omega H_\nu^{(2)}(ka)}{\Omega H_\nu^{(1)}(ka)} F(\nu)e^{iq_1(\nu)}\,d\nu,\end{aligned} \tag{2.293}$$

where $F(\nu)$ is given by (2.287) and where

$$q_1(\nu) = \sqrt{(k\rho)^2 - \nu^2} + \sqrt{(kr)^2 - \nu^2} - \nu\left(\operatorname{Arc\,cos}\frac{\nu}{k\rho} + \operatorname{Arc\,cos}\frac{\nu}{kr}\right) - \frac{\pi}{2}. \tag{2.294}$$

Now the asymptotic expansion of $H_\nu^{(1,2)}(ka)$ is dependent on whether $R_e\nu < ka$ or $R_e\nu > ka$. For $R_e\nu > ka$, the Debye asymptotic expansion of $H_\nu^{(1)}(ka)$ corresponds to its expression valid in region II of figure 2.30(a) and is given by

$$H_\nu^{(1)}(ka) \approx -\sqrt{\frac{2}{\pi\, ka \sin\gamma}}\; e^{i\frac{\pi}{4}}\, e^{-i\,ka\,(\sin\gamma - \gamma\cos\gamma)}, \tag{2.295}$$

where $\cos\gamma = \dfrac{\nu}{ka}$. Similarly, for $R_e\nu > ka$, the Debye asymptotic expansion of $H_\nu^{(2)}(ka)$ is given by its expression valid in region IV of figure 2.30(b), which reads

$$H_\nu^{(2)}(ka) \approx \sqrt{\frac{2}{\pi\, ka \sin\gamma}}\; e^{i\frac{\pi}{4}}\, e^{-i\,ka\,(\sin\gamma - \gamma\cos\gamma)}. \tag{2.296}$$

Hence for $R_e\nu > ka$ we have

$$H_\nu^{(1)}(ka) \approx -H_\nu^{(2)}(ka), \tag{2.297}$$

and the resulting integrand resembles that of equation (2.286), but now $q(\nu)$ is given by

$$q(\nu) = \sqrt{(k\rho)^2 - \nu^2} + \sqrt{(kr)^2 - \nu^2} + \nu\left(|\theta| - \operatorname{Arc}\cos\frac{\nu}{k\rho} - \operatorname{Arc}\cos\frac{\nu}{kr}\right) - \frac{\pi}{2}.$$

The stationary phase points are the solutions of the following equation:

$$q'(\nu) = |\theta| - \left(\operatorname{Arc}\cos\frac{\nu}{k\rho} + \operatorname{Arc}\cos\frac{\nu}{kr}\right) = 0. \tag{2.298}$$

When ν_s is real, the saddle point condition (2.298) admits a graphical interpretation similar to the one shown in figure 2.34, but now we must have $SQ > SH$. It is also noted that the restriction $R_e\nu_s / k > a$ defines the rays that connect S to Q without touching the cylinder. Moreover, since

$$q''(\nu_s) = \frac{1}{\sqrt{(k\rho)^2 - \nu_s^2}} + \frac{1}{\sqrt{(kr)^2 - \nu_s^2}} > 0,$$

the SDP makes an angle $\frac{\pi}{4}$ with the real ν-axis.

Near the stationary point ν_s, we deform the contour C_s into the SDP and apply the saddle point method. According to the general formula (2.59), giving the leading term of the asymptotic expansion of the integral (2.293) for $\nu > ka$, we obtain

$$I_2 = \frac{e^{i\frac{\pi}{4}}}{\sqrt{8\pi k}} \frac{\exp\left[i\left[\sqrt{(kr)^2 - \left(\frac{\nu_s}{k}\right)^2} + \sqrt{(k\rho)^2 - \left(\frac{\nu_s}{k}\right)^2}\right]\right]}{\left[\sqrt{(kr)^2 - \left(\frac{\nu_s}{k}\right)^2} + \sqrt{(k\rho)^2 - \left(\frac{\nu_s}{k}\right)^2}\right]^{1/2}}. \tag{2.299}$$

According to (2.298), we have

$$\cos(|\theta|) = \frac{\nu_s^2}{k^2 r\rho} - \sqrt{1 - \left(\frac{\nu_s}{k\rho}\right)^2}\sqrt{1 - \left(\frac{\nu_s}{kr}\right)^2}$$

and

$$\left|\overrightarrow{SQ}\right|^2 = r^2 + \rho^2 - 2r\rho\cos\theta = \left(\sqrt{r^2 - \left(\frac{\nu_s}{k}\right)^2} + \sqrt{\rho^2 - \left(\frac{\nu_s}{k}\right)^2}\right). \tag{2.300}$$

Hence for $R_e\nu > ka$,

$$I_2 = \frac{e^{i\frac{\pi}{4}}}{\sqrt{8\pi k}} \frac{e^{iSQ}}{\sqrt{SQ}}, \tag{2.301}$$

which corresponds to the incident or direct field restricted to the illuminated region.

For $R_e\nu < ka$, by writing (2.285) in the form

$$H_\nu^{(1)}(x) \approx h(x,\nu)\, e^{i\psi(x,\nu)},$$
$$H_\nu^{(2)}(x) \approx h(x,\nu)\, e^{-i\psi(x,\nu)},$$

where

$$h(x,\nu) = \sqrt{\frac{2}{\pi}}\left(x^2 - \nu^2\right)^{-1/4},$$
$$\psi(x,\nu) = \sqrt{\left(x^2 - \nu^2\right)} - \nu\, Arc\cos\frac{\nu}{x} - \frac{\pi}{4}, \tag{2.302}$$

we obtain

$$H_\nu^{\prime(1)}(x) \approx i\, h(x,\nu)\, \psi'(x,\nu)\, e^{i\psi(x,\nu)}$$
$$H_\nu^{\prime(2)}(x) \approx -i\, h(x,\nu)\, \psi'(x,\nu)\, e^{-i\psi(x,\nu)}$$

for x and $|\nu|$ large and $|\nu - x| > O\left(|\nu|^{1/3}\right)$. Hence, recalling that $\Omega = k\left(\frac{\partial}{\partial x} + iZ\right)$, we have

$$\frac{\Omega H_\nu^{(2)}(x)}{\Omega H_\nu^{(1)}(x)} \approx -\frac{\psi'(x,\nu) - Z}{\psi'(x,\nu) + Z} e^{-2i\psi(x,\nu)}. \tag{2.303}$$

Inserting (2.303), with $x = ka$, into (2.293) yields

$$I_2 \approx \frac{i}{4\pi} \int_{C_s} \frac{\psi'(x,\nu) - Z}{\psi'(x,\nu) + Z} F(\nu)\, e^{i\bar{q}(\nu)} d\nu, \tag{2.304}$$

where

$$\begin{aligned}\bar{q}(\nu) &= q(\nu) - 2\psi(ka,\nu) \\ &= \sqrt{(k\rho)^2 - \nu^2} + \sqrt{(kr)^2 - \nu^2} - 2\sqrt{(ka)^2 - \nu^2} \\ &\quad + \nu\left[\theta - \left(\text{Arc}\cos\frac{\nu}{k\rho} + \text{Arc}\cos\frac{\nu}{kr} - 2\text{Arc}\cos\frac{\nu}{ka}\right)\right] + \frac{\pi}{2}.\end{aligned} \tag{2.305}$$

The integrand of (2.304) has a stationary point verifying the following:

$$\bar{q}'(\nu) = \theta - \left(\text{Arc}\cos\frac{\nu}{k\rho} + \text{Arc}\cos\frac{\nu}{kr} - 2\text{Arc}\cos\frac{\nu}{ka}\right) = 0. \tag{2.306}$$

If we let $\nu = ka \sin \theta_R$, then (2.306) transforms in the following equation:

$$\theta - \left[\text{Arc}\cos\left(\frac{a}{\rho}\sin\theta_R\right) + \text{Arc}\cos\left(\frac{a}{r}\sin\theta_R\right)\right] + \pi - 2\theta_R = 0. \tag{2.307}$$

Since the term in brackets is close to π for $\frac{a}{\rho} \ll 1$ and $\frac{a}{r} \ll 1$, we see that when the observation point lies outside the shadow region, equation (2.307) has a unique solution given by $\theta/2 + \varepsilon$, where $0 < \varepsilon \ll 1$. When the observation point approaches the shadow boundary, we have $\theta_R \to \frac{\pi}{2}$. In this instance, the direct and reflected waves merge. It corresponds to a coalescence of two saddle points, therefore requiring a more sophisticated asymptotic evaluation of the integral. In the transition region, the contribution of the integral I_2 is of lower order and needs a different asymptotic expansion of the Hankel functions. In section 3.2, we will show that by using a uniform expansion of the Hankel functions for $R_e\nu \le ka$, the result in the deep shadow will be given by the series of creeping wave modes corresponding to the shortest path length between the source and the observation point in the shadow region. These creeping wave modes were absent in the asymptotic evaluation of the integral I_1. We now come back to the evaluation of I_2 when the observation point lies outside the shadow region and the transition region.

Since

$$q''(\nu) = \frac{1}{\sqrt{(k\rho)^2 - \nu^2}} + \frac{1}{\sqrt{(kr)^2 - \nu^2}} - \frac{2}{\sqrt{(ka)^2 - \nu^2}} < 0, \tag{2.308}$$

the tangent to the SDP at the saddle point $\nu_R = ka \sin \theta_R$ makes an angle $-\frac{\pi}{4}$ with the real ν-axis.

By applying the steepest descent method to the integral (2.304), we obtain

$$I_2 = \frac{i}{4\pi}\frac{\psi'(ka,\nu_R) - Z}{\psi'(ka,\nu_R) + Z} F(\nu_R)\, e^{i\bar{q}(\nu_R)} \sqrt{\frac{2\pi}{|q''(\nu_R)|}}\, e^{-\pi/4}, \tag{2.309}$$

where $\nu_R = ka \sin \theta_R$.

According to (2.302), we obtain

$$\psi'(ka,\nu_R) = \frac{x}{\sqrt{x^2 - \nu_R^2}} - \frac{\nu_R^2}{x^2\sqrt{1 - \frac{\nu_R^2}{x^2}}} = \sqrt{1 - \left(\frac{\nu_R}{x}\right)^2} = \cos\theta_R. \tag{2.310}$$

Hence the coefficient

$$R = \frac{\psi'(ka,\nu_R) - Z}{\psi'(ka,\nu_R) + Z} = \frac{\cos\theta_R - Z}{\cos\theta_R + Z} \tag{2.311}$$

is the reflection coefficient of the surface of a circular cylinder, and θ_R is the angle of reflection.

The phase $\bar{q}(\nu_R)$ is given by (2.305), which according to (2.306) reads

$$\bar{q}(\nu_R) = \sqrt{(k\rho)^2 - \nu_R^2} + \sqrt{(kr)^2 - \nu_R^2} - 2\sqrt{(ka)^2 - \nu_R^2}.$$

Referring to figure 2.35, we see that

$$\rho^2 = a^2 + (SM)^2 + 2a(SM)\cos\theta_R,$$

$$r^2 = a^2 + (QM)^2 + 2a(QM)\cos\theta_R,$$

where M is the point of reflection.

Hence

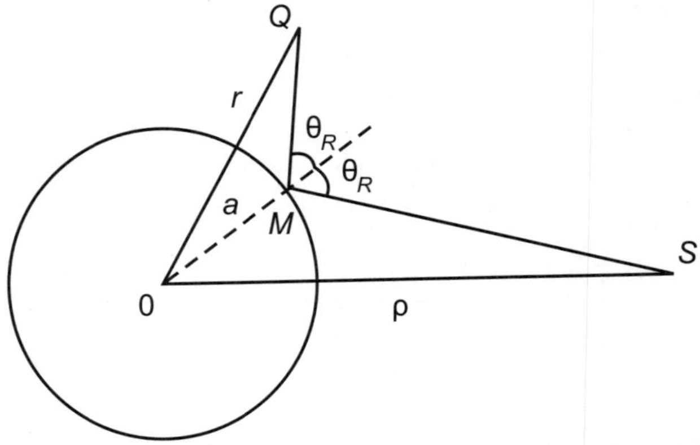

Figure 2.35. Reflected ray and reflection angle θ_R.

$$\left(k\rho\right)^2 - \nu_R^2 = k^2\left(a^2\cos^2\theta_R + \left(SM\right)^2 + 2a\left(SM\right)\cos\theta_R\right) = k^2\left(SM + a\cos\theta_R\right)^2. \quad (2.312)$$

Similarly

$$\left(kr\right)^2 - \nu_R^2 = k^2\left(QM + a\cos\theta_R\right)^2 \quad (2.313)$$

and consequently

$$\bar{q}\left(\nu_R\right) = k\left(SM + MQ\right), \quad (2.314)$$

which corresponds to the phase of the reflected wave at the observation point Q.

We still must express the factor:

$$\mathrm{T} = \frac{F\left(\nu_R\right)}{\sqrt{q''\left(\nu_R\right)}} = \left(\frac{2}{\sqrt{\left(ka\right)^2 - \nu_R^2}} - \frac{1}{\sqrt{\left(k\rho\right)^2 - \nu_R^2}} - \frac{1}{\sqrt{\left(kr\right)^2 - \nu_R^2}}\right)^{-1/2}\left[\left(k\rho\right)^2 - \nu_R^2\right]^{-1/4}\left[\left(kr\right)^2 - \nu_R^2\right]^{-1/4}.$$

According to (2.312) and (2.313), we obtain immediately

$$\mathrm{T} = \frac{1}{\sqrt{k}}\left(\frac{a\cos\theta_R}{2SM\,QM + a\,SM\cos\theta_R + a\,QM\cos\theta_R}\right)^{1/2}.$$

By separating the divergence factor of the incident field at the reflection point M given by $(SM)^{-1/2}$, the factor T can also be written in the form

$$\begin{aligned} \mathrm{T} &= \frac{1}{\sqrt{k}}\frac{1}{\sqrt{SM}}\left(\frac{a\,SM\cos\theta_R}{(2SM+a\cos\theta_R)QM+a\,SM\cos\theta_R}\right)^{1/2} \\ &= \frac{1}{\sqrt{k}}\frac{1}{\sqrt{SM}}\left(\left(\frac{a\,SM\cos\theta_R}{2SM+a\cos\theta_R}\right)\left(\frac{1}{QM+\dfrac{a\,SM\cos\theta_R}{2SM+a\cos\theta_R}}\right)\right)^{1/2}. \end{aligned} \tag{2.315}$$

Let

$$\rho_c = \frac{SM\,a\cos\theta_R}{2SM+a\cos\theta_R}. \tag{2.316}$$

We know from appendix D that ρ_c is the radius of curvature of the reflected wave front. Hence

$$\mathrm{T} = \frac{1}{\sqrt{k}}\frac{1}{\sqrt{SM}}A, \quad A = \sqrt{\frac{\rho_c}{(\rho_c+QM)}}, \tag{2.317}$$

where A is the divergence factor of the reflected wave.

Inserting (2.311), (2.314), and (2.317) into (2.309), we obtain

$$I_2 = \frac{e^{i\frac{\pi}{4}}}{\sqrt{8\pi k}}\frac{e^{ik\,SM}}{\sqrt{SM}}\frac{\cos\theta_R - Z}{\cos\theta_R + Z}\sqrt{\frac{\rho_c}{\rho_c+MQ}}\,e^{ik\,MQ} = p_i(M)R\sqrt{\frac{\rho_c}{\rho_c+MQ}}\,e^{ik\,MQ}, \tag{2.318}$$

where $p_i(M)$ is the incident field at the reflection point M. The integral I_2 therefore gives the reflected field $p_R(Q)$ at the observation point $Q(r, \theta)$, and the integral $p_2(r, \theta)$ given by (2.283) is equal to the sum of the incident field and the reflected field at Q:

$$p_2(r,\theta) = p_i(r,\theta) + p_R(r,\theta). \tag{2.319}$$

The formulas (2.301) and (2.318) are not valid when $|\theta|$ tends to π.

Assuming $\rho >> a$, $r >> a$, it follows from (2.307) that $\theta_R \cong \frac{|\theta|}{2}$. Hence for $\theta = \pi$ we have $\nu_R \cong ka$. But in this case, the Debye asymptotic expansion of the Hankel function that has been used to derive (2.307) is no longer valid. A similar conclusion can be deduced from (2.298). When ν_R is close to ka, the Debye asymptotic expansion of the Hankel functions $H_\nu^{(1,2)}(ka)$ can be approximated by Airy functions that provide the means of passing smoothly from the region $|\nu - ka| > O\left((ka)^{1/3}\right)$, wherein the Debye formulas are valid, to the region $|\nu - ka| \leq O\left((ka)^{1/3}\right)$. Expansions of this type, limited to their dominant order, are given by (2.328). The transition from one region to another occurs at the shadow boundary. That boundary is therefore a transition region in which one passes from the GA formulation outside the shadow region to the creeping wave formulation in the deep shadow region. Uniform solutions through the shadow regions are derived in chapter 3, section 3.3.

2.3.1.3. Expression of the Diffraction Coefficient

According to (2.275) for Q in the shadow region and (2.280) for Q outside the shadow region, the dominant term of the asymptotic expansion of the diffracted field is given by

$$p^d(Q) = -\frac{i}{2k}\frac{e^{ik(r_1+r_2)}}{\sqrt{r_1 r_2}}\sum_{n=1}^{\infty}\left(e^{i\frac{\nu_n}{a}\ell} + e^{i\frac{\nu_n}{a}\ell'}\right)\frac{\Omega H_{\nu_n}^{(2)}(ka)}{\left[\frac{\partial}{\partial \nu}\Omega H_{\nu_n}^{(1)}(ka)\right]_{\nu=\nu_n}}, \tag{2.320}$$

where ℓ and ℓ' are the arc lengths followed by the creeping rays on the cylinder and where r_1 and r_2 are defined in figure 2.32.

Our objective is to determine the diffraction coefficient by comparing (2.320) to the general expression of the field diffracted by a smooth convex surface derived in chapter 1, section 1.9, which for a circular cylinder excited by a line source parallel to the generatrices, reads

$$p^d(Q) = D_s\frac{e^{ikr_1}}{\sqrt{r_1}}\left[\sum_{n=1}^{\infty}\left(D_{sr,cr}^{(n)}D_{cr,sr}^{(n)}\right)\left(e^{i\frac{\nu_n}{a}\ell} + e^{i\frac{\nu_n}{a}\ell'}\right)\right]D_s\frac{e^{ikr_2}}{\sqrt{r_2}}, \tag{2.321}$$

where D_s is the coefficient corresponding to the excitation of a space wave by a line source of unit strength, and $D_{sr,cr}$ is the excitation coefficient of a creeping wave by a space wave. Also, $D_{cr,sr}$ is the radiation coefficient of a creeping wave into a space wave. Since the field radiated by a line source at an observation point at the distance r from the source is given by $i/4\,H_0^{(1)}(kr)$, we obtain the following for kr large:

$$\frac{i}{4}H_0^{(1)}(kr) \approx \frac{e^{i\frac{\pi}{4}}}{\sqrt{8\pi k}}\frac{e^{ikr}}{\sqrt{r}} = D_s\frac{e^{ikr}}{\sqrt{r}}. \tag{2.322}$$

Comparing (2.321) to (2.320), we obtain

$$D^{(n)}_{sr,cr}\,D^{(n)}_{cr,sr} = -4\pi \frac{\Omega H^{(2)}_{\nu_n}(ka)}{\frac{\partial}{\partial\nu}\left[\Omega H^{(1)}_{\nu_n}(ka)\right]_{\nu=\nu_n}} = -4\pi R_n. \tag{2.323}$$

The Wronskian determinant between $H^{(1)}_{\nu_n}(ka)$ and $H^{(2)}_{\nu_n}(ka)$ is given by

$$H^{(1)}_{\nu_n}(ka)H^{(2)\prime}_{\nu_n}(ka) - H^{(1)\prime}_{\nu_n}(ka)H^{(2)}_{\nu_n}(ka) = -\frac{4i}{\pi\,ka}, \tag{2.324}$$

and since ν_n are the zeros of $\Omega H^{(1)}_{\nu_n}(ka)$, the following equation is verified:

$$H^{(1)\prime}_{\nu_n}(ka) + iZ\,H^{(1)}_{\nu_n}(ka) = 0. \tag{2.325}$$

Inserting (2.325) into (2.324) and (2.323) gives

$$H^{(2)\prime}_{\nu_n}(ka) + i\,Z\,H^{(2)}_{\nu_n}(ka) = -\frac{4\,i}{\pi\,ka\,H^{(1)}_{\nu_n}(ka)}$$

and

$$R_n = -\frac{4i}{Q_n}, \tag{2.326}$$

where

$$Q_n = \pi\,ka\left[H^{(1)}_{\nu_n}(ka)\left(\frac{\partial}{\partial\nu}H^{(1)\prime}_{\nu}(ka)\right)_{\nu=\nu_n} - H^{(1)\prime}_{\nu_n}(ka)\left(\frac{\partial}{\partial\nu}H^{(1)}_{\nu}(ka)\right)_{\nu=\nu_n}\right]. \tag{2.327}$$

We now assume that the zeros of (2.325) are located in the vicinity of the Stokes line h_1, and we replace the Hankel function $H^{(1)}_{\nu}(x)$ and its derivatives with the dominant term of their asymptotic expansions, which is valid in a neighborhood $|\nu - ka| = O\left(|\nu|^{1/3}\right)$ of this line, obtained by Olver [2.18] and given by (see also Abramowitz and Stegun [2.5])

$$H_{\nu_n}^{(1)}(ka) \approx 2^{4/3}\left(\nu_n\right)^{-1/3} e^{-i\frac{\pi}{3}}\left[A_i(y)\right]_{x=ka}$$

$$\left[\frac{\partial}{\partial x} H_{\nu_n}^{(1)}(x)\right]_{x=ka} \approx 2^{4/3}\left(\nu_n\right)^{-1/3} e^{-i\frac{\pi}{3}}\left[A_i'(y)\frac{\partial y}{\partial x}\right]_{x=ka}$$

$$\left[\frac{\partial}{\partial \nu} H_{\nu_n}^{(1)}(ka)\right]_{\nu=\nu_n} \approx 2^{4/3}\left(\nu_n\right)^{-1/3} e^{-i\frac{\pi}{3}}\left[A_i'(y_n)\left(\frac{\partial y}{\partial \nu}\right)_{\nu=\nu_n}\right]_{x=ka}$$

$$\left[\frac{\partial}{\partial \nu} H_{\nu_n}^{(1)'}(ka)\right]_{\nu=\nu_n} \approx 2^{4/3}\left(\nu_n\right)^{-1/3}\left[y_n A_i(y_n)\left(\frac{\partial y}{\partial \nu}\right)_{\nu=\nu_n}\frac{\partial y}{\partial x}\right]_{x=ka}, \tag{2.328}$$

where

$$y = 2^{1/3} e^{-i\frac{\pi}{3}} x^{-1/3}(x-\nu), \quad \frac{\partial y}{\partial x} \approx 2^{1/3} e^{-i\frac{\pi}{3}} x^{-1/3}, \quad \frac{\partial y}{\partial \nu} \approx -2^{1/3} e^{-i\frac{\pi}{3}} x^{-1/3}$$

and where $A_i(y)$ is the Airy function defined by

$$A_i(y) = \int_0^{\infty} \cos\left(\frac{1}{3}\tau^3 + y\tau\right) d\tau, \tag{2.329}$$

verifying

$$A_i''(y) = y A_i(y).$$

Inserting (2.328) into (2.327) yields

$$Q_n = \pi ka\, 2^{\frac{10}{3}} \nu_n^{-2/3} e^{2i\frac{\pi}{3}} (ka)^{-2/3}\left\{\left[A_i'(y_n)\right]^2 - y_n\left[A_i(y_n)\right]^2\right\}_{x=ka}$$

and

$$R_n = -\frac{2^{-4/3}}{\pi}(ka)^{1/3} e^{-2i\frac{\pi}{3}}\left(\left[A_i'(y_n)\right]^2 - y_n\left[A_i(y_n)\right]^2\right)_{x=ka}^{-1}.$$

Finally, the diffraction coefficient $D_n^2 = D_{sr,cr}^{(n)} D_{cr,sr}^{(n)}$ is given by

$$D_n^2 = \frac{2^{2/3} e^{-i\frac{\pi}{6}} (ka)^{1/3}}{\left(\left[A_i'(y_n)\right]^2 - y_n\left[A_i(y_n)\right]^2\right)_{x=ka}}. \tag{2.330}$$

This diffraction coefficient is valid for an impedance boundary condition, although the impedance does not figure explicitly in the formula. As a matter of fact, the impedance comes indirectly into the formula through the zeros of the characteristic equation (2.325) that, after approximating the Hankel function $H_\nu^{(1)}(x)$ and its derivative $H_\nu^{(1)'}(x)$ by the relations (2.328), transforms into the following equation:

$$A_i'(-\xi) + i\,Z\left(\frac{ka}{2}\right)^{1/3} e^{i\frac{\pi}{3}}\, A_i(-\xi) = 0. \tag{2.331}$$

The solution ξ_n is related to ν_n by

$$\xi_n = 2^{1/3}\, e^{-i\frac{\pi}{3}} (ka)^{-1/3} (\nu_n - ka) = -y_n. \tag{2.332}$$

For a soft boundary condition, we have $\xi_n = \eta_n$ and $A_i(-\eta_n) = 0$. Hence (2.330) gives

$$D_n^2 = \frac{2^{2/3}\, e^{-i\frac{\pi}{6}} (ka)^{1/3}}{\left[A_i'(-\eta_n)\right]^2}. \tag{2.333}$$

Likewise, for a hard boundary condition we have $\xi_n = \bar{\eta}_n$ and $A_i'(-\bar{\eta}_n) = 0$ and consequently

$$D_n^2 = \frac{2^{2/3}\, e^{-i\frac{\pi}{6}} (ka)^{1/3}}{\bar{\eta}_n \left[A_i'(-\bar{\eta}_n)\right]^2}. \tag{2.334}$$

2.3.1.4. Noncircular Convex Cylinder

According to (2.321), if we assume that the diffraction processes at Q and Q' are symmetric, which means that $D_{sr,cr}^{(n)}(Q) = D_{cr,sr}^{(n)}(Q') D_s$, then we get the correct result for the diffraction coefficient of a circular cylinder, equal to $D_n^2 D_s$. In addition, this relation satisfies the theorem of reciprocity. Because $D_{sr,cr}^{(n)}\, D_{cr,sr}^{(n)}$ depend only on the radius of the circular cylinder and because, at high frequencies, the diffraction process depends only on the local geometry of the surface, a way to extend these coefficients to a noncircular convex cylinder at normal incidence consists of writing

$$D_{sr,cr}^{(n)}(Q) = D_n(a_1)\sqrt{D_s}, \quad D_{cr,sr}^{(n)}(Q') = D_n(a_2)/\sqrt{D_s}, \tag{2.335}$$

where (a_1) and (a_2) are the radii of curvature of the cylindrical surface at Q and Q', respectively, and where $D_n(a_i)$, $i = (1, 2)$ is given by (2.330) with (a) replaced by (a_1) or (a_2).

Taking (2.335) into account in the formula (1.238) derived in chapter 1, section 1.9, we obtain the following (as expected):

$$\frac{D_{cr,\,sr}^{(n)}(Q)}{D_{sr,\,cr}^{(n)}(Q)} = \frac{D_{cr,\,sr}^{(n)}(Q')}{D_{sr,\,cr}^{(n)}(Q')} = \frac{1}{D_s}. \tag{2.336}$$

Hence for a noncircular convex cylinder submitted to the pressure field emitted by a line source of unit strength parallel to the generatrices of the cylinder, (2.321) takes the more general form

$$\begin{aligned} p^d(Q) = {} & D_S \frac{e^{ik\,r_1}}{\sqrt{r_1}} \sum_{n=1}^{\infty} D_n(a_1) D_n(a_2) e^{ik\,s_1}\, e^{i\int_{P_1}^{P_2} \frac{b_n(s)}{\rho_s} ds}\, D_S \frac{e^{ik\,r_2}}{\sqrt{r_2}} \\ & + D_S \frac{e^{ik\,r_1'}}{\sqrt{r_1'}} \sum_{n=1}^{\infty} D_n(a_1') D_n(a_2') e^{ik\,s_1'}\, e^{i\int_{P_1}^{P_2} \frac{b_n(s)}{\rho_s} ds}\, D_S \frac{e^{ik\,r_2'}}{\sqrt{r_2'}}, \end{aligned} \tag{2.337}$$

where a_1, a_2, a_1', a_2' are the local radii of curvature of the cross section of the cylinder at P_1, P_2, P_1', P_2', respectively; where P_1 and P_2 are the excitation and launching points of the creeping wave encircling the cylinder in the trigonometric direction; and where P_1' and P_2' are the excitation and launching points of the creeping wave encircling the cylinder in the opposite direction. We have also stated that

$$\nu_n = k\rho + b_n, \tag{2.338}$$

with b_n depending on the local radius of curvature ρ of the surface. Also, s_1 and s_1' are the arc lengths P_1P_2 and $P_1'P_2'$, respectively, followed by the creeping waves on the surface of the cylinder. The other parameters are $r_1 = SP_1$, $r_2 = P_2Q$, $r_1' = SP_1'$, and $r_2' = P_2'Q$.

2.3.1.5. Field at an Observation Point Located on the Surface of a Circular Cylinder

We consider the case where the observation point Q is located on the surface of a circular cylinder. We suppose that Q is situated in the shadow region of the source. Consequently, the incident and reflected fields are absent, and the total field is equal to the diffracted field given by (2.264). We assume that the source is far away from the cylinder ($\rho >> a$). Then $H_\nu^{(1)}(k\rho)$ in (2.264) can be replaced by its Debye asymptotic expansion (2.274). Taking (2.273) into account, the dominant term of the asymptotic expansion of the diffracted field is given by

$$p^d(a,\theta) = \sqrt{\frac{\pi}{8k}}\, e^{-\frac{i\pi}{4}} \frac{e^{ik\left(\rho^2-a^2\right)^{\frac{1}{2}}}}{\left(\rho^2-a^2\right)^{\frac{1}{4}}} \sum_{n=1}^{\infty} \frac{e^{i\nu_n\theta} + e^{i\nu_n(2\pi-\theta)}}{1-e^{2i\pi\nu_n}}\, e^{-i\nu_n \operatorname{Arc\,cos}\left(\frac{\nu_n}{k\rho}\right)} \frac{\Omega H_{\nu_n}^{(2)}(ka)}{\left[\frac{\partial}{\partial\nu}\Omega H_\nu^{(1)}(ka)\right]_{\nu=\nu_n}} H_\nu^{(1)}(ka). \tag{2.339}$$

The physical interpretation of (2.339) in terms of creeping rays is similar to that presented in section 2.3.1.2 for an observation point far away from the cylinder. For each value of ν_n, two series of creeping waves reach the point Q, one encircling the cylinder in the positive trigonometric direction and the other in the opposite direction. Using the notations

$$r_1 = \sqrt{\rho^2 - a^2}, \quad \phi = \operatorname{Arc\,cos}\frac{a}{\rho}, \quad \ell = a(\theta-\phi), \quad \ell' = a(2\pi-\theta-\phi), \tag{2.340}$$

equation (2.339) becomes

$$p^d(a,\theta) = \sqrt{\frac{\pi}{8k}}\, e^{-\frac{i\pi}{4}} \frac{e^{ikr_1}}{\sqrt{r_1}} \sum_{n=1}^{\infty}\sum_{m=0}^{\infty} \left[\left(e^{i\frac{\nu_n}{a}(\ell+2m\pi a)} + e^{i\frac{\nu_n}{a}(\ell'+2m\pi a)}\right)\right] \frac{\Omega H_{\nu_n}^{(2)}(ka)}{\left[\frac{\partial}{\partial\nu}\Omega H_\nu^{(1)}(ka)\right]_{\nu=\nu_n}} H_\nu^{(1)}(ka), \tag{2.341}$$

and by taking (2.322) and (2.323) into account, equation (2.341) may also be written as

$$p^d(a,\theta) = D_s \frac{e^{ikr_1}}{\sqrt{r_1}} \sum_{n=1}^{\infty} D_{sr,cr}^{(n)} \sum_{m=0}^{\infty} \left[\left(e^{i\frac{\nu_n}{a}(\ell+2m\pi a)} + e^{i\frac{\nu_n}{a}(\ell'+2m\pi a)}\right)\right] A^{(n)}, \tag{2.342}$$

where $A^{(n)}$ is the attachment coefficient given by

$$A^{(n)} = \frac{i}{4} D_{cr,sr}^{(n)} H_{\nu_n}^{(1)}(ka). \tag{2.343}$$

Replacing $H_\nu^{(1)}(ka)$ with its asymptotic expansion given by (2.328), we finally obtain

$$A^{(n)} = \frac{i}{\sqrt{2}} \frac{\left(\frac{ka}{2}\right)^{-1/6} e^{i\frac{\pi}{12}} A_i(\bar{y}_n)}{\left[A_i'(\bar{y}_n)\right]^2 - y_n\left[A_i(\bar{y}_n)\right]^2}, \tag{2.344}$$

where $\bar{y}_n = 2^{1/3} e^{-i\frac{\pi}{3}} (ka)^{-1/3} (ka - \nu_n)$.

For $m = 0$, equation (2.342) reduces to

$$p^d(a,\theta) = D_s \frac{e^{ikr_1}}{\sqrt{r_1}} \sum_{n=1}^{\infty} D^{(n)}_{sr,cr} \left[\left(e^{i\frac{\nu_n}{a}\ell} + e^{i\frac{\nu_n}{a}\ell'}\right)\right] A^{(n)}. \tag{2.345}$$

The physical interpretation of (2.345) is the following: Each value of n has two corresponding creeping rays that have traveled less than one time around the cylinder to reach the observation point Q. The first creeping ray is excited at P, with an excitation coefficient $D^{(n)}_{sr,cr}$. It then travels an arc length ℓ along the cylinder, with a propagation constant ν_n/a, to the observation point Q where the field is given by applying the attachment coefficient $A^{(n)}$ (fig. 2.36).

The second creeping ray follows the path $SP'Q$ and travels an arc length ℓ' with the same propagation constant ν_n. Both coefficients $D^{(n)}_{sr,cr}$ and $A^{(n)}$ depend on ka and on the surface impedance through the value of the root ν_n of the following equation:

$$\frac{\partial}{\partial x} H^{(1)}_{\nu}(x) + iZ\, H^{(1)}_{\nu}(x) = 0. \tag{2.346}$$

These coefficients remain unchanged when the incident field is a plane wave, as can be readily seen by recognizing that the factor $D_s e^{ikr_1} / \sqrt{r_1}$ in (2.342) corresponds to the incident field emitted by a line source of unit strength.

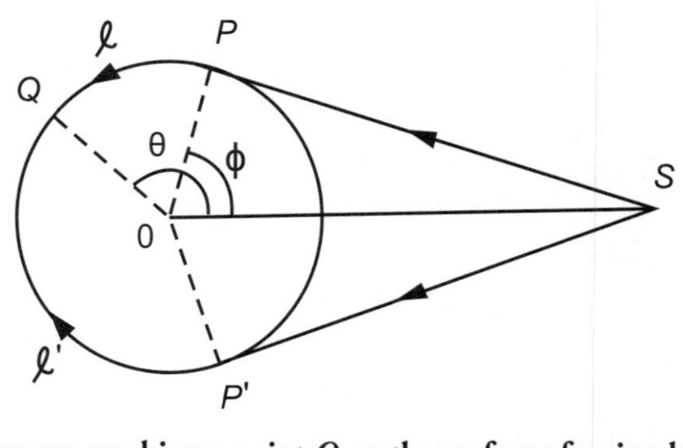

Figure 2.36. Creeping rays reaching a point Q on the surface of a circular cylinder.

We now present some analytical and numerical results concerning the position of the zeros of (2.346) in the upper half of the complex ν-plane. We recall that these zeros are the poles of the integrand of the exact integral form of the solution.

2.3.1.6. Position of the Poles in the Complex ν-Plane

From the results in section 2.3.1.1, the poles of the integrand of the general solution of the diffraction problem are the zeros in the complex ν-plane of equation (2.346).

We consider first the special cases of hard $(Z = 0)$ and soft $(Z = \infty)$ boundary conditions, in which (2.346) reduces to $H_\nu^{(1)'}(ka) = 0$ and $H_\nu^{(1)}(ka) = 0$, respectively. According to the asymptotic forms of the Hankel function $H_\nu^{(1)}(ka)$ and its derivative when $|\nu - ka| = O\left(|\nu|^{1/3}\right)$ and ka large, given by (2.328), we see that the zeros ν_n and $\bar{\nu}_n$ of $H_\nu^{(1)}(ka)$ and $H_\nu^{(1)'}(ka)$ are given in first approximation by the zeros η_n and $\bar{\eta}_n$ of the Airy functions $A_i(-\eta)$ and $A_i'(-\eta)$, respectively, where

$$\eta = -2^{1/3}\, e^{-i\frac{\pi}{3}} (ka)^{-1/3} (ka - \nu). \tag{2.347}$$

According to (2.347), we have

$$\begin{aligned} \nu_n &\approx ka + \left(\frac{ka}{2}\right)^{1/3} e^{i\frac{\pi}{3}}\, \eta_n, \\ \bar{\nu}_n &\approx ka + \left(\frac{ka}{2}\right)^{1/3} e^{i\frac{\pi}{3}}\, \bar{\eta}_n, \end{aligned} \tag{2.348}$$

and η_n and $\bar{\eta}_n$ are tabulated in reference [2.20].

The first few zeros are

$$\begin{aligned} &\eta_1 = 2.338, \quad \eta_2 = 4.088, \quad \eta_3 = 5.521, \quad \eta_4 = 6.787, \quad \eta_5 = 7.944, \\ &\bar{\eta}_1 = 1.019, \quad \bar{\eta}_2 = 3.248, \quad \bar{\eta}_3 = 4.820, \quad \bar{\eta}_4 = 6.163, \quad \bar{\eta}_5 = 7.372. \end{aligned}$$

They are located in the first quadrant of the complex ν-plane, along a straight line making an angle $\pi/3$ with the real ν-axis.

Equation (2.347) is only true when $\nu \cong ka$, and therefore, the number of terms that can be taken is limited. However, when ka is large, only a few terms are necessary in practical applications. We consider now the case where $Z = i\gamma$ is purely imaginary (γ taking positive or negative values). Then when $|\gamma|$ passes through zero, equation (2.346) is verified at the zeros $\bar{\nu}_n$ of $H_\nu^{(1)'}(ka)$.

Similarly, when $|\gamma|$ is infinite, equation (2.346) is verified at the zeros ν_n of $H_\nu^{(1)}(ka)$. Hence when γ varies from $-\infty$ to $+\infty$, the zeros ν_n of (2.346) follow a sinusoidal trajectory passing through the zeros of $H_\nu^{(1)}(ka)$ and $H_\nu^{(1)'}(ka)$, as shown in figure 2.37(a). Moreover, when $|\gamma|$ or $|\gamma|^{-1}$ is small, it is possible to calculate the zeros of (2.346) by making a Taylor series expansion with respect to the zeros of $H_\nu^{(1)}(ka)$ and $H_\nu^{(1)'}(ka)$, respectively, and to show that the sinusoidal trajectory has a continuous tangent and curvature at the points $\bar{\nu}_n$ and ν_n in the complex ν-plane [2.21]. It can also be shown [2.21] that the trajectory of the zero ν_1 crosses the real ν-axis at the point $\nu_1 = ka\sqrt{1+\gamma^2}$.

For $Z = |Z|e^{i\theta}$ complex, the trajectories of the zeros of (2.346) also pass through the poles $\bar{\nu}_n$ and ν_n, when $|Z|$ or $|Z|^{-1}$ is equal to zero. When $|Z|$ varies from zero to infinity, θ being constant between two successive poles $\bar{\nu}_n$ and ν_n but changing its sign at each pole $\bar{\nu}_n$ or ν_n,

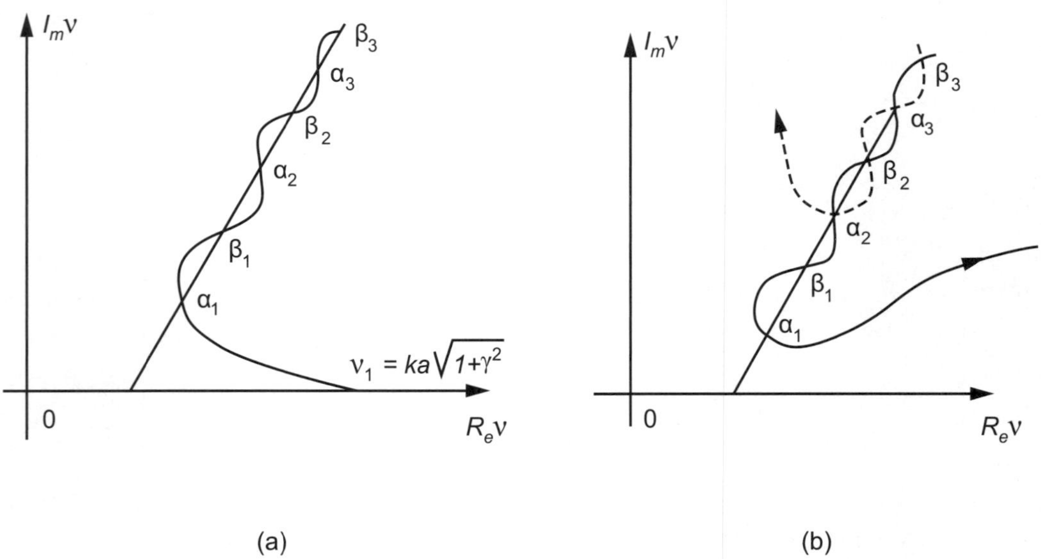

Figure 2.37. Trajectories of the zeros of equation (2.346) for (a) purely imaginary values of Z and (b) complex values $Z = |Z|e^{i\theta}$, when $|Z|$ varies from zero to infinity.

the trajectory of the zeros of the equation (2.346) is again a sinusoidal curve passing through $\bar{\nu}_n$ and ν_n with a continuous tangent and curvature, as shown in figure 2.37(b). However, for certain values of θ, the trajectory may leave the vicinity of the Stokes line and enter into a region where the Hankel function can be represented by its Debye asymptotic expansion. Especially the curve followed by ν_1 does not tend to the real ν-axis.

It is possible to give a physical interpretation of these results. When $Z = i\gamma$ is purely imaginary and γ takes large values, the creeping wave corresponding to the zero ν_1 transforms progressively into a surface wave that travels around the cylinder with a negligible attenuation. However, when Z is complex and ν_1 enters into the Debye region, the corresponding creeping wave transforms into a leaky wave.

2.3.1.7. Oblique Incidence

In this section we consider a plane wave obliquely incident on a circular cylinder. According to the translation invariance, the pressure field at an observation point $M(\rho, \theta, z)$ in the deep shadow region is given by

$$p(\rho,\theta,z) = P(\rho,\theta;k_t)\, e^{ik_z z}, \tag{2.349}$$

where $k_t = k \sin \beta_0$ and $k_z = k \cos \beta_0$, β_0 being the angle of incidence, and where $P(\rho, \theta; k_t)$ verifies the Helmholtz equation for normal incidence. The boundary conditions on the surface read

$$\frac{\partial}{\partial \rho} p(\rho,\theta,z) + ikZ\, p(\rho,\theta,z) = 0\ , \quad \rho = a,$$

which according to (2.349), can also be written as

$$\frac{\partial}{\partial \rho} P(\rho,\theta;k_t) + ik_t \frac{Z}{\sin \beta_0} P(\rho,\theta;k_t) = 0, \quad \rho = a.$$

Consequently, the expression of $P(\rho, \theta; k_t)$ in (2.349) can be deduced from the solution of the diffraction problem for normal incidence by replacing Z with $Z' = Z / \sin \beta_0$. Moreover, for an incident plane wave, $P(\rho, \theta; k_t)$ is obtained by replacing the factor $D_s / \sqrt{r_1} \exp(ikr_1)$, corresponding to the pressure field emitted by a line source of unit strength, with the incident field p^i at the excitation point of each creeping wave. Hence $P(\rho, \theta; k_t)$ is known. As in (2.321), if we neglect the contribution of the creeping waves reaching the observation point after having encircled the cylinder m times with $m \geq 1$, then the solution $p(\rho, \theta, z)$ is defined by the contribution of two creeping rays. If we project these rays on a plane perpendicular to the cylinder axis, then we obtain a representation of the rays similar to that shown in figure 2.31, with S removed to infinity. Let $p_1(\rho, \theta, z)$ be the contribution of one of these creeping rays, touching the cylinder at glancing incidence at Q and leaving the cylinder at Q', as shown in figure 2.38. Then according to (2.349) and (2.321), we obtain

$$p_1(\rho,\theta,z) = p^i(Q)\sum_{n=1}^{\infty} D^{(n)}_{sr,\,cr}(a,k_t)\, D^{(n)}_{cr,sr}(a,k_t)\, e^{i\frac{\nu_n(k_t)}{a}\ell_\perp}\, e^{ik_z z}\, D_s(k_t)\frac{e^{ik_t\rho_1}}{\sqrt{\rho_1}}, \quad (2.350)$$

where $\ell_\perp = arc\,Q\,Q'_\perp$, $\rho_1 = Q'_\perp M_\perp$, $z = M_\perp M$, Q being the excitation point of the creeping wave and $Q'_\perp$ and $M_\perp$ being the projection of the launching point Q' of the creeping ray and the projection of the observation point M, respectively, on the XY plane passing through Q.

Also,

$$\frac{\nu_n(k_t)}{a} = k_t + \frac{b_n}{a}(a,k_t),$$

where $I_m b_n/a$ is the attenuation coefficient, depending on the boundary conditions. By noting that

$$\ell_\perp = \ell\sin\beta_0, \quad z = (\ell + Q'M)\cos\beta_0,$$

where $\ell = \text{arc}\,Q\,Q'$, the expression (2.350) can also be written as follows:

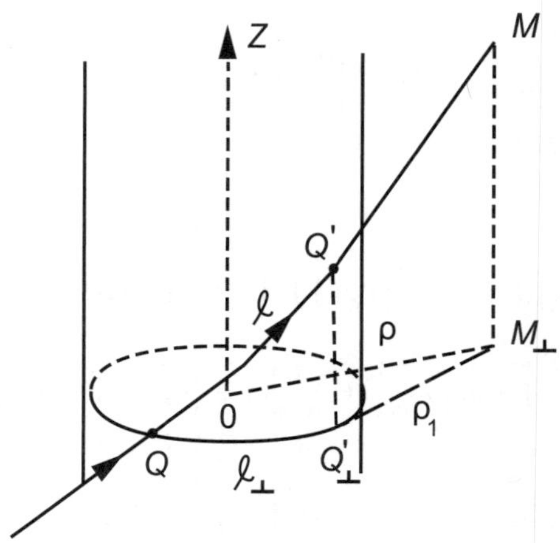

Figure 2.38. Geometrical parameters for oblique incidence.

$$p_1(\rho,\theta,z)=\frac{p^i(Q)}{\sqrt{\sin\beta_0}}D^{(n)}_{sr,cr}(a,k_t)\,D^{(n)}_{cr,sr}(a,k_t)\,e^{\frac{i b_n(a,k_t)}{a}\ell_\perp}\,e^{ik\ell}\,D_s(k_t)\frac{e^{ikQ'M}}{\sqrt{Q'M}}. \tag{2.351}$$

For a soft or hard surface, it follows from (2.348) that

$$b_n(a,k_t)=\left(\frac{k_t a}{2}\right)^{1/3}e^{i\frac{\pi}{3}}\xi_n, \tag{2.352}$$

where $\xi_n=\eta_n$ for a soft surface and $\xi_n=\bar{\eta}_n$ for a hard surface, η_n and $\bar{\eta}_n$ being, respectively, the zeros of the Airy function $A_i(-\xi)$ and its derivative $A_i'(-\xi)$.

The geodesic QQ' followed by the creeping wave is a helix with a constant principal curvature $\bar{\rho}$. According to Euler's theorem, we have

$$a=\bar{\rho}\sin^2\beta_0. \tag{2.353}$$

Hence (2.352) yields

$$\frac{b_n(a,k_t)}{a}=\frac{q_n(\bar{\rho},k)}{\sin\beta_0}, \tag{2.354}$$

where

$$q_n(\bar{\rho},k)=\frac{1}{\bar{\rho}}\left(\frac{k\bar{\rho}}{2}\right)^{1/3}e^{i\frac{\pi}{3}}\xi_n, \tag{2.355}$$

in which ξ_n has the same meaning as in (2.352). It follows from (2.354) that

$$\frac{b_n(a,k_t)}{a}\ell_\perp=q_n(\bar{\rho},k)\ell=\frac{m\ell}{\bar{\rho}}\xi_n e^{i\frac{\pi}{3}}, \tag{2.356}$$

where

$$m=\left(\frac{k\bar{\rho}}{2}\right)^{1/3}$$

is the Fock parameter along the geodesic.

Returning to (2.351), we observe that for oblique incidence, the diffraction coefficient without the attenuation factor is given by

$$D_c^{(n)}=\frac{1}{\sqrt{\sin\beta_0}}D^{(n)}_{sr,cr}(a,k_t)\,D^{(n)}_{cr,sr}(a,k_t)=\frac{1}{\sqrt{\sin\beta_0}}D_n^2(a,k_t),$$

where $D_n^2(a, k_t)$ is given by (2.330). For soft and hard boundary conditions, this coefficient can be readily expressed in terms of the Fock parameter m by replacing k_t with $k \sin \beta_o$ and by taking (2.353) into account. The result is given by

$$\frac{D_n^2(a, k_t)}{\sqrt{\sin \beta_0}} = \sqrt{\sin \beta_0}\, d_n^2(\bar{\rho}, k)$$

with

$$d_n^2(\bar{\rho}, k) = 2m e^{-i\pi/6} \times \begin{cases} (\bar{\eta}_n)^{-1}\left[A_i(-\bar{\eta}_n)\right]^{-2}, & \text{hard} \\ \left[A_i'(-\eta_n)\right]^{-2}, & \text{soft}. \end{cases} \tag{2.357}$$

Inserting (2.356) and (2.357) into (2.351) and noting that $D_s(k_t) = 1/\sqrt{\sin \beta_0}\, D_s(k)$ where $D_s(k) = (8\pi k)^{-1/2} \exp\left(+i\frac{\pi}{4}\right)$, we obtain the final expression for the contribution to the diffracted field by a circular cylinder at oblique incidence, due to the creeping ray $QQ'M$,

$$p_1(M) = p^i(Q) \sum_{n=1}^{\infty} d_n^2(\bar{\rho}, k)\, e^{ik\ell}\, e^{i\alpha_n \frac{m\ell}{\bar{\rho}}}\, D_s \frac{e^{ik\, Q'M}}{\sqrt{Q'M}}, \tag{2.358}$$

where $D_s = D_s(k)$ and $\alpha_n = e^{i\frac{\pi}{3}} \eta_n$ or $\alpha_n = e^{i\frac{\pi}{3}} \bar{\eta}_n$ for soft and hard boundary conditions, respectively.

We can observe that all the terms in (2.358) involve only the parameters of the ray trajectory. In the case of a general convex cylindrical surface, the radius of curvature $\bar{\rho}$ of the geodesic is not constant. However, by invoking the local nature of the diffraction phenomena at high frequencies, (2.358) can be adapted to this case by replacing $d_n^2(\bar{\rho}, k)$ with the product

$$d_n\left[\bar{\rho}(Q), k\right] d_n\left[\bar{\rho}(Q'), k\right] \tag{2.359}$$

and the attenuation coefficient $q_n(\bar{\rho}, k)$ with

$$\int_Q^{Q'} q_n\left[\bar{\rho}(\ell), k\right] d\ell = \alpha_n \int_Q^{Q'} \frac{m(\ell)}{\bar{\rho}(\ell)} d\ell. \tag{2.360}$$

For a general convex cylindrical soft or hard surface at oblique incidence, we therefore obtain

$$p_1(M) = p^i(Q) \sum_{n=1}^{\infty} d_n(Q) d_n(Q')\, e^{ik\ell}\, e^{i\alpha_n \int_Q^{Q'} \frac{m\, d\ell}{\bar{\rho}}}\, D_s \frac{e^{ik\, Q'M}}{\sqrt{Q'M}}. \tag{2.361}$$

For an impedance boundary condition on a circular cylinder, the contribution of the creeping ray $QQ'M$ to the diffracted field, at oblique incidence, is still given by (2.358) with $\alpha_n = e^{i\pi/3}\xi_n$, but now ξ_n $(n = 1,\infty)$ are the solutions of (see equation 2.331)

$$A_i'(-\xi) + iZ'\left(\frac{k_t a}{2}\right)^{1/3} e^{i\frac{\pi}{3}} A_i(-\xi) = 0, \quad Z' = \frac{Z}{\sin\beta_0},$$

which since $k_t a = k\bar{\rho}\sin^3\beta_0$, can also be written as

$$A_i'(-\xi) + iZme^{i\pi/3} A_i(-\xi) = 0\,, \quad m = \left(\frac{k\bar{\rho}}{2}\right)^{1/3}. \qquad (2.362)$$

Hence the zeros of (2.362) depend on the impedance Z and the radius of curvature of the geodesic followed by the creeping wave.

Consequently, since

$$\xi = (\nu - k_t a)\left(\frac{k_t a}{2}\right)^{-1/3} e^{-i\frac{\pi}{3}},$$

we have

$$\nu_n = k_t a + \left(\frac{k_t a}{2}\right)^{1/3} e^{+i\frac{\pi}{3}} \xi_n$$

and

$$\frac{b_n(a,k_t)}{a}\ell_\perp = \frac{\ell}{\bar{\rho}}\left(\frac{k\bar{\rho}}{2}\right)^{1/3} e^{i\frac{\pi}{3}}\xi_n = \frac{m\ell}{\bar{\rho}} e^{i\frac{\pi}{3}}\xi_n. \qquad (2.363)$$

The diffraction coefficient $d_n^2(\bar{\rho},k)$ in (2.358) is also modified. According to (2.330), it is now given by

$$d_n^2(\bar{\rho},k) = 2me^{-i\frac{\pi}{6}}\left(\left[A_i'(-\xi_n)\right]^2 - y_n\left[A_i(-\xi_n)\right]^2\right)^{-1}. \qquad (2.364)$$

Formula (2.358) can also be extended to a general convex impedance cylinder by replacing $d_n^{(2)}(\bar{\rho},k)$ with the product

$$d_n\left[\bar{\rho}(Q),k\right]d_n\left[\bar{\rho}(Q'),k\right], \qquad (2.365)$$

where d_n is given by (2.364) and by substituting $\alpha_n \dfrac{m\ell}{\bar{\rho}}$ with the expression $\alpha_n \int_Q^{Q'} \dfrac{md\ell}{\bar{\rho}}$, where $\alpha_n(\ell) = \xi_n(\ell)\exp\left(\dfrac{i\pi}{3}\right)$, ξ_n being the zeros of (2.362), which now depend on the position ℓ along the geodesic.

2.3.2. *THREE-DIMENSIONAL CONVEX SURFACE WITH HARD, SOFT, OR IMPEDANCE BOUNDARY CONDITIONS*

Formulas for the diffracted field associated with a creeping ray following a geodesic without torsion on a three-dimensional convex surface have been proposed by Kouyoumjian [2.22]. They are founded on the work by Keller and Levy [2.23], Franz and Klante [2.24], Hong [2.25], and Voltmer [2.26].

When the transverse radius of curvature ρ_t of the surface in a plane orthogonal to the geodesic is large compared to the radius of curvature of the geodesic ($\rho_t >> \rho_g$), formula (2.361) can be extended to a three-dimensional convex surface by adapting the divergence factor of the diffracted rays. Hence, for a hard or soft surface,

$$p(M) = p^i(Q)\sum_{n=1}^{\infty} d_n(Q)\, d_n(Q')\, e^{i\alpha_n \int_Q^{Q'} \frac{md\ell}{\rho_g}}\, D_s\, e^{ikQ'M} \sqrt{\frac{\rho}{Q'M(\rho + Q'M)}} \sqrt{\frac{d\Sigma(Q)}{d\Sigma(Q')}}, \qquad (2.366)$$

where d_n is given by (2.357) for hard and soft boundary conditions and $\alpha_n = \xi_n \exp\left(i\pi/3\right)$, where ξ_n are the zeros $\overline{\eta_n}$ and η_n of $A_i'(-\xi)$ and $A_i(-\xi)$, respectively. Also, ρ has the same meaning as in (1.224), and $\left[d\Sigma(Q)/d\Sigma(Q')\right]^{1/2}$ is the divergence factor of the creeping rays along the surface. For an impedance boundary condition, d_n is given by (2.364), and ξ_n are the zeros of

$$A_i'(-\xi) + iZm\, e^{i\pi/3}\, A_i(-\xi) = 0.$$

Therefore, they depend on the local radius of curvature of the geodesic. Hence $p_1(M)$ is given by (2.366) with the phase term $\int_Q^{Q'} \alpha_n(\ell)\dfrac{md\ell}{\rho_g}$.

When $\rho_t \cong \rho_g$, the transverse curvature ρ_t and the first and second derivatives of ρ_g appear in the phase term and yield small corrections of order $\left(k\rho_g\right)^{-1/3}$ in this term.

For a hard boundary, for example, the phase term reads (after [2.26] or [2.22])

$$\alpha_n \int_Q^{Q'} \frac{md\ell}{\rho_g} \left\{ 1 + \frac{1}{m^2} \left[\alpha_n \left(\frac{1}{60} - \frac{2}{45} \rho_g \ddot{\rho}_g + \frac{4}{135} \dot{\rho}_g^2 \right) - \frac{1}{\alpha_n^2} \left(\frac{1}{10} + \frac{\rho_g}{4\rho_t} - \frac{\rho_g \ddot{\rho}_g}{60} + \dot{\rho}_g^2 \right) \right] \right\}. \tag{2.367}$$

In the case of a hard sphere, the term $\frac{\rho_g}{4\rho_t}$ in the correction term needs to be introduced in (2.366) for moderately small values of $k\rho_g$—typically $k\rho_g \approx 5$. This term is absent in the formula for a soft surface, which is therefore less sensitive to this correction term.

When $\rho_g / \rho_t \approx (k\rho_g)^{1/3}$, the correction term in (2.367) is of order 1 and modifies the phase even for large values of $k\rho_g$. Finally, when $\rho_g / \rho_t \approx (k\rho_g)^{2/3}$, the term with ρ_t in (2.367) appears to be of the same order as the leading term. In this case, (2.367) loses its asymptotic character and may give wrong results.

More recently, new formulas for the amplitude and propagation parameters of creeping waves have been obtained by Andronov and Bouche [2.27–2.28], who applied the boundary layer method. The basic ideas of this method can be found in the work of Fock [2.29] and Buchal and Keller [2.30]. A comprehensive presentation of it is given in [2.31] and [2.32]. The main results obtained by Andronov and Bouche concern the case of a moderately elongated body characterized by $\rho_g / \rho_t \approx (k\rho_g)^{1/3}$ and that of a strongly elongated body characterized by $\rho_g / \rho_t \approx (k\rho_g)^{2/3}$.

In the case of a moderately elongated object, they have shown that the diffraction coefficient is modified. New formulas for this coefficient can be found in [2.27]. In the case of a strongly elongated object, they have shown that the transverse curvature leads to a total modification of the creeping wave asymptotics. This topic is still a matter of research. Preliminary results for a body of revolution are given in [2.33].

Besides the general formula (2.366) that, except in some special cases mentioned previously, gives satisfactory results in most practical applications, another important formula can be derived from (2.361) for a geodesic with torsion. The method is to locally approximate, along the geodesic, the actual surface by a cylindrical surface defined in such a way that locally the radius of curvature of the geodesic on the cylindrical surface is identical to that on the actual surface, and the angle between the geodesic and the principal radii of curvature is the same for both surfaces.

Let $\beta(\ell)$ be the angle between the geodesic and the principal direction corresponding to the largest principal radius of curvature of the surface. We suppose that along the geodesic, ρ_t is large compared to ρ. Then the generalization of (2.366) to a geodesic with torsion, yields

$$p(M) = p^i(Q)\sum_{n=1}^{\infty} d_n(Q)\, d_n(Q')\, e^{ik\ell} \exp\left(i\int_Q^{Q'} \frac{\alpha_n(\ell) m d\ell}{\rho_g} \right) D_s \sqrt{\frac{d\Sigma(Q)}{d\Sigma(Q')}} \sqrt{\frac{\rho}{Q'M(\rho + Q'M)}}\, e^{ikQ'M}. \tag{2.368}$$

For a hard or soft surface, $\alpha_n(\ell)$ is independent of ℓ and equal to $\xi_n \exp\left(i\frac{\pi}{3}\right)$, where $\xi_n = \bar{\eta}_n$ and $\xi_n = \eta_n$ are the zeros of $A_i'(-\xi)$ and $A_i(-\xi)$, respectively. However, for an impedance surface, according to (2.362), $\alpha_n(\ell)$ depends on ℓ through the value of $m = m(\ell)$. In this last case, the zeros of (2.362) need to be calculated at each sampling point along the geodesic.

It is important to observe that (2.368) does not depend explicitly on the angle $\beta(\ell)$. This result has already been established in (2.361) for a cylindrical convex surface. As a consequence, formulas (2.366) and (2.368) are identical.

2.3.3. HOLLOW ELASTIC CIRCULAR CYLINDER

We consider a hollow elastic circular cylinder embedded in a fluid. The cylinder is closed and the medium occupying the interior domain is vacuum. A plane sound wave of circular frequency ω is incident on the cylinder. This canonical problem is representative, in practical applications, of a closed shell embedded in a dense fluid like water, with a low density fluid inside the shell like air. Typical applications are underwater mines and submarine vessels.

2.3.3.1. General Solution for Normal Incidence

The solution of this problem is classical and can be found in [2.34] for the more general case of a fluid occupying the interior domain. The axis of the hollow circular cylinder, of outer radius *a* and inner radius *b*, is taken to coincide with the *z*-axis of the coordinate system, and the plane sound wave is incident along the negative *x*-axis (fig. 2.39).

The fluid outside the shell is labeled by 1 and the elastic shell material by 2. The fluid medium is described by the density ρ_1, the sound velocity C_1, and the propagation constant $k_1 = \omega / C_1$. The elastic medium is characterized by the density ρ_2 and the Lamé elastic constants λ, μ. The longitudinal and transverse elastic wave velocities inside the shell medium are given by

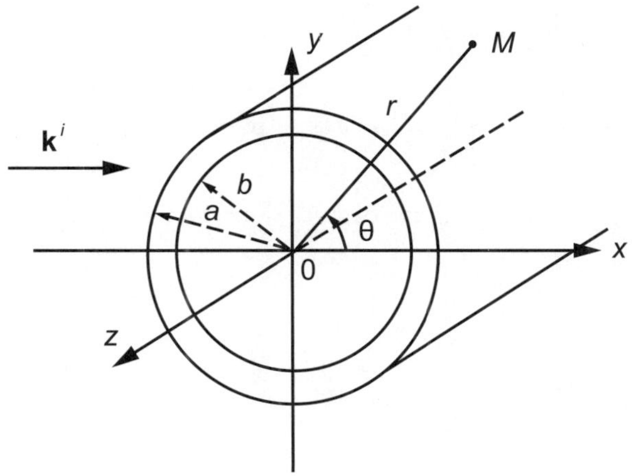

Figure 2.39. Cylinder geometry.

$$C_\ell^2 = (\lambda + 2\mu) / \rho_2, \quad C_t^2 = \mu / \rho_2, \tag{2.369}$$

and the corresponding propagation constants are

$$k_\ell = \frac{\omega}{C_\ell}, \quad k_t = \frac{\omega}{C_t}. \tag{2.370}$$

Let us denote by r, θ with $x = r\cos\theta$, $y = r\sin\theta$, the polar coordinates of the observation point. In medium 1, the total pressure field p is taken as $p_1 = p_i + p_s$, where p_i is the incident field and p_s is the scattered field. The given incident plane wave pressure can be written as

$$p_i = P_0 \sum_{n=0}^{\infty} i^n \varepsilon_n J_n(k_1 r) \cos n\theta, \tag{2.371}$$

where $\varepsilon_0 = 1$ for $n = 0$ and $\varepsilon_n = 2$ for $n \neq 0$. A time factor $\exp(-i\omega t)$ is not explicitly written.

The scattered field is an outgoing wave. By symmetry, it must be even in θ. It is a solution of the homogeneous Helmholtz equation and can be written as

$$p_s = P_0 \sum_{n=0}^{\infty} i^n \varepsilon_n b_n H_n^{(1)}(k_1 r) \cos n\theta. \tag{2.372}$$

According to (A.9) applied to harmonic waves, the displacement $\mathbf{u}^{(1)}$ is given by

$$\mathbf{u}^{(1)} = \frac{1}{\rho_1 \omega^2} \nabla_p. \tag{2.373}$$

In medium 2, the displacement vector $\mathbf{u} = \mathbf{u}^{(\ell)} + \mathbf{u}^{(t)}$ is written as

$$\mathbf{u} = \nabla\psi + \nabla\times\mathbf{A}\ ,\ \mathbf{u}^{(\ell)} = \nabla\psi\ ,\ \ \mathbf{u}^{(t)} = \nabla\times\mathbf{A}\ , \tag{2.374}$$

where the scalar potential ψ describing the compressional waves satisfies

$$\Delta\psi - \frac{1}{C_\ell^2}\frac{\partial^2\psi}{\partial t^2} = 0, \tag{2.375}$$

and the vector potential **A** describing the shear waves satisfies

$$\Delta\mathbf{A} - \frac{1}{C_t^2}\frac{\partial^2\mathbf{A}}{\partial t^2} = 0. \tag{2.376}$$

Since the direction of propagation of the incident wave is normal to the cylinder axis and the problem is infinite in the z direction, we must have $u_z = 0$. Hence

$$A_r = A_\theta = 0, \quad A_z \neq 0. \tag{2.377}$$

The solutions of equations (2.375) and (2.376) with the proper symmetry in θ are

$$\psi = P_0 \sum_{n=0}^{\infty} i^n \varepsilon_n \left[c_n J_n(k_\ell r) + d_n H_n^{(1)}(k_\ell r)\right]\cos n\theta, \tag{2.378}$$

$$A_z = P_0 \sum_{n=0}^{\infty} i^n \varepsilon_n \left[e_n J_n(k_t r) + f_n H_n^{(1)}(k_t r)\right]\sin n\theta. \tag{2.379}$$

At $r = a$ the following boundary conditions need to be satisfied (see appendix B).

- The normal component of stress is continuous:

$$\sigma_{rr}^{(1)} = p = \sigma_{rr}^{(\ell)} + \sigma_{rr}^{(t)}. \tag{2.380}$$

- The normal component of displacement is continuous:

$$u_r^{(1)} + u_r^{(\ell)} + u_r^{(t)} = 0. \tag{2.381}$$

- The tangential components of shearing stress are zeros:

$$\sigma_{r\theta}^{(\ell)} + \sigma_{r\theta}^{(t)} = 0. \tag{2.382}$$

Moreover, at the free boundary $r = b$, the normal and tangential components of stress are zero:

$$\sigma_{rr}^{(\ell)} + \sigma_{rr}^{(t)} = 0, \tag{2.383}$$

$$\sigma_{r\theta}^{(\ell)} + \sigma_{r\theta}^{(t)} = 0. \tag{2.384}$$

Since $\mathbf{u}^{(\ell)}$ and $\mathbf{u}^{(t)}$ do not depend on z, the conditions $\sigma_{rz}^{(\ell)} + \sigma_{rz}^{(t)} = 0$ and $\sigma_{\theta z}^{(\ell)} + \sigma_{\theta z}^{(t)} = 0$ at $r = a$ and $r = b$ are identically satisfied.

Expressing the stress components by displacements, by way of (B19), the boundary conditions lead to five linear equations in the five unknowns b_n, c_n, d_n, e_n, f_n from equations (2.380) to (2.384), which are solved by Cramer's rule. Since the unknown b_n appears only in (2.380) and (2.381), the general form of the determinant D_2 of the system is

$$D_{2n} = \begin{vmatrix} \alpha_{11} & \alpha_{12} & \alpha_{13} & \alpha_{14} & \alpha_{15} \\ \alpha_{21} & \alpha_{22} & \alpha_{23} & \alpha_{24} & \alpha_{25} \\ 0 & \alpha_{32} & \alpha_{33} & \alpha_{34} & \alpha_{35} \\ 0 & \alpha_{42} & \alpha_{43} & \alpha_{44} & \alpha_{45} \\ 0 & \alpha_{52} & \alpha_{53} & \alpha_{54} & \alpha_{55} \end{vmatrix} = \alpha_{11} B_n^{(1)} - \alpha_{21} B_n^{(2)}, \tag{2.385}$$

where $B_n^{(1)}$ and $B_n^{(2)}$ are the minors of D_{2n} corresponding to α_{11} and α_{21}, respectively.

The solution for the coefficient b_n is given by

$$b_n = \frac{D_{1n}}{D_{2n}}, \tag{2.386}$$

where D_{1n} is obtained by replacing the first column of D_{2n} with the terms β_1 and β_2, corresponding to the contribution of the incident pressure field to equations (2.380) and (2.381), respectively. Since D_{1n} has the same structure as D_{2n}, with β_1 and β_2 substituted for α_{11} and α_{21}, respectively, we have

$$D_{1n} = \beta_1 B_n^{(1)} - \beta_2 B_n^{(2)}. \tag{2.387}$$

The elements of D_{2n} are listed in appendix F, noting that

$$\alpha_{11} = -\left(\frac{x_1}{k_1}\right)^2 H_n^{(1)}(x_1)\,, \qquad \alpha_{21} = -x_1\, H_n^{(1)'}(x_1), \tag{2.388}$$

$$\beta_1 = \left(\frac{x_1}{k_1}\right)^2 J_n(x_1) \;, \qquad \beta_2 = x_1 \, J_n'(x_1), \tag{2.389}$$

where $x_1 = k_1 a$.

The solution (2.386) does not change if we multiply the same row in D_{1n} and D_{2n} by the same factor. By multiplying the first row by $(\rho_1 / \rho_2)\, k_t^2$, (2.386) and (2.385) can be written as

$$D_1(n, x_1) = \frac{\rho_1}{\rho_2} x_t^2 \, J_n(x_1) B_n^{(1)} - x_1 \, J_n'(x_1) B_n^{(2)}, \tag{2.390}$$

$$D_2(n, x_1) = -\frac{\rho_1}{\rho_2} x_t^2 \, H_n^{(1)}(x_1) B_n^{(1)} + x_1 \, H_n^{(1)'}(x_1) B_n^{(2)}, \tag{2.391}$$

where $x_t = k_t a$, and where we have stated $D_{1n} = D_1(n, x_1), D_{2n} = D_2(n, x_1)$.

By introducing the notations

$$\Omega_n = \frac{\partial}{\partial r} - \frac{\rho_1}{\rho_2} \frac{x_t^2}{x_1} k_1 \frac{B_n^{(1)}}{B_n^{(2)}} = \frac{\partial}{\partial r} + ik_1 \, \eta_n, \tag{2.392}$$

where

$$\eta_n = i \frac{\rho_1}{\rho_2} \frac{x_t^2}{x_1} \frac{B_n^{(1)}}{B_n^{(2)}}, \tag{2.393}$$

the solution for the total pressure field in the fluid (medium 1) is given by

$$\begin{aligned} p_t(r,\theta) &= p_i(r,\theta) + p_s(r,\theta) \\ &= P_0 \sum_{n=0}^{\infty} i^n \varepsilon_n \left[J_n(k_1 r) - \frac{\Omega_n \, J_n(k_1 a)}{\Omega_n \, H_n^{(1)}(k_1 a)} H_n^{(1)}(k_1 r) \right] \cos n\theta. \end{aligned} \tag{2.394}$$

2.3.3.2. Watson Transformation

The Watson transformation has been introduced in section 2.3.1.1, where it has been shown that

$$\sum_{n=-\infty}^{+\infty} f(n) = \frac{i}{2} \int_C \frac{e^{-i\pi\nu} f(\nu)}{\sin \pi\nu} d\nu, \tag{2.395}$$

where C is the contour defined in figure 2.28. For the series (2.394), the relation (2.395) takes the form

$$\sum_{n=0}^{\infty} \varepsilon_n \, f(n) = i \, \boldsymbol{P} \int_{\bar{C}} \frac{e^{-i\pi\nu} f(\nu)}{\sin \pi\nu} d\nu, \tag{2.396}$$

where $\bar{C}$ surrounds the positive real axis in the clockwise sense, and the principal value $\mathcal{P}$ is to be taken at the origin where only half of the contribution of the pole needs to be taken into account (fig. 2.40).

Applied to the sum of (2.371) and (2.372), the relation (2.396) gives

$$p = p_i + p_s = i\,P_0\;\mathcal{P}\int_{\bar{C}} \frac{e^{-iv\frac{\pi}{2}}\cos v\theta}{\sin v\pi}\left[J_v(k_1 r) - \frac{\Omega_v J_v(k_1 a)}{\Omega_v H_v^{(1)}(k_1 a)} H_v^{(1)}(k_1 r)\right]dv, \quad (2.397)$$

where

$$\Omega_v = \frac{\partial}{\partial r} + ik_1\,\eta_v, \quad \eta_v = i\frac{\rho_1}{\rho_2}\frac{x_t^2}{x_1}\frac{B_v^{(1)}}{B_v^{(2)}}, \quad (2.398)$$

in which $B_v^{(1)}$ and $B_v^{(2)}$ are the complex extensions of $B_n^{(1)}$ and $B_n^{(2)}$, respectively.

The difference between (2.397) and (2.260) is the localization of the zeros of the denominator $\Omega_v H_v^{(1)}(k_1 a)$ in the complex v-plane. Numerical computations of the zeros have been performed in the past for a water-filled aluminum shell in water [2.35] and have shown that for x_1 large, unlike the integrand in (2.260), the denominator of (2.397) may have zeros in the second and forth quadrants, other than those of sin $v\pi$. In order to avoid these zeros, we proceed as in [2.35] and close the contour $-\bar{C}$ by adding C_0 ,C', and C_∞, as shown in figure 2.41.

The demonstration of the convergence and vanishing of the integral in the infinite portions of the contour are shown as in 2.3.1.1, and the integral over $\bar{C}$ equals the sum of the integrals over C_0 and C'. The contribution of the integral over C' has no physical sense. It represents a sort of background integral that is usually small, and in some special cases, it is equal to zero. It decreases when $k_1 a$ increases and can be neglected for $k_1 a$ large. Some numerical verifications of this behavior for an aluminum cylinder have been performed in [2.36].

By neglecting the background integral, (2.397) transforms into

$$p = i\,P_0\int_{C_0} \frac{e^{-iv\frac{\pi}{2}}\cos v\theta}{\sin v\pi}\left[J_v(k_1 r) - \frac{\Omega_v J_v(k_1 a)}{\Omega_v H_v^{(1)}(k_1 a)} H_v^{(1)}(k_1 r)\right]dv. \quad (2.399)$$

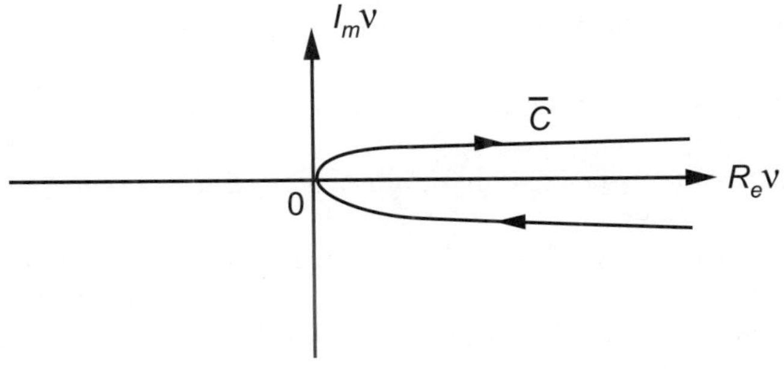

Figure 2.40. Original integration contour for the Watson transformation.

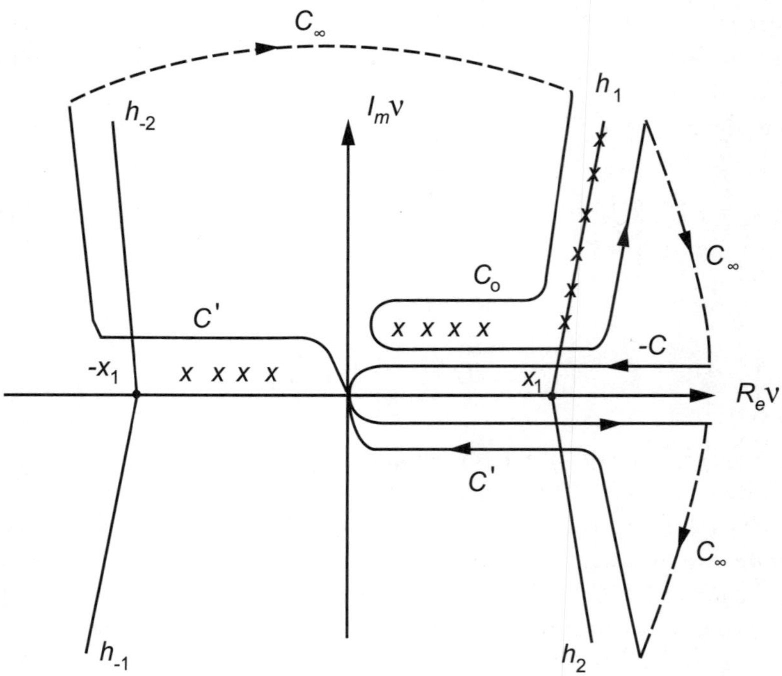

Figure 2.41. Modified contour for the Watson transformation.

When the observation point lies in the shadow region of the incident wave, the integral can be replaced by a series of residues having the creeping wave character that correspond to poles in the complex ν-plane, given by the zeros of $\Omega_\nu\, H_\nu^{(1)}(k_1 a)$. We have shown in section 2.3.1.1 that for soft or rigid cylinders, one has the zeros of $H_\nu^{(1)}(k_1 a)$ and $H_\nu^{(1)'}(k_1 a)$, which are located on the Stokes line h_1 in figure 2.41. These zeros give rise to creeping waves that encircle the cylinder with speeds somewhat below the sound speed in the surrounding fluid, and these waves are strongly attenuated. A similar result is obtained for an impedance cylinder when the surface impedance has a small imaginary part, in which case the poles remain close to those of the soft and rigid cylinder. These creeping waves are of Franz type [2.37]; they enter and leave the cylinder surface tangentially.

For elastic cylindrical shells embedded in a fluid with vacuum in the interior domain, the denominator is a combination of two minors of a 5×5 determinant containing Bessel and Hankel functions. The zeros of the denominator may be split into two parts $\{F\}$ and $\{R\}$.

The poles corresponding to $\{F\}$ have a real part $\geq k_1 a$. They are of the Franz type. For a not-too-thin shell, these poles are found numerically, very near the zeros of the rigid case. The poles corresponding to the category $\{R\}$ are located near the real axis, with a real part smaller than $k_1 a$. These poles are new and correspond to Rayleigh-type circumferential waves of elastic character that propagate with high speeds and little attenuation.

Separating the path $C_0 = C_1 + C_2$, as in figure 2.42, the total pressure in region 1 is

$$p = p_1 + p_2 = i P_0 \left(\int_{C_1} F(\nu) d\nu + \int_{C_2} F(\nu) d\nu \right),$$

where

$$\begin{aligned} F(\nu) &= \left[J_\nu(k_1 r) - \frac{\Omega_\nu J_1(k_1 a)}{\Omega_\nu H_\nu^{(1)}(k_1 a)} H_\nu^{(1)}(k_1 r) \right] e^{-i\nu\frac{\pi}{2}} \frac{\cos \nu\,\theta}{\sin \nu\,\pi} \\ &= \frac{1}{2}\left[H_\nu^{(2)}(k_1 r) - \frac{\Omega_\nu H_\nu^{(2)}(k_1 a)}{\Omega_\nu H_\nu^{(1)}(k_1 a)} H_\nu^{(1)}(k_1 r) \right] e^{-i\nu\frac{\pi}{2}} \frac{\cos \nu\,\theta}{\sin \nu\,\pi}. \end{aligned} \qquad (2.400)$$

The contour C_1 has no infinite branches. We can therefore directly apply the residue theorem to the integral p_1, which gives the contribution of the poles $\{R\}$:

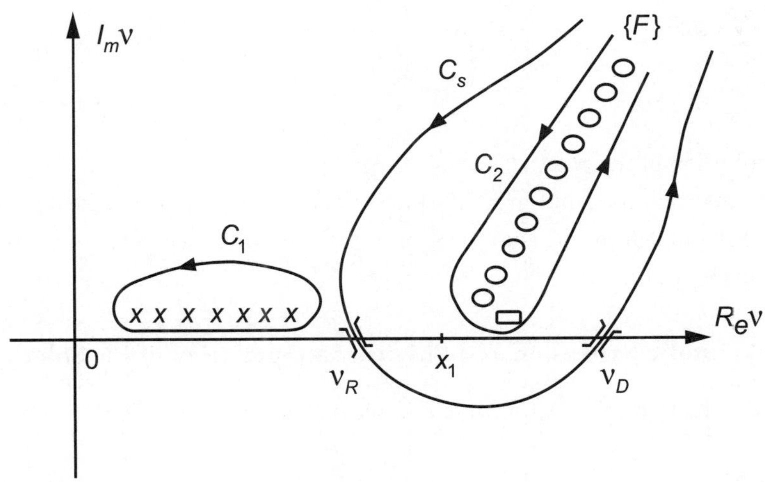

Figure 2.42. Final integration contours for the Watson transformation.

$$p_1 = \pi P_0 \sum_{\ell=1}^{N} e^{-i\nu_\ell \frac{\pi}{2}} \frac{\cos \nu_\ell \theta}{\sin \pi \nu_\ell} \frac{\Omega_{\nu_\ell} H_{\nu_\ell}^{(2)}(k_1 a)}{\left[\frac{\partial}{\partial \nu} \Omega_\nu H_\nu^{(1)}(k_1 a)\right]_{\nu=\nu_\ell}} H_{\nu_\ell}^{(1)}(k_1 r). \tag{2.401}$$

The integral p_2 is treated as in section 2.3.1.2 by applying the Imai transformation (2.279), which we can write as

$$\cos \nu \theta = e^{i\nu\pi} \cos \nu(\pi - \theta) - i e^{i\nu(\pi-\theta)} \sin \pi \nu, \tag{2.402}$$

permitting to split off the geometric direct and reflected waves

$$p_2 = p_c + p_g, \tag{2.403}$$

where the integral

$$p_g = \frac{1}{2} P_0 \int_{C_2} e^{i\nu\left(\frac{\pi}{2}-\theta\right)} \left[H_\nu^{(2)}(k_1 r) - \frac{\Omega_\nu H_\nu^{(2)}(k_1 a)}{\Omega_\nu H_\nu^{(1)}(k_1 a)} H_\nu^{(1)}(k_1 r) \right] d\nu \tag{2.404}$$

contains the geometric waves and where

$$\begin{aligned} p_c &= \frac{1}{2} i P_0 \int_{C_2} e^{i\nu\frac{\pi}{2}} \frac{\cos \nu(\pi - \theta)}{\sin \nu\theta} \left[H_\nu^{(2)}(k_1 r) - \frac{\Omega_\nu H_\nu^{(2)}(k_1 a)}{\Omega_\nu H_\nu^{(1)}(k_1 a)} H_\nu^{(1)}(k_1 r) \right] d\nu \\ &= \pi P_0 \sum_{n=1}^{\infty} e^{i\nu_n \frac{\pi}{2}} \frac{\cos \nu_n(\pi - \theta)}{\sin \nu_n \theta} \frac{\Omega_{\nu_n} H_{\nu_n}^{(2)}(k_1 a)}{\left[\frac{\partial}{\partial \nu} \Omega_\nu H_\nu^{(1)}(k_1 a)\right]_{\nu=\nu_n}} H_{\nu_n}^{(1)}(k_1 r) \end{aligned} \tag{2.405}$$

gives the contribution of the poles $\{F\}$ of the integral p_c.

Observing that the integrand in (2.404) has no poles on the real ν-axis, we can proceed as in section 2.3.1.2 and deform the contour C_2 into the contour C_s of figure 2.42, passing through the saddle points ν_R and ν_D.

2.3.3.3. Asymptotic Expansion and Physical Interpretation of the Solution

We first consider the integral (2.404), which we separate in two terms $p_g = p_{g_1} + p_{g_2}$, where

$$p_{g_1} = \frac{1}{2} P_0 \int_{C_s} e^{i\nu\left(\frac{\pi}{2}-\theta\right)} H_\nu^{(2)}(k_1 r)\, d\nu, \tag{2.406}$$

$$p_{g_2} = -\frac{1}{2} P_0 \int_{C_s} e^{i\nu\left(\frac{\pi}{2}-\theta\right)} \frac{\Omega_\nu H_\nu^{(2)}(k_1 a)}{\Omega_\nu H_\nu^{(1)}(k_1 a)} H_\nu^{(1)}(k_1 r)\, d\nu. \tag{2.407}$$

In order to asymptotically evaluate the integrals (2.406) and (2.407), we proceed as in section 2.3.1.2. In both integrals, the Hankel functions $H_\nu^{(1)}(k_1 a)$ and $H_\nu^{(2)}(k_1 a)$ are replaced by their Debye asymptotic expansions (2.285). Then (2.406) gives

$$p_{g_1} \cong \frac{P_0 e^{i\frac{\pi}{4}}}{\sqrt{2\pi}} \int_{C_s} \left[(k_1 r)^2 - \nu^2\right]^{-1/4} e^{i\left[\nu\left(\frac{\pi}{2}-\theta+\operatorname{Arc}\cos\frac{\nu}{k_1 r}\right)-\sqrt{(k_1 r)^2-\nu^2}\right]} d\nu = \frac{P_0 e^{i\frac{\pi}{4}}}{\sqrt{2\pi}} \int_{C_s} F(\nu) e^{iq(\nu)}\, d\nu \tag{2.408}$$

with

$$q(\nu) = \nu\left(\frac{\pi}{2} - \theta + \operatorname{Arc}\cos\frac{\nu}{k_1 r}\right) - \sqrt{(k_1 r)^2 - \nu^2}, \quad F(\nu) = \left[(k_1 r)^2 - \nu^2\right]^{-1/4}.$$

The stationary points are given by the roots of the following equation:

$$q'(\nu) = \frac{\pi}{2} - \theta + \operatorname{Arc}\cos\frac{\nu}{k_1 r} = 0. \tag{2.409}$$

This equation has only one real stationary point, given by

$$\nu_D = k_1 r \sin\theta, \tag{2.410}$$

and since $q''(\nu_D) = -(k_1 r \sin\theta)^{-1} < 0$, the SDP Γ makes an angle $\alpha = -\pi/4$ with the positive real ν-axis.

Applying the general formula (2.59), giving the leading term of the asymptotic expansion of an integral

$$\int_\Gamma F(\nu) e^{iq(\nu)} d\nu \cong F(\nu_s) e^{iq(\nu_s)} \sqrt{\frac{2\pi}{|q''(\nu_s)|}}\, e^{i\alpha}, \tag{2.411}$$

we obtain readily

$$p_{g_1} = P_0 e^{-ik_1 r|\cos\theta|}. \tag{2.412}$$

Observing that the Debye expansions (2.285) imply $\alpha_1 = \text{Arc}\cos\left(\frac{\nu}{k_1 r}\right) > 0$, it follows that (2.412) is only valid for $\theta > \frac{\pi}{2}$ (see fig. 2.34, with the source removed to infinity).

Equation (2.412) can also be written as

$$p_{g_1} = P_0\, e^{ik_1 r\cos\theta}, \qquad \theta > \frac{\pi}{2},$$

which is the incident field at the observation point (r, θ).

Similarly, the integral p_{g_2} is approximated by

$$p_{g_2} \cong -\frac{P_0}{\sqrt{2\pi}}\, e^{-i\frac{\pi}{4}} \int_{C_s} \frac{\Omega_\nu H_\nu^{(2)}(k_1 a)}{\Omega_\nu H_\nu^{(1)}(k_1 a)} \left[(k_1 r)^2 - \nu^2\right]^{-1/4} e^{i\left[\nu\left(\frac{\pi}{2}-\theta-\text{Arc}\cos\frac{\nu}{k_1 r}\right)+\sqrt{(k_1 r)^2-\nu^2}\right]} d\nu.$$

(2.413)

We further replace $H_\nu^{(1,2)}(ka)$ with their Debye expansions. Those depend on whether $R_e\nu < ka$ or $R_e\nu > ka$. Following the demonstration given in section 2.3.1.2, we obtain the relation

$$H_\nu^{(1)}(ka) \approx -H_\nu^{(2)}(ka) \tag{2.414}$$

for $R_e\nu > ka$, using (2.295) and (2.296), and consequently we obtain

$$p_{g_2} \cong -\frac{P_0}{\sqrt{2\pi}}\, e^{-i\frac{\pi}{4}} \int_{C_s} \left[(k_1 r)^2 - \nu^2\right]^{-1/4} e^{i\left[\nu\left(\frac{\pi}{2}-\theta-\text{Arc}\cos\frac{\nu}{k_1 r}\right)+\sqrt{(k_1 r)^2-\nu^2}\right]} d\nu, \tag{2.415}$$

which is of the form

$$p_{g_2} \cong \frac{P_0}{\sqrt{2\pi}}\, e^{-i\frac{\pi}{4}} \int_{C_s} F(\nu)\, e^{iq(\nu)}\, d\nu \tag{2.416}$$

with

$$q(\nu) = \nu\left(\frac{\pi}{2} - \theta - \text{Arc}\cos\frac{\nu}{k_1 r}\right) + \sqrt{(k_1 r)^2 - \nu^2}. \tag{2.417}$$

Since

$$q'(\nu) = \frac{\pi}{2} - \theta - \operatorname{Arc\,cos}\frac{\nu}{k_1 r} = \frac{\pi}{2} - \theta - \alpha_1,$$

the integrand of (2.415) has a stationary point given by $\alpha_1 = \frac{\pi}{2} - \theta$ or $\nu_D = k_1 r \sin\theta$, but since $\alpha_1 > 0$, we must now have $\theta < \frac{\pi}{2}$.

Furthermore,

$$q''(\nu_D) = (k_1 r \cos\theta)^{-1} > 0.$$

Hence the tangent to the SDP at $\nu = \nu_D$ makes an angle $\alpha = \frac{\pi}{4}$ with the positive real ν-axis. Applying the steepest descent method, using (2.411), we obtain

$$p_{g_2} = P_0\, e^{ikr\cos\theta}, \quad \theta < \frac{\pi}{2}, \tag{2.418}$$

which is the incident pressure field in the domain $\theta < \frac{\pi}{2}$. Consequently, by adding (2.418) to (2.412), we obtain the incident pressure field in the whole domain $(0, \pi)$ of variation of θ. The analysis excludes the line $\theta = \frac{\pi}{2}$, although the solution remains valid there.

For $R_e(\nu) < ka$, the dominant terms of the Debye asymptotic expansions of $H_\nu^{(1,2)}(ka)$ and $H_\nu^{(1,2)'}(ka)$ are given by the relations in front of (2.302), where $\psi(x, \nu)$ is defined by (2.302). Taking (2.398) into account, it follows that

$$\frac{\Omega_\nu H_\nu^{(2)}(x_1)}{\Omega_\nu H_\nu^{(1)}(x_1)} \cong -\frac{\psi'(x_1,\nu) - \eta_\nu}{\psi'(x_1,\nu) + \eta_\nu}\, e^{-2i\psi(x_1,\nu)} \tag{2.419}$$

and

$$p_{g_2} \cong \frac{1}{\sqrt{2\pi}} P_0\, e^{-i\frac{\pi}{4}} \int_{C_s} e^{i\nu\left(\frac{\pi}{2}-\theta\right)} \frac{\psi'(x_1,\nu) - \eta_\nu}{\psi'(x_1,\nu) + \eta_\nu} \left[(k_1 r)^2 - \nu^2\right]^{-1/4} e^{i\left[-2\psi(x_1,\nu) - \nu \operatorname{Arc\,cos}\frac{\nu}{k_1 r} + \sqrt{(k_1 r)^2 - \nu^2}\right]} d\nu. \tag{2.420}$$

This integral has the same form as (2.416), with

$$q(\nu) = \nu\left(\frac{\pi}{2} - \theta - \text{Arc}\cos\frac{\nu}{k_1 r} + 2\text{Arc}\cos\frac{\nu}{k_1 r}\right) - 2\sqrt{(k_1 a)^2 - \nu^2} + \sqrt{(k_1 r)^2 - \nu^2} + \frac{\pi}{2}. \tag{2.421}$$

The integrand of p_{g_2} has a stationary point ν_R, verifying

$$q'(\nu_R) = \frac{\pi}{2} - \theta - \text{Arc}\cos\frac{\nu_R}{k_1 r} + 2\text{Arc}\cos\frac{\nu_R}{k_1 a} = 0. \tag{2.422}$$

If we let $\nu_R = k_1 a \sin\theta_R$, then (2.422) yields

$$\frac{\pi}{2} - \theta - \text{Arc}\cos\left(\frac{a}{r}\sin\theta_R\right) + \pi - 2\theta_R = 0.$$

Since $\frac{a}{r} << 1$, this equation has a unique solution given by $\theta_R = \frac{\pi - \theta}{2} + \varepsilon$, where $0 < \varepsilon << 1$ when the observation point lies outside the shadow region. The comments about this solution are similar to those given in section 2.3.1.2 after the equation (2.307).

Applying the steepest descent method to the integral (2.420) and noting that

$$q''(\nu_R) = \frac{1}{k_1 r}\frac{1}{\sqrt{1 - \left(\frac{a}{r}\right)^2 \sin^2\theta_R}} - \frac{2}{k_1 a \cos} < 0,$$

$$\psi'(k_1 a, \nu_R) = \cos\theta_R,$$

$$q(\nu_R) = \sqrt{(k_1 r)^2 - \nu_R^2} - 2\sqrt{(k_1 a)^2 - \nu_R^2} + \frac{\pi}{2},$$

$$\frac{\psi'(ka, \nu_R) - \eta_{\nu_R}}{\psi'(ka, \nu_R) + \eta_{\nu_R}} = \frac{\cos\theta_R - \eta(\theta_R)}{\cos\theta_R + \eta(\theta_R)} = R,$$

where R is the reflection coefficient and $\eta(\theta_R) = \eta_{\nu R}$, we obtain

$$p_{g_2} \cong P_0\left(\frac{2}{a\cos\theta_R} - \frac{1}{\sqrt{r^2 - a^2\sin^2\theta_R}}\right)^{-1/2} e^{ik_1\left(\sqrt{r^2 - a^2\sin^2\theta_R} - 2a\cos\theta_R\right)}\left(r^2 - a^2\sin^2\theta_R\right)^{-1/4}. \tag{2.423}$$

As in section 2.3.1.2, it is possible to give a geometrical interpretation of this relation.

Referring to figure 2.43, we have

$$\sqrt{r^2 - a^2 \sin^2 \theta_R} - 2a \cos \theta_R = MQ - a \cos \theta_R, \tag{2.424}$$

and since $\exp(-k_1 a \cos\theta_R)$ is the phase at M, the phase of p_{g2} is equal to the phase at the reflection point M augmented by the phase along the reflected ray from M to the observation point Q.

Likewise, the divergence factor in (2.423) yields

$$A = \sqrt{\frac{a \cos \theta_R}{2MQ + a \cos \theta_R}} = \sqrt{\frac{\rho_{cR}}{\rho_{cR} + MQ}}, \tag{2.425}$$

where ρ_{cR} is the radius of curvature of the reflected wave front, which for an incident plane wave reads

$$\rho_{cR} = \frac{a \cos \theta_R}{2}.$$

Inserting (2.424) and (2.425) into (2.423) yields the final result:

$$p_{g_2} \cong p_i(M) \frac{\cos \theta_R - \eta(\theta_R)}{\cos \theta_R + \eta(\theta_R)} \sqrt{\frac{\rho_{cR}}{\rho_{cR} + MQ}} \, e^{ik\,MQ}, \tag{2.426}$$

where $p_i(M)$ is the incident field at the reflection point M. We recognize in (2.426) the reflected field $p_R(Q)$ at the observation point $Q(r,\theta)$. Therefore, the integral (2.404) gives the sum of the incident field and the reflected field:

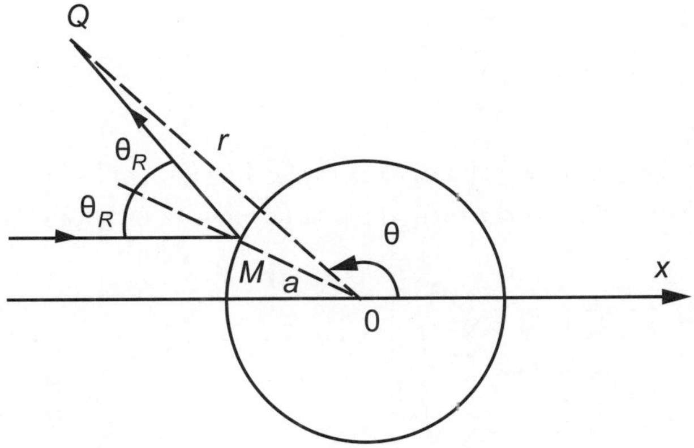

Figure 2.43. Ray geometry for the reflected field.

$$p_g(r,\theta) = p_i(r,\theta) + p_R(r,\theta). \tag{2.427}$$

The validity of (2.426) is limited by the validity of the Debye asymptotic expansions of the Hankel functions $H_\nu^{(1)}(ka)$ and $H_\nu^{(2)}(ka)$. Those are not valid for $\theta \cong \pi$ or $\theta_R \cong \frac{\pi}{2}$, where $\nu_R \cong k_1 a$. When $|\nu - k_1 a| \leq 0\left(|ka|^{1/3}\right)$, they can be replaced by Airy functions, as in (2.328). The GA formulation is therefore not valid close to the shadow boundary, which constitutes a transition region to the creeping wave formulation. Uniform solutions through the shadow boundary when a plane wave strikes an elastic shell are derived in chapter 3, section 3.4.

2.3.3.4. Field Diffracted by Creeping Waves

For an observation point lying outside the shadow region, the creeping wave contribution to the scattered field is given by the residue series (2.405), where ν_n are the zeros of the characteristic equation

$$\frac{\partial}{\partial r} H_\nu^{(1)}(k_1 r) + ik_1 \ \eta_\nu \ H_\nu^{(1)}(k_1 r) = 0 \ , \quad r = a. \tag{2.428}$$

Owing to the complexity of the expression of the equivalent impedance η_ν, which according to (2.398) depends on the quotient of two 4×4 determinants, the elements of which are linear combinations of Bessel and Hankel functions of order ν and argument $k_1 a$, $k_\ell a$, or $k_t a$, the zeros of (2.428) can only be found numerically. However, for a not-too-thin elastic shell, $|\eta_\nu|$ remains small compared to unity, and consequently the zeros of (2.428) are found near the zeros of the rigid case. Hence for $r >> a$ we can make the hypothesis

$$|k_1 r - \nu_n| >> |\nu_n|^{1/3}$$

and replace the Hankel function $H_\nu^{(1)}(k_1 r)$ with the leading term of its Debye asymptotic expansion. Thus, far from the surface of the shell, the residue series (2.405) can be approximated by

$$p_c(r,\theta) \cong e^{-3i\frac{\pi}{4}} \sqrt{\frac{2\pi}{k_1}} P_0 \frac{e^{ik_1\sqrt{r^2-a^2}}}{\left(r^2 - a^2\right)^{1/4}} \sum_{n=1}^{\infty} \frac{e^{i\nu_n\left(\frac{3\pi}{2}+\psi-\phi_1\right)} + e^{i\nu_n\left(\frac{3\pi}{2}-\psi-\phi_1\right)}}{1 - e^{2i\nu_n \pi}} \frac{\Omega_{\nu_n} H_{\nu_n}^{(2)}(k_1 a)}{\left[\frac{\partial}{\partial \nu} \Omega_\nu H_\nu^{(1)}(k_1 a)\right]_{\nu=\nu_n}}, \tag{2.429}$$

where $\psi = \pi - \theta$ and $\phi_1 = \text{Arc}\cos\dfrac{\nu_n}{k_1\rho}$.

Since $I_m \nu_n > 0$, we can replace $\left(1-e^{2\pi i\nu_n}\right)^{-1}$ with

$$\left(1-e^{2\pi i\nu_n}\right)^{-1} = \sum_{m=0}^{\infty} e^{2im\nu_n}. \tag{2.430}$$

With this modification, (2.429) has a simple physical interpretation in terms of creeping waves similar to those obtained in section 2.3.1.2.

For $m = 0$, each term in the residue series (2.429) may be interpreted as being associated with an incident wave that strikes the cylindrical shell at a glancing angle, travels along the surface of the cylindrical shell, and sheds energy continually during its progress, due to the imaginary part of ν_n. The associated interpretation in terms of geometrical rays is shown in figure 2.44.

It is evident from figure 2.44 that $r_1 = \sqrt{r^2 - a^2}$ and that for an observation Q located outside the shadow region, we have

$$\begin{aligned} \frac{3\pi}{2} + \psi - \phi_1 &= \frac{\ell}{a}, \\ \frac{3\pi}{2} - \psi - \phi_1 &= \frac{\ell'}{a}, \end{aligned} \tag{2.431}$$

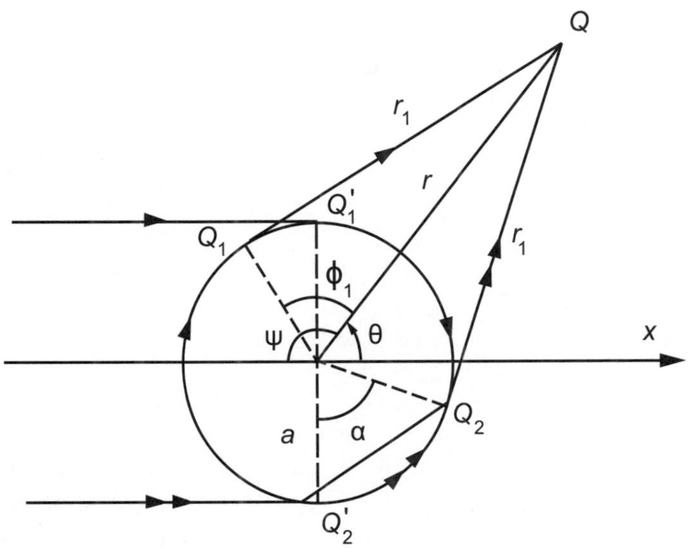

Figure 2.44. Geometrical interpretation of the creeping wave solution.

where ℓ is the arc length Q'_1Q_1 in the clockwise rotation, and ℓ' is the arc length Q'_2Q_2 in the trigonometric rotation. When the observation point is in the shadow region, the arc length Q'_1Q_1 is longer than one tour around the cylinder. In this case, the creeping ray corresponding to the shortest path from Q_1 to Q'_1 is absent in (2.429), which must be replaced by the residue series corresponding to the solution before the application of the Imai transform given by (2.399) integrated over C_2. The solution (2.429) and figure 2.44 correspond to plane wave incidence. When the cylindrical shell is struck by the pressure field emitted by a line source parallel to the cylinder axis, the geometrical interpretation of the solution is identical to that which is given in figure 2.32.

Proceeding as in section 2.3.1.2, we obtain the following for the field, due to the dominant creeping rays:

$$\bar{p}_c(r,\theta) \cong e^{-3i\frac{\pi}{4}} P_0 \sqrt{\frac{2\pi}{k_1}} \sum_{n=1}^{\infty} e^{\frac{i\,\nu_n}{a}(\ell+\ell')} R_n \frac{e^{ik r_1}}{\sqrt{r_1}}, \tag{2.432}$$

where

$$R_n = \frac{\Omega_{\nu_n} H^{(2)}_{\nu_n}(k_1 a)}{\left[\frac{\partial}{\partial \nu} \Omega_\nu H^{(1)}_\nu (k_1 a)\right]_{\nu=\nu_n}}. \tag{2.433}$$

Comparing (2.432) to the general expression (1.240) derived in chapter 1, section 1.9, which for an incident plane wave striking a circular cylindrical shell at normal incidence reads

$$p(M) = P_0 \sum_{n=1}^{\infty} D^{(n)}_{sr,cr} D^{(n)}_{cr,sr} e^{\frac{i\,\nu_n}{a}(\ell+\ell')} D_s \frac{e^{ik r_1}}{\sqrt{r_1}} \tag{2.434}$$

$$\text{with } D_s = \frac{e^{i\frac{\pi}{4}}}{\sqrt{8\pi k}}, \tag{2.435}$$

we obtain

$$D_n^2 = D^{(n)}_{sr,cr} D^{(n)}_{cr,sr} = -4\pi R_n. \tag{2.436}$$

When the zeros of the characteristic equation (2.428) are located in the vicinity of the Stokes line h_1, the Hankel functions can be approximated for *ka* large by Airy functions, as in section 2.3, equation (2.328). Moreover, the formulas (2.323) to (2.327) are still valid and show that R_n does not explicitly depend on the equivalent impedance η_{ν_n}. Consequently, the diffraction coefficient is still given by (2.330). The impedance η_{ν_n} comes indirectly in the expression of D_n^2 through the solution ν_n of the characteristic equation (2.428).

2.3.3.5. Noncircular Convex Cylindrical Shell

The generalization of (2.432) to a noncircular convex cylindrical shell follows the same rules as in section 2.3.1.4 and leads to a result that is identical to (2.337).

2.3.3.6. Oblique Incidence for Creeping Waves

In this section we consider a plane wave obliquely incident on a circular cylindrical shell. According to the translation invariance, the pressure field at an observation point $M(\rho, \theta, z)$ in the fluid of propagation constant k_1, is given by

$$p_s(\rho,\theta,z) = p_1(\rho,\theta;k_{1\perp})\, e^{ik_z z},$$

where $k_{1\perp} = k_1 \sin\beta_0$, $k_z = k_1 \cos\beta_0$, β_0 being the acute angle between the incident ray and the oz axis, and where $p_1(r,\theta;k_{1\perp})$ verifies the homogeneous Helmholtz equation for normal incidence. Hence at oblique incidence the formulas (2.371) and (2.372) giving the pressure in the fluid and those giving the displacements in the elastic medium, particularly the scalar and vector potentials, are multiplied by $\exp(ik_z z)$.

In order to solve the problem, we need to apply the boundary conditions that are given by the five conditions in (2.380) to (2.384), augmented by the following two conditions:

$$\sigma_{rz}^{(\ell)} + \sigma_{rz}^{(t)} = 0\,, \quad r = a\,,b$$

or

$$\sigma_{\theta z}^{(\ell)} + \sigma_{\theta z}^{(t)} = 0\,, \quad r = a,\ b.$$

These must be verified at $r = a$ and $r = b$ because $\mathbf{u}^{(\ell)}$ and $\mathbf{u}^{(t)}$ now depend on z and also because among the three components of the vector potential $\mathbf{A}$, only two are necessary.

Applying the boundary conditions leads to seven linear equations of seven unknowns, the solution of which, for the amplitude b_n of the scattered field, is given by (2.386) where D_{1n} and D_{2n} are now determinants of dimensions 7×7 having the same structure as (2.385). Especially the first column has only two elements that differ from zero (see appendix F), given by

$$\alpha_{11} = -\frac{\rho_1}{\rho_2} x_t^2\, H_n^{(1)}(x_{1\perp}), \qquad \alpha_{21} = -x_{1\perp} H_n^{(1)'}(x_{1\perp}),$$

$$\beta_1 = \frac{\rho_1}{\rho_2} x_t^2\, J_n(x_{1\perp}), \qquad \beta_2 = x_{1\perp}\, J_n'(x_{1\perp}).$$

By expressing the determinants D_{1n} and D_{2n} with respect to the minors $\bar{B}_n^{(1)}$ and $\bar{B}_n^{(2)}$ corresponding to the elements of the first column of D_{1n} and D_{2n}, we obtain, as in (2.385) and (2.387),

$$D_{2n} = \alpha_{11}\,\bar{B}_n^{(1)} - \alpha_{21}\,\bar{B}_n^{(2)},$$
$$D_{1n} = \beta_1\,\bar{B}_n^{(1)} - \beta_2\,\bar{B}_n^{(2)},$$

where now $\bar{B}_n^{(1)}$ and $\bar{B}_n^{(2)}$ are of dimensions 6×6. The unknown b_n may therefore be written in the form

$$b_n = -\frac{\bar{\Omega}_n\, J_n\left(k_{1\perp} a\right)}{\bar{\Omega}_n\, H_n^{(1)}\left(k_{1\perp} a\right)},$$

where

$$\bar{\Omega}_n = \frac{\partial}{\partial r} - k_{1\perp}\frac{\rho_1}{\rho_2}\frac{x_t^2}{x_{1\perp}}\frac{\bar{B}_n^{(1)}}{\bar{B}_n^{(2)}} = \frac{\partial}{\partial r} + ik_{1\perp}\bar{\eta}_n$$

and

$$\bar{\eta}_n = i\frac{\rho_1}{\rho_2}\frac{x_t^2}{x_1}\frac{\bar{B}_n^{(1)}}{\bar{B}_n^{(2)}} \tag{2.437}$$

with $x_1 = k_1 a$, $x_t = k_t a$.

Consequently, the expression of $p_1\left(r,\theta;k_{1\perp}\right)$ can be deduced from the solution of the diffraction problem for normal incidence by replacing the equivalent impedance η_n by $\bar{\eta}_n$, which is a rather complicated function of β_0 (see appendix F).

Proceeding as in section 2.3.1.7 and taking into account the solution obtained for normal incidence, it can be rapidly shown that the contribution of the creeping wave $QQ'M$ (fig. 2.38) is given by (2.358) in which $\alpha_n = y_n \exp\left(i\pi/3\right)$, where y_n are the zeros of the equation

$$A_i'(-y) + i\bar{\eta}(y)\, m\sin\beta_0 e^{i\pi/3}\, A_i(-y) = 0, \tag{2.438}$$

where $\bar{\eta}(y) = \bar{\eta}_{v(y)}$ is given by (2.437) and

$$v(y) = k\bar{\rho}\sin^3\beta_0 + m\sin\beta_0\, e^{i\pi/3}\, y,$$

where $\bar{\rho}$ is the radius of curvature of the geodesic.

As in section 2.3.1.6, this result can be readily extended to a noncircular cylindrical shell at oblique incidence by taking into account that α_n is now a function of the abscissa ℓ on the geodesic since $\alpha_n(\ell) = y_n(\ell)e^{i\pi/3}$ and $y_n(\ell)$ are the zeros of (2.438), where $\overline{\eta}(y)$ and m depend on the local radius of curvature.

2.3.3.7. Elastic Surface Waves

The Rayleigh-type circumferential waves obtained in section 2.3.3.2 are also known as elastic surface waves. They correspond to poles located near the real axis with a real part smaller than $k_1 a$. The residue at these poles is given by the series of elastic surface wave modes (2.401). The propagation constants of these modes are given by the poles ν_n of the complex function

$$r(\nu) = \frac{\Omega_\nu H_\nu^{(2)}(x_1)}{\Omega_\nu H_\nu^{(1)}(x_1)}, \tag{2.439}$$

where $x_1 = k_1 a$.

Replacing the Hankel functions $H_\nu^{(1)}(x_1)$ and $H_\nu^{(2)}(x_1)$ and their derivatives with respect to x_1 by the dominant term of their Debye asymptotic expansions, we obtain, as in (2.51),

$$r(\nu) \cong -\frac{\psi'(x_1,\nu) - \eta_\nu}{\psi'(x_1,\nu) + \eta_\nu}\, e^{-2i\psi(x_1,\nu)}, \tag{2.440}$$

where

$$\psi(x_1,\nu) = \sqrt{x_1^2 - \nu^2} - \nu \operatorname{Arccos}\frac{\nu}{x} - \frac{\pi}{4}, \tag{2.441}$$

$$\psi'(x_1,\nu) = \sqrt{1 - \left(\frac{\nu}{x}\right)^2}.$$

Hence

$$r(\nu) = -R(\nu)\, e^{-2i\psi(x_1,\nu)}, \tag{2.442}$$

where

$$R(\nu) = \frac{\sqrt{1 - \left(\dfrac{\nu}{x_1}\right)^2} - \eta_\nu}{\sqrt{1 - \left(\dfrac{\nu}{x_1}\right)^2} + \eta_\nu} \tag{2.443}$$

is the analytical extension in the complex ν-plane of the reflection coefficient.

Coming back to the integral form of the solution for the elastic surface waves

$$p_1 = \frac{i P_0}{2} \int_{C_1} e^{-i\nu\frac{\pi}{2}} \frac{\cos\nu\theta}{\sin\nu\pi} \left[H_\nu^{(2)}(k_1 r) - \frac{\Omega_\nu H_\nu^{(2)}(x_1)}{\Omega_\nu H_\nu^{(1)}(x_1)} H_\nu^{(1)}(k_1 r) \right] d\nu \tag{2.444}$$

and replacing $r(\nu)$ by its expression (2.442), we obtain the following for the residue series:

$$p_1 = -P_0\sqrt{2\pi} \sum_{\ell=1}^{N} e^{-i\nu_\ell\frac{\pi}{2}} \frac{\cos\nu_\ell\theta}{\sin\nu_\ell\pi} \frac{\sqrt{1-\left(\frac{\nu_\ell}{x_1}\right)^2} - \eta(\nu_\ell)}{\left[\frac{\partial}{\partial\nu}\left(\sqrt{1-\left(\frac{\nu}{x_1}\right)^2} + \eta(\nu)\right)\right]_{\nu=\nu_\ell}} \tag{2.445}$$

$$\times \frac{e^{i\left[\psi(k_1 r,\nu_\ell) - 2\psi(x_1,\nu_\ell)\right]}}{\left[(k_1 r)^2 - \nu_\ell^2\right]^{1/4}},$$

where ν_ℓ are the poles of $R(\nu)$ and N are the number of poles inside the contour C_1 (fig. 2.42).

By stating

$$\nu_\ell = k_1 a \sin\theta_\ell, \quad \theta_\ell = \theta'_\ell + i\theta''_\ell \tag{2.446}$$

and assuming that $\theta''_\ell << \theta'_\ell$, it is possible to give a geometrical interpretation of (2.445) in harmony with the Fermat principle applied to surface waves (see the end of section 1.3).

At once, we simplify equation (2.445) by using the transformation $\nu = k_1 a \sin\theta$ and noting that

$$\frac{\partial}{\partial\nu} = \frac{1}{k_1 a\cos\theta}\frac{\partial}{\partial\theta},$$

which gives

$$\left[\frac{\partial}{\partial\nu}\left(\sqrt{1-\left(\frac{\nu}{x_1}\right)^2} + \eta(\nu)\right)\right]_{\nu=\nu_\ell} = -\frac{1}{k_1 a\cos\theta_\ell}\left[\sin\theta_\ell - \left(\frac{\partial\eta}{\partial\theta}\right)_{\theta=\theta_\ell}\right],$$

and since θ_ℓ verifies

$$\cos\theta_\ell = -\eta(\theta_\ell),$$

we obtain

$$p_1 = -i P_0 \sqrt{8\pi}\, k_1 a \sum_{\ell=1}^{N} e^{i\nu_\ell \frac{\pi}{2}} \frac{e^{i\nu_\ell\theta} + e^{-i\nu_\ell\theta}}{1 - e^{2i\nu_\ell\pi}} \frac{\cos^2\theta_\ell}{\sin\theta_\ell - \left[\frac{\partial}{\partial\theta}\eta(\theta)\right]_{\theta=\theta_\ell}} \frac{e^{i\left[\psi(k_1 r, \nu_\ell) - 2\psi(x_1, \nu_\ell)\right]}}{\left[(k_1 r)^2 - \nu_\ell^2\right]^{1/4}}. \tag{2.447}$$

Furthermore, according to (2.441), the phase term $\psi(k_1 r - \nu_\ell) - 2\psi(x_1, \nu_\ell)$ yields

$$\psi(k_1 r - \nu_\ell) - 2\psi(x_1, \nu_\ell) = k_1\sqrt{r^2 - a^2\sin^2\theta_\ell} + \nu_\ell(\pi - 2\theta_\ell - \beta) - 2k_1 a\cos\theta_\ell + \frac{\pi}{4}, \tag{2.448}$$

where

$$\beta = \beta' + i\beta'' = \text{Arc}\cos\left(\frac{a}{r}\sin\theta_\ell\right). \tag{2.449}$$

Equation (2.449) is equivalent to the following system:

$$\cos\beta'\, ch\beta'' = \frac{a}{r}\sin\theta'_\ell,$$

$$\sin\beta'\, sh\beta'' \cong -\frac{a}{r}\theta''_\ell\cos\theta'_\ell,$$

the solution of which is given by

$$\cos\beta' \cong \frac{a}{r}\sin\theta'_\ell\ , \quad sh\beta'' \cong -\frac{a}{r}\theta''_\ell\cos\theta'_\ell$$

and consequently

$$\beta' \cong \text{Arc}\cos\left(\frac{a}{r}\sin\theta'_\ell\right), \quad \beta'' = -\text{Arg}\, sh\left(\frac{a}{r}\theta''_\ell\cos\theta'_\ell\right). \tag{2.450}$$

Now in order to give a geometrical interpretation of the solution, we introduce the ray path $Q_\ell Q'_\ell M$, as shown in figure 2.45, which corresponds to the ℓth elastic surface wave launched at Q_ℓ propagating along the arc $Q_\ell Q'_\ell$ and radiating to the observation point M along the space ray $Q'_\ell M$. A second surface ray is launched at $\bar{Q}_\ell$ and surrounds the cylinder in the opposite direction.

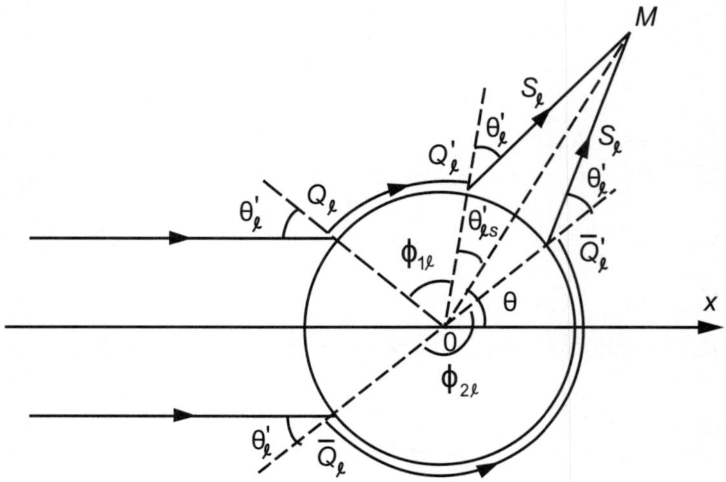

Figure 2.45. Geometrical representation of the elastic surface rays.

It can be readily shown that

$$\sqrt{r^2 - a^2 \sin^2 \theta'_\ell} = s_\ell + a\cos\theta'_\ell, \tag{2.451}$$

$$\sqrt{r^2 - a^2 \sin^2 \theta_\ell} \cong \sqrt{r^2 - a^2 \sin^2 \theta'_\ell} + i\frac{a^2\theta''_\ell \sin\theta'_\ell}{\sqrt{r^2 - a^2 \sin^2 \theta'_\ell}} = s_\ell + a\cos\theta'_\ell + i\frac{a^2\theta''_\ell \sin\theta'_\ell}{s_\ell + a\cos\theta'_\ell}, \tag{2.452}$$

where $s_\ell = Q'_\ell M$.

Since $\theta''_\ell << \theta'_\ell$ and $s_\ell >> a$, the imaginary part in (2.452) can be neglected. Hence

$$\sqrt{r^2 - a^2 \sin^2 \theta_\ell} \cong s_\ell + a\cos\theta'_\ell. \tag{2.453}$$

We also neglect β'' in (2.450). In the triangle $OQ'_\ell M$, we have

$$\frac{a}{r}\sin\theta'_\ell = \sin\left(\theta'_\ell - \theta'_{\ell s}\right), \tag{2.454}$$

where the real angle $\theta'_{\ell s}$ is defined in figure 2.45. Consequently, since

$$\cos\beta' = \frac{a}{r}\sin\theta'_\ell,$$

we have

$$\beta' = \frac{\pi}{2} - \theta'_\ell + \theta'_{\ell s}. \tag{2.455}$$

Inserting (2.453) and (2.455) into (2.448) yields

$$\psi(k_1 r, \nu_\ell) - 2\psi(x_1, \nu_\ell) = k_1 s_\ell - k_1 a \cos\theta'_\ell + \nu_\ell\left(\frac{\pi}{2} - \theta'_\ell - \theta'_{\ell s}\right) + 2k_1 a\left(\theta''_\ell\right)^2 \cos\theta'_\ell + \frac{\pi}{4}. \tag{2.456}$$

Finally, by neglecting the term of order $\left(\theta''_\ell\right)^2$ in (2.456), we obtain the following expression for the contribution of the surface waves encircling the circular cylindrical shell:

$$p_1 = D_s^{-1} \sum_{n=1}^{N} P^i(Q_\ell) \frac{a \cos^2\theta'_\ell}{\sin\theta'_\ell - \left(\frac{\partial \eta}{\partial \theta}\right)_{\theta=\theta'_\ell}} \frac{e^{i\nu_\ell \phi_{1\ell}} + e^{-i\nu_\ell \phi_{2\ell}}}{1 - e^{2i\pi \nu_\ell}} \frac{e^{i k_1 s_\ell}}{\sqrt{s_\ell + a\cos\theta'_\ell}}, \tag{2.457}$$

where

$$P^i(Q_\ell) = P_0 \, e^{-ik_1 a \cos\theta'_\ell}, \tag{2.458}$$

$$D_s = \frac{e^{i\frac{\pi}{4}}}{\sqrt{8\pi k_1}} \tag{2.459}$$

and (fig. 2.45)

$$\phi_{1\ell} = \pi - \theta - \theta'_\ell - \theta'_{\ell s}, \quad \phi_{2\ell} = \pi + \theta - \theta'_\ell - \theta'_{\ell s}. \tag{2.460}$$

By writing

$$\left(1 - e^{2i\nu_\ell \pi}\right)^{-1} = \sum_{m=0}^{\infty} e^{2im\pi\nu_\ell}, \tag{2.461}$$

we see that for $m = 0$, (2.457) represents the field of N surface rays $Q_\ell Q'_\ell M$ and $\bar{Q}_\ell \bar{Q}'_\ell M$ $(\ell = 1 \ldots N)$, excited at the critical angle θ'_ℓ and encircling the elastic shell in opposite directions.

The propagation constants $k_{s\ell} = \nu_\ell / a$ are given by the solutions of the characteristic equation

$$\sqrt{1-\left(\frac{\nu}{x_1}\right)^2}+\eta(\nu)=0, \tag{2.462}$$

where $x_1 = k_1 a$. They are also the poles of the reflection coefficient, located in the first quadrant of the complex ν-plane. Numerical solutions of (2.462) show that $\nu_\ell'' << \nu_\ell'$ (see the end of chapter 3). Hence the propagation constants $k_{s\ell} = k'_{s\ell} + ik''_{s\ell}$ of the elastic surface waves verify

$$k'_{s\ell} = k_1 \sin\theta'_\ell, \quad k''_{s\ell} = k_1\,\theta''_\ell \cos\theta'_\ell. \tag{2.463}$$

Using the representation introduced in chapter 1, section 1.3 of a local inhomogeneous plane wave defined by a complex wave number $\mathbf{K}_\ell = \mathbf{K}'_\ell + i\,\mathbf{K}''_\ell$ with $\mathbf{K}'_\ell \cdot \mathbf{K}''_\ell = 0$ and $K'^2_\ell - K''^2_\ell = k_1^2$, we see that according to (2.463) we have

$$\mathbf{K}'_\ell \cdot \mathbf{s} = K'_\ell \sin\theta'_\ell = k_1 \sin\theta'_\ell,$$

$$\mathbf{K}''_\ell \cdot \mathbf{s} = K''_\ell \cos\theta'_\ell = k_1\,\theta''_\ell \cos\theta'_\ell,$$

where $\mathbf{s}$ is the unit vector tangent to the cylinder at an arbitrary point N along $Q_\ell Q'_\ell$. Hence

$$K'_\ell = k_1, \quad K''_\ell = k_1\,\theta''_\ell. \tag{2.464}$$

Consequently, the ℓth order elastic surface wave is a local inhomogeneous plane wave propagating at each point N on the arc $Q_\ell Q'_\ell$ along $\mathbf{K}'_\ell$ and decaying exponentially in the direction of the vector $\mathbf{K}''_\ell$ (see fig. 2.46).

The elastic surface wave is therefore a leaky wave continuously shading rays in the direction $\mathbf{K}'_\ell$, making an angle θ'_ℓ with the normal to the cylinder. However, since the attenuation of the elastic surface wave is very small, it is necessary to take into account the contribution of the rays encircling the cylinder many times. The series (2.461) is therefore not used in the computation of the field. The latter is obtained by directly calculating the second member of (2.457), which can also be written as

$$p_1 = D_s^{-1} \sum_{n=1}^{N} P^i(Q_\ell) D_\ell^2 \frac{e^{i\nu_\ell \phi_{1\ell}} + e^{-i\nu_\ell \phi_{2\ell}}}{1-e^{2i\pi\nu_\ell}} A\, e^{ik_1 s_\ell},$$

where A is the divergence factor of the radiated wave given by

$$A = \sqrt{\frac{\rho_\ell}{s_\ell + \rho_\ell}},$$

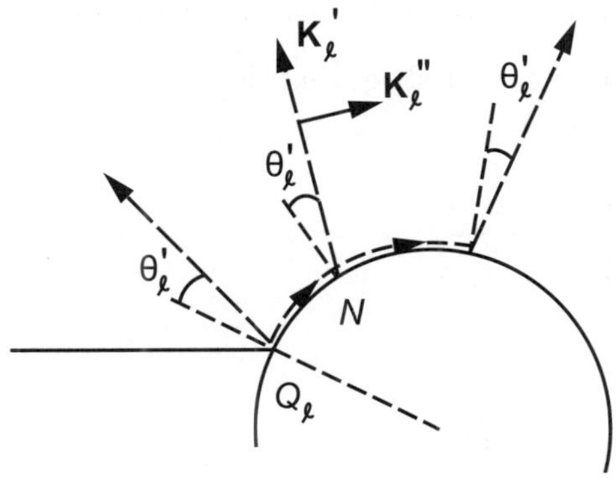

Figure 2.46. Schematic representation of an elastic surface wave shading a local inhomogeneous plane wave.

and $\rho_\ell = a\cos\theta'_\ell$ is the distance of Q'_ℓ to the caustic of the radiated rays. This formula can be explained by observing that the directions in which the elastic surface waves shed energy in the fluid have envelopes or caustics that form circles concentric to the cylinder, the radii of which are equal to $a\sin\theta'_\ell$. Hence the distance from an arbitrary point Q on the surface to the caustic is $\rho_\ell = a\cos\theta'_\ell$, and the divergence factor $\left(s_\ell + a\cos\theta'_\ell\right)^{-1/2}$ in (2.457) is equal to $\left(s_\ell + \rho_\ell\right)^{-1/2}$.

The coefficient

$$D_\ell^2 = \frac{1}{D_s}\frac{\sqrt{a\cos^3\theta'_\ell}}{\sin\theta'_\ell - \left(\dfrac{\partial\eta}{\partial\theta}\right)_{\theta=\theta'_\ell}} \tag{2.465}$$

is usually written in the form

$$D_\ell^2 = D^{(\ell)}_{sr,su}\, D^{(\ell)}_{su,sr}, \tag{2.466}$$

where $D^{(\ell)}_{sr,su}$ is the excitation coefficient of the ℓth surface ray by a space ray and where $D^{(\ell)}_{su,sr}$ is the radiation coefficient of the ℓth surface ray into a space ray. According to (2.465), we have

$$D^{(\ell)}_{sr,su} = D^{(\ell)}_{su,sr} = \frac{1}{\sqrt{D_S}}\frac{\left(a\cos^3\theta'_\ell\right)^{1/4}}{\sqrt{\sin\theta'_\ell - \left(\dfrac{\partial\eta}{\partial\theta}\right)_{\theta=\theta'_\ell}}}. \tag{2.467}$$

2.3.3.8. Oblique Incidence for Elastic Surface Waves

We have seen in section 2.3.3.6 that at oblique incidence the amplitude b_n of the mode of order n of the scattered field is given by

$$b_n = -\frac{\bar{\Omega}_n J_n(k_{1\perp}a)}{\bar{\Omega}_n H_n^{(1)}(k_{1\perp}a)},$$

where

$$\bar{\Omega}_n = \frac{\partial}{\partial r} + ik_{1\perp}\bar{\eta}_n$$

and

$$\bar{\eta}_n = i\frac{\rho_1}{\rho_2}\frac{x_t^2}{x_1}\frac{\bar{B}_n^{(1)}}{\bar{B}_n^{(2)}} \tag{2.468}$$

with $x_1 = k_1 a$, $x_t = k_t a$. Consequently, the expression $p_1(\rho,\theta;k_{1\perp})$ can be deduced from the solution of the diffraction problem for normal incidence by replacing the equivalent impedance η_n with $\bar{\eta}_n$.

We limit our analysis to one surface wave. The formulas obtained are the same for all other elastic surface waves, the only difference being the value of the propagation constant, which also defines the angles of excitation and radiation.

According to (2.457) and (2.468), for the pole $\nu_1 = k_{1\perp}\sin\theta_{1\perp}$ of the reflection coefficient in the complex ν-plane at normal incidence, we obtain the following expression for the field of the corresponding elastic surface wave at oblique incidence:

$$p_1(M) = p_i(Q_1)\sqrt{8\pi k_{1\perp}}\, e^{-i\frac{\pi}{4}} \frac{a\cos^2\theta_t'}{\sin\theta_t' - \left(\frac{\partial\bar{\eta}}{\partial\theta}\right)_{\theta=\theta_t'}} e^{ik_{s\perp}\ell_\perp} \frac{e^{ik_{1\perp}s_\perp} e^{ik_z z}}{\sqrt{s_\perp + a\cos\theta_t'}}, \tag{2.469}$$

where M is the observation point and where Q_1 is the launching point of the elastic surface wave $s_\perp = Q_\perp' M_\perp$, $\ell_\perp = \text{arc}\, Q_1 Q_\perp'$, $z = MM_\perp$ (see fig. 2.47).

The transverse propagation constant of the elastic surface wave is given by

$$k_{s\perp} = k_{1\perp}\sin\theta_t, \quad \theta_t = \theta_t' + i\,\theta_t''. \tag{2.470}$$

Some elementary geometrical calculations give

$$\begin{aligned} k_{1\perp}\cos\theta_t' &= k_1\cos\theta', \\ k_{1\perp}\sin\theta_t' &= k_1\sqrt{\sin^2\beta - \cos^2\theta'}. \end{aligned} \tag{2.471}$$

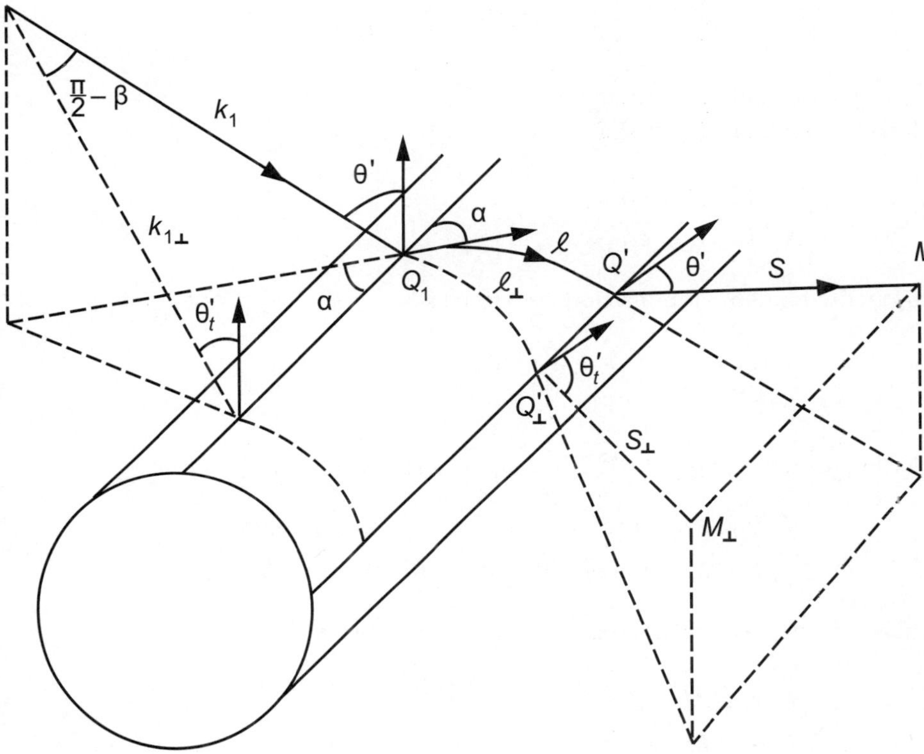

Figure 2.47. Geometrical representation of the rays corresponding to transverse and oblique incidence.

Furthermore, in accordance with the Fermat principle for surface waves (see chapter 1.3),

$$k_s = k_1 \sin\theta, \quad \theta = \theta' + i\,\theta'', \tag{2.472}$$

where k_s is the propagation constant of the surface wave. The real part of the phase of the surface wave follows a geodesic Q_1Q' on the surface, where Q_1 is the point on the surface where the incident ray makes an angle θ' with the normal to the surface at that point. Likewise, the diffracted ray $Q'M$ makes an angle θ' with the normal to the surface at Q'. Moreover, the tangent to the geodesic at Q_1 makes an acute angle α with the generatrix of the cylinder passing through Q_1, verifying

$$k_{1\perp} \sin\theta'_t = k_1 \sin\theta' \sin\alpha. \tag{2.473}$$

Hence

$$\ell_{\perp} = \ell \sin\alpha, \quad z = s\cos\beta + \ell\cos\alpha, \quad s_{\perp} = s\sin\beta, \tag{2.474}$$

where ℓ is the length of the arc of the geodesic Q_1Q' and $s = Q'M$.

It follows from (2.471) and (2.473) that

$$\sin\alpha = \frac{\sqrt{\sin^2\beta - \cos^2\theta'}}{\sin\theta'}. \tag{2.475}$$

The phase term in (2.469) can therefore be written as

$$e^{ik_1 s}\, e^{i\left(k_{s\perp}\ell_{\perp} + k_1\cos\beta\cos\alpha\right)}. \tag{2.476}$$

By neglecting the terms of order $\theta_t''^2$, since θ_t'' is small, we obtain

$$\begin{aligned} R_e k_{s\perp} &\cong k_{1\perp}\sin\theta_t' = k_1\sin\beta\sin\theta_t' = k_1\sin\theta'\sin\alpha, \\ I_m k_{s\perp} &\cong k_{1\perp}\cos\theta_t'\, sh\,\theta_t'' = k_1\cos\theta'\, sh\theta_t''. \end{aligned} \tag{2.477}$$

Similarly, by neglecting the terms of order θ''^2, we obtain

$$\begin{aligned} R_e k_s &\cong k_1\sin\theta', \\ I_m k_s &\cong k_1\cos\theta'\, sh\,\theta''. \end{aligned} \tag{2.478}$$

Then

$$k_{s\perp}\ell_{\perp} + k_1\cos\beta\cos\alpha = \left(R_e k_s\right)\ell\left(\sin^2\alpha + \frac{\cos\beta\cos\alpha}{\sin\theta'}\right) + ik_1\ell\sin\alpha\cos\theta'\, sh\theta_t'',$$

and since it follows from (2.475) that

$$\sin^2\alpha + \frac{\cos\beta\cos\alpha}{\sin\theta'} = 1,$$

we finally obtain the following for the phase term (2.476):

$$e^{ik_1 s}\, e^{i\left(R_e k_s\right)\ell}\, e^{-k_1\ell\sin\alpha\cos\theta'\, sh\theta_t''}. \tag{2.479}$$

The divergence factor in (2.469) reads as follows, according to (2.471):

$$\sqrt{\frac{a\cos\theta_t'}{s_{\perp} + a\cos\theta_t'}} = \sqrt{\frac{\rho_c}{s+\rho_c}}, \tag{2.480}$$

where $\rho_c = a \cos \theta' / \sin^2 \beta$ is the distance to the caustic of the radiated wave.

It can be readily verified that the caustic for oblique incidence is a cylinder of radius $a \sin \theta_t'$. Finally, the pressure field at the observation point M, due to the elastic surface wave launched at Q_1 by a plane wave obliquely incident on a circular cylindrical shell, is given by

$$p_1(M) = p_i(Q_1) D_1^2 \, e^{i k_s' \ell} \, e^{-\gamma \ell} \, e^{i k_1 s} \sqrt{\frac{\rho_c}{s + \rho_c}}, \tag{2.481}$$

where $k_s' = R_e k_s$ and where γ is an attenuation coefficient given by

$$\gamma = k_1 \sin\alpha \cos\theta' \, sh\theta_t''. \tag{2.482}$$

The diffraction coefficient D_1^2 is given by

$$D_1^2 = \frac{1}{D_{s\perp}} \frac{\sqrt{a \cos^3 \theta_t'}}{\sin\theta_t' - \left(\frac{\partial \bar{\eta}}{\partial \theta}\right)_{\theta = \theta_t'}} = \frac{1}{D_s} \frac{\sqrt{\rho_c \cos^3 \theta'}}{\sin\theta_t' - \left(\frac{\partial \bar{\eta}}{\partial \theta}\right)_{\theta = \theta_t'}}, \tag{2.483}$$

where

$$D_{s\perp} = \frac{e^{i\frac{\pi}{4}}}{\sqrt{8\pi k_{1\perp}}} = \frac{1}{\sqrt{\sin\beta}} \frac{e^{i\frac{\pi}{4}}}{\sqrt{8\pi k_1}} = \frac{1}{\sqrt{\sin\beta}} D_s.$$

2.3.3.9. Noncircular Cylindrical Shell at Normal and Oblique Incidence

The extension of (2.481) to a noncircular cylindrical shell is straightforward. By writing (2.483) as a product of an excitation coefficient and a radiation coefficient, verifying the reciprocity theorem, we obtain

$$(D_1)^2 = D^{(1)}_{sr,su}(Q_1) \, D^{(1)}_{su,sr}(Q_2), \tag{2.484}$$

where Q_1 and Q_2 are the points of excitation and radiation of the elastic surface wave and where

$$D^{(1)}_{sr,su}(Q_1) = \frac{1}{\sqrt{D_s}} \frac{\left(\rho_1 \cos^3 \theta_1'\right)^{1/4}}{\sqrt{\sin\theta_{1t}' - \left(\frac{\partial \bar{\eta}}{\partial \theta}\right)_{\theta = \theta_{1t}'}}}, \quad D^{(1)}_{su,sr}(Q_2) = \frac{1}{\sqrt{D_s}} \frac{\left(\rho_2 \cos^3 \theta_2'\right)^{1/4}}{\sqrt{\sin\theta_{2t}' - \left(\frac{\partial \bar{\eta}}{\partial \theta}\right)_{\theta = \theta_{2t}'}}}. \tag{2.485}$$

In (2.485), ρ_1 and ρ_2 are the radii of curvature of the geodesic followed by the elastic surface wave at Q_1 and Q_2, respectively. The angles θ'_{1t} and θ'_{2t} are defined by

$$R_e\, \nu_1^{(1)} = k_{1\perp} a_1 \sin\theta'_{1t}, \quad R_e\, \nu_1^{(2)} = k_{1\perp} a_2 \sin\theta'_{2t}, \quad k_{1\perp} = k_1 \sin\beta, \tag{2.486}$$

where $\nu_1^{(1,2)}$ is the first root of the characteristic equation (2.462) at Q_1 and Q_2 and where $a_1 = \rho_1 / \sin^2\beta$ and $a_2 = \rho_2 / \sin^2\beta$.

With these notations, the expression of the field at oblique incidence is given by

$$p_1(M) = p_i(Q_1)\, D^{(1)}_{sr,su}(Q_1)\, D^{(1)}_{su,sr}(Q_2)\, e^{i\int_{Q_1}^{Q_2} k'_s(\ell)d\ell}\, e^{-\int_{Q_1}^{Q_2} \gamma(\ell)d\ell} \sqrt{\frac{\rho_c}{s+\rho_c}}\, e^{ik_1 s}, \tag{2.487}$$

where ρ_c is the distance of Q_2 to the caustic of the radiated wave and where

$$k'_s(\ell) = R_e k_s(\ell) = k_1 \sin\theta'(\ell). \tag{2.488}$$

By using the geometrical relation derived from figure 2.47,

$$\cos\beta = \cos\alpha \sin\theta', \tag{2.489}$$

and the relation (2.473), we obtain

$$\sin\theta'(\ell) = \left[\cos^2\beta + \sin^2\beta \sin^2\theta_t(\ell_\perp)\right]^{1/2} \tag{2.490}$$

and

$$k'_s(\ell) = k_1 \sin\theta'(\ell) = \left[k_1^2 \cos^2\beta + k_{1\perp}^2 \sin^2\theta'_t(\ell_\perp)\right]^{1/2}, \tag{2.491}$$

where

$$k_1 \sin\theta'_t(\ell_\perp) = \frac{R_e\, \nu_1(\ell_\perp)}{a(\ell_\perp)} \tag{2.492}$$

and where $\ell_\perp$ is the abscissa along the projected geodesic $\ell_\perp = \ell \sin\alpha$.

Likewise, by using the geometrical relation

$$\tan(\alpha) = \sin\theta_t' \tan\beta, \tag{2.493}$$

we obtain

$$\gamma(\ell) = k_1 \sin\alpha(\ell)\cos\theta_t'(\ell) sh\theta_t''(\ell_\perp) \frac{k_1 \tan\beta \sin\theta_t'(\ell_\perp)}{\sqrt{1+\tan^2\beta \sin^2\theta_t'(\ell_\perp)}} \sin\beta \cos\theta_t'(\ell_\perp) sh\theta_t''(\ell_\perp), \tag{2.494}$$

where

$$k_1 \sin\beta \cos\theta_t'(\ell_\perp) sh\theta_t''(\ell_\perp) = \frac{I_m \nu_1(\ell_\perp)}{a(\ell_\perp)}.$$

The values of $\nu_1(\ell_\perp)$ are calculated by solving the characteristic equation $\cos\theta = -\overline{\eta}(\theta)$ at a series of sampling points along the projected geodesic.

The first step for applying (2.487) is to find the points Q_1 and Q_2. For normal plane wave incidence, we denote by $\theta_g'(\ell)$ the acute angle between the incident direction and the normal to the surface, and we denote by $a(\ell)$ the radius of curvature at abscissa ℓ. The angle $\theta_g'(\ell)$ is a geometrical quantity that can be determined if we know the geometry of the cross section of the convex cylindrical shell and the incident direction. We locally replace the original shell at abscissa ℓ with a circular cylindrical shell of exterior radius $a(\ell)$ that has the same thickness as the original shell. Solving the characteristic equation for this cylinder, we obtain a characteristic angle $\theta(\ell) = \theta'(\ell) + i\theta''(\ell)$. The value of ℓ for which $\theta'(\ell) = \theta_g'(\ell)$ defines the position of Q_1. A similar procedure can be applied for the determination of Q_2.

For oblique incidence, the points Q_1 and Q_2 are derived by applying the procedure for normal incidence to $a(\ell_\perp)$ and by using the geometrical relations presented previously.

2.3.4. THREE-DIMENSIONAL CONVEX SHELL

2.3.4.1. General Solution for Creeping Waves

A general solution for the field of a creeping wave following a torsionless geodesic on a three-dimensional convex shell can be derived by using the same rules presented in section 2.3.2.

Let $QQ'M$ be a creeping ray, where Q is the excitation point of the creeping ray and where Q' is the point where the creeping ray sheds energy along the space ray $Q'M$ to the observation point M. The contribution to the diffracted field in the deep shadow region of this ray is given by

$$p(M) = p^i(Q) \sum_{n=1}^{\infty} D_n(Q) D_n(Q') e^{ik\ell} e^{i\int_Q^{Q'} \frac{\alpha_n(\ell) m d\ell}{\rho_g}} D_s \sqrt{\frac{d\Sigma(Q)}{d\Sigma(Q')}} \sqrt{\frac{\rho}{Q'M(\rho + Q'M)}} e^{ikQ'M}, \tag{2.495}$$

where ρ has the same meaning as in (1.224) and where the diffraction coefficient D_n is given by (2.330). The computation of α_n needs to solve $\Omega_\nu H_\nu^{(1)}(x) = 0$ at each sampling point along the geodesic. Formula (2.495) is a good approximation of the solution when the transverse radius of curvature ρ_t is large compared to the radius of curvature ρ_g of the geodesic.

For a geodesic with torsion, we choose a local cylindrical approximation. We denote by ℓ the abscissa of a point Q on the geodesic. Let $a_1(Q)$ and $a_2(Q)$ be the principal radii of curvature at Q with $a_1(Q) > a_2(Q)$. The geodesic makes an angle $\alpha(\ell)$ with the principal direction $\mathbf{t}_1$ corresponding to $a_1(Q)$. If $\rho_g(Q)$ denotes the radius of curvature of the geodesic at Q, then we have

$$\frac{1}{\rho_g} = \frac{\sin^2\alpha}{a_2} + \frac{\cos^2\alpha}{a_1}. \tag{2.496}$$

If we approximate the surface at Q by a circular cylinder of radius $a_2(Q)$, having an axis parallel to the principal direction $\mathbf{t}_1$, then the radius of the geodesic on its surface, in the direction α, would be different from $\rho_g(Q)$. In order to keep the value of ρ_g, we have to choose a radius $a(Q)$ for the local cylinder, verifying

$$\frac{\sin^2\alpha}{a} = \frac{1}{\rho_g}. \tag{2.497}$$

With this choice for the local cylindrical approximation, the formulas established in section 2.3.3.6 can be applied locally. Notably, at each position $Q(\ell)$ on the geodesic, we can determine the angle $\beta(\ell) = \alpha(\ell)$ between the incident ray and the principal direction $\mathbf{t}_1(\ell)$. Then, knowing $a(\ell)$ and $\beta(\ell)$, the value of $\nu_1(\ell)$ can be calculated by solving the characteristic equation (2.438) for a circular cylinder. The expression for the creeping wave field follows from (2.358) and (2.361) with $d_n(Q) = d_n[\rho_g(Q), k]$. It is given by

$$p(M) = p^i(Q) \sum_{n=1}^{\infty} d_n(Q)\, d_n(Q')\, e^{ik\ell}\, e^{i\int_Q^{Q'} \frac{\alpha_n(\ell)\, m\, d\ell}{\rho_g}}\, D_s \sqrt{\frac{d\Sigma(Q)}{d\Sigma(Q')}} \sqrt{\frac{\rho}{Q'M(\rho + Q'M)}}\, e^{ikQ'M}. \tag{2.498}$$

Equation (2.498) is identical to (2.366). However, the computation of $\alpha_n(\ell)$, $d_n(Q)$, and $d_n(Q')$ are more complex. Particularly, the computation of $\alpha_n(Q)$ needs the resolution of the characteristic equation (2.438) at each sampling point along the geodesic.

2.3.4.2. General Solution for Elastic Surface Waves

We consider first the case where the elastic surface waves follow a torsionless geodesic. Then, according to Fermat's principle, the source *S*, the observation *P*, and the geodesic followed by the elastic surface wave are in the same plane. Hence the search of the ray trajectory follows the same rules as in the case of a noncircular cylindrical surface at normal incidence. If the transverse radius of curvature ρ_t of the surface in a plane orthogonal to the geodesic is large compared to the radius of curvature of the geodesic ($\rho_t >> \rho_g$), then the formula (2.481) can be extended to a three-dimensional convex shell by adapting the divergence factor of the diffracted rays (the excitation and radiation coefficients remaining unchanged). Hence the field radiated by the elastic surface wave corresponding to a pole ν_1 is given by

$$p_1(M) = p_i(Q_1)\, D_{sr,su}(Q)\, D_{su,sr}(Q_2)\, e^{i\int_{Q_1}^{Q_2} k_s'(\ell)d\ell}\, e^{-\int_{Q_1}^{Q_2}\gamma(\ell)d\ell} \sqrt{\frac{d\Sigma(Q_1)}{d\Sigma(Q_2)}} \sqrt{\frac{\rho_1\,\rho_2}{(\rho_1+s)(\rho_2+s)}}\, e^{ik_1 s}, \tag{2.499}$$

where $\left[d\Sigma(Q_1)/d\Sigma(Q_2)\right]^{1/2}$ is the divergence factor of the elastic surface rays and where ρ_1, ρ_2 is the principal radii of curvature of the diffracted wave front at Q_2. The other parameters in (2.499) are identical to their homologues for normal incidence on a noncircular convex cylindrical shell.

When the surface wave follows a geodesic with torsion, the main difficulty is the determination of the ray path, which depends on the solution of the characteristic equation. However, at a first approximation we can ignore the effect of the curvature and solve the characteristic equation on a flat surface. Then the excitation and radiation angles θ_1' and θ_2' are equal and can be determined a priori. In the frame of this approximation, it is possible to determine the ray trajectory by applying the Fermat principle (see chapter 1, section 1.3). Numerical algorithms using an approximation of the surface by bicubic spline functions have been reported in the literature [2.38]. Once the ray trajectory is known, the field can be calculated by using a local cylindrical approximation of the surface along the geodesic followed by the ray, similar to that described in section 2.3.4.1.

2.4. CONCAVE SURFACE

2.4.1. *INTRODUCTION*

We have seen in chapter 1, section 1.6 that the field reflected by a smooth convex or concave rigid, soft, or impedance surface has the same asymptotic structure as the incident field. However, on a concave surface, if we limit our analysis to the dominant term, then an incident doubly curved GA wave gives rise, generally at a fixed observation point *M*, to two doubly curved reflected GA waves and also, depending on the position of *M*, to multiple reflected rays. Outside the caustics of single or multiple reflected rays, the reflected field associated with each category of waves follows the laws of GA and may be calculated by the formulas derived in chapter 1, section 1.6.

When the incident field is generated by a line source or a point source located near the concave surface, other diffraction phenomena occur that contribute to the diffracted field. New types of surface waves appear, known as whispering gallery waves. Such waves are special solutions of the Helmholtz equation that cannot be constructed by the ray method. One way to solve this problem is to derive a solution to the homogeneous Helmholtz equation near the surface, by the boundary layer method, and to construct a superposition of these solutions that describes a surface wave propagating from a line source or a point source. This last operation needs a matching procedure at the caustic boundary layer of the rays emanating radially from the source, which is rather involved and has not been performed until now.

Another way to solve this problem is to find a solution of an etalon or model problem and then extend this solution to the general case. The simplest model is a line source near a circular boundary, where the variables are separable.

2.4.2. SOLUTION OF THE CANONICAL PROBLEM OF A LINE SOURCE PARALLEL TO THE GENERATRIX OF A CONCAVE CIRCULAR CYLINDER

A solution to this problem has been obtained separately by Wasylkiwskyj [2.39] and Babich and Buldyrev [2.40]. A complete analysis of the solution with hybrid ray-mode representations is given in [2.41]. We follow here the method developed in [2.39] and extend it to an impedance surface.

2.4.2.1. Derivation of the Solution

Let Σ be an angular sector of a circular cylinder of radius a, delimited by two radial perfectly absorbing planes P_1 and P_2. We suppose that the angular sector is less than π. A line source S' and an observation point S of polar coordinates (ρ', ϕ) and (ρ, ϕ) are located inside the concave cylindrical cavity delimited by the planes P_1 and P_2 (fig. 2.48).

Inside the domain delimited by Σ and the planes P_1 and P_2, the Green function of the problem verifies the Helmholtz equation in cylindrical coordinates:

$$\left(\frac{1}{\rho}\frac{\partial}{\partial\rho}\rho\frac{\partial}{\partial\rho}+\frac{1}{\rho^2}\frac{\partial^2}{\partial\phi^2}+k^2\right)G(\mathbf{\rho},\mathbf{\rho}')=-\frac{1}{\rho'}\delta(\rho-\rho')\delta(\phi-\phi'). \tag{2.500}$$

We first consider the case of the hard boundary condition, which reads

$$\frac{\partial}{\partial\rho}G(\rho,\rho')\bigg|_{\rho=a}=0. \tag{2.501}$$

We seek a solution of (2.500) and (2.501) in the form of angular modes (see [2.16], chapter 3.2):

$$G(\mathbf{\rho},\mathbf{\rho}')=\sum_{\nu} g_\phi(\phi,\phi';\nu)\psi_\nu(\rho)\psi_\nu(\rho'), \tag{2.502}$$

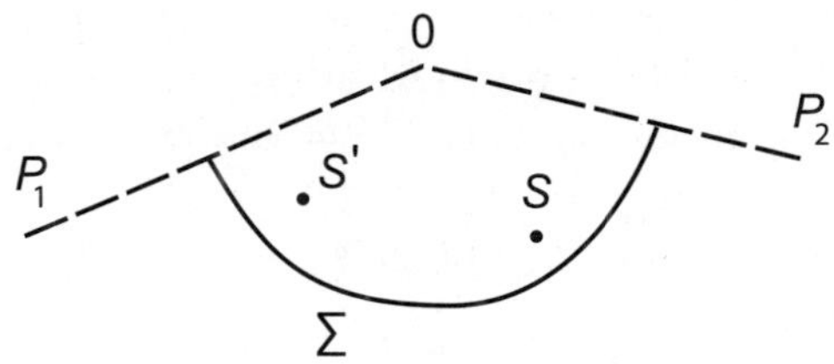

Figure 2.48. Line source within a circular cavity bounded by planar absorbers.

where the functions $\psi_\nu(\rho)$ form a complete set of functions in the domain $0 \le \rho \le a$.

Due to the presence of the absorbing planes P_1 and P_2, the Green function g_ϕ verifies

$$\left(\frac{d^2}{d\phi^2} + \nu^2\right) g_\phi(\phi, \phi'; \nu) = -\delta(\phi - \phi'). \tag{2.503}$$

Inserting (2.502) into (2.500) and taking (2.503) into account, we see that the functions $\psi_\nu(\rho)$ verify

$$\left(\rho \frac{d}{d\rho} \rho \frac{\partial}{\partial \rho} + k^2 \rho^2\right) \psi_\nu(\rho) = \nu^2 \psi_\nu(\rho). \tag{2.504}$$

To this equation, we must add the completeness condition:

$$\rho' \delta(\rho - \rho') = \sum_\nu \psi_\nu(\rho) \psi_\nu(\rho'). \tag{2.505}$$

According to (2.501), the boundary conditions on $\psi_\nu(\rho)$ read

$$\frac{\partial \psi_\nu}{\partial \rho}\bigg|_{\rho=a} = 0. \tag{2.506}$$

Moreover, for $\rho = 0$, the eigenfunctions of (2.504) must take finite values:

$$|\psi_\nu(0)| < \infty. \tag{2.507}$$

Now since the circumferential waves have a reflection coefficient equal to zero on the planes P_1 and P_2, the solution of (2.503) gives

$$g_\phi(\phi, \phi'; \nu) = \frac{e^{\pm i\nu|\phi - \phi'|}}{\mp 2i\nu}, \tag{2.508}$$

since, due to the perfectly absorbing planes, no waves propagating toward the source can exist. The circumferential waves (2.508) tend to zero when $|\phi - \phi'|$ tends to infinity, if $I_m \nu$ is positive when the sign (+) is chosen in the exponent of (2.508) or if $I_m \nu$ is negative when the sign (–) is chosen.

In order to find the eigenfunctions g_ρ of (2.504) satisfying (2.505) and (2.506), we use the relation (see [2.16] section 3.3a)

$$\sum_{\nu} \psi_\nu(\rho)\psi_\nu(\rho') = \frac{1}{2\pi i}\oint g_\rho(\rho,\rho';\mu)\,d\mu, \tag{2.509}$$

where μ is a complex variable and where g_ρ is the Green function in the ρ domain verifying

$$\left(\frac{1}{\rho}\frac{d}{d\rho}\rho\frac{d}{d\rho} + k^2 - \frac{\mu}{\rho^2}\right)g_\rho(\rho,\rho';\mu) = -\frac{1}{\rho}\delta(\rho-\rho') \tag{2.510}$$

and the conditions

$$\frac{d}{d\rho}g(\rho)\Big|_{\rho=a} = 0,\quad g_\rho\left|(0,\rho';\mu)\right| < \infty. \tag{2.511}$$

The integration contour in (2.509) encircles all the singularities of g_ρ. We observe that (2.510) is a Bessel equation, the general solution of which is a linear combination of $J_{\sqrt{\mu}}(k\rho)$ and $H^{(1)}_{\sqrt{\mu}}(k\rho)$. We note that $J_{\sqrt{\mu}}(k\rho)$ is not singular at $\rho = 0$ and that $H^{(1)}_{\sqrt{\mu}}(k\rho)$, with the time dependence exp($-i\omega t$) chosen in this book, represents traveling waves propagating toward the boundary $\rho = a$. By taking the reciprocity theorem into account and applying the boundary conditions (2.511), we readily obtain

$$g_\rho(\rho,\rho';\mu) = a(\mu)J_{\sqrt{\mu}}(k\rho_<)\left[J_{\sqrt{\mu}}(k\rho_>)\frac{H^{(1)'}_{\sqrt{\mu}}(ka)}{J'_{\sqrt{\mu}}(ka)} - H^{(1)}_{\sqrt{\mu}}(k\rho_>)\right], \tag{2.512}$$

where $\rho_< = \min(\rho, \rho')$ and $\rho_> = \max(\rho, \rho')$. The parameters $a(\mu)$ can be determined by multiplying both sides of (2.510) by ρ and integrating from $\rho = \rho' + \Delta$ to $\rho = \rho' - \Delta$, with $\Delta \to 0$. This operation yields

$$\frac{\partial}{\partial\rho}g_\rho\left(\rho^+,\rho'\right) - \frac{\partial}{\partial\rho}g_\rho\left(\rho^-,\rho'\right) = -\frac{1}{\rho'} \tag{2.513}$$

and, according to (2.512),

$$a(\mu)\left[J_{\sqrt{\mu}}(k\rho')H^{(1)'}_{\sqrt{\mu}}(k\rho') - J'_{\sqrt{\mu}}(k\rho')H^{(1)}_{\sqrt{\mu}}(k\rho')\right] = \frac{1}{k\rho'}. \tag{2.514}$$

The expression contained within brackets is a Wronskian expression of the Bessel and Hankel functions, equal to $2i / \pi k\rho'$. Hence (2.514) yields $a(\mu) = -i\pi / 2$, which is independent of μ and

$$g_\rho(\rho,\rho';\mu) = -\frac{i\pi}{2} J_{\sqrt{\mu}}(k\rho_<)\left[J_{\sqrt{\mu}}(k\rho_>)\frac{H^{(1)'}_{\sqrt{\mu}}(ka)}{J'_{\sqrt{\mu}}(ka)} - H^{(1)}_{\sqrt{\mu}}(k\rho_>)\right]. \tag{2.515}$$

The eigenfunctions are obtained by inserting (2.515) into (2.509) and integrating g_ρ in the complex μ-plane, over a circle centered at the origin, by taking all the singularities of g_ρ into account. These singularities are simple poles corresponding to the zeros of the denominator. In addition, there is a branch point at $\mu = 0$.

Since for fixed ka, $J'_{\sqrt{\mu}}(ka)$ decreases monotically as a function of $\sqrt{\mu}$ for $\sqrt{\mu} > ka$, it has only a finite number of zeros located on the real μ-axis Moreover, to ensure boundedness of g_ρ at $\rho = 0$, the properties of the cylinder functions at $\rho = 0$, $J_{\sqrt{\mu}}(k\rho) \approx (k\rho)^{\sqrt{\mu}}$, require that $R_e\sqrt{\mu} > 0$. By choosing the branch cut along the negative real μ-axis, we deform the circular path around the branch cut, as well as around the simple poles on the positive real μ-axis (fig. 2.49).

Then the integral on the right-hand side of (2.509), in which g_ρ is replaced by (2.515), is transformed in a residue series arising from the pole contributions and in an integral over the path P around the branch cut. The spectrum therefore comprises a finite member of discrete eigenfunctions or modes associated with a continuous spectrum contained in the integral over P, provided that the sum in (2.509) is itself interpreted as an integral.

Let $\nu = \sqrt{\mu}$. Since on the Riemann sheet on which $R_e\sqrt{\mu} > 0$ we have $-\pi < \arg\mu < \pi$, it follows that

$$-\frac{\pi}{2} < \arg\nu < \frac{\pi}{2}, \tag{2.516}$$

and consequently $R_e\nu > 0$. Hence, with the integration variable transformation $\nu^2 = \mu$ on the right-hand side of (2.509), the contour integral over P is transformed into an integral along the

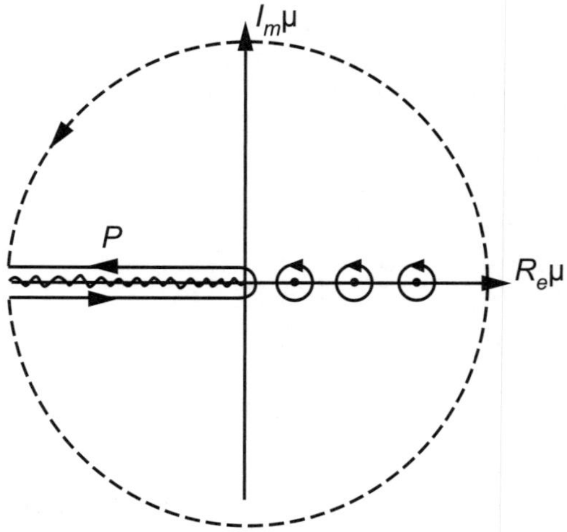

Figure 2.49. Path of integration in the complex μ-plane.

contour P' following the imaginary axis on the side $R_e \nu > 0$, and the poles of $J'_{\sqrt{\mu}}(ka)$ in the complex plane μ are also the poles of $J'_{\nu}(ka)$ in the complex plane ν (fig. 2.50).

It remains to introduce the weighting functions g_φ to construct, with the aid of (2.502), the Green function $G(\mathbf{\rho},\mathbf{\rho}')$. Therefore, for the contribution of the integral along P', we obtain

$$-\frac{1}{2\pi}\int_{-i\infty+\varepsilon}^{0} e^{-i\nu|\phi-\phi'|} g_\rho(\rho,\rho';\nu)d\nu + \frac{1}{2\pi}\int_{0}^{i\infty+\varepsilon} e^{i\nu|\phi-\phi'|} g_\rho(\rho,\rho';\nu)d\nu, \tag{2.517}$$

where ε is an infinitely small positive member.

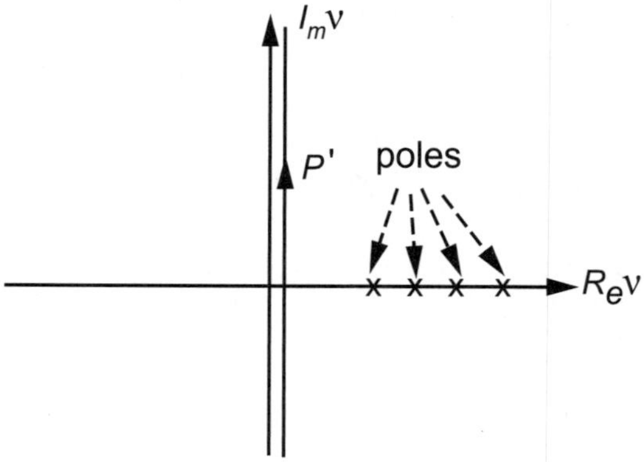

Figure 2.50. Path of integration and poles in the complex ν-plane.

By changing ν into $-\nu$ in the first integral (2.517), it is possible to transform the sum of the two integrals in (2.517) into one integral along a contour C' around the positive imaginary ν-axis going from $-\varepsilon + i\infty$ to $\varepsilon + i\infty$. However, since the integrand is discontinuous through the imaginary axis and since the integrand decreases exponentially along the infinite branches of C', we must introduce a branch cut along the positive imaginary ν-axis, as shown in figure 2.51.

Finally, (2.517) can be written

$$\frac{1}{2\pi}\int_{C'} e^{i\nu|\phi-\phi'|}\,\tilde{g}_\rho(\rho,\rho';\nu)\,d\nu \tag{2.518}$$

with

$$\tilde{g}_\rho(\rho,\rho';\nu)\,d\nu = \begin{cases} g_\rho(\rho,\rho';\nu) & , \quad R_e\nu \geq 0 \\ g_\rho(\rho,\rho';-\nu) & , \quad R_e\nu < 0, \end{cases} \tag{2.519}$$

where g_ρ is given by (2.515) with $\sqrt{\mu} = \nu$.

The term (2.518) gives the contribution of the integral along C'. The Green function $G(\boldsymbol{\rho},\boldsymbol{\rho}')$ is obtained by adding this term to the sum of the residues at the poles. It is also possible to include the contribution of the poles in a unique integral by deforming the branch of the contour C' corresponding to $R_e\nu > 0$ into a branch C'', as shown in figure 2.52.

Moreover, since there are no other singularities in the complex ν-plane and since the integrand of (2.518) tends to zero when $|\nu|$ tends to infinity in the second quadrant, it is possible to deform the branch $R_e\nu < 0$ of C' into a contour C''' along the real axis, as shown in figure 2.53. By denoting C as the sum of the contours C'' and C''', we obtain the following integral form of the Green function:

$$G(\boldsymbol{\rho},\boldsymbol{\rho}') = \frac{1}{2\pi}\int_{C} e^{i\nu|\phi-\phi'|}\,\tilde{g}_\rho(\rho,\rho';\nu)\,d\nu. \tag{2.520}$$

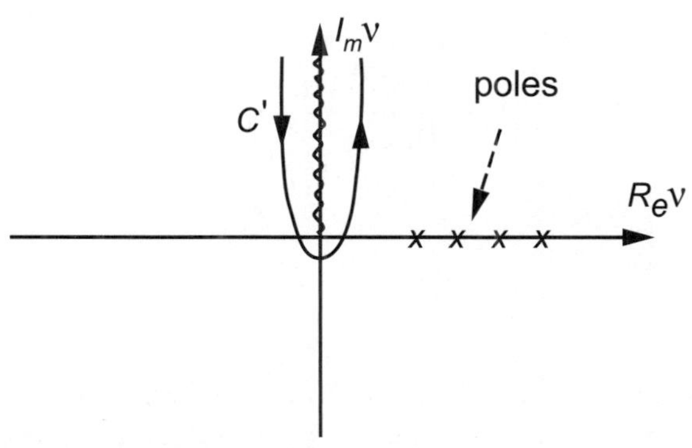

Figure 2.51. Contour deformation and branch cut.

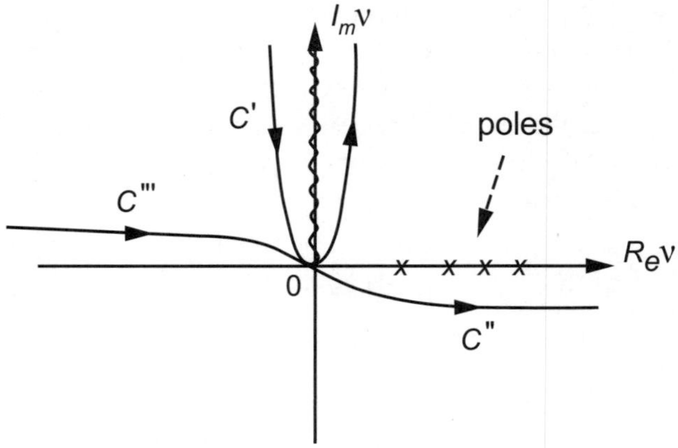

Figure 2.52. Contour deformation of C' into C'' and C'''.

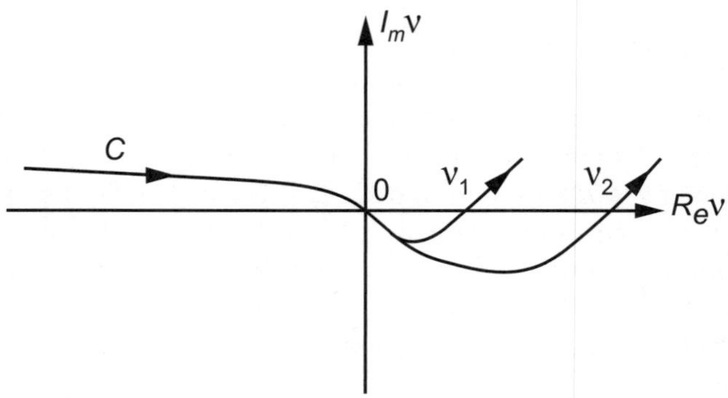

Figure 2.53. Deformation of the integration path through the stationary points ν_1 and ν_2.

The solution (2.520) has been obtained by Wasylkiwskyj [2.39]. It can also be written in the following form:

$$G(\boldsymbol{\rho},\boldsymbol{\rho}') = \frac{\pi}{2}\sum_n e^{i\nu_n|\phi-\phi'|} J_{\nu_n}(k\rho_<) \frac{H^{(1)'}(ka)}{\left[\dfrac{\partial}{\partial\nu} J_\nu'(ka)\right]_{\nu=\nu_n}} J_{\nu_n}(k\rho_>)$$

$$+\frac{1}{2}\int_{C'} e^{i\nu|\phi-\phi'|}\, \tilde{g}_\rho(\rho,\rho';\nu)\, d\nu. \tag{2.521}$$

The residue series comprises a finite number of terms, which are called whispering gallery modes. Their angular constants of propagation ν_n are the solutions of the characteristic equation

$$J_\nu'(ka) = 0. \tag{2.522}$$

The residual integral can also be written as

$$\frac{1}{2\pi}\int_0^{i\infty} e^{i\nu|\phi-\phi'|}\left[g_\rho(\rho,\rho';\nu) - g_\rho(\rho,\rho';-\nu)\right]d\nu. \tag{2.523}$$

Since the exponential function decreases rapidly along the imaginary axis when $|\phi-\phi'|$ is large, the integral (2.523) is only important when $|\phi-\phi'|$ is small, which is verified, for instance, in the special case when the source and the observation point are close together and near the surface.

The representation (2.521) of the Green function allows one to easily calculate the field due to a line source in the presence of a concave surface having a circular cross section. However, this representation of the solution is not adequate for its generalization to an arbitrary concave surface. In particular, the relation between (2.521) and the GA field when S' and S are moved away from the surface is difficult to establish. For this reason, the integral representation (2.520) will be used subsequently.

For a soft boundary condition, the derivation of the solution follows the same steps, and the solution is still given by (2.518) to (2.520). However, due to the boundary condition for a soft surface, we now have

$$G(\boldsymbol{\rho},\boldsymbol{\rho}')\Big|_{\rho=a} = 0, \quad \psi_\nu(a) = 0 \tag{2.524}$$

and consequently

$$g_\rho(\rho,\rho';\nu) = -\frac{i\pi}{2}J_\nu(k\rho_<)\left[\frac{H_\nu^{(1)}(ka)}{J_\nu(ka)}J_\nu(k\rho_>) - H_\nu^{(1)}(k\rho_>)\right]. \tag{2.525}$$

For an impedance boundary condition, we have

$$\left(\frac{d}{d\rho} - ikZ\right)\psi_\nu(\rho) = 0\,, \quad \rho = a, \tag{2.526}$$

where Z is the surface impedance. By using the notation

$$\Omega = \frac{d}{d\rho} - ikZ, \tag{2.527}$$

we obtain

$$g_{\rho}(\rho,\rho';\nu) = -\frac{i\pi}{2} J_{\nu}(k\rho_{<})\left\{\left[\frac{\Omega H_{\nu}^{(1)}(k\rho)}{\Omega J_{\nu}(k\rho)}\right]_{\rho=a} J_{\nu}(k\rho_{>}) - H_{\nu}^{(1)}(k\rho_{>})\right\}. \tag{2.528}$$

Equation (2.528) contains (2.515) and (2.525) by letting Z take the particular values $Z = 0$ and $Z = \infty$, respectively.

2.4.2.2. Extraction of the GA Contributions

We start with the identity

$$\frac{1}{\Omega J_{\nu}(ka)} = \frac{2}{\Omega H_{\nu}^{(2)}(ka)} \frac{1}{1-R}, \quad R = -\frac{\Omega H_{\nu}^{(1)}(ka)}{\Omega H_{\nu}^{(2)}(ka)}, \tag{2.529}$$

which can be readily established by applying the relation between Bessel and Hankel functions given by

$$J_{\nu}(x) = \frac{H_{\nu}^{(1)}(x) + H_{\nu}^{(2)}(x)}{2}, \tag{2.530}$$

which implies

$$\Omega J_{\nu}(x) = \frac{\Omega H_{\nu}^{(1)}(x) + \Omega H_{\nu}^{(2)}(x)}{2}. \tag{2.531}$$

Following Babich and Buldyrev [2.40], in (2.528) we replace the Bessel functions with Hankel functions and apply (2.529). This operation yields

$$g(\rho,\rho';\nu) = \frac{i\pi}{2} \Omega H_{\nu}^{(2)}(ka) \frac{F_{\nu}(\rho_{>},a) J_{\nu}(k\rho_{<})}{2\Omega J_{\nu}(ka)} = \frac{i\pi}{2} \frac{F_{\nu}(\rho_{>},a) J_{\nu}(k\rho_{<})}{1-R} \tag{2.532}$$

where

$$F_{\nu}(\rho,a) = H_{\nu}^{(1)}(k\rho) + R H_{\nu}^{(2)}(k\rho). \tag{2.533}$$

By taking the identity

$$2J_{\nu}(k\rho_{<}) = F_{\nu}(\rho_{<},a) + (1-R) H_{\nu}^{(2)}(k\rho_{<}) \tag{2.534}$$

into account, we finally obtain

$$g(\rho,\rho';\nu)=\frac{i\pi}{4}\ \frac{F_\nu(\rho_>,a)F_\nu(\rho_<,a)}{1-R}+\frac{i\pi}{4}\ F_\nu(\rho_>,a)H_\nu^{(2)}(k\rho_<). \quad (2.535)$$

Now in (2.520) we deform the integration path C into the path $C_0 = C' + C''$, which for $R_e\nu < 0$ follows the positive imaginary axis and for $R_e\nu > 0$ is identical to C (see fig. 2.52). Then the original integral is written as the sum of three terms:

$$\begin{aligned}G(\boldsymbol{\rho},\boldsymbol{\rho}')&=\frac{i}{8}\int_{C_0} F_\nu(\rho_>,a)H_\nu^{(2)}(k\rho_<)e^{i\nu|\phi-\phi'|}\,d\nu\\&+\frac{i}{8}\int_0^\infty \frac{F_\nu(\rho_>,a)F_\nu(\rho_<,a)}{1-R}e^{i\nu|\phi-\phi'|}\,d\nu\\&+\frac{i}{8}\int_{i\infty}^0 e^{2i\pi\nu}\frac{F_\nu(\rho_>,a)F_\nu(\rho_<,a)}{1-e^{2i\pi\nu}R}e^{i\nu|\phi-\phi'|}\,d\nu.\end{aligned} \quad (2.536)$$

In the first integral on the right-hand side of (2.536), in which C_0 can be replaced by C, we have taken into account that $H_\nu^{(2)}(k\rho_<)F_\nu(\rho_>,a)$ is invariant when we change ν in $-\nu$, since

$$H_{-\nu}^{(1,2)}(x)=e^{\pm i\nu\pi}\,H_\nu^{(1,2)}(x). \quad (2.537)$$

The two other integrals come from the first term on the right-hand side of (2.35), which has been separated into two parts corresponding to $R_e\nu > 0$ and $R_e\nu < 0$. For $R_e\nu < 0$, ν has been replaced by $-\nu$ in accordance with (2.519).

It can be readily shown that if we replace, in the last integral on the right-hand side of (2.536), the Hankel functions with their Debye asymptotic approximations given by (2.295) and (2.296), the integrand has no stationary point in the second quadrant of the complex ν-plane.

The dominant term of its asymptotic expansion is given by the end point contribution at $\nu = 0$. This term can be calculated by the integration of parts and gives a contribution of order k^{-1}. It is of the same order as the parasitic diffraction at the origin, introduced by the perfectly absorbing planes used in the derivation of the solution of the diffraction problem. Therefore, it can be neglected compared to the principal contribution, which is of order $k^{-1/2}$.

In the same way, by using the Debye asymptotic approximation of the Hankel functions in the integrand of the first integral on the right-hand side of (2.536), it can be shown that the integrand has two stationary points situated on the real axis that correspond to a direct ray and a reflected ray.

By deforming the path of integration into the SDP passing through these points (see fig. 2.53), we obtain the corresponding contributions to the field by applying the steepest descent method. The existence of the stationary points is submitted to the following conditions:

$$0 \leq |\phi - \phi'| \leq \frac{\pi}{2} - \text{Arc}\sin\left(\frac{\rho_<}{\rho_>}\right) \text{ for the direct ray,}$$

$$0 \leq |\phi - \phi'| \leq \tau_1, \quad \tau_1 = \frac{\pi}{2} + \text{Arc}\sin\left(\frac{\rho_<}{\rho_>}\right) - 2\text{Arc}\sin\left(\frac{\rho_<}{a}\right) \text{ for the reflected ray.}$$

If the source S' and the observation point S are both on the surface ($\rho_> = \rho_< = a$), then these rays exist only when $|\phi - \phi'| = 0$ so that S' and S coalesce, and their contributions are strictly zero since $F_\nu(a,a) = 0$. If S' and S are in the boundary layer defined by $\frac{kh}{m} \leq 1$, where $h = a - \rho'$ or $a - \rho$ and m is the Fock parameter $m = (ka/2)^{1/3}$, then h is a small quantity of order $k^{-2/3}$, and consequently the value of τ_1 is close to zero. Hence when the source and the observation point are close to the surface, the direct and reflected rays exist only when S is close to S', and their contributions are negligibly small since $F_\nu(a, a-h)$ is negligibly small.

Later on we limit our investigation to the case where the source and the observation point are both close to the surface. Then according to the previous analysis, the contribution of the first integral on the right-hand side of (2.536) can be neglected. Finally, only the second integral is of importance, and we have

$$G(\mathbf{\rho},\mathbf{\rho}') \cong \frac{i}{8}\int_0^\infty \frac{F_\nu(\rho_>,a)F_\nu(\rho_<,a)}{1-R} e^{i\nu|\phi-\phi'|}\,d\nu. \tag{2.538}$$

This integral does not change at the principal order if we replace the contour of integration by the contour C_1 shown in figure 2.54, which is the equivalent of adding (2.538) to the integral

$$\frac{i}{8}\int_{-i\infty}^{0} \frac{F_\nu(\rho_>,a)F_\nu(\rho_<,a)}{1-R} e^{i\nu|\phi-\phi'|}\,d\nu, \tag{2.539}$$

which has no stationary point on the contour of integration. Thus its contribution reduces to the end contribution, which is of lower order and can therefore be neglected.

The Green function (2.538) may therefore be approximated by

$$G(\mathbf{\rho},\mathbf{\rho}') \cong \frac{i}{8}\int_{C_1} \frac{F_\nu(\rho_>,a)F_\nu(\rho_<,a)}{1-R} e^{i\nu|\phi-\phi'|}\,d\nu. \tag{2.540}$$

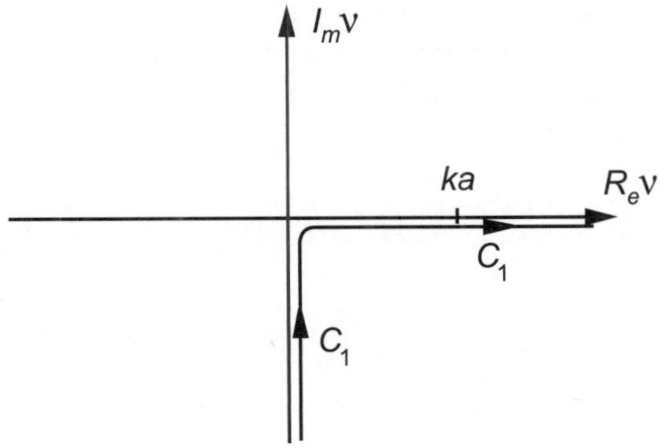

Figure 2.54. Integration path C_1.

This expression of the Green function comes from Babich and Buldyrev [2.40]. It differs from the one obtained in (2.520) by Wasylkiwskyj [2.39] by a term of order k^{-1}. But since Wasylkiwskyj's solution contains a parasitic term of the same order due to the diffraction by the edge of the absorbing planes at the origin, delimiting the domain of integration, both solutions are equivalent for the principal term, of order $k^{-1/2}$, of their asymptotic expansions, but they are not adequate for calculating higher-order terms.

To extract the GA field from (2.540), we decompose $(1-R)^{-1}$ as follows:

$$\frac{1}{1-R} = \sum_{n=0}^{N} R^n + \frac{R^{N+1}}{1-R}. \tag{2.541}$$

According to (2.541), the Green function is separated into two parts, G_1 and G_2, given by

$$G_1(\boldsymbol{\rho},\boldsymbol{\rho}') = \frac{i}{8}\sum_{n=0}^{N} \int_{C_1} R^n F_\nu(\rho_>,a)F_\nu(\rho_<,a)\, e^{i\nu|\phi-\phi'|}\, d\nu, \tag{2.542}$$

$$G_2(\boldsymbol{\rho},\boldsymbol{\rho}') = \frac{i}{8}\int_{C_1} F_\nu(\rho_>,a)F_\nu(\rho_<,a)\frac{R^{N+1}}{1-R} e^{i\nu|\phi-\phi'|}\, d\nu. \tag{2.543}$$

Observing that the integrand of (2.542) has no pole on the real axis, it can be shown that under some conditions, which will be specified later on in this chapter, $G_1(\boldsymbol{\rho},\boldsymbol{\rho}')$ gives the contributions of the direct field and the field, due to single and multiple reflections up to the order N, whereas $G_2(\boldsymbol{\rho},\boldsymbol{\rho}')$ gives a residual contribution that cannot be represented by GA waves.

The technique for deriving the GA contributions consists of replacing the Hankel functions in (2.542) with the principal term of their Debye asymptotic expansions and calculating the integral over G_1 by the steepest descent method.

We first consider the term $N = 0$ in (2.542). The contributions of this term are given by

$$G_1^{(0)}(\boldsymbol{\rho},\boldsymbol{\rho}') = \frac{i}{8}\int_{C_1} F_\nu(\rho_>,a)F_\nu(\rho_<,a)\, e^{i\nu|\phi-\phi'|}\,d\nu$$
$$= \frac{i}{8}\int_{C_1}\left(\sum_{j=1}^{4}\phi_j\right)e^{i\nu|\phi-\phi'|}\,d\nu, \tag{2.544}$$

where

$$\phi_1 = H_\nu^{(1)}(k\rho_>)\,H_\nu^{(1)}(k\rho_<)\ ,\quad \phi_2 = R\,H_\nu^{(1)}(k\rho_>)\,H_\nu^{(2)}(k\rho_<),$$
$$\phi_3 = R\,H_\nu^{(2)}(k\rho_>)\,H_\nu^{(1)}(k\rho_<),\quad \phi_4 = R^2\,H_\nu^{(2)}(k\rho_>)\,H_\nu^{(2)}(k\rho_<). \tag{2.545}$$

Replacing the Hankel functions with the first term of their Debye asymptotic expansion, we obtain the following expression for the phase of the term $\phi_1\exp\left[i\nu|\phi-\phi'|\right]$:

$$q(\nu) = \nu|\phi-\phi'| + \sqrt{(k\rho_>)^2-\nu^2} + \sqrt{(k\rho_<)^2-\nu^2} - \nu\left(\text{Arc}\cos\frac{\nu}{k\rho_>} + \text{Arc}\cos\frac{\nu}{k\rho_<}\right). \tag{2.546}$$

The saddle point condition reads

$$q'(\nu) = |\phi-\phi'| - \left(\text{Arc}\cos\frac{\nu}{k\rho_>} + \text{Arc}\cos\frac{\nu}{k\rho_<}\right) = 0. \tag{2.547}$$

If we denote by

$$\alpha_1 = \text{Arc}\cos\frac{\nu}{k\rho_>},\quad \alpha_2 = \text{Arc}\cos\frac{\nu}{k\rho_<}, \tag{2.548}$$

then we have $\nu_s = k\rho_>\cos\alpha_1 = k\rho_<\cos\alpha_2$. For ν real, the saddle point condition admits the graphical interpretation shown in figure 2.55.

We observe that a real solution is possible only if $S'H < S'S$, which is equivalent to $|\phi-\phi'| > \dfrac{\pi}{2} - \text{Arc}\cos\dfrac{\rho_<}{\rho_>}$. By taking (2.547) into account in (2.546), we see that the phase $q(\nu_s)$ is given by

$$q(\nu_s) = k(\rho_>\sin\alpha_1 + \rho_<\sin\alpha_2) = k\,S'S.$$

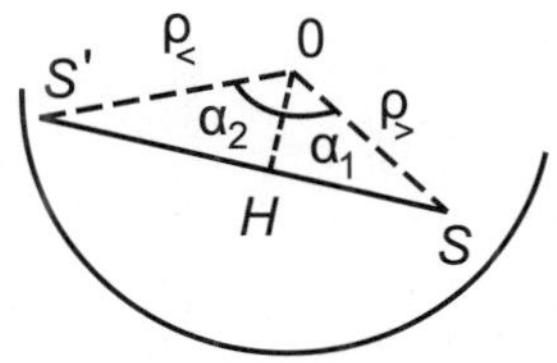

Figure 2.55. Graphical interpretation of the saddle point condition for ϕ_1.

The contribution of ϕ_1 therefore corresponds to the direct ray $S'S$. Likewise, the Debye approximation for ϕ_2 and ϕ_3, respectively, leads to the following saddle point conditions:

$$q'(v_s) = |\phi - \phi'| - \text{Arc}\cos\frac{v}{k\rho_>} + \text{Arc}\cos\frac{v}{k\rho_<} - 2\text{Arc}\cos\frac{v}{ka} = 0. \tag{2.549}$$

$$q'(v_s) = |\phi - \phi'| - \text{Arc}\cos\frac{v}{k\rho_<} + \text{Arc}\cos\frac{v}{k\rho_>} - 2\text{Arc}\cos\frac{v}{ka} = 0. \tag{2.550}$$

By denoting

$$\alpha_1 = \text{Arc}\cos\frac{v}{k\rho_>}, \quad \alpha_2 = \text{Arc}\cos\frac{v}{k\rho_<}, \quad \alpha_3 = \text{Arc}\cos\frac{v}{ka}, \tag{2.551}$$

we have

$$v = k\rho_> \cos\alpha_1 = k\rho_< \cos\alpha_2 = ka\cos\alpha_3 \tag{2.552}$$

and

$$|\phi - \phi'| = \alpha_1 - \alpha_2 + 2\alpha_3,$$
$$|\phi - \phi'| = -\alpha_1 + \alpha_2 + 2\alpha_3$$

for the saddle point conditions (2.549) and (2.550), respectively.

The corresponding graphical interpretation is shown in figure 2.56, where θ_1 and θ_2 are the angles of reflection of the rays $S'Q_1S$ and $S'Q_2S$.

The saddle point condition corresponding to ϕ_4 reads

$$q'(v_s) = |\phi - \phi'| + \text{Arc}\cos\frac{v}{k\rho_>} + \text{Arc}\cos\frac{v}{k\rho_<} - 4\text{Arc}\cos\frac{v}{ka} = 0 \tag{2.553}$$

or

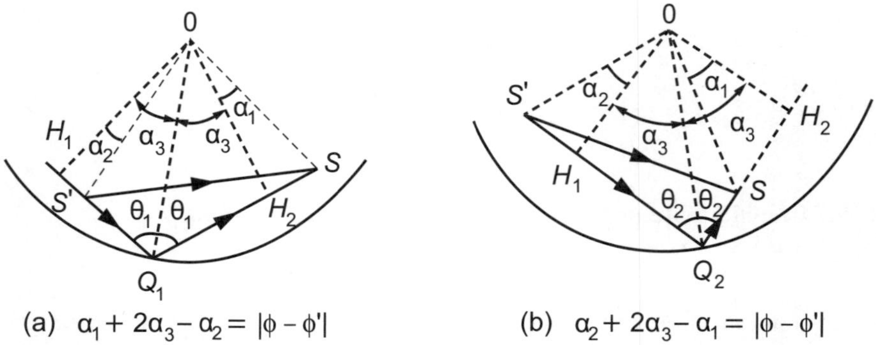

Figure 2.56. Graphical interpretation of the saddle point conditions corresponding to (a) ϕ_2 and (b) ϕ_3.

$$|\phi - \phi'| = 4\alpha_3 - \alpha_1 - \alpha_2, \tag{2.554}$$

where α_1, α_2, and α_3 are defined by (2.551). The graphical interpretation of (2.554) is shown in figure 2.57. It leads to a doubly reflected ray.

For $N = 1$, the term $R\phi_1$ gives another singly reflected ray corresponding to the saddle point condition:

$$|\phi - \phi'| = 2\alpha_3 - \alpha_1 - \alpha_2.$$

However $R\phi_2$ and $R\phi_3$ give doubly reflected rays, and $R\phi_4$ gives a triply reflected ray. We observe that when N augments by one unit, the order of reflection of each term ϕ_j (j = 1, 2, 3, 4) also augments by one unit. Now it remains to derive the conditions under which the GA approximation can be applied. It is clear that this depends on the validity of the Debye approximation,

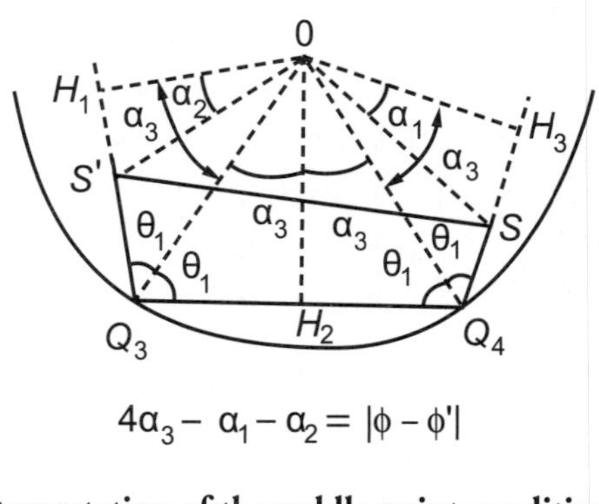

Figure 2.57. Graphical interpretation of the saddle point condition corresponding to ϕ_4.

which imposes $|\nu - ka| \geq (ka)^{1/3}$. If we consider the saddle point conditions corresponding to the terms ϕ_j (j = 1, 2, 3, 4), then we observe that when $|\phi - \phi'|$ is small compared to π, the dominant contributions to the integral for $N = 0$ comes from ν, close to ka. Hence the Debye approximation remains valid when the angular distance between the source and the observation point is not too small.

In order to specify this rule more carefully, we consider the rays emanating from a source located on the concave surface, which undergo one reflection on the surface. These rays have an envelope or caustic C_1 that is real and divides the space above the concave surface into two domains denoted by regions 1 and 2 (fig. 2.58).

The field in the neighborhood of the caustic, in region 2, expressed in curvilinear coordinates s, n, where s is the abscissa along the caustic C_1 and n is the normal distance of the observation point to the caustic, is given by (see chapter 4, equation (4.26))

$$u = A(s)e^{iks}\, A_i(-\nu), \tag{2.555}$$

where $A_i(x)$ is the Airy function, $A(s)$ is an amplitude factor, and $\nu = kn\,/\,m$, where m is the Fock parameter $m = (k\rho_c\,/\,2)^{1/3}$ and ρ_c is the radius of curvature of the caustic at the abscissa s. The caustic boundary layer is defined by $\nu \leq 1$ or $n \leq m/k$. In the shadow region of the caustic (region 1), the GA theory gives a zero field, but this is only an approximation. The true field given by (2.555) is an evanescent field. In region 2, corresponding to the simply reflected rays, the field

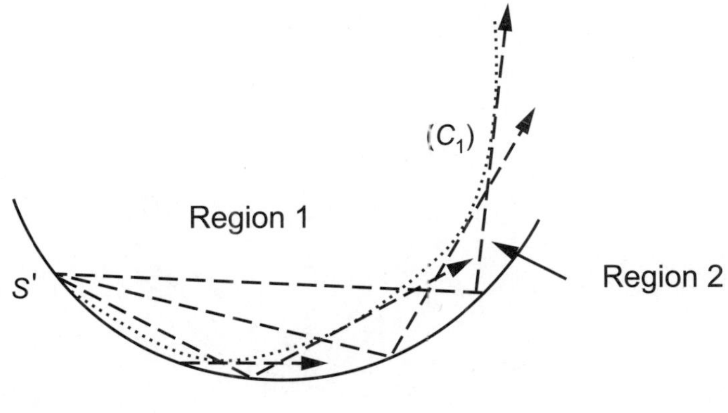

Figure 2.58. Caustic of singly reflected rays.

is correctly represented by the GA field, except in the caustic boundary layer. The condition $n > m/k$ defines the minimum distance ℓ_m from the source, along the concave surface, where the GA approximation of the field given by the simply reflected rays is valid.

Likewise, the doubly reflected rays are tangent to a caustic C_2. When the order of reflections augments, the corresponding caustics approach the concave surface as illustrated in figure 2.59.

As a result, the minimum distance where the GA laws are valid increases with the order of reflection.

A minimum distance also exists for the direct ray. When the observation point is located on the surface, the saddle point condition (2.547) defines a saddle point $\nu_s = ka \sin w_s$, where

$$w_s = \frac{\pi}{2} - \frac{|\phi - \phi'|}{2}.$$

Hence when $|\phi - \phi'|$ decreases ν_s tends to ka, and at a certain angular distance, $|\nu - ka|$ becomes smaller than $(ka)^{1/3}$. Consequently, the Debye asymptotic approximation of the Hankel functions is no longer valid.

It follows from the preceding discussion that it is not possible to replace the Hankel functions throughout with their Debye asymptotic expansions. The correct procedure consists of extracting from (2.540) the direct field or the direct and reflected field and treating the remainder integral in a different way, which we will describe in the following section.

2.4.2.3. Expression of the Remainder Integral

It is well known that for $ka >> 1$, the Debye expansions of $H_\nu^{(1,2)}(ka)$ remain valid very close to $\nu = ka$. Consequently, the most important contribution to (2.540) comes from the neighborhood of $\nu = ka$. This argument leads to the following change of variables:

$$\nu = ka + m\xi, \quad m = \left(\frac{ka}{2}\right)^{1/3}. \tag{2.556}$$

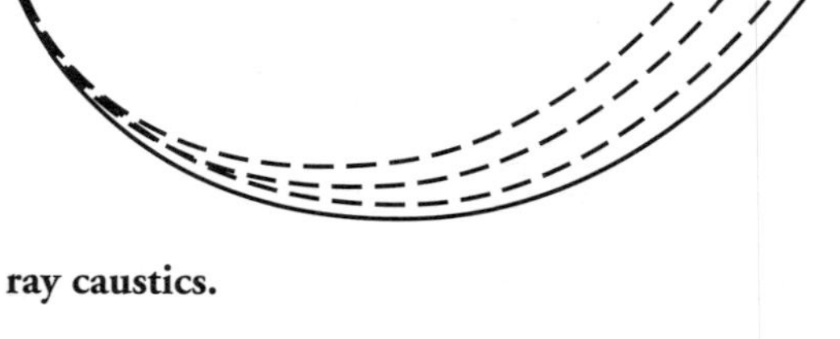

Figure 2.59. Reflected ray caustics.

For $ka >> 1$ and $\xi \leq 1$, the Bessel and Hankel functions may be replaced by their asymptotic expansions, valid for $|\nu - ka| \leq (ka)^{1/3}$ (see equations (3.197) and (3.199)):

$$J_\nu(ka) \cong \left(m\sqrt{\pi}\right)^{-1} V(\xi),$$
$$H_\nu^{(1,2)}(ka) \cong \mp i\left(m\sqrt{\pi}\right)^{-1} \mathrm{w}_{1,2}(\xi). \tag{2.557}$$

Moreover, it is also possible to derive similar equations for $J_\nu(k\rho)$ and $H_\nu^{(1,2)}(k\rho)$ by observing that

$$\nu = k(a+n) + m\left(\xi - \frac{kn}{m}\right) = k\rho + m(\xi - h), \quad h = \frac{kn}{m}$$
$$= k(a+n') + m\left(\xi - \frac{kn'}{m}\right) = k\rho' + m(\xi - h'), \quad h' = \frac{kn'}{m}. \tag{2.558}$$

Consequently, for $h \leq 1$ and $h' \leq 1$, it follows that

$$J_\nu(k\rho) \cong \left(m\sqrt{\pi}\right)^{-1} V(\xi - h), \; J_\nu(k\rho') \cong \left(m\sqrt{\pi}\right)^{-1} V(\xi - h'),$$
$$H_\nu^{(1,2)}(k\rho) \cong \mp i\left(m\sqrt{\pi}\right)^{-1} \mathrm{w}_{1,2}(\xi - h), \; H_\nu^{(1,2)}(k\rho') \cong \mp i\left(m\sqrt{\pi}\right)^{-1} \mathrm{w}_{1,2}(\xi - h'). \tag{2.559}$$

We consider first the case where no GA contribution has been extracted from (2.540). By inserting (2.557) through (2.559) into (2.540) and by taking (2.529) into account, we obtain

$$G(\boldsymbol{\rho}, \boldsymbol{\rho}') = \frac{e^{ik\ell}}{4\pi m} S(\sigma, h, h'), \tag{2.560}$$

where ℓ is the distance along the concave surface between the projections S_1' and S_1 of the source S' and the observation point S on the surface and $\sigma = m\ell / a$. The function $S^{(0)}$, where the superscript (0) means that no GA contribution has been extracted, is given by

$$S^{(0)}(\sigma, h, h') = \frac{1}{4} \int_{C_\xi} e^{i\sigma\xi} F_\xi(h) F_\xi(h') \left(\frac{\Omega_\xi \mathrm{w}_2(\xi)}{\Omega_\xi V(\xi)}\right) d\xi, \tag{2.561}$$

where

$$F_\xi(x) = w_1(\xi - x) - \frac{\Omega_\xi w_1(\xi)}{\Omega_\xi w_2(\xi)} w_2(\xi - x) \tag{2.562}$$

and

$$\Omega_\xi = \frac{\partial}{\partial \xi} - im\,Z. \tag{2.563}$$

In (2.540), the contour C_1 can be deformed into the contour C_1' passing through *ka*, as shown in figure 2.60(a). Furthermore, since $V(\xi)$ has its fastest growth on the straight line,

$$\xi = |\xi| \exp\left(-2i\,\pi/3\right)$$

and decays algebraically when $\xi = |\xi| e^{i\pi/3}$, and since its fastest decay is on the real axis and is exponential (see page 408 in reference [2.40] or page 514 in reference [2.32]), it is advantageous to deform the integration path in the complex ξ-plane into a straight line $\exp\left(-2i\,\pi/3\right), \exp\left(i\,\pi/3\right)$ passing through the origin (fig. 2.60[b]). This path is designated by C_ξ.

We now consider the term $N = 0$ in (2.542) and (2.543). The remainder integral (2.543) corresponds to the case where the contributions of the direct ray $S'S$; the singly reflected rays $S'Q_1S$, $S'Q_2S$; and the doubly reflected ray $S'Q_3Q_4S$ (see fig. 2.56 and 2.57) have been subtracted from (2.540). Applying the transformation (2.556), we obtain the following for the reminder integral of order $N = 0$:

$$G_2^{(0)}(\boldsymbol{\rho},\boldsymbol{\rho}') = \frac{e^{ik\ell}}{16 m\pi} \int_{C_\xi} e^{i\sigma\xi} F_\xi(h) F_\xi(h') \frac{\Omega_\xi\, \mathrm{w}_2(\xi)}{\Omega_\xi\, V(\xi)} \frac{\Omega_\xi\, \mathrm{w}_1(\xi)}{\Omega_\xi\, \mathrm{w}_2(\xi)} d\xi. \tag{2.564}$$

Likewise, the remainder integral of order N is given by

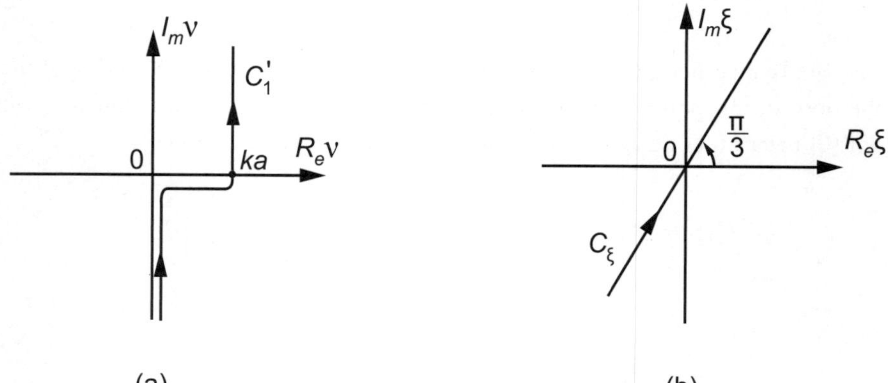

(a) (b)

Figure 2.60. Path deformations in the complex v- and ξ-planes.

$$G_2^{(N)}(\boldsymbol{\rho},\boldsymbol{\rho}') = \frac{e^{ik\ell}}{16m\pi}(-1)^{N+1}\int_{C_\xi} e^{i\sigma\xi}F_\xi(h)F_\xi(h')\frac{\Omega_\xi \, \mathrm{w}_2(\xi)}{\Omega_\xi V(\xi)}\left[\frac{\Omega_\xi \, \mathrm{w}_1(\xi)}{\Omega_\xi \, \mathrm{w}_2(\xi)}\right]^{N+1} d\xi, \quad (2.565)$$

which can also be written as

$$G_2^{(N)}(\boldsymbol{\rho},\boldsymbol{\rho}') = (-1)^{N+1}\frac{e^{ik\ell}}{4\pi m}S^{(N+1)}(\sigma,h,h'), \quad (2.566)$$

where $S^{(N+1)}(\sigma, h, h')$ is a Fock-type function given by

$$S^{(N+1)}(\sigma,h,h') = \frac{1}{4}\int_{C_\xi} e^{i\sigma\xi}F_\xi(h)F_\xi(h')\frac{\Omega_\xi \, \mathrm{w}_2(\xi)}{\Omega_\xi V(\xi)}\left[\frac{\Omega_\xi \, \mathrm{w}_1(\xi)}{\Omega_\xi \, \mathrm{w}_2(\xi)}\right]^{N+1} d\xi. \quad (2.567)$$

When the source S' and the observation point S are both located on the surface of the boundary, the rays $S'Q_1S$, $S'Q_2S$, and $S'Q_3Q_4S$ merge with the direct ray $S'S$. For a hard boundary, we then find for $N = 0$ four times the field of the direct ray in free space, and the total Green function is decomposed as follows:

$$G_h^{(0)}(\sigma,0,0) = e^{i\frac{\pi}{4}}\sqrt{\frac{2}{\pi k}}\frac{e^{ikD_0}}{\sqrt{D_0}} - \frac{e^{ik\ell}}{4\pi m}S^{(1)}(\sigma,0,0), \quad (2.568)$$

where D_0 is the rectilinear distance from the source to the observation point.

The factor four is in agreement with the fact that the field radiated by a source on a hard boundary is multiplied by two and that the field observed on the surface of a hard boundary is also multiplied by two.

Numerical results for the modulus $\left|G_h^{(0)}(\sigma,0,0)\right|$ are given in figure 2.61 and 2.62 for $ka = 8$ and $ka = 80$, respectively, and for σ varying from 0 to 7.

In these figures, we observe two regimes: a transitional regime for $0 < \sigma \leq 3.5$ and an oscillatory regime for $3.5 < \sigma < 7$. If we compare these results to the exact solution, then we observe that (2.568) gives accurate values for $\sigma > 3.5$. For σ smaller than 3, the accuracy of the results decreases progressively, independently of the value of ka. Hence $\sigma \cong 3$ is the shortest distance where the contribution of the direct ray can be separated from the global solution. Beneath this value, the GA approximation is no longer valid, and the global solution (2.560) must be retained. However, when $\sigma > 3$ the global solution (2.560) deviates from the exact results and convergence problems appear, due to the slow decrease of the integrand when $|\xi|$ augments on the branch of the integration path, situated in the third quadrant of the complex ξ-plane. In this case, the approximation, which consists of replacing the Hankel functions with their asymptotic expansions (2.557), is no longer valid, and the integral (2.560) no longer represents the exact solution. In particular, it does not give the oscillations of the solution when $\sigma > 3.5$. This result highlights the importance of the decomposition of the solution in a GA contribution and a remainder term.

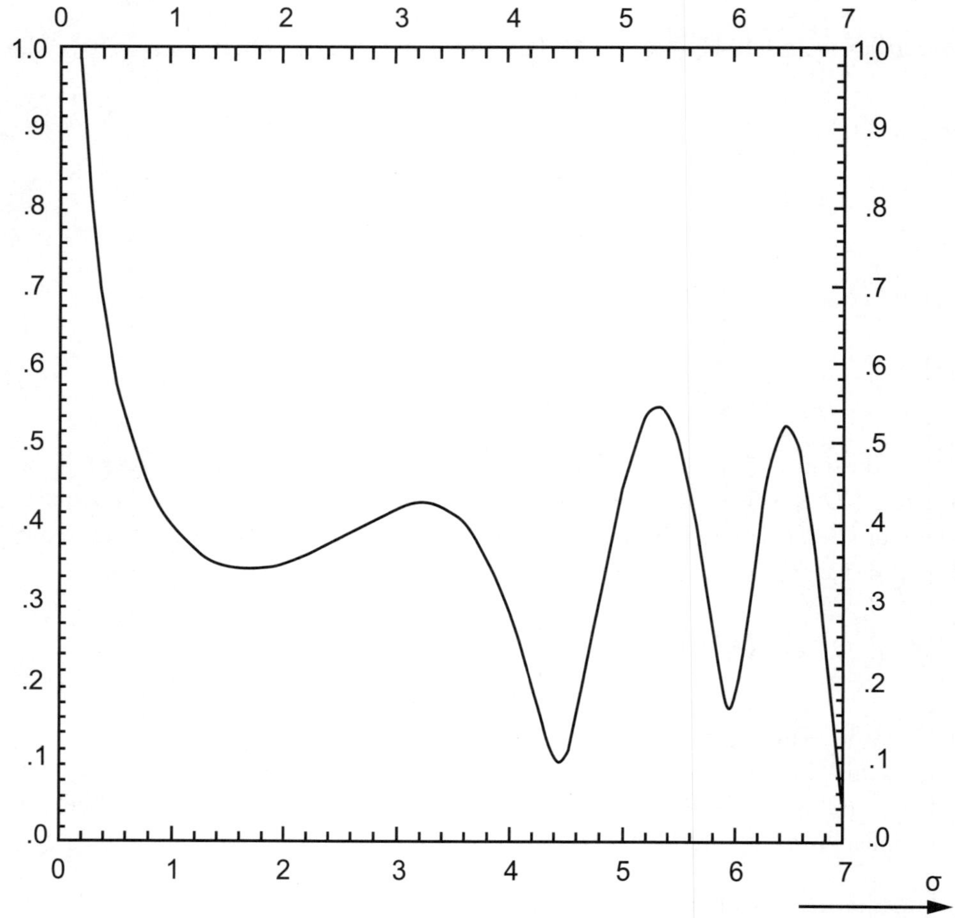

Figure 2.61. Source and observation point on a concave hard circular cylinder modulus of $G_h(\sigma, 0, 0)$ versus σ, ka = 8.

For $\sigma \geq 6$, the contribution of the reflected ray can be extracted from the integral form of the solution, in addition to the contribution of the direct ray. The remainder is then given by the canonical integral corresponding to $N = 1$. A general rule for the choice of N has been derived in [2.42] and is given by $\frac{\sigma}{\sigma_c} - 1 < N + 1 \leq \frac{\sigma}{\sigma_c}$, where $\sigma_c \cong 3$.

Additional results for a source and an observation point located on either a hard or soft boundary may be found in references [2.42] and [2.40]. For an impedance boundary, the number of rays to be extracted from the integral form of the solution follows the same rule as for a hard or soft boundary.

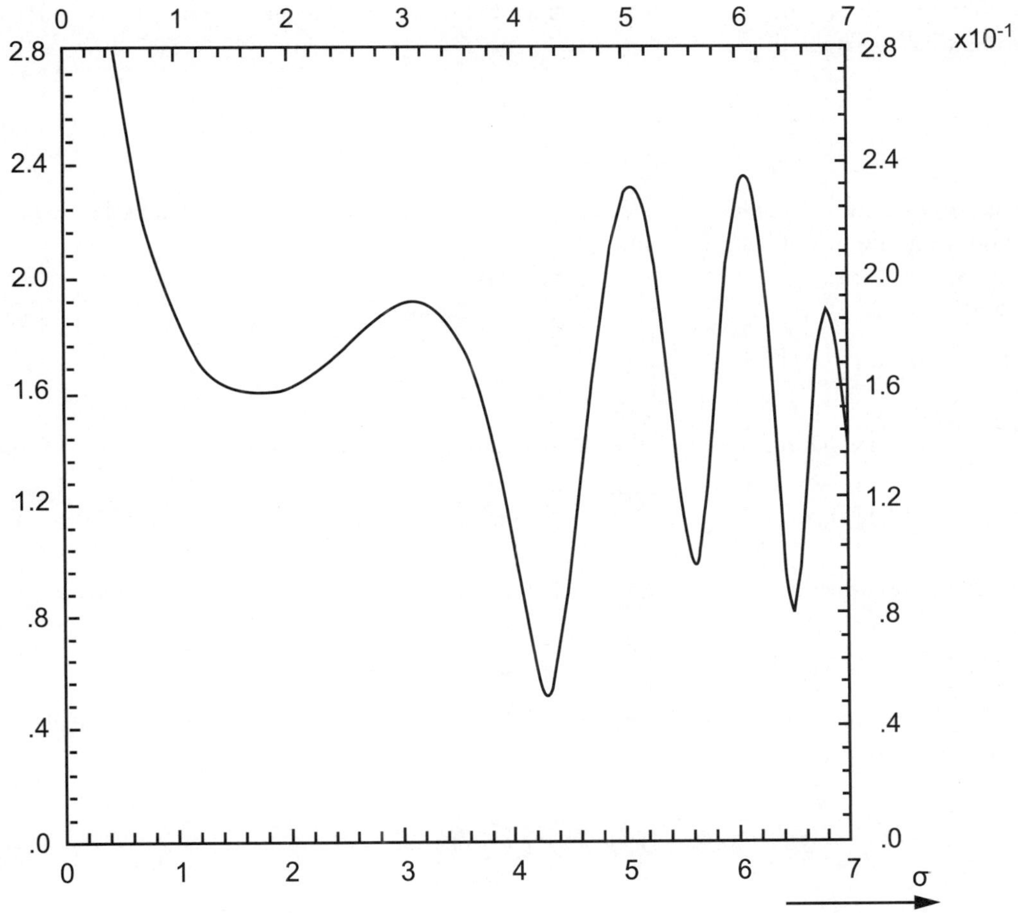

Figure 2.62. Source and observation point on a concave hard circular cylinder modulus of $G_h(\sigma, 0, 0)$ versus σ, ka = 80.

2.4.3. EXTENSION OF THE SOLUTION TO MORE GENERALIZED SITUATIONS

If the cylindrical surface has an arbitrary smooth concave cross section, then the Green function of the problem can be obtained by extending the solution of the canonical problem derived in section 2.4.2 by invoking the principle of localization, which is valid when the local radius of curvature is large compared to the wavelength. Particularly, the Green function can be decomposed into a GA term and a remainder term. The generalization of the GA term is standard. We therefore focus our investigation on the remainder term.

Let ℓ be the curvilinear abscissa along the surface measured from the projection S_p' of the source to the projection S_p of the observation point on the boundary. The canonical integral

(2.567) depends only on the parameter σ, which can be extended to an arbitrary concave surface in the usual way by setting

$$\sigma = \int_{S'_p}^{S_p} \frac{m(\ell)}{\rho(\ell)} d\ell, \qquad (2.569)$$

where $\rho(\ell)$ is the local radius of curvature of the cross section of the cylindrical boundary, which is also the geodesic joining S'_p to S_p and

$$m(\ell) = \left[\frac{k\rho(\ell)}{2}\right]^{1/3}. \qquad (2.570)$$

Furthermore, in order to satisfy the theorem of reciprocity, we replace m in the factor in front of $S^{(N+1)}$ in (2.566) with $\sqrt{m(\rho')m(\rho)}$, where ρ' and ρ are the radii of curvature of the cross section of the surface at S'_p and S_p, respectively. The conservation of energy in the boundary layer also introduces a factor $\left(\frac{\rho}{\rho'}\right)^{1/6} = \sqrt{\frac{m(\rho)}{m(\rho')}}$.

Finally, for an arbitrary smooth concave cylindrical surface excited by a line source parallel to the generatrices, the Green function takes the form

$$G(\boldsymbol{\rho},\boldsymbol{\rho}') = G_1^{(N)}(\boldsymbol{\rho},\boldsymbol{\rho}') + G_2^{(N)}(\boldsymbol{\rho},\boldsymbol{\rho}'), \qquad (2.571)$$

where $G_1^{(N)}(\boldsymbol{\rho},\boldsymbol{\rho}')$ corresponds to the GA contribution of the direct and reflected rays of order 1 to N and where the remainder contribution is given by

$$G_2^{(N)}(\boldsymbol{\rho},\boldsymbol{\rho}') = \frac{e^{ik\ell}}{4\pi\, m(\rho')} S^{(N+1)}(\sigma, h, h'), \qquad (2.572)$$

where the canonical integral $S^{(N+1)}(\sigma, h, h')$ is given by (2.567), with σ and m given by (2.569) and (2.570), respectively.

A generalization of (2.571) and (2.572) to a point source in front of an arbitrary smooth concave cylindrical surface is obtained by extending the solution of the canonical problem of a point source in front of a concave circular cylinder. The analytical formulation of the three-dimensional problem from a knowledge of the two-dimensional one is straightforward (see [2.16]). Explicit formulas for the decomposition of the Green function in a GA contribution and a remainder integral are given in [2.42].

A heuristic generalization of the solution given in [2.42] to a point source in front of a concave smooth three-dimensional surface is also possible by extending the integration path of the integral in (2.569) to an integration path along the geodesic, joining S'_p and S_p and by taking into account the divergence factor along this geodesic.

REFERENCES

[2.1] MacDonald, H. M. 1902. *Electric Waves*. Cambridge: Cambridge University Press.

[2.2] Sommerfeld, A. 1928. "Asymtotische Darstellung von Formeln aus Beugungstheorie des Lichtes." *Journal für die Reine und Angewandte Mathematik* 158: 199–208.

[2.3] ———. 1966. *Partial Differential Equations in Physics*. New York: Academic Press.

[2.4] Meixner, J. 1972. "The Behavior of Electromagnetic Fields at Edges." *IEEE Transactions on Antennas and Propagation* 20 (4): 442–46.

[2.5] Abramowitz, M., and I. A. Stegun. 1964. *Handbook of Mathematical Functions*. New York: Dover.

[2.6] Maliuzhinets, G. D. 1958. "Excitation, Reflection and Emission of Surface Waves from a Wedge with Given Face Impedances." *Soviet Physics—Doklady* 3: 752–55.

[2.7] ———. 1960. "Das Sommerfeldsche Integral und die Lösung von Beugunsaufgaben in Winkelgebieten." *Annalen der Physik* 7: 107–12 .

[2.8] Vaccaro, V. 1980. "The Generalized Reflection Method in Electromagnetism." *AEÜ—International Journal of Electronics and Communications* 34 (12): 493–500.

[2.9] Gradshteyn, I. S., and I. M. Ryzhik. 1980. *Table of Integrals, Series, and Products*. San Diego: Academic Press.

[2.10] Senior, T. B. A., and J. L. Volakis. 1995. *Approximate Boundary Conditions in Electromagnetics*. London: IEEE.

[2.11] Bernard, J. M. L. 1987. "Diffraction by a Metallic Wedge Covered with a Dielectric Material." *Wave Motion* 9: 543–61.

[2.12] Michaeli, A. 1988. "Extension of Maliuzhinets' Solution to Arbitrary Permittivity and Permeability of Coatings on a Perfectly Conducting Wedge." *Electronics Letters* 24 (20): 1291–92.

[2.13] Viktorov, I. A. 1967. *Rayleigh and Lamb Waves: Physical Theory and Applications*. New York: Plenum Press.

[2.14] De Billy, M. 1991. "Finite Sample Effects on the Diffracted Signal by a Right-Angle Wedge." *Journal Acoustique* 4: 563–73.

[2.15] Bowman, J. J., T. B. A. Senior, and P. L. Uslenghi. 1987. *Acoustic and Electromagnetic Scattering by Simple Shapes*. New York: Hemisphere.

[2.16] Felsen, L. B., and N. Marcuvitz. 1972. *Radiation and Scattering of Waves*. New York: Prentice-Hall.

[2.17] Watson, G. N. 1918. "The Diffraction of Electric Waves by the Earth." *Procedings of the Royal Society* 95: 83–99.

[2.18] Olver, F. W. J. 1974. *Asymptotic and Special Functions*. San Diego: Academic Press.

[2.19] Imai, I. 1954. "Die Beugung Elektromagnetischer Wellen an einem Kreiszylinder." *Zeitschrift für Physik* 137: 31–48.

[2.20] Miller, J. C. P. 1946. *The Airy Integral*. Cambridge: Cambridge University Press.

[2.21] Helstrom, C. W. 1963. "Scattering from a Cylinder Coated with Dielectric Material." In *Electromagnetic Theory and Antennas*, edited by E. C. Jordan. New York: Pergamon.

[2.22] Kouyoumjian, R. G. 1975. "The Geometrical Theory of Diffraction and Its Application." Chap. 6 in *Numerical and Asymptotic Techniques in Electromagnetics*, edited by R. Mittra. Berlin: Springer-Verlag.

[2.23] Keller, J. B., and B. R. Levy. 1959. "Decay Exponents and Diffraction Coefficients for Surface Waves on Surfaces of Non Constant Curvature." *IEEE Transactions on Antennas and Propagation* 7 (5): 52–61.

[2.24] Franz, W., and K. Klante. 1959. "Diffraction by Surfaces of Variable Curvature." *IEEE Transactions on Antennas and Propagation* 7 (5): 58–70.

[2.25] Hong, S. 1967. "Asymptotic Theory of Electromagnetic and Acoustic Diffraction by Smooth Convex Surfaces of Variable Curvature." *Journal of Mathematical Physics* 8 (6): 1223–32.

[2.26] Voltmer, D. R. 1970. "Diffraction by Doubly Curved Convex Surfaces." PhD diss., Ohio State University.

[2.27] Andronov, I., and D. Bouche. 1994. "Computation of the Second-Order Term for the Propagation Parameter of Creeping Waves by Boundary Layer Method." *Annales des Télécommunications* 49 (3–4): 199–204.

[2.28] ———. 1994. "Asymptotics of Creeping Waves on a Strongly Prolate Body." *Annales des Télécommunications* 49 (3–4): 205–10.

[2.29] Fock, V. R. 1960. *Electromagnetic Wave Propagation and Diffraction Problems*. Oxford: Pergamon.

[2.30] Buchal, R. N., and J. B. Keller. 1960. "Boundary Layer Problems in Diffraction Theory." *Communications On Pure and Applied Mathematics* 13: 35.

[2.31] Babich, W. M., and N. Ya. Kirpichnikova. 1979. *The Boundary Layer Method in Diffraction Problems*. Berlin: Springer-Verlag.

[2.32] Bouche, D., F. Molinet, and R. Mittra. 1997. *Asymptotic Methods in Electromagnetics*. Berlin: Springer-Verlag.

[2.33] Molinet, F., I. Andronov, and D. Bouche. 2005. *Asymptotic and Hybrid Methods in Electromagnetics*. London: IEEE.

[2.34] Dodlittle, R. D., and H. Überall. 1966. "Sound Scattering by Elastic Cylindrical Shells." *Journal of the Acoustical Society of America* 39: 272–75.

[2.35] Ugincius, P., and H. Überall. 1968. "Creeping-Wave Analysis of Acoustic Scattering by Elastic Cylindrical Shells." *Journal of the Acoustical Society of America* 50: 1025–35.

[2.36] Brill, D., and H. Überall. 1971. "Acoustic Waves Transmitted Through Solid Elastic Cylinders." *Journal of the Acoustical Society of America* 50: 921–39.

[2.37] Franz, W. 1957. *Theorie der Beugung Elektromagnetischer Wellen*. Berlin: Springer-Verlag.

[2.38] Yang, Y., A. N. Norris, and L. S. Couchman. 1996. "Ray Tracing Over Smooth Elastic Shells of Arbitrary Shape." *Journal of the Acoustical Society of America* 99 (1): 55–64.

[2.39] Wasylkiwskyj, W. 1975. "Diffraction by a Concave Perfectly Conducting Circular Cylinder." *IEEE Transactions on Antennas and Propagation* 23 (4): 480–92.

[2.40] Babich, V. M., and V. S. Buldyrev. 1991. *Asymptotic methods of short wavelength diffraction*. Berlin: Springer-Verlag.

[2.41] Ishihara, T., L. B. Felsen, and A. Green. 1978. "High-Frequency Fields Excited by a Line Source Located on a Perfectly Conducting Concave Cylindrical Surface." *IEEE Transactions on Antennas and Propagation* 26 (6): 757–67.

[2.42] Felsen, L. B., and T. Ishihara. 1979. "High-Frequency Surface Fields Excited by a Point Source on a Concave Perfectly Conducting Cylindrical Boundary." *Radio Science* 14 (2): 205–16.

Chapter 3

UNIFORM ASYMPTOTIC THEORY OF ACOUSTIC WAVE DIFFRACTION

3.1. INTRODUCTION

The asymptotic solutions to the canonical problems described in chapter 2 are valid outside the caustic regions and the transition regions at the shadow boundaries of the direct and reflected rays. In chapter 1 we saw how these asymptotic solutions can be used to construct the dominant term of the asymptotic expansion valid outside the caustic and transition regions of more general problems (curved wedge, smooth curved convex, or concave surfaces).

In the present chapter we seek asymptotic solutions that are uniformly valid inside and outside the transition regions of the incident and reflected fields for the general problems considered in chapter 1. Several methods leading to such solutions are based on the fact that it is possible to construct a uniform solution once the nonuniform solutions inside and outside the transition regions are known (see reference [3.1]). In our case, the main problem is to find the inner solution, since the outer solution is known. For this we can use techniques based on a stretching of the coordinates in order to make their variations smooth. These techniques are also known as the boundary layer method (see references [3.2] and [3.3]). However, these techniques can be quite tedious in regions where multiple boundary layers intersect (fig. 3.1).

We can use a direct method leading to solutions that, although less general, are extremely useful in their applications. The direct method consists of choosing, as an ansatz for the uniform solution, a solution having the same structure as the uniform solution of the corresponding canonical problem.

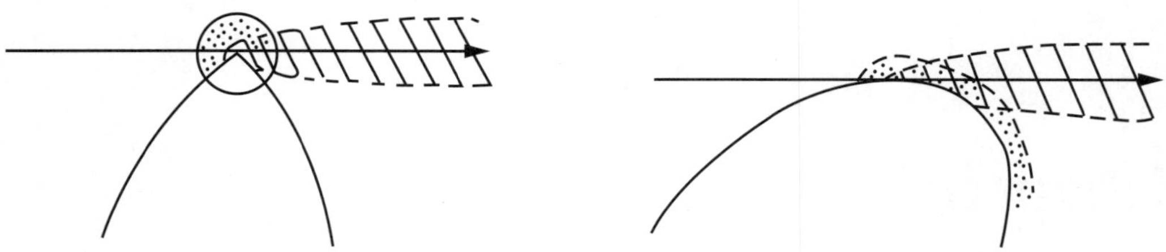

Figure 3.1. Intersection of (a) a line caustic boundary layer and (b) a surface caustic boundary layer with a shadow boundary layer.

The ansatz generally contains special functions characterized by the type of singularities responsible for the nonuniformity of the outer expansion, when the variable enters into the inner domain. We will see that the amplitude of the field in the vicinity of a shadow boundary involves a Fresnel function. In order to satisfy the matching conditions with the known solution of the outer problem, the technique consists of properly modifying the arguments of the special function as well as the amplitude of the terms in the representation.

We will implement this method to construct uniform solutions that are uniformly valid as the observation point traverses through the shadow boundaries of the general problems considered in chapter 1.

3.2. THE THREE-DIMENSIONAL CONVEX WEDGE

3.2.1. UNIFORM SOLUTION FOR THE ACOUSTIC FIELD SCATTERED BY A CONVEX THREE-DIMENSIONAL HARD OR SOFT WEDGE

In chapter 1, section 1.8, we derived a nonuniform solution for the acoustic field scattered by a three-dimensional, impenetrable wedge. For a hard or soft wedge, this solution is valid when the observation point is located far from the edge in region 1, is outside the shadow boundaries 2 and 2′ of the incident and reflected fields, respectively, and is away from the transition zones 3 and 3′, where the diffracted rays in space are transformed into creeping rays (fig. 3.2).

The solution is given in terms of a diffraction coefficient that is itself obtained in chapter 2, section 2.2, for hard and soft boundary conditions. This diffraction coefficient is derived from the dominant term of the asymptotic expansion of the exact solution of the diffraction problem of a wedge with planar faces that are tangent to the wedge with curved faces at the point of diffraction. The derived coefficient provides the dominant term of the asymptotic expansion of the field diffracted by the edge. The total field is obtained by adding the incident and reflected fields to the field diffracted by the edge. If we limit the expansion of the total field up to $0\left(k^{-1/2}\right)$, then

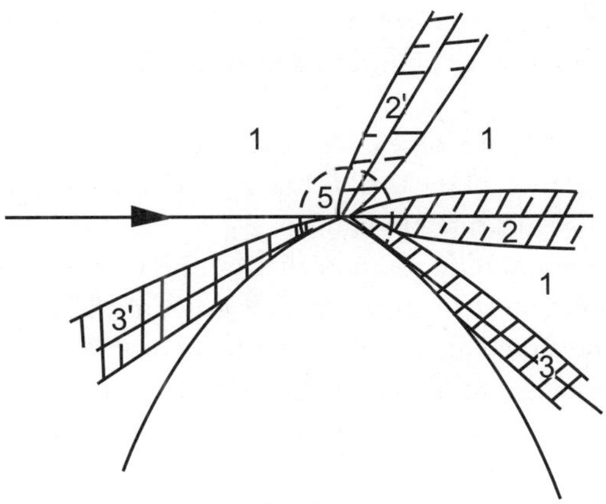

Figure 3.2. Region of validity of the nonuniform solution for a wedge.

only the dominant terms of $0(k^o)$ of the incident and reflected fields appear in the expansion, which takes the form

$$P(M) = e^{ikS^i(M)} p_o^i(M) U(-\varepsilon_i) + e^{ikS^R(M)} p_o^R(M) U(-\varepsilon_R) + e^{ikS^d(M)} \frac{p_o^d}{\sqrt{k}} + 0(k^{-1}), \qquad (3.1)$$

where p_o^i, p_o^R, and p_o^d are, respectively, the dominant terms of the incident, reflected, and diffracted fields and where U is the Heaviside function whose argument is $\varepsilon_i = +1$ or -1, depending on which observation point is in the shadow or the sound zone of the incident pressure field, respectively. Likewise, $\varepsilon_R = +1$ or $\varepsilon_R = -1$, respectively, if the observation point M is in the shadow or the sound zone of the reflected field. The functions $S^i(M)$, $S^R(M)$, and $S^d(M)$ represent the phase of the incident, reflected, and refracted fields, respectively—the phase reference point being some arbitrary point in space.

The solution (3.1) is not valid in the transition zones 2 and 2′ of the direct and reflected field, nor is it valid in the zones 3 and 3′, but outside these zones it provides an appropriate solution in each subdomain of region 1.

In chapter 1, section 1.6 and in appendix D, we have derived the general formula for the dominant term p_o^i of the asymptotic expansion of the field reflected by a curved surface defined by its principal radii of curvature. It is given in terms of the incident field at the point of reflection, a divergence factor depending on the principal radii of curvature of the reflected wave front, and a reflection coefficient R depending on the boundary conditions. For a hard or soft wedge, R is given by +1 and −1, respectively.

A general expression for the dominant p_o^d term of the asymptotic expansion of the diffracted field is given in chapter 1, section 1.8. The expression is presented in terms of (1) the incident field at the diffraction point; (2) a divergence factor depending on the radius of curvature of the diffracted wave front in the plane of diffraction, defined by the tangent to the edge at the diffraction point and the direction of observation; and (3) a diffraction coefficient, the expression of which depends on the boundary conditions and is given in chapter 2, section 2.2 for a hard and a soft wedge.

The different terms of (3.1) are therefore completely determined so that this formula can be directly used for the computation of the total field scattered by a three-dimensional wedge in each subdomain of the region 1, represented in figure 3.2.

Next, we will seek an asymptotic solution that can be valid through the transition regions 2 and 2′ and yield the solution defined by (3.1) in each of the subdomains of the region 1. Following the direct method presented in the introduction of this chapter, we must first come back to the exact solution of the canonical problem of the wedge with planar faces, derived in chapter 2, section 2.2, and extract a uniform asymptotic solution from it.

3.2.1.1. Uniform Asymptotic Solution of a Hard or Soft Wedge with Planar Faces Submitted to a Planar Sound Wave

The exact solution of the problem of plane wave scattering by a hard or soft wedge with planar faces has been derived in chapter 2, section 2.2.2. For an incident plane wave of unit amplitude, the solution is explicitly given by

$$p_{s_h}(\rho,\phi,z)=\frac{e^{ikz\cos\beta_o}}{4\pi in}\int_{L-L'}e^{-ik\rho\sin\beta_o\cos\xi}\left[\cot\left(\frac{\xi+\phi-\phi'}{2n}\right)\mp\cot\left(\frac{\xi+\phi+\phi'}{2n}\right)\right]d\xi, \qquad (3.2)$$

where β_o is the angle of the Keller cone $\left(0\le\beta_o\le\frac{\pi}{2}\right)$; ϕ and ϕ' are the angles between the projections of the opposite $-\vec{k}^i$ of the incident wave vector and the diffracted wave vector $\vec{k}^d$, respectively, onto the plane *xoy* and the *ox* axis; and $n\pi$ is the outer angle of the wedge (fig. 3.3).

The integration contour in the complex plane of the angular variable ξ is composed of the curves L and L' depicted in figure 2.4. After closing the contour with the steepest descent paths (SDP) $-\Gamma$ and $-\Gamma'$ passing through the saddle points π and $-\pi$ of the integrand and applying the residue theorem, the integral in (3.2) can be replaced by the integrals along $-\Gamma$ and $-\Gamma'$, to which one must add the residues at the poles of the integrand. The poles of the integrand are located at the values of ξ, where the cotangents in (3.2) go to zero. Namely,

$$\begin{aligned}\xi+\phi-\phi'&=2nN\pi\\ \xi+\phi+\phi'&=2nN\pi\end{aligned}\qquad N=0,\pm1,\pm2,\dots\,. \qquad (3.3)$$

These poles are located on the real axis. In addition, it is easily verified that there exist at most two poles in the interval $(-\pi,\pi)$, and they correspond to the incident and reflected fields.

A shadow boundary corresponds to the situation where one of the poles crosses one of the SDP. As this pole crosses the contour from the interior, the field associated with it (incident or reflected) disappears suddenly, and this creates a discontinuity in the representation of the field

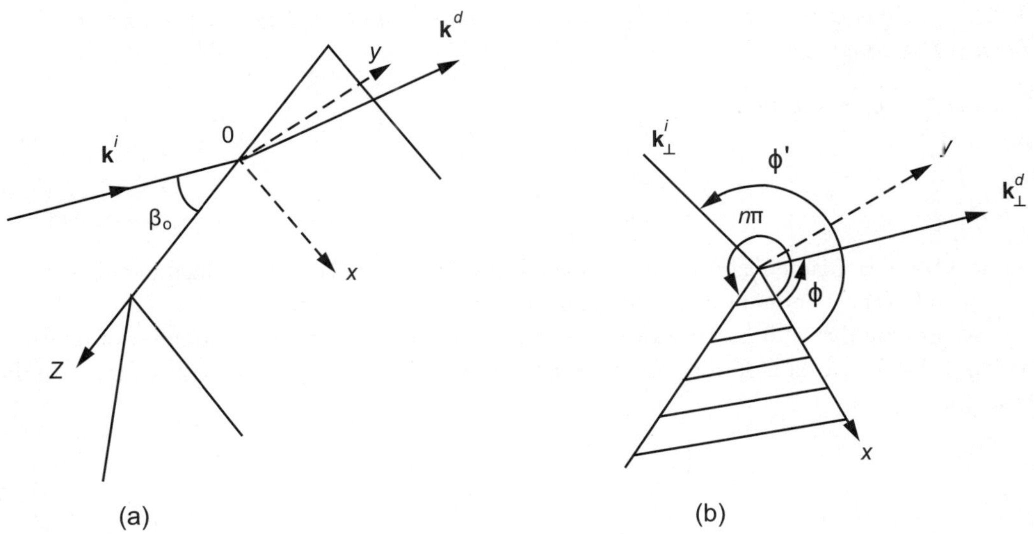

Figure 3.3. Geometry of the scattering from a wedge with plane wave incidence, where (a) is the coordinate system and (b) is the projection of incident and diffracted rays on a plane perpendicular to the edge.

that is not present in the initial expression of the integral form of the solution given by (3.2). Consequently, there is a similar jump with an opposite sign of the integral along the contour $-\Gamma$ or $-\Gamma'$, which has been crossed by this pole. This problem was not encountered in chapter 2, section 2.2, since the saddle point approximations of the integrals along the SDP Γ and Γ' have only been performed in the case where the poles are not close to the saddle points and therefore had no effect on them.

We now consider the case of an integral whose integrand is oscillatory in nature that has an isolated pole close to a saddle point—our objective being to search for a uniform asymptotic expansion. When this pole moves close to the SDP, the uniformity of the solution can be assured by staying on the same side of this path. If we do this, we can obtain two different uniform expansions, depending on which side of the SDP contains the pole. These two expansions do not match. There is a jump discontinuity that must equal the residue of the pole.

In order to obtain a uniform asymptotic expansion possessing these properties, two different methods have been proposed in scientific literature. The first of these is the Pauli method [3.4], which has been generalized by Clemmow [3.5]. We call this the Pauli-Clemmow method. The second method comes from Van der Waerden [3.6]. A comparative study of the two uniform expansions resulting from these methods has been carried out by Hutchins and Kouyoumjian [3.7].

We are now going to review these two methods. We also refer the reader to Bleistein [3.8] and to chapter 4 of Felsen and Marcuvitz [3.9] for additional information.

3.2.1.1.1. Asymptotic Evaluation of an Integral with a First Order Saddle Point and an Isolated Pole Singularity

Consider a simple integral of the type

$$I(\Omega,\alpha)=\int_{\gamma} g(z,\alpha)e^{\Omega q(z)}dz, \tag{3.4}$$

where $\Omega>0$ is a large real number compared to unity, α is a complex-valued parameter, and $g(z, \alpha)$ and $q(z)$ are holomorphic functions of z.

We assume that $q(z)$ has an isolated stationary point at z_s, which is a solution of $q'(z) = 0$ with $q''(z_s)\neq 0$. The path of integration is the SDP traversing through z_s. We also assume that the function

$$g(z,\alpha)=\frac{f(z,a)}{z-z_o} \tag{3.5}$$

has a simple pole at $z = z_o(\alpha)$ and that $z_o\rightarrow z_s$ as $\alpha\rightarrow\alpha_s$. Next, the integral $I(\Omega,\alpha)$ is transformed into a "canonical form," as in chapter 2, section 2.2, equation (2.53), by applying the transformation

$$q(z)=\tau(s)=q(z_s)-s^2 \tag{3.6}$$

to (3.4), which transforms γ into the real axis of the complex s-plane, the stationary point z_s into $s = 0$, and the pole into $s = s_1(\alpha)$ (fig. 3.4).

The integral (3.4) then reads

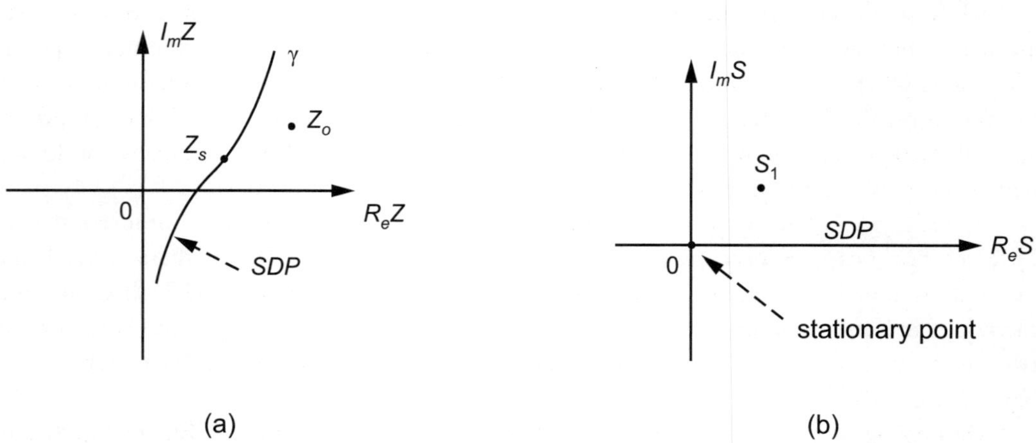

Figure 3.4. Canonical transformation.

$$I(\Omega,\alpha)=\int_{-\infty}^{+\infty} G(s,\alpha)e^{-\Omega s^2}\, ds, \tag{3.7}$$

where

$$G(s,\alpha)=g(z,\alpha)\,\frac{dz}{ds}\,e^{\Omega q(z_s)}. \tag{3.8}$$

Since $g(z, \alpha)$ has a pole singularity at $z = z_o$, it follows that $\mathbf{G}(s, \alpha)$ has a pole singularity at $s = s_1(\alpha)$.

In the Pauli-Clemmow method, one sets

$$G(s,\alpha)=\frac{T(s,\alpha)}{s-s_1(\alpha)}, \tag{3.9}$$

where $T(s, \alpha)$ is holomorphic. In the Van der Waerden method, the pole singularity is separated by using

$$G(s,\alpha)=\frac{A(\alpha)}{s-s_1(\alpha)}+H(s,\alpha), \tag{3.10}$$

where A is a constant and $H(s, \alpha)$ is holomorphic.

We will see that both methods are equivalent to the order considered in (3.1), namely, $\Omega^{-1/2}$. However, the Pauli-Clemmow method, in which the derivation of the final formula is much more involved than in the Van der Waerden method, provides the complete asymptotic expansion in a direct manner. Thus at this point one might raise a question regarding the utility of the Pauli-Clemmow method. We will see that the answer to this question has two parts in the case of a soft or hard wedge with planar faces. First, the pole crosses the SDP at $s = 0$. In the particular situation when the incident field is a planar wave, the Pauli-Clemmow method permits us to extract a first-order term from the complete asymptotic expansion. This is much easier to compute numerically and is exact when $s_1 \to 0$. It remains accurate in the transition zone. Second, the asymptotic expansion obtained with the Van der Waerden method remains important in more general situations, encountered later in sections 3.3 and 3.4, where poles cross the SDP at points away from the saddle point (e.g., impedance wedge, impedance-shaped shell) or where other terms of $0\left(\Omega^{-1/2}\right)$ are generated in the transition region. Terms of the same order appear, for instance, when the derivative of the incident field in the direction normal to the shadow boundary does not vanish.

For this reason, we will present both methods in detail.

The Pauli-Clemmow Method

Since the function T(s) in (3.9) is holomorphic, it can be expanded as a Taylor series in the neighborhood of $s = 0$:

$$\mathrm{T}(s) = (s - s_1)G(s) = \sum_{n=o}^{\infty} A_n s^n, \tag{3.11}$$

where the explicit α dependence has been suppressed. It follows

$$G(s) = \sum_{n=o}^{\infty} \frac{A_n s^n}{s - s_1} = \sum_{n=o}^{\infty} A_n \frac{s^n (s + s_1)}{s^2 - s_1^2}$$

and

$$I(\Omega) = \sum_{n=o}^{\infty} A_n \int_{-\infty}^{+\infty} \frac{e^{-\Omega s^2}}{s^2 - s_1^2} s^n (s + s_1) ds. \tag{3.12}$$

The integrals occurring in (3.12) are of the form

$$\int_{-\infty}^{+\infty} \frac{e^{-\Omega s^2}}{s^2 - s_1^2} s^m \, ds = \begin{cases} 0, m = 2p+1 \\ J_p, m = 2p \end{cases} \quad (p = 0,1,2,\ldots),$$

where

$$J_p = \int_{-\infty}^{+\infty} \frac{e^{-\Omega s^2}}{s^2 - s_1^2} s^{2p} \, ds. \tag{3.13}$$

Zero-Order Terms

Thus

$$\begin{aligned} J_o = \int_{-\infty}^{+\infty} \frac{e^{-\Omega s^2}}{s^2 - s_1^2} ds &= e^{-\Omega s_1^2} \int_{-\infty}^{+\infty} ds \int_{\Omega}^{\infty} e^{-\xi(s^2 - s_1^2)} d\xi \\ &= \sqrt{\pi}\, e^{-\Omega s_1^2} \int_{\Omega}^{\infty} \frac{e^{\xi s_1^2}}{\sqrt{\xi}} d\xi. \end{aligned} \tag{3.14}$$

Changing the integration variable by stating

$$s_1\sqrt{\xi} = \pm\, t\, e^{\frac{i\pi}{4}} \tag{3.15}$$

and using the notation $X = -i\Omega s_1^2$, we obtain

$$J_o = -\frac{1}{s_1^2}\sqrt{\frac{\pi}{\Omega}}\, F(X), \tag{3.16}$$

where $F(X)$ is known as the Kouyoumjian transition function. It is given by

$$F(X) = -2i\sqrt{X}\, e^{-ix}\int_{\sqrt{x}}^{\infty} e^{it^2}\, dt\,, \quad -\frac{\pi}{2} < \arg X < \frac{3\pi}{2}. \tag{3.17}$$

This function is defined on the upper Riemann sheet of the complex X-plane, with a cut along the negative imaginary axis as shown in figure 3.5 and with the following choice for $\sqrt{X}$:

$$\begin{aligned} \sqrt{X} &= e^{-\frac{i\pi}{4}}\sqrt{\Omega}\, s_1, \quad R_e X > 0, \\ \sqrt{X} &= e^{\frac{3i\pi}{4}}\sqrt{\Omega}\, s_1, \quad R_e X < 0. \end{aligned} \tag{3.18}$$

Inserting (3.16) into (3.12) and substituting for A_o its expression $A_o = T(o) = -s_1 G(o)$, given by (3.11), it follows that

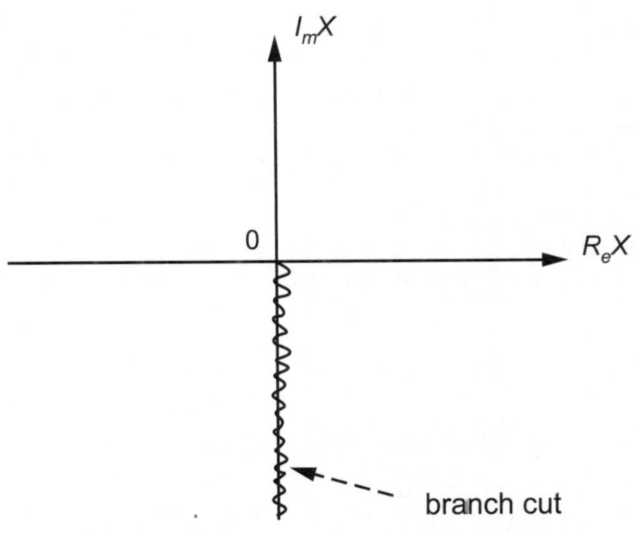

Figure 3.5. Branch cut in the complex X-plane uniquely defining Kouyoumjian's transition function for a complex argument.

$$I(\Omega) = \sqrt{\frac{\pi}{\Omega}}\, G(o) F\left(-i\Omega s_1^2\right). \tag{3.19}$$

We will now verify that $I(\Omega)$, given by (3.19), has a jump discontinuity as the pole crosses the contour γ and compensates exactly for the residue of the integrand of (3.4).

When the pole s_1 crosses the SDP γ, we have, by definition of γ,

$$I_m q(z_o) = I_m q(z_s)$$

and, according to (3.6),

$$I_m\left(-s_1^2\right) = 0, \quad s_1 \in \gamma, \tag{3.20}$$

which is equivalent to

$$R_e\left(-i\Omega s_1^2\right) = 0, \quad s_1 \in \gamma. \tag{3.21}$$

On the other hand, denoting the residue of the integrand of (3.4) by R_1, we get

$$R_1 = \mathrm{T}(s_1) e^{-\Omega s_1^2}. \tag{3.22}$$

In order for R_1 to converge as $\Omega \to \infty$, it is necessary for $R_e\, s_1^2 > 0$, which leads to the condition that $I_m\left(-i\Omega s_1^2\right) < 0$. Hence it follows that when the pole crosses γ, the complex variable X crosses the branch cut in the complex X-plane and $\sqrt{X}$ exhibits a jump discontinuity that also produces a jump discontinuity of $F(X)$.

Since $R_e\, s_1^2 > 0$ and $I_m\, s_1^2 = 0$ on γ, it follows that $I_m\, s_1 = 0$ on γ and

$$\sqrt{X} = \pm e^{-\frac{i\pi}{4}} \sqrt{\Omega}\, s_1 = \pm\Delta, \quad s_1 \in \gamma, \quad R_e X = 0^{\pm}. \tag{3.23}$$

We can now calculate the jump of $F(X)$ when $R_e X$ passes from 0^- to 0^+:

$$\delta F(X) = 2i\, \Delta e^{-iX}\left(-\int_{-\Delta}^{\infty} e^{it^2} dt - \int_{\Delta}^{\infty} e^{it^2} dt\right) = -2i\, \Delta e^{-iX} \sqrt{\pi}\, e^{\frac{i\pi}{4}}.$$

By substituting for Δ its expression (3.23), it follows that

$$\delta F\left(-i\Omega s_1^2\right) = -2i\sqrt{\pi\Omega}\, e^{-\Omega s_1^2}\, s_1 \tag{3.24}$$

and finally, according to (3.20),

$$\delta I(\Omega) = 2\pi i\, \mathrm{T}(o) e^{-\Omega s_1^2}.$$

Comparing this expression to (3.22), we see that when s_1 crosses the SDP at the stationary point $s = 0$, the jump of the zero-order term of the asymptotic expansion of the integral (3.4) exactly compensates for the loss of the residue of the integrand of (3.4).

Furthermore, when the pole is far from the stationary point, the function $F(X)$ for Ω large can be replaced by its asymptotic expansion, which with the notation $X = \Omega a$ is given by

$$F(\Omega a) = 1 - i\frac{1}{2\Omega a} - \frac{3}{4}\frac{1}{(\Omega a)^2} + 0\left(\Omega^{-3}\right), \tag{3.25}$$

where

$$a = -i s_1^2 = i\left[q(z_o) - q(z_s)\right]. \tag{3.26}$$

The analytical details for the demonstration of the formula (3.25) are provided later on in this chapter.

Substituting the right-hand side of the first term of (3.25) into (3.19), we see that far from the SDP, we recover the first term I_{nu} of the nonuniform asymptotic expansion of the integral in (3.4). It is given by

$$I_{nu}(\Omega) = \sqrt{\frac{\pi}{\Omega}}\, G(o) + 0\left(\Omega^{-3/2}\right). \tag{3.27}$$

In view of this result, we see that the solution given in (3.19) is uniform with respect to the position of the pole on each side of the SDP γ and that on γ, it exhibits a jump discontinuity that equals the residue of the pole of the integrand, provided that the pole crosses γ at the stationary point $s = 0$. An immediate consequence is that the expression

$$S_u(\Omega) = 2\pi i\, \mathrm{T}(o) e^{-\Omega s_1^2}\, U\left[-R_e\left(-i\Omega s_1^2\right)\right] + \sqrt{\frac{\pi}{\Omega}}\, G(o) F(\Omega a) + 0\left(\Omega^{-3/2}\right), \tag{3.28}$$

where U denotes the Heaviside function, is an asymptotic solution that is uniformly valid regardless of the position of the pole with respect to the SDP.

For the demonstration of the formula (3.25), we start with the identity

$$\int_{\sqrt{X}}^{\infty} e^{it^2}dt = \frac{1}{2}\int_{-\infty}^{+\infty} e^{it^2}dt - \frac{1}{2}\int_{\sqrt{-X}}^{\sqrt{X}} e^{it^2}dt. \tag{3.29}$$

The second member of (3.29) can also be written as

$$\frac{1}{2}e^{\frac{i\pi}{4}}\sqrt{\pi} - \frac{1}{2}\sqrt{\Omega}\int_{-\sqrt{a}}^{+\sqrt{a}} e^{i\Omega u^2}du. \tag{3.30}$$

Consider now the integral

$$J(\Omega) = \int_{-\sqrt{a}}^{+\sqrt{a}} e^{i\Omega u^2}du. \tag{3.31}$$

This integral has an oscillatory integrand and can therefore be evaluated for large Ω, by its asymptotic expansion. The stationary point is at $u = 0$, and the SDP is given by $I_m\left(i\Omega u^2\right) = R_e\,\Omega u^2 = 0$, which gives $u_1 = \pm u_2$, where $u = u_1 + iu_2$. The convergence for $\left|\Omega u^2\right|$ large imposes $R_e\left(i\Omega u^2\right) = I_m\left(\Omega u^2\right) < 0$. Hence the SDP is given by $u_1 = u_2$. Integration along the steepest path gives

$$\int_{\infty e^{-\frac{3i\pi}{4}}}^{\infty e^{\frac{i\pi}{4}}} e^{i\,\Omega u^2}du = e^{\frac{i\pi}{4}}\int_{-\infty}^{+\infty} e^{-\Omega u^2}du = \sqrt{\frac{\pi}{\Omega}}\,e^{\frac{i\pi}{4}}. \tag{3.32}$$

To this contribution, we must add the end point contributions, which can be derived by successive integrations by parts of the integral (3.31). We have

$$e^{i\,\Omega u^2}du = \frac{d}{du}\frac{e^{i\,\Omega u^2}}{2\,i\Omega u} + \frac{e^{i\,\Omega u^2}}{2\,i\Omega u^2}du,$$

$$\frac{e^{i\,\Omega u^2}}{2\,i\Omega u^2}du = \frac{d}{du}\frac{e^{i\,\Omega u^2}}{\left(2\,i\Omega\right)^2 u^3} + \frac{3e^{i\,\Omega u^2}}{\left(2\,i\Omega\right)^2 u^4},$$

$$\frac{3e^{i\,\Omega u^2}}{\left(2\,i\Omega\right)^2 u^4} = \frac{d}{du}\frac{3e^{i\,\Omega u^2}}{\left(2\,i\Omega\right)^3 u^5} + \frac{15e^{i\,\Omega u^2}}{\left(2\,i\Omega\right)^3 u^6}.$$

Hence by integrating the terms we obtain the following for the end point contributions $\mathrm{End}\left(J(\Omega)\right)$ of $J(\Omega)$:

$$\text{End}\left(\int_{-\sqrt{a}}^{+\sqrt{a}} e^{i\Omega u^2} du\right) \cong e^{i\,\Omega a}\left(\frac{1}{i\Omega\sqrt{a}} + \frac{2}{(2\,i\Omega)^2\, a^{3/2}} + \frac{6}{(2\,i\Omega)^3\, a^{5/2}}\right) + 0\left(\Omega^{-4}\right). \tag{3.33}$$

Inserting (3.32) and (3.33) into (3.30) gives

$$\int_{\sqrt{X}}^{\infty} e^{it^2} dt \cong e^{i\,\Omega a}\left(-\frac{1}{2i\sqrt{\Omega a}} + \frac{1}{4(\Omega a)^{3/2}} - \frac{3}{8}\frac{i}{(\Omega a)^{5/2}}\right) + 0\left(\Omega^{-7/2}\right). \tag{3.34}$$

Finally, by multiplying (3.34) by $-2i\sqrt{\Omega a}\ \exp(-i\Omega a)$, we obtain the asymptotic expansion of the function $F(\Omega a)$ for Ω large, given by (3.25).

Higher-Order Terms

We come back to the integral J_p given by (3.13) and write it in the form

$$J_p = \int_{-\infty}^{+\infty} \frac{e^{-\Omega s^2}}{s^2 - s_1^2} s^{2p}\, ds = (-1)^p \frac{d^p}{d\Omega^p} \int_{-\infty}^{+\infty} \frac{e^{-\Omega s^2}}{s^2 - s_1^2} ds. \tag{3.35}$$

The identity (3.35) can be verified by iterating the identity

$$J_{p+1} = -\frac{d}{d\Omega} J_p. \tag{3.36}$$

Starting from the expression J_1, which reads

$$J_1 = \int_{-\infty}^{+\infty} \frac{e^{-\Omega s^2}}{s^2 - s_1^2} s^2\, ds = \int_{-\infty}^{+\infty} e^{-\Omega s^2} ds + s_1^2\, J_o, \tag{3.37}$$

and applying (3.37) gives

$$\begin{aligned} J_2 &= \int_{-\infty}^{+\infty} s^2\, e^{-\Omega s^2} ds + s_1^2\, J_1 \\ &- - - - - - - - - - - - - - - \\ J_p &= \int_{-\infty}^{+\infty} s^{2p-2}\, e^{-\Omega s^2} ds + s_1^2\, J_{p-1}. \end{aligned} \tag{3.38}$$

The integral representations in (3.38) provide contributions of a higher order than $\Omega^{-1/2}$. If we denote by $\overline{J}_p$ the terms of order $\Omega^{-1/2}$ of J_p, then we get

$$\overline{J}_p = s_1^{2(p-1)} \overline{J}_1 = s_1^{2p-1} J_1, \tag{3.39}$$

since $\overline{J}_1 = J_1$.

Now coming back to (3.12), which we write in the form

$$I(\Omega) = \sum_{n=o}^{\infty} A_n \int_{-\infty}^{+\infty} \frac{e^{-\Omega s^2}}{s^2 - s_1^2} s^n (s + s_1) ds = \sum_{p=o}^{\infty} s_1 A_{2p} J_p + A_{2p+1} J_{p+1}, \tag{3.40}$$

and using (3.39), we obtain the following expression for $I(\Omega)$, restricted to the terms of $0\left(\Omega^{-1/2}\right)$ that we denote by $\overline{I}(\Omega)$:

$$\begin{aligned} \overline{I}(\Omega) &= A_o s_1 J_o + A_1 J_1 + J_1 \sum_{p=1}^{\infty} A_{2p} s_1^{2p-1} + A_{2p+1} s_1^{2p} \\ &= A_o s_1 J_o + A_1 J_1 + \frac{J_1}{s_1} \sum_{p=1}^{\infty} A_{2p} s_1^{2p} + A_{2p+1} s_1^{2p+1} \\ &= A_o s_1 J_o + \frac{J_1}{s_1}\left[\mathrm{T}(s_1) - \mathrm{T}(o)\right]. \end{aligned} \tag{3.41}$$

It is therefore possible to formally calculate the sum of the terms of the series (3.40) corresponding to each order $\Omega^{-(n+1/2)}$ $(n \geq 0)$. Suppose now that we want to derive the results for $p = 1$ in (3.40), limited to the order $\Omega^{-1/2}$. From (3.40) and the fact that $\overline{J}_1 = J_1$, we obtain

$$\overline{I}(\Omega) = A_o s_1 J_o + A_1 J_1 + A_2 s_1 J_1 = A_o s_1 J_o + J_1 (A_1 + s_1 A_2),$$

where J_o is given by (3.16), and according to (3.37),

$$J_1 = \sqrt{\frac{\pi}{\Omega}}\left[1 - F\left(-i\Omega s_1^2\right)\right]. \tag{3.42}$$

We note that to derive the results for the order $p = 1$, we need to calculate the expansion coefficients A_n up to the order 2. Due to the complexity of the integrand in (3.2), an explicit derivation of the coefficients A_1 and A_1 is quite cumbersome, and for $p > 1$, the Pauli-Clemmow approach ceases to be practicable.

Coming back to (3.40) and inserting the expressions of J_o and J_1, given by (3.16) and (3.42), into (3.41), we are led to the result

$$\bar{I}(\Omega) = \sqrt{\frac{\pi}{\Omega}} \frac{1}{s_1} \left\{ \mathrm{T}(s_1)\left[1 - F\left(-i\Omega s_1^2\right)\right] - \mathrm{T}(o) \right\}. \tag{3.43}$$

We know from (3.24) that regardless of where the pole crosses the SDP, the jump of $F(X)$ from $R_e X < 0$ to $R_e X < 0$, where $X = -i\Omega s_1^2$, is given by

$$\delta F\left(-i\Omega s_1^2\right) = -2i\pi\sqrt{\pi\Omega}\; e^{-\Omega s_1^2}\, s_1. \tag{3.44}$$

Hence it follows that the jump of $\bar{I}(\Omega)$ is given by

$$\delta \bar{I}(\Omega) = 2i\pi\,\mathrm{T}(s_1) e^{-\Omega s_1^2} = R_1, \tag{3.45}$$

where R_1 is the residue of the integrand of (3.4). A consequence of this result is that the expression

$$\bar{S}_u(\Omega) = 2\pi i R_1 U\left[-R_e\left(-i\Omega s_1^2\right)\right] + \sqrt{\frac{\pi}{\Omega}} \left[\frac{\mathrm{T}(s_1) - \mathrm{T}(o)}{s_1} - \frac{\mathrm{T}(s_1)}{s_1} F\left(-i\Omega s_1^2\right)\right] \tag{3.46}$$

is uniform as a function of location of the pole, regardless of where the pole crosses the SDP. In contrast, the uniformity of (3.28) is achieved only if the pole crosses the path of steepest descent at the stationary point.

By writing (3.43) in the form

$$\bar{I}(\Omega) = \sqrt{\frac{\pi}{\Omega}} \left\{ \frac{\mathrm{T}(s_1) - \mathrm{T}(o)}{s_1} \left[1 - F\left(-i\Omega s_1^2\right)\right] - \frac{\mathrm{T}(o)}{s_1} F\left(-i\Omega s_1^2\right) \right\} \tag{3.47}$$

and comparing it with (3.19), we find that the latter is an incomplete asymptotic expansion to the order $\left(\Omega^{-1/2}\right)$, the missing term being

$$\sqrt{\frac{\pi}{\Omega}}\; \frac{\mathrm{T}(s_1) - \mathrm{T}(o)}{s_1} \left[1 - F\left(-i\Omega s_1^2\right)\right]. \tag{3.48}$$

According to (3.25), this term vanishes when the pole is far from the SDP. However, it may not be negligible in the vicinity of this path, since it is proportional to the derivative of the function $T(s_1)$ when s_1 tends to zero.

The conditions under which this term is negligible can be established by expanding $T(s_1)$ and $T(o)$ in the vicinity of the stationary point, in Taylor series with respect to s_1. Starting with the expression of $T(s)$, given per (3.9) and (3.8),

$$\mathrm{T}(s) = (s - s_1) G(s) = (s - s_1) g(z) \frac{dz}{ds} e^{\Omega q(z_s)}. \tag{3.49}$$

Since $g(z)$ has a simple pole at $z = z_1$, it can be expanded in a Laurent series in the vicinity of z_1:

$$g(z) = \frac{A}{z - z_1} + B + C(z - z_1) + D(z - z_1)^2 + \cdots, \tag{3.50}$$

where $A, B, C, D \ldots$ are constants depending on z_1. According to (3.6), we get

$$\frac{dz}{ds} = -\frac{2s}{q'(z)}, \quad s = \sqrt{q(z_s) - q(z)}. \tag{3.51}$$

Expanding $q(z) - q(z_s)$ in a Taylor series, we obtain

$$q(z) - q(z_s) = \frac{(z - z_s)^2}{2} q''(z_s) + \frac{(z - z_s)^3}{6} q^{(3)}(z_s) + 0(z - z_s)^4$$

and

$$s = \sqrt{-(q(z) - q(z_s))} = (z - z_s)\sqrt{-\frac{q''(z_s)}{2}}\left[1 + \frac{(z - z_s)}{6}\frac{q^{(3)}(z_s)}{q''(z_s)} + 0(z - z_s)^2\right]. \tag{3.52}$$

Similarly,

$$q'(z) = (z - z_s)q''(z_s)\left[1 + \frac{z - z_s}{2}\frac{q^{(3)}(z_s)}{q''(z_s)} + 0(z - z_s)^2\right]. \tag{3.53}$$

Hence it follows that

$$\frac{dz}{ds} = -\frac{2s}{q'(z)} = \sqrt{-\frac{2}{q''(z_s)}}\left[1 - \frac{1}{3}(z - z_s)\frac{q^{(3)}(z_s)}{q''(z_s)} + 0(z - z_s)^2\right], \tag{3.54}$$

$$s - s_1 = \sqrt{-\frac{q''(z_s)}{2}}\left\{z - z_1 + \frac{q^{(3)}(z_s)}{6q''(z_s)}(z - z_1)(z + z_1 - 2z_s) + 0\left[(z - z_s)^3 - (z_1 - z_s)^3\right]\right\}, \tag{3.55}$$

and finally, by inserting (3.55) and (3.54) into (3.49) and taking (3.50) into account, it follows that

$$\mathrm{T}(s) = A e^{\Omega q(z_s)} \left\{ 1 - \frac{1}{6} \frac{q^{(3)}(z_s)}{q''(z_s)} \left[z - z_s - (z_1 - z_s) \right] + 0(z - z_s)^2 \right\}$$

$$+ B e^{\Omega q(z_s)} (s - s_1) \sqrt{-\frac{2}{q''(z_s)}} \left[1 - \frac{1}{3}(z - z_s) \frac{q^{(3)}(z_s)}{q''(z_s)} + 0(z - z_s)^2 \right] + \cdots$$

and

$$\frac{\mathrm{T}(s_1) - \mathrm{T}(o)}{s_1} = A e^{\Omega q(z_s)} \left[-\frac{1}{6} \frac{q^{(3)}(z_s)}{q''(z_s)} \frac{z_1 - z_s}{s_1} + 0(z_1 - z_s) \right]$$

$$- B e^{\Omega q(z_s)} \left[1 - \frac{1}{3}(z - z_s) \frac{q^{(3)}(z_s)}{q''(z_s)} + 0(z_1 - z_s)^2 \right] + \ldots. \tag{3.56}$$

Equation (3.56) shows that if $q^{(3)}(z_s) = 0$ and $B = 0$, then the missing term in (3.48) tends to zero when $s_1 \to 0$ and remains negligible in a neighborhood of $s = 0$. The latter must include the transition region so that the Pauli-Clemmow method, limited to the zero-order term, can still be a good approximation in that region. It can be verified directly in (3.2) that these conditions are satisfied in the case of plane wave incidence. They are also satisfied in the case of an incident cylindrical wave, when the distance of the line source to the edge is large compared to the wavelength, as can be verified on the integral form of the solution given by (2.42).

The Van der Waerden Method

In this method, we begin with the expression (3.10) for $G(s)$. Inserting (3.10) into (3.7), one gets

$$I(\Omega) = \int_{-\infty}^{+\infty} \frac{A}{s - s_1} e^{-\Omega s^2} ds + \int_{-\infty}^{+\infty} H(s) e^{-\Omega s^2} ds. \tag{3.57}$$

The first integral in (3.57), which we denote by $I_1(\Omega)$, can be evaluated to yield

$$I_1 = A \int_{-\infty}^{+\infty} \frac{s + s_1}{s^2 - s_1^2} e^{-\Omega s^2} ds = A s_1 \int_{-\infty}^{+\infty} \frac{e^{-\Omega s^2}}{s^2 - s_1^2} ds = A s_1 J_o,$$

where J_o is given by (3.16). Hence it follows that

$$I_1(\Omega) = -\frac{A}{s_1} \sqrt{\frac{\pi}{\Omega}} F(-i\Omega s_1^2). \tag{3.58}$$

Since $H(s)$ is a holomorphic function, an asymptotic evaluation to $0\left(\Omega^{-1/2}\right)$ of the second integral, which we denote by $I_2(\Omega)$, gives, according to (2.56),

$$I_2(\Omega) = H(o)\int_{-\infty}^{+\infty} e^{-\Omega s^2}\, ds + \frac{H''(o)}{2}\int_{-\infty}^{+\infty} s^2 e^{-\Omega s^2}\, ds + \cdots$$
$$= H(o)\sqrt{\frac{\pi}{\Omega}} + 0\left(\Omega^{-3/2}\right), \tag{3.59}$$

and since

$$G(o) = -\frac{A}{s_1} + H_o,$$

we obtain

$$I_2 = \left(G(o) + \frac{A}{s_1}\right)\sqrt{\frac{\pi}{\Omega}} + 0\left(\Omega^{-3/2}\right). \tag{3.60}$$

Now according to (3.9) and (3.10), we have

$$G(o) = -\frac{\mathrm{T}(o)}{s_1}, \quad A = \mathrm{T}(s_1). \tag{3.61}$$

Hence after grouping together the terms we get

$$I_1(\Omega) + I_2(\Omega) = I(\Omega) = \sqrt{\frac{\pi}{\Omega}}\frac{1}{s_1}\left\{\mathrm{T}(s_1)\left[1 - F\left(-i\Omega s_1^2\right)\right] - \mathrm{T}(o)\right\} + 0\left(\Omega^{-3/2}\right) = \bar{I}(\Omega) + 0\left(\Omega^{-3/2}\right), \tag{3.62}$$

where $\bar{I}(\Omega)$ is $I(\Omega)$ restricted to the terms of $0\left(\Omega^{-1/2}\right)$.

The expression (3.62) is seen as identical to (3.43). Thus the Van der Waerden method provides the complete asymptotic expansion of (3.7) to $0\left(\Omega^{-1/2}\right)$ in a direct manner.

Historically, the Pauli-Clemmow method was limited to the zero-order ($p = 0$) term, whereas Van der Waerden derived the complete asymptotic expansion. The fact that the summation of the higher-order terms in Pauli-Clemmow's procedure leads to Van der Waerden's results has been proved more recently by Rojas [3.10].

The solution given in (3.62) can be written in an alternative form that is more convenient for the construction of an ansatz for the solution of the diffraction problem of a wedge with curved faces. To derive this new form, we define

$$V(s_1) = -\frac{1}{s_1}\frac{1}{2\pi i}\sqrt{\frac{\pi}{\Omega}}\, F\left(-i\Omega s_1^2\right)^{e^{\Omega s_1^2}}, \tag{3.63}$$

$$\hat{V}(s_1) = -\frac{1}{s_1}\frac{1}{2\pi i}\sqrt{\frac{\pi}{\Omega}}\, e^{\Omega s_1^2}. \tag{3.64}$$

Coming back to (3.25), we see that $\hat{V}(s_1)$ is the first term of the asymptotic expansion of $V(s_1)$. Using (3.27) together with $T(o) = -s_1 G(o)$, we can write (3.62) in the form

$$\bar{I}(\Omega) = 2\pi i R_1\left[V(s_1) - \hat{V}(s_1)\right] + I_{nu}(\Omega), \tag{3.65}$$

where $R_1 = T(s_1)$ is the residue of the integrand of (3.7).

We have seen in section 3.2.1.1 that the function $F(X)$, where $X = -i\Omega s_1^2$, exhibits a jump discontinuity given by (3.24) when $R_e X$ passes from 0^- to 0^+. Hence it follows that $V(s_1)$ also has a jump discontinuity, which according to (3.24) and (3.63) is equal to

$$V^+(s_1) - V^-(s_1) = 1, \tag{3.66}$$

where the functions V^+ and V^- correspond to $R_e X = 0^+$ and $R_e X = 0^-$, respectively.

A consequence of this result is that the expression

$$\bar{S}_u(\Omega) = 2\pi i R_1\left[V(s_1) - \hat{V}(s_1)\right] + 2\pi i R_1 U\left[-R_e\left(-i\Omega s_1^2\right)\right] + I_{nu}(\Omega) \tag{3.67}$$

is equivalent to (3.46) and is uniform as a function of the location of the pole.

According to (3.66), $V(s_1)$ can be written as

$$V(s_1) = V^+(s_1) - U\left[-R_e\left(-i\Omega s_1^2\right)\right]. \tag{3.68}$$

Inserting the first equation (3.18) into (3.63) and using (3.17), the equation (3.68) can be written as

$$V(s_1) = \bar{F}(\tau) - U\left[R_e\left(-i\Omega s_1^2\right)\right], \tag{3.69}$$

where

$$F(\tau) = \frac{e^{-\frac{i\pi}{4}}}{\sqrt{\pi}}\int_{\tau}^{\infty} e^{it^2}\, dt \tag{3.70}$$

and

$$\tau = e^{-\frac{i\pi}{4}}\sqrt{\Omega}\, s_1. \tag{3.71}$$

Thus the solution (3.65) can be written as

$$\overline{I}(\Omega) = 2\pi i R_1 \left[\overline{F}(\tau) - \overset{\wedge}{\overline{F}}(\tau) \right] - 2\pi i R_1 U\left[-R_e\left(-i\Omega s_1^2 \right) \right], \tag{3.72}$$

where $\overset{\wedge}{\overline{F}}(\tau)$ is obtained by inserting (3.71) into (3.64). Its expression is given by

$$\overset{\wedge}{\overline{F}}(\tau) = \frac{1}{2\tau\sqrt{\pi}} e^{i\left(\tau^2 + \frac{\pi}{4}\right)}. \tag{3.73}$$

Finally, by adding the contribution of the pole to (3.72), the uniform solution (3.67) can be written as

$$\overline{S}_u(\Omega) = 2\pi i R_1 \left[\overline{F}(\tau) - \overset{\wedge}{\overline{F}}(\tau) \right] + I_{nu}(\Omega). \tag{3.74}$$

We will see that for a plane wave incidence on a hard or soft wedge with planar faces, τ is real. In this case, when $|\tau| \to \infty$, we get the following from (3.70):

$$\overline{F}(\tau) \cong U(-\tau) + \overset{\wedge}{\overline{F}}(\tau) + 0\left(\Omega^{-3/2} \right). \tag{3.75}$$

Hence, if the pole S is far from the saddle point, then the uniform solution (3.72) reduces to

$$S_u(\Omega) = 2\pi R_1 U(-\tau) + I_{nu}(\Omega) + 0\left(\Omega^{-3/2} \right). \tag{3.76}$$

This completes the derivation of the various uniform solutions.

In summary, we have presented two methods for obtaining a uniform asymptotic expansion of an integral whose integrand is oscillatory and has an isolated pole close to an isolated stationary point of the integrand.

The first method, labeled the Pauli-Clemmow method, gives a series representation of the term of the asymptotic expansion corresponding to order $\Omega^{-1/2}$. Only the zero-order term of this series is retained, and an asymptotic expansion, given by (3.28), is derived that is uniformly valid regardless of the position of the pole with respect to the SDP. This uniform asymptotic solution provides accurate numerical results when the pole crosses the SDP at the stationary point. This condition is verified by the integrals occurring in the exact solution of the diffraction problem of a plane wave striking a hard or soft wedge with planar faces.

The second method, attributed to Van der Waerden, directly gives the complete expression of the asymptotic expansion of the integral to the order $\Omega^{-1/2}$. Again, from this result, a uniform expansion (see equation (3.46)) has been derived that remains uniformly valid regardless of the position of the pole. However, unlike the asymptotic solution obtained by the Pauli-Clemmow method, the pole is not constrained to cross the SDP at the stationary point. Instead, it may cross this path at an arbitrary point. Of course, the effect of the pole is more important when it crosses the SDP at the stationary point. In this case, we have shown that the difference between the solutions given by equations (3.28) and (3.46) is negligibly small. Moreover, experience shows that the numerical calculation of the expression (3.46) is very sensitive, since the terms inside the brackets vanish to the first order in s_1. As $s_1 \to 0$, one is forced to accurately extract the second-order terms in order to use this expression for numerical computations. This problem does not occur in the Pauli-Clemmow method. For these and other reasons that will be presented in section 3.2.2.2, the Van der Waerden method is typically not used in practice, aside from some special situations where the Pauli-Clemmow method fails to give the correct result. Such a situation occurs when the derivative of the incident field, in the direction perpendicular to the ray incident on the edge of the wedge, is large. In this case, it is necessary to add a correction term to the Pauli-Clemmow solution, known as the slope diffraction term, whereas no modification is needed in the Van der Waerden solution, which contains the correct terms. There is, however, a theoretical interest in comparing both solutions that justifies our effort to present both methods. In the following section, we turn to the application of the solutions (3.28) and (3.46) to the diffraction of a plane wave by a soft or hard wedge with planar faces.

3.2.1.1.2. Application of Uniform Theories to the Problem of Diffraction by a Wedge with Planar Faces

We start with the exact solution for the diffraction problem of an incident plane wave striking a soft or hard wedge with planar faces, given by (3.2). The integrand of (3.2) has two stationary points $\xi_s = \pm\pi$. Consequently, the field diffracted by the wedge is given by

$$P_{s_b}^{D}(\rho,\phi,z) = -\frac{e^{ikz\cos\beta_o}}{4\pi in}\int_{\Gamma+\Gamma'} e^{-ik\rho\sin\beta_o\cos\xi}\left[\cot\left(\frac{\xi+\phi-\phi'}{2n}\right) \mp \cot\left(\frac{\xi+\phi+\phi'}{2n}\right)\right]d\xi, \quad (3.77)$$

where Γ and Γ' are the SDP passing through $\xi_s = \pm\pi$ (see fig. 2.4). Equation (3.77) is of the form

$$P_{s_b}^{D} = \int_{\Gamma+\Gamma'} e^{-\Omega q(\xi)}\left[g(\xi,\bar{\beta}) \mp g\left(\xi,\overset{+}{\beta}\right)\right]d\xi, \quad (3.78)$$

where $\overset{\pm}{\beta} = \phi \pm \phi'$, $\Omega = k\rho\sin\beta_o$, and

$$q(\xi) = -i\cos\xi, \quad (3.79)$$

$$g(\xi,\beta) = -\frac{e^{ikz\cos\beta_o}}{4\pi in}\cot\left(\frac{\xi+\beta}{2n}\right). \quad (3.80)$$

The Pauli-Clemmow Method

The uniform solution (3.28), given by the zero-order term of the method of Pauli-Clemmow, is represented by the term given in (3.19). Thus we need to calculate $G(o)$ and the argument $a = -i\Omega s_1^2$ of the transition function F to construct the uniform solution in the Pauli-Clemmow method.

Using the expression (3.26) at the saddle points $\xi_s = \pm\pi$, we obtain

$$a = i\left[q(\xi_o) - q(\xi_s)\right] = 1 + iq(\xi_o), \quad (3.81)$$

and since, according to (3.3),

$$\xi_o = -\beta + 2nN\pi, \quad N = 0, \pm 1, \pm 2, \ \dots, \quad (3.82)$$

we have

$$a(\beta) = 1 + iq(-\beta + 2nN\pi). \quad (3.83)$$

In the transition regions of the shadow and reflection boundaries, a pole is close to the saddle point. The value of N associated with the pole closest to the saddle point $\xi_s = \pi$ is the integer N^+, which most nearly satisfies the equation

$$2nN^+\pi - \beta = \pi. \quad (3.84)$$

Similarly, the value of N associated with the pole closest to the saddle point $\xi_s = -\pi$ is the integer N^-, which most nearly satisfies the equation

$$2n N^- \pi - \beta = -\pi. \tag{3.85}$$

The parameter $a(\beta)$, therefore, takes two values $\overset{\pm}{a}(\beta)$, defined by

$$\overset{\pm}{a}(\beta) = 1 + iq\left(-\beta + 2n N^{\pm}\pi\right) = 1 + \cos\left(-\beta + 2n N^{\pm}\pi\right). \tag{3.86}$$

Using (3.8) and replacing z by ξ and $g(z, \alpha)$ by $g(\xi, \beta)$, we get

$$G(o) = g(\xi_s, \beta)\; e^{\Omega q(\xi_s)} \left(\frac{d\xi}{ds}\right)_{s=o}, \tag{3.87}$$

and since

$$q(\xi) - q(\xi_s) = -s^2 \cong \frac{(\xi - \xi_s)^2}{2} q''(\xi_s)$$

or

$$(\xi - \xi_s) = s\sqrt{\frac{-2}{q''(\xi_s)}},$$

we have

$$\left(\frac{d\xi}{ds}\right)_{s=o} = \sqrt{\frac{-2}{q''(\xi_s)}} = \sqrt{\left|\frac{2}{q''(\xi_s)}\right|}\; e^{i\alpha} = \sqrt{2}\, e^{i\alpha}, \tag{3.88}$$

where α is the phase of $d\xi / ds$, which is identical to the phase of $d\xi$ since ds is real. From the values of the angles of the tangent to the contours Γ and Γ' at $\xi_s = \pi$ and $\xi_s = -\pi$, respectively (see fig. 2.4), we obtain $\alpha = \frac{3\pi}{4}$ when $\xi = \pi$ and $\alpha = -\frac{\pi}{4}$ when $\xi_s = -\pi$.

Finally, for each value of β in (3.78), we obtain two values for $G(o)$, corresponding to $\xi_s = \pi$ and $\xi_s = -\pi$, respectively:

$$\overset{\pm}{G}(o, \beta) = \mp \frac{e^{\frac{i\pi}{4}}\sqrt{2}}{4\pi n} \cot\left(\frac{\pm\pi + \beta}{2n}\right) e^{ik\left(z\cos\beta_o + \rho\sin\beta_o\right)}. \tag{3.89}$$

Summing the asymptotic evaluations of the integrals along the SDP Γ and Γ' gives

$$P_{s_b}^{D} = \sqrt{\frac{\pi}{\Omega}} \left\{ G^{+}\left(o, \overset{-}{\beta}\right) F\left[\Omega \overset{+}{a}\left(\overset{-}{\beta}\right)\right] + G^{-}\left(o, \overset{-}{\beta}\right) F\left[\Omega \overset{-}{a}\left(\overset{-}{\beta}\right)\right] \right.$$
$$\left. \mp \left(G^{+}\left(o, \overset{+}{\beta}\right) F\left[\Omega \overset{+}{a}\left(\overset{+}{\beta}\right)\right] + G^{-}\left(o, \overset{+}{\beta}\right) F\left[\Omega \overset{-}{a}\left(\overset{+}{\beta}\right)\right]\right) \right\}, \tag{3.90}$$

and by taking (3.89) into account, we obtain the final result:

$$P_{s_b}^{D}(\rho, \phi, z) = -\frac{e^{\frac{i\pi}{4}}}{2n\sqrt{2\pi k \rho \sin \beta_o}} e^{ik(z \cos \beta_o + \rho \sin \beta_o)}$$
$$\left\{ \cot\left(\frac{\pi + \overset{-}{\beta}}{2n}\right) F\left[K \overset{+}{a}\left(\overset{-}{\beta}\right)\right] + \cot\left(\frac{\pi - \overset{-}{\beta}}{2n}\right) F\left[K \overset{-}{a}\left(\overset{-}{\beta}\right)\right] \right. \tag{3.91}$$
$$\left. \mp \left(\cot\left(\frac{\pi + \overset{+}{\beta}}{2n}\right) F\left[K \overset{+}{a}\left(\overset{+}{\beta}\right)\right] + \cot\left(\frac{\pi - \overset{+}{\beta}}{2n}\right) F\left[K \overset{-}{a}\left(\overset{+}{\beta}\right)\right]\right) \right\},$$

where $K = k\rho \sin \beta_o$.

Equation (3.91) represents the contribution of the diffracted field to the uniform solution. The total field obtained by adding the incident field and the reflected field to (3.91) is continuous through the shadow and reflection boundaries.

The transition function F, defined by (3.17), tends to zero when its argument tends to zero. Since the cotangent function tends to infinity, the product

$$\cot\left(\frac{\pm\pi + \beta}{2n}\right) F\left[K \overset{\pm}{a}(\beta)\right]$$

is undetermined. However, this difficulty can easily be removed by replacing $F(X)$ with its small argument expansion. When $|X|$ is small, we directly obtain the following from (3.17) by stating $u = \sqrt{X}$ and developing $F(u^2)$ in a Taylor series in the neighborhood of $u = 0$:

$$F(X) \cong \left[\sqrt{\pi X} - 2X e^{-\frac{i\pi}{4}} - \frac{2}{3} X^2 e^{\frac{i\pi}{4}} + 0\left(X^3\right) \right] e^{-i\left(\frac{\pi}{4}+X\right)}. \tag{3.92}$$

For example, a reflection boundary associated with the wedge face $\phi = 0$ occurs at $\phi = \pi - \phi'$. Hence it follows from (3.85) that $N^- = 0$ and

$$\bar{a}\left(\overset{+}{\beta}\right) = 1 + \cos \overset{+}{\beta}.$$

In the transition region, close to the reflection boundary, we have on one side of the boundary, $\phi + \phi' = \pi + \varepsilon$, $\varepsilon > 0$, where ε is a small angle. Then $\bar{a}\left(\overset{+}{\beta}\right) = 1 - \cos \varepsilon$ and by using (3.92), we obtain the following from a simple calculation:

$$-\frac{e^{\frac{i\pi}{4}}}{2n\sqrt{2\pi K}} \cot\left(\frac{\pi - \overset{+}{\beta}}{2n}\right) F\left[K \bar{a}\left(\overset{+}{\beta}\right)\right] = \frac{1}{2}.$$

For $\varepsilon < 0$, the result would be $-\frac{1}{2}$. Hence we find a jump of 1 for the term when this observation direction crosses the reflection boundary, as expected.

In (3.77) we have assumed that the incident field is equal to 1 at the diffraction point. Hence the expression (3.91) of the diffracted field can also be written in the following form:

$$P_{s_b}^{D}(\rho,\phi',z) = p_{s_b}^{i}(0,\phi',0) D_{s_b} \frac{e^{ik(z\cos\beta_o + \rho\sin\beta_o)}}{\sqrt{\rho / \sin\beta_0}}, \tag{3.93}$$

where D_{s_b} is the diffraction coefficient. By using the notations

$$\overset{\pm}{d}(\beta,n) = -\frac{e^{-\frac{i\pi}{4}}}{n\sqrt{2\pi k}} \frac{1}{2} \cot\left(\frac{\pi \pm \beta}{2n}\right) \tag{3.94}$$

introduced by Pathak and Kouyoumjian [3.11], the diffraction coefficient can be written as

$$\begin{aligned} D_{s_b} = \frac{1}{\sin\beta_o} & \overset{+}{d}\left(\overset{-}{\beta},n\right) F\left[K \overset{+}{a}\left(\overset{-}{\beta}\right)\right] + \overset{-}{d}\left(\overset{-}{\beta},n\right) F\left[K \overset{-}{a}\left(\overset{-}{\beta}\right)\right] \\ & \mp \left(\overset{+}{d}\left(\overset{+}{\beta},n\right) F\left[K \overset{+}{a}\left(\overset{+}{\beta}\right)\right] + \overset{-}{d}\left(\overset{+}{\beta},n\right) F\left[K \overset{-}{a}\left(\overset{+}{\beta}\right)\right] \right) \Big\}. \end{aligned} \tag{3.95}$$

For large values of $K\overset{\pm}{a}(\beta)$, we know from (3.25) that $F\left[K\overset{\pm}{a}(\beta)\right]$ tends to unity. When $F = 1$ in all four terms in (3.95), we recover the Keller's nonuniform diffraction coefficient established in chapter 2, equation (2.76).

The Van der Waerden Method

The unifom solution derived by using the Van der Waerden method was given in (3.74). To apply it to the integral in (3.77), we need to calculate the residue R_1 and the parameter τ. Let us assume that the incident field makes an angle $\phi' < \pi$ with the OX face (see fig. 3.3). In this case, the integrand (3.2) has two poles located at $\xi_1 = \phi' - \phi$ and $\xi_2 = -(\phi + \phi')$. The corresponding residues at these poles are given by

$$R_1 = \frac{1}{2\pi i}\exp\left\{ik\left[z\cos\beta_o - \rho\sin\beta_o\cos(\phi-\phi')\right]\right\}, \tag{3.96}$$

$$R_2 = \mp\exp\left\{ik\left[z\cos\beta_o - \rho\sin\beta_o\cos(\phi+\phi')\right]\right\}. \tag{3.97}$$

The signs (–) and (+) in (3.97) correspond to the soft or hard case, respectively. We observe that $2\pi R_1$ and $2\pi R_2$ are identical to the direct and reflected field in their domain of existence. However, since the residues are defined by (3.96) and (3.97), regardless of whether the poles ξ_1 and ξ_2 are located on the real axis of the complex ξ-plane, we must find a significance for R_1 and R_2 outside the domain of existence of the incident and reflected fields. This poses no difficulty for the residue R_1, because it is identical to the field that exists at the observation point in the absence of the wedge. To find a significance for the residue R_2, we enlarge the face from which the reflection occurs and extend the standard definition of the reflected field by means of the continuity of the phase as shown in figure 3.6.

Since the observation point lies outside the wedge, we restrict the domain of definition of the extended incident and reflected fields to the exterior region of the wedge.

The parameter τ is defined in (3.71) and is related to the phase factor in (3.4) by the quantity s_1, given by (3.26),

$$\tau^2 = -i\Omega s_1^2 = i\Omega\left[q(z_o) - q(z_s)\right]. \tag{3.98}$$

Using (3.2), we obtain

$$\tau_1^2 = 2k\rho\sin\beta_o\cos^2\left(\frac{\phi-\phi'}{2}\right), \tag{3.99}$$

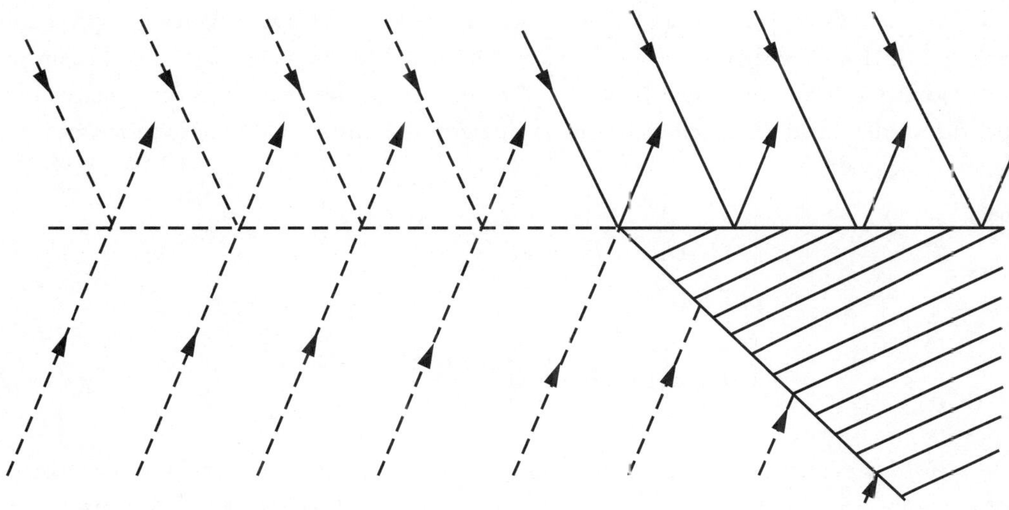

Figure 3.6. Extension of the field reflected on a straight wedge outside its physical domain.

$$\tau_2^2 = 2k\rho \sin\beta_o \cos^2\left(\frac{\phi+\phi'}{2}\right), \tag{3.100}$$

where τ_1 and τ_2 correspond to the poles ξ_1 and ξ_2. In (3.2) we have assumed that the phase reference is at the diffraction point. Hence it follows that the phase of the diffracted field at the observation point $M(\rho, \phi, z)$ is

$$kS^d(M) = k\rho \sin\beta_o + kz\cos\beta_o. \tag{3.101}$$

Since the phase $S^i(M)$ of the direct field at M is given by the exponent in (3.96), we obtain

$$k\left[S^i(M) - S^d(M)\right] = 2k\rho \sin\beta_o \cos^2\left(\frac{\phi-\phi'}{2}\right) = \tau_1^2. \tag{3.102}$$

Similarly, if $S^R(M)$ denotes the phase of the extended reflected field at M, given by the exponent in (3.97), we obtain

$$k\left[S^R(M) - S^d(M)\right] = 2k\rho \sin\beta_o \cos^2\left(\frac{\phi+\phi'}{2}\right) = \tau_2^2. \tag{3.103}$$

We now observe that (3.102) and (3.103), respectively, correspond to the detour that the phase makes along the path followed by the diffracted ray compared with the phase along the path followed by the direct and reflected rays.

In view of this property, τ_1 and τ_2 have been termed the "detour parameter" by Lee and Deschamps [3.14]. The sign of these parameters can be defined by imposing the constraint on $U(-\tau)$, occurring in (3.76), that it be set equal to 1 when the observation point is reached by either the incident field or the reflected field of geometrical acoustics (GA) and to zero elsewhere. Therefore, we get

$$\tau_1 = \varepsilon_1(M)\sqrt{2k\rho \sin\beta_o}\left|\cos\left(\frac{\phi-\phi'}{2}\right)\right|, \tag{3.104}$$

$$\tau_2 = \varepsilon_2(M)\sqrt{2k\rho \sin\beta_o}\left|\cos\left(\frac{\phi+\phi'}{2}\right)\right|, \tag{3.105}$$

where $\varepsilon_1(M)$ is set equal to 1 if M is in the shadow region of the GA incident field, while it equals −1 if the point is in the zone reached by the GA incident field. Likewise, $\varepsilon_2(M)$ is equal to 1 if the observation point is in the GA shadow of the reflected field and −1 if it is in the region where the reflected GA field exists.

Inserting (3.96) and (3.97), together with (2.75), into (3.74) leads to the final result:

$$\begin{aligned} P_{s_b}(\rho,\phi,z) &= e^{ik\left[z\cos\beta_o - \rho\sin\beta_o\cos(\phi-\phi')\right]}\left[\overline{F}(\tau_1) - \hat{\overline{F}}(\tau_1)\right] \\ &\mp e^{ik\left[z\cos\beta_o - \rho\sin\beta_o\cos(\phi+\phi')\right]}\left[\overline{F}(\tau_2) - \hat{\overline{F}}(\tau_2)\right] \\ &+ D_{s_b}\frac{e^{ik(z\cos\beta_o + \rho\sin\beta_o)}}{\sqrt{\rho/\sin\beta_o}}, \end{aligned} \tag{3.106}$$

where D_{s_b} is the nonuniform diffraction coefficient given by (2.76). Equation (3.106) is the uniform asymptotic solution given by the Van der Waerden method. It represents the total field.

One of the difficulties in employing (3.106) comes from the fact that at the shadow and reflection boundaries, τ tends to zero and $\hat{F}(\tau)$ tends to infinity. Since D_s and D_h tend also to infinity, the first-order terms of their expansions for small τ must compensate each other, and one must extract accurately the second-order terms for numerical calculations.

Another difficulty is related to the fact that the method used to extend the reflected field fails on a caustic of reflected rays. This situation occurs on a wedge with planar faces at the image points of a line source parallel to the edge or at the image point of a point source illuminating the edge of the wedge (fig. 3.7).

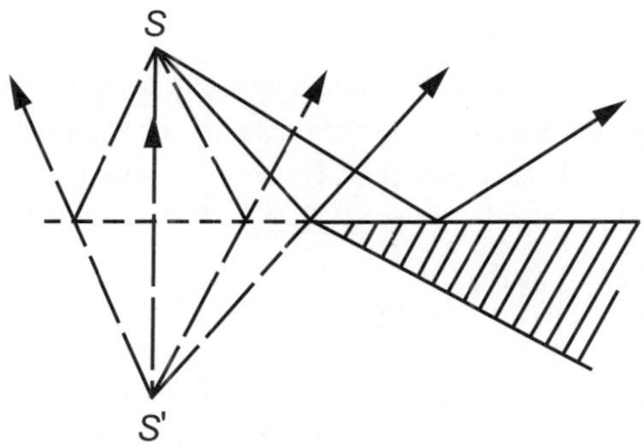

Figure 3.7. **Critical situation for line source or point source illuminations of a wedge with planar faces.**

3.2.1.2. Uniform Asymptotic Solutions for a Soft or Hard Wedge with Curved Faces

We consider an incident acoustic ray field (see chapter 1, section 1.6) limited to its principal order, which we know is identical to the GA field, falling on a soft or hard wedge with curved faces. A nonuniform asymptotic expansion limited to $0\left(k^{-1/2}\right)$ of the solution to this problem is given by (3.1). This solution is valid in the subdomains labeled 1 in figure 3.2. Our objective is now to construct a uniform asymptotic expansion valid through the transition regions 2 and 2′ of figure 3.2 that reduces to the solution (3.1) in the subdomains of region 1.

To construct the uniform solution, we follow the direct method described in the introduction of this chapter that consists of adapting the uniform expansions derived from a wedge with planar faces to the curved wedge problem. Since we have obtained two uniform solutions depending on the procedure used to derive the asymptotic expansion of an integral having a pole close to the saddle point, we will also obtain two different uniform solutions for the curved wedge problem.

The first one, founded on the method of Pauli-Clemmow, was first derived by Kouyoumjian and Pathak [3.12] who called it the uniform theory of diffraction (UTD) solution. The second one, founded on the method of Van der Waerden and on the work of Lewis and Boersma [3.13], was derived by Lee and Deschamps [3.14] and is called the uniform asymptotic theory (UAT) solution.

In both approaches, the technique consists of adapting the geometrical parameters in the straight wedge solution in order to enforce the continuity of the solution through the shadow and reflection boundaries and to recover the nonuniform solution (3.1) in the subdomains of region 1, away from the transition regions corresponding to these boundaries.

3.2.1.2.1. The UTD Solution

We consider a wedge with planar faces that is locally tangent to the wedge with curved faces at the diffraction point Q. By adapting the divergence factor of the diffracted wave to the case of a curved wedge illuminated by a locally planar wave, we can write (3.93) in the form

$$P_{s_b}^{D}\left(M\right)=P_{s_b}^{i}\left(Q\right)D_{s_b}\sqrt{\frac{\rho_e}{s\left(\rho_e+s\right)}}\,e^{iks}, \tag{3.107}$$

where $s = |\mathbf{QM}|$, ρ_e is the radius of curvature of the diffracted wave in the plane of diffraction (see chapter 1, section 1.8, equation (1.199)), and $P^i(Q)$ is the incident field at the diffraction point Q.

In order for the diffracted field to compensate for the jump discontinuity in the incident and reflected fields at the crossing of the respective shadow boundaries, we adapt the arguments of the transition functions $F(X)$ in (3.95). There are four transition functions in (3.95), the arguments of which are different since $a(\beta)$ takes four different values. However, all these arguments depend on the same parameter $K = kL$, where $L = \rho \sin \beta_o$ is a geometrical parameter. The functions $a(\beta)$ depend on the angular positions of the shadow and reflection boundaries and can therefore not be modified. However, the parameter L can be adapted. Since the diffraction coefficient (3.95) has four jump discontinuities, depending on the angular positions of the shadow and reflection boundaries, we state our ansatz in the form given by (3.107), where

$$\begin{aligned} D_{s_b} = \frac{1}{\sin\beta_o}\Bigg\{ & \overset{+}{d}\left(\overset{-}{\beta},n\right)F\left[kL^{i}\,\overset{+}{a}\left(\overset{-}{\beta}\right)\right]+\overset{-}{d}\left(\overset{-}{\beta},n\right)F\left[kL^{i}\,\overset{-}{a}\left(\overset{-}{\beta}\right)\right] \\ & \mp\left(\overset{+}{d}\left(\overset{+}{\beta},n\right)F\left[kL^{ro}\,\overset{+}{a}\left(\overset{+}{\beta}\right)\right]+\overset{-}{d}\left(\overset{+}{\beta},n\right)F\left[kL^{rn}\,\overset{-}{a}\left(\overset{+}{\beta}\right)\right]\right)\Bigg\} \end{aligned} \tag{3.108}$$

and where L^i, L^o, L^{rn} are three unknown parameters. The parameter L^i is associated with the incident wave, while the parameters L^{ro} and L^{rn} are associated with the reflected waves on the faces $\phi = 0$ and $\phi = n\pi$, respectively. To determine these parameters, we identify the jump discontinuity in the transition function $F(X)$ with the jump discontinuity of the incident or the reflected fields. According to (3.24), we have

$$\delta F\left(kLa\right)\approx 2e^{-\frac{i\pi}{4}}\sqrt{\pi\, kLa}\;,a\to 0.$$

Hence from (3.108), (3.94), and (3.86), the discontinuity in D_{s_b} can be written

$$\delta D_{s_b}=\begin{cases}-\dfrac{1}{\sin\beta_0}\sqrt{L^{i}} \\ \pm\dfrac{1}{\sin\beta_0}\sqrt{L^{r}},\end{cases} \tag{3.109}$$

where L^r is used to denote either L^{ro} or L^{rn}. According to the general expressions for the incident and reflected fields established in chapter 1, sections 1.5 and 1.6, it is possible to express these fields with respect to a reference wave front passing through the diffraction point Q. The jumps of these fields are then given by

$$\delta P^i(M) = -P^i(Q)\sqrt{\frac{\rho_1^i\,\rho_2^i}{(\rho_1^i + s)(\rho_2^i + s)}}\,e^{iks}, \tag{3.110}$$

$$\delta P_{s_b}^R(M) = \pm P^i(Q)\sqrt{\frac{\rho_1^r\,\rho_2^r}{(\rho_1^r + s)(\rho_2^r + s)}}\,e^{iks}. \tag{3.111}$$

The jump discontinuity of the diffracted field is obtained by inserting (3.109) into (3.107). Equating the jump discontinuity of the diffracted field with the jump discontinuity of the incident or reflected fields yields

$$L^i = \frac{s(\rho_e^i + s)\rho_1^i\,\rho_2^i \sin^2\beta_0}{\rho_e^i(\rho_1^i + s)(\rho_2^i + s)}, \tag{3.112}$$

$$L^r = \frac{s(\rho_e^r + s)\rho_1^r\,\rho_2^r \sin^2\beta_0}{\rho_e^r(\rho_1^r + s)(\rho_2^r + s)}, \tag{3.113}$$

where ρ_e^i and ρ_e^r correspond to the value of ρ_e in the plane of incidence and reflection, respectively.

To summarize, the uniform UTD solution for a three-dimensional curved wedge is given by (3.107) with D_{s_b} defined by (3.108). The geometrical parameters L^i and L^r are given by (3.112) and (3.113). The parameter L^i depends on the principal radii of curvature ρ_1^i and ρ_2^i of the incident wave front at Q, on the radius of curvature ρ_e^i of the diffracted wave front at Q in the plane of incidence defined by the tangent to the edge and the incident ray, and on the distance to the observation point $s = |\mathbf{QM}|$. The parameter L^{ro} depends on ρ_e^{ro} and s and on the principal radii of curvature ρ_1^{ro} and ρ_2^{ro} at Q of the wave reflected on the curved face $\phi = 0$. Likewise, the parameter L^{rn} depends on ρ_e^{rn} and s and on the principal radii of curvature ρ_1^{rn} and ρ_2^{rn} at Q of the wave reflected on the curved face $\phi = n\pi$. The total field obtained by adding the incident and reflected GA fields to (3.107) is continuous, as their respective shadow boundaries are crossed.

3.2.1.2.2. The UAT Solution

The UAT solution derived by Lee and Deschamps [3.14] relies on an ansatz for the total field that is constructed from the uniform solution (3.106) of the wedge with planar faces. By introducing in (3.106) the divergence factor of the diffracted wave corresponding to a curved wedge submitted to a local plane wave, established in chapter 1, section 1.8, and by generalizing the detour parameters τ_1 and τ_2 for the three-dimensional curved wedge, we can write (3.106) in the form

$$P_{s_b}(M) = P^i(M)\left[\bar{F}(\tau_i) - \hat{\bar{F}}(\tau_i)\right] + P^R_{s_b}(M)\left[\bar{F}(\tau_R) - \hat{\bar{F}}(\tau_R)\right] + D_{s_b}\, P^i(Q)\sqrt{\frac{\rho_e}{s(\rho_e + s)}}\, e^{iks}, \tag{3.114}$$

where D_{s_b} is the nonuniform diffraction coefficient given by (2.76), ρ_e is the radius of curvature of the diffracted wave in the plane of diffraction, and s is the distance $|\mathbf{QM}|$ from the diffraction point Q to the observation point M.

According to (3.102) and (3.103), the detour parameters extended to the three-dimensional case read

$$\tau_i = \varepsilon_i(M)\sqrt{k\left|S^i(M) - S^d(M)\right|}, \tag{3.115}$$

$$\tau_R = \varepsilon_R(M)\sqrt{k\left|S^R(M) - S^d(M)\right|}, \tag{3.116}$$

where $\varepsilon_i = 1$ if the observation point M is in the shadow region of the incident field, and $\varepsilon_i = -1$ if it is in the region where the incident field exists. Likewise, $\varepsilon_R = 1$ if the point M is in the shadow region of the reflected field, and $\varepsilon_R = -1$ in the region where the reflected field exists.

The field $P^R_{s_b}(M)$ is an extension of the GA reflected field in the region where the latter does not exist, through the continuation of the phase and the amplitude. In order to perform this extension, we elongate each of the curved faces by enforcing the continuity of the tangent plane and the principle curvatures. Then we prolong the reflected field on the real surfaces using the one reflected on the fictitious surfaces that we have just defined. Finally, this field must be extended to the domain located on the other side of these surfaces by the continuation of the phase and the amplitude.

Figures 3.8(a) and 3.8(b) provide an illustration of this procedure for the field reflected from the face $\phi = 0$. A similar picture can be provided for the reflected field from the face $\phi = n\pi$.

The extended reflected field satisfies the laws of GA along the fictitious rays. It is therefore not defined on the caustics of these rays. In figure 3.8(a), the caustic of the fictitious reflected rays lies inside the wedge. Since the observation point can only be outside the wedge, it is always away from this caustic. However, the caustic of the reflected rays depends on the direction of

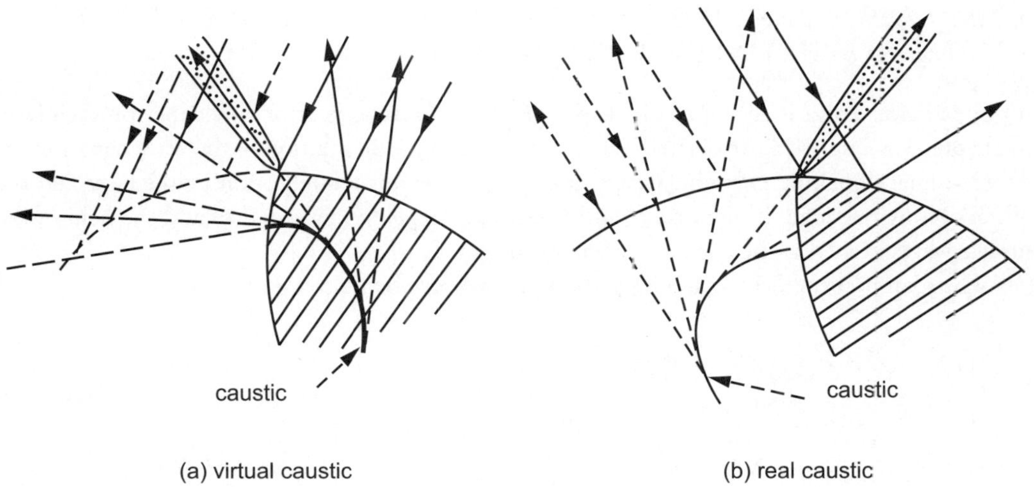

Figure 3.8. Extension of the field reflected on a curved wedge outside its physical domain and the transition zone of the diffracted field, where (a) represents virtual caustic and (b) represents real caustic.

propagation of the incident field. In figure 3.8(b), the caustic of the reflected rays lies outside the wedge. This situation is a drawback for the UAT solution.

Outside the transition zones and the caustics of the fictitious rays, the ansatz (3.114) yields the solution (3.1). This result is a direct consequence of (3.76). It can be shown that the ansatz (3.114) leads to a continuous field through the shadow boundaries of the direct and reflected fields. Moreover, not only the total field but also its derivatives are continuous across the shadow boundaries. A demonstration of these properties is given in reference [3.1].

On the other hand, the continuity of the derivatives of the field are not generally satisfied by the UTD solution. Specifically, as mentioned in section 3.2.2.1, the terms that depend on the derivative of the incident field in the direction normal to the shadow boundary are not present in the UTD solution. To remedy this defect, Hwang and Kouyoumjian [3.15] have added a corrective term to the UTD solution, which they call the "slope diffraction coefficient." In contrast, the UAT solution does not exhibit a similar behavior, and it correctly yields this term.

In addition, in the UAT solution, one finds a term that depends on the derivative of the reflected field in the direction normal to the shadow boundary. This term plays a role that is similar to that of the derivative of the incident field and makes it necessary to introduce in the UTD formulation a slope diffraction coefficient of the reflected field.

However, as far as practical applications are concerned, UTD augmented by the slope diffraction terms is more convenient to use than the UAT, since it is necessary neither to extend the surface nor to search for fictitious rays when the UTD is used.

3.2.2. UNIFORM ASYMPTOTIC SOLUTIONS FOR THE ACOUSTIC FIELD SCATTERED BY A CONVEX IMPEDANCE WEDGE

An impedance wedge is an impenetrable wedge with a constant surface impedance on each face. In chapter 1, section 1.8, we derived a nonuniform asymptotic solution of the wave equation for a three-dimensional impenetrable wedge, taking into account the reflection by the faces and the diffraction by the edge. In the case of an impedance wedge, we also need to add the field of the surface waves excited at the edge to these contributions.

The total pressure field at an observation point M can be written in the form

$$\begin{aligned} P(M) &= e^{ikS^i(M)} p_0^i(M)U(-\varepsilon_i) + e^{ikS^R(M)} p_0^R(M)U(-\varepsilon_R) \\ &\quad + e^{ikS^d(M)} \frac{p_0^d(M)}{\sqrt{k}} + \sum_{n=1}^{N} e^{ikS_n^{sw}(M)} p_{on}^{sw}(M)U(-\varepsilon_n^{sw}) + o\left(k^{-1/2}\right), \end{aligned} \tag{3.117}$$

where N is the number of surface waves and where the index o means that only the dominant term of each species of waves is retained. As shown in chapter 1, the dominant term of the incident, reflected, and surface waves is of order $k^0 = 1$, whereas the dominant term of the diffracted field is of order $k^{-1/2}$. Apart from the term concerning the surface waves, the notations used in (3.117) for all other terms are identical to those of equation (3.1). The function p_{on}^{sw} and S_n^{sw} are, respectively, the amplitude and phase of the nth surface wave, which is a local inhomogeneous plane wave. The argument ε_n^{sw} of the Heaviside function is $\varepsilon_n^{sw} = 1$ or -1, depending on whether the observation point is in the shadow or the sound zone of the nth surface wave.

Altogether the terms in (3.117), which do not depend on the surface waves, look similar to those of (3.1). Their explicit expressions are different because the reflection coefficient depends on the surface impedance and also because the diffraction coefficient is no longer Keller's diffraction coefficient but Maliuzhinets's diffraction coefficient, given by (2.203).

The term describing the contribution of the surface waves can be calculated by using the general formulas giving the complex amplitude and phase of a surface wave, established in chapter 1, section 1.9, together with the excitation coefficient of a surface wave by an anisotropic line source, derived later in chapter 5, section 5.3.1.

Formula (3.117) is a nonuniform asymptotic solution for the acoustic field scattered by a convex impedance wedge. This solution is valid outside the transition regions at the shadow boundaries of the different types of waves considered in (3.117). Away from these transition regions at large distance from the edge ($kr >> 1$), it gives an accurate solution for the field scattered by a three-dimensional impedance wedge. In order to construct a uniform asymptotic solution valid in the transition regions and reducing to (3.117) outside these regions, we proceed as in section 3.2.1. Since the shadow boundaries of the direct and reflected fields also exist on a straight impedance wedge, the method consists of first establishing a uniform solution for this canonical problem and then adapting the geometrical parameters of this solution such that, outside the transition regions, it yields the first three terms of the solution defined in (3.117). The same procedure cannot be applied at the shadow boundaries of the surface waves or at the shadow boundaries of the diffracted rays since such boundaries do not exist on a straight wedge. This problem, involving surface diffraction and described by elastic surface waves and creeping waves, will be solved by a different technique in chapter 5.

Next we come back to the exact solution of the canonical problem of the impedance wedge with planar faces, derived in section 2.2.2, and extract a uniform asymptotic solution from it.

3.2.2.1. Uniform Asymptotic Solution of an Impedance Wedge with Planar Faces

We have seen in chapter 2, section 2.2.2 that, when a plane wave strikes an impedance wedge with planar faces, the exact solution for the total field in the fluid outside the wedge is given by

$$P(r,\varphi) = \frac{1}{2\pi i} \int_{\gamma} e^{-ik\,r\cos\alpha}\, \sigma(\alpha+\varphi)\psi(\alpha+\varphi)\,d\varphi, \tag{3.118}$$

where γ is the Sommerfeld contour shown in figure 2.14 and where $\sigma(\alpha)$ and $\psi(\alpha)$ are given by (2.150) and (2.137), respectively. For the asymptotic evaluation of the solution (3.118), the contour γ has been closed by the SDP Γ and Γ' passing through the stationary points $\alpha = \pi$ and $\alpha = -\pi$, respectively (see fig. 2.15).

By applying the residue theorem, the integral in (3.118) can be written in the form

$$\int_{\gamma} F(\alpha)d\alpha = \int_{\Gamma+\Gamma'} F(\alpha)d\alpha + 2\pi i \sum_{p=1}^{N} R(\alpha_p),$$

where $F(\alpha) = \sigma(\alpha + \varphi)\psi(\alpha + \varphi)$ and $R(\alpha_p)$ are the residues at the poles α_p of $F(\alpha)$, located inside the closed contour $\gamma - \Gamma - \Gamma'$. The method for obtaining the asymptotic expansion of the solution (3.118) for $kr >> 1$ consists of evaluating the integrals along the paths Γ and Γ' by the method of steepest descent. We know from chapter 2, section 2.2.2 that the poles of $\sigma(\alpha + \varphi)$ are located on the real axis and correspond to the incident and reflected fields, whereas those of $\psi(\alpha + \varphi)$ correspond to the surface waves.

3.2.2.1.1. Uniform Asymptotic Solution Valid through the Shadow Boundaries of the Incident and Reflected Fields

We consider first the poles of $\sigma(\alpha + \varphi)$. When one of these poles approaches the saddle points π or $-\pi$, the classical method of steepest descent is no longer applicable. Especially, the Taylor series expansion (2.183) becomes invalid. We have seen in section 3.2.1 that in this case, two different methods have been developed for treating the singularity of the integrand: the method of Pauli-Clemmow and the method of Van der Waerden. Each of these methods has its own advantages.

However, referring to our comments at the end of section 3.2.2.2, the UTD solution (which is founded on the Pauli-Clemmow method) is easier to implement in a computer program for practical applications.

In order to apply the method of Pauli-Clemmow, we must first decompose the integrand into a sum of terms such that each of them presents only a single pole in the domain of variation of the incident and observation angles. For this purpose, we come back to the equation (3.77), giving the exact expression of the field diffracted by a hard or soft wedge with planar faces. We observe that the signs $\pm$ in the integrand of (3.77) can be replaced by the reflection coefficient R since for a hard wedge $R = 1$ and for a soft wedge $R = -1$. We recognize therefore, along each contour Γ and Γ', the term corresponding to the incident field and the term corresponding to the reflected field. Each of these terms has only a single pole in the domain of variation of the incidence and observation angles.

In the same way, it is possible to decompose the integrand of (3.118). For this purpose, we refer the angles to one of the faces of the wedge and use the relations in (2.192). With these notations, the diffracted field is given by

$$P^d\left(M\right)=\frac{P_0^i}{2\pi i}\int_{\Gamma+\Gamma'} e^{-ikr\cos\alpha}\,\sigma_1\left(\alpha-\phi,\,\phi'\right)\frac{\psi\left(\alpha+\dfrac{n\pi}{2}-\phi\right)}{\psi\left(\dfrac{n\pi}{2}-\phi'\right)}d\alpha, \tag{3.119}$$

where

$$\sigma_1\left(\alpha-\phi,\,\phi'\right)=\frac{1}{n}\frac{\sin\dfrac{\phi'}{n}}{\cos\left(\dfrac{\alpha-\phi}{n}\right)-\cos\dfrac{\phi'}{n}}. \tag{3.120}$$

Following the method of Tiberio et al. [3.16], we start with the identity

$$\begin{aligned}Q&=R_n\left(\alpha+\frac{n\pi}{2}-\phi,\frac{n\pi}{2}-\phi';\,\theta\right)\sigma_1\left(\alpha-\phi,\,\phi'\right)\\&=\frac{1}{2n}\left\{\cot\left[\frac{\alpha-\left(\phi-\phi'\right)}{2n}\right]+\Gamma_n\left(\phi',\theta\right)\cot\left[\frac{\alpha-\left(\phi+\phi'\right)}{2n}\right]\right\},\end{aligned} \tag{3.121}$$

where

$$R_n\left(\alpha,\alpha';\theta\right)=\frac{\cos\dfrac{\alpha}{n}+\sin\dfrac{\theta}{n}}{\cos\dfrac{\alpha'}{n}+\sin\dfrac{\theta}{n}}, \tag{3.122}$$

$$\Gamma_n(\phi',\theta)=\frac{\sin\frac{\phi'}{n}-\sin\frac{\theta}{n}}{\sin\frac{\phi'}{n}+\sin\frac{\theta}{n}}. \tag{3.123}$$

The identity (3.121) can be easily verified by noting that

$$R_n\left(\alpha+\frac{n\pi}{2}-\phi\,,\frac{n\pi}{2}-\phi';\,\theta\right)=\frac{\sin\left(\frac{\phi-\alpha}{n}\right)+\sin\frac{\theta}{n}}{\sin\frac{\phi'}{n}+\sin\frac{\theta}{n}}=1+\frac{\sin\left(\frac{\phi-\alpha}{n}\right)-\sin\frac{\phi'}{n}}{\sin\frac{\phi'}{n}+\sin\frac{\theta}{n}}.$$

Hence

$$Q=\frac{1}{n}\left[\frac{\sin\frac{\phi'}{n}}{\cos\left(\frac{\alpha-\phi}{n}\right)-\cos\frac{\phi'}{n}}+\frac{\sin\frac{\phi}{n}}{\sin\frac{\phi'}{n}+\sin\frac{\theta}{n}}\quad\frac{\sin\left(\frac{\phi-\alpha}{n}\right)-\sin\frac{\phi'}{n}}{\cos\left(\frac{\phi-\alpha}{n}\right)-\cos\frac{\phi'}{n}}\right]$$

and since

$$\frac{\sin\frac{\phi}{n}}{\sin\frac{\phi'}{n}+\sin\frac{\theta}{n}}=\frac{1}{2}\left[1+\Gamma_n(\phi',\theta)\right],$$

we obtain

$$Q=\frac{1}{2n}\left[\frac{\sin\frac{\phi'}{n}+\sin\left(\frac{\phi-\alpha}{n}\right)}{\cos\left(\frac{\phi-\alpha}{n}\right)-\cos\frac{\phi'}{n}}+\Gamma_n(\phi',\theta)\frac{\sin\left(\frac{\phi-\alpha}{n}\right)-\sin\frac{\phi'}{n}}{\cos\left(\frac{\phi-\alpha}{n}\right)-\cos\frac{\phi'}{n}}\right],$$

and by replacing the sum of trigonometric functions by products of such functions, we obtain readily the formula (3.121) for Q.

According to (3.121), we have

$$\sigma_1(\alpha-\phi,\,\phi')=\frac{1}{2n}\frac{\sin\frac{\phi'}{n}+\sin\frac{\theta}{n}}{\sin\left(\frac{\phi-\alpha}{n}\right)-\sin\frac{\theta}{n}}\quad\left[\cot\left(\frac{\alpha-\bar{\beta}}{2n}\right)+\Gamma_n(\phi',\theta)\cot\left(\frac{\alpha-\overset{+}{\beta}}{2n}\right)\right], \tag{3.124}$$

where $\overset{-}{\beta} = \phi - \phi'$ and $\overset{+}{\beta} = \phi + \phi'$.

Inserting (3.124) into (3.119) yields

$$P^d(M) = \frac{P_0^i}{4\pi n i} \int_{\Gamma} e^{-ikr\cos\alpha} \frac{\psi\left(\alpha + \frac{n\pi}{2} - \phi\right)}{\psi\left(\frac{n\pi}{2} - \phi'\right)} \frac{\sin\frac{\phi'}{n} + \sin\frac{\theta_0}{n}}{\sin\left(\frac{\phi-\alpha}{n}\right) + \sin\frac{\theta_0}{n}} \left[\cot\left(\frac{\alpha - \overset{-}{\beta}}{2n}\right) + \Gamma_n(\phi', \theta_0)\cot\left(\frac{\alpha - \overset{+}{\beta}}{2n}\right)\right] d\alpha$$
$$+ \frac{P_0^i}{4\pi n i} \int_{\Gamma'} e^{-ikr\cos\alpha} \frac{\psi\left(\alpha + \frac{n\pi}{2} - \phi\right)}{\psi\left(\frac{n\pi}{2} - \phi'\right)} \frac{\sin\frac{\phi'}{n} + \sin\frac{\theta_n}{n}}{\sin\left(\frac{\phi-\alpha}{n}\right) + \sin\frac{\theta_n}{n}} \left[\cot\left(\frac{\alpha - \overset{-}{\beta}}{2n}\right) + \Gamma_n(\phi', \theta_n)\cot\left(\frac{\alpha - \overset{+}{\beta}}{2n}\right)\right] d\alpha, \tag{3.125}$$

where $\theta_{0,n}$ are related to the impedances $Z_{0,n}$ of the faces 0 and n by $\sin\theta_{0,n} = Z_{0,n}$ and where we have assigned the value $\theta = \theta_0$ to the contour Γ and the value $\theta = \theta_n$ to the contour Γ'.

In a second step, we transform the integrands in (3.125) in order that the reflection coefficients $\Gamma_1(\phi', \theta_0)$ and $\Gamma_2(\phi', \theta_n)$ on the faces 0 and n come forth. To this end, we apply the relation (2.115), expressed with the notation $\Phi = n\pi / 2$. Noting that

$$\Gamma_1(\phi', \theta_0) = \frac{\sin\phi' - \sin\theta_0}{\sin\phi' + \sin\theta_0}, \quad \Gamma_1(\phi', \theta_n) = \frac{\sin(n\pi - \phi') - \sin\theta_n}{\sin(n\pi - \phi') + \sin\theta_n}, \tag{3.126}$$

we obtain, by applying (2.115),

$$\psi\left(\frac{n\pi}{2} - \phi'\right) = \psi\left(\frac{n\pi}{2} + \phi'\right) \frac{\sin\phi' + \sin\theta_0}{-\sin\phi' + \sin\theta_0} = \frac{-\psi\left(\frac{n\pi}{2} + \phi'\right)}{\Gamma_1(\phi', \theta_0)} \tag{3.127}$$

and

$$\psi\left(\frac{n\pi}{2}-\phi'\right)=-\psi\left(\phi'-\frac{3n\pi}{2}\right)\frac{\sin(\phi'-n\pi)-\sin\theta_n}{\sin(\phi'-n\pi)+\sin\theta_n}=\frac{-\psi\left(\phi'-\frac{3n\pi}{2}\right)}{\Gamma_1(\phi',\theta_n)}. \tag{3.128}$$

Inserting (3.127) and (3.128) into (3.125), we finally obtain

$$\begin{aligned}
P^d(M) = &-\frac{P_0^i}{4\pi n i}\int_{\Gamma} e^{-ikr\cos\alpha}\left\{\frac{\psi\left(\alpha+\frac{n\pi}{2}-\phi\right)}{\psi\left(\frac{n\pi}{2}-\phi'\right)}\frac{\sin\frac{\phi'}{n}+\sin\frac{\theta_0}{n}}{\sin\left(\frac{\phi-\alpha}{n}\right)+\sin\frac{\theta_0}{n}}\cot\left(\frac{\alpha-\bar{\beta}}{2n}\right)\right.\\
&\left.-\frac{\psi\left(\alpha+\frac{n\pi}{2}-\phi\right)}{\psi\left(\frac{n\pi}{2}+\phi'\right)}\frac{\sin\frac{\phi'}{n}-\sin\frac{\theta_0}{n}}{\sin\left(\frac{\phi-\alpha}{n}\right)+\sin\frac{\theta_0}{n}}\Gamma_1(\phi',\theta_0)\cot\left(\frac{\alpha-\overset{+}{\beta}}{2n}\right)\right\}d\alpha\\
&-\frac{P_0^i}{4\pi n i}\int_{\Gamma'} e^{-ikr\cos\alpha}\left\{\frac{\psi\left(\alpha+\frac{n\pi}{2}-\phi\right)}{\psi\left(\frac{n\pi}{2}-\phi'\right)}\frac{\sin\frac{\phi'}{n}+\sin\frac{\theta_n}{n}}{\sin\left(\frac{\phi-\alpha}{n}\right)+\sin\frac{\theta_n}{n}}\cot\left(\frac{\alpha-\bar{\beta}}{2n}\right)\right.\\
&\left.-\frac{\psi\left(\alpha+\frac{n\pi}{2}-\phi\right)}{\psi\left(\varphi'-\frac{3n\pi}{2}\right)}\frac{\sin\frac{\phi'}{n}-\sin\frac{\theta_n}{n}}{\sin\left(\frac{\phi-\alpha}{n}\right)+\sin\frac{\theta_n}{n}}\Gamma_1(\phi',\theta_n)\cot\left(\frac{\alpha-\overset{+}{\beta}}{2n}\right)\right\}d\alpha.
\end{aligned} \tag{3.129}$$

It can be readily verified that the behavior of the integrands in (3.129) is similar to that of (3.77)—namely, that when the poles $\alpha_p = \overset{+}{\beta}$ and $\alpha_p = \bar{\beta}$ tend toward one of the saddle points $\alpha_s = \pi$ or $\alpha_s = -\pi$, the coefficients in front of the cotangents in the first and third line of (3.129) tend to 1, whereas the coefficients in front of the cotangents in the second and forth line tend to $\Gamma_1(\phi', \theta_0)$ and $\Gamma_1(\phi', \theta_n)$, respectively.

It follows from this analysis that the second member of (3.129) and consequently the second member of (3.125) are in the desired form for the application of Pauli-Clemmow's method. The dominant term of the asymptotic expansion is therefore given by (3.91), which in the present case, for normal incidence, takes the form

$$P^d(M) = -P_0^i \frac{e^{\frac{i\pi}{4}} e^{ikr}}{2n\sqrt{2\pi kr}} \left\{ \frac{\psi\left(\pi + \frac{n\pi}{2} - \phi\right)}{\psi\left(\frac{n\pi}{2} - \phi'\right)} \frac{\sin\frac{\phi'}{n} + \sin\frac{\theta_0}{n}}{\sin\left(\frac{\phi - \pi}{n}\right) + \sin\frac{\theta_0}{n}} \right.$$

$$\left(\cot\left(\frac{\pi - \overset{-}{\beta}}{2n}\right) F\left[K\overset{-}{a}\left(\overset{-}{\beta}\right)\right] + \Gamma_n(\phi', \theta_0) \cot\left(\frac{\pi - \overset{+}{\beta}}{2n}\right) F\left[K\overset{-}{a}\left(\overset{+}{\beta}\right)\right]\right)$$

$$+ \frac{\psi\left(-\pi + \frac{n\pi}{2} - \phi\right)}{\psi\left(\frac{n\pi}{2} - \phi'\right)} \frac{\sin\frac{\phi'}{n} + \sin\frac{\theta_n}{n}}{\sin\left(\frac{\phi + \pi}{n}\right) + \sin\frac{\theta_n}{n}}$$

$$\left.\left(\cot\left(\frac{\pi + \overset{-}{\beta}}{2n}\right) F\left[K\overset{+}{a}\left(\overset{-}{\beta}\right)\right] + \Gamma_n(\phi', \theta_n) \cot\left(\frac{\pi + \overset{+}{\beta}}{2n}\right) F\left[K\overset{+}{a}\left(\overset{+}{\beta}\right)\right]\right) \right\}, \tag{3.130}$$

where $K = kr$ and where the functions $F(x)$ and $\overset{\pm}{a}\left(\overset{\pm}{\beta}\right)$ are defined by equations (3.17) and (3.86), respectively, in section 3.2.1.

By setting

$$\overset{\pm}{d}(\beta, n) = \frac{-e^{\frac{i\pi}{4}}}{n\sqrt{2\pi k}} \frac{1}{2} \cot\left(\frac{\pi \pm \beta}{2n}\right), \tag{3.131}$$

the formula (3.130) can also be written in the form

$$P^d(M) = P^i(O)\, D(\phi, \phi'; \theta_0, \theta_n; kr) \frac{e^{ikr}}{\sqrt{r}}, \tag{3.132}$$

where D is the diffraction coefficient given by (3.133).

When the observation point leaves the transition region of the incident or the reflected wave, the argument of the transition function $F(x)$ in (3.133) becomes of the order $K = kr$, which is supposed to be large. Then $F(x)$ will be replaced by the dominant term of its asymptotic expansion for $|x|$ large, which according to (3.25), is equal to 1. Replacing $F(x)$ with 1 in (3.133) and using the identity (3.124), where σ_1 is defined by (3.120), we recover the expression (2.193), giving the diffraction coefficient of an impedance wedge, which is valid outside the transition regions:

$$D(\phi,\phi';\theta_0,\theta_n;kr)=\frac{\psi\left(\pi+\frac{n\pi}{2}-\phi\right)}{\psi\left(\frac{n\pi}{2}-\phi'\right)}\frac{\sin\frac{\phi'}{n}+\sin\frac{\theta_0}{n}}{\sin\left(\frac{\phi-\pi}{n}\right)+\sin\frac{\theta_0}{n}}$$

$$\left(\overset{-}{d}\left(\overset{-}{\beta},n\right)F\left[K\overset{-}{a}\left(\overset{-}{\beta}\right)\right]+\Gamma_n(\phi',\theta_0)\overset{-}{d}\left(\overset{+}{\beta},n\right)F\left[K\overset{-}{a}\left(\overset{+}{\beta}\right)\right]\right)$$

$$+\frac{\psi\left(-\pi+\frac{n\pi}{2}-\phi\right)}{\psi\left(\frac{n\pi}{2}-\phi'\right)}\frac{\sin\frac{\phi'}{n}+\sin\frac{\theta_n}{n}}{\sin\left(\frac{\phi+\pi}{n}\right)+\sin\frac{\theta_n}{n}}$$

$$\left(\overset{+}{d}\left(\overset{-}{\beta},n\right)F\left[K\overset{+}{a}\left(\overset{-}{\beta}\right)\right]+\Gamma_n(\phi',\theta_n)\overset{+}{d}\left(\overset{+}{\beta},n\right)F\left[K\overset{+}{a}\left(\overset{+}{\beta}\right)\right]\right). \tag{3.133}$$

3.2.2.1.2. Explicit Form of the Diffraction Coefficient for Numerical Applications

In formula (3.133) the function $\psi(\alpha)$ appears with three different arguments. According to (2.137), $\psi(\alpha)$ is a product of four Maliuzhinets functions $\psi_\Phi(\alpha)$ with different arguments. Therefore the calculation of the diffraction coefficient (3.133) requires 12 calculations of the special Maliuzhinets function $\psi_\Phi(\alpha)$, given by (2.139). Hence we need to perform twelve integrations.

It is possible to reduce the numerical operations to one integration. For this purpose, we make use of equations (E.4) to (E.6) and (E.9) to (E.10) of appendix E. If we replace all the functions $\psi_\Phi(\alpha)$ in (E.6) with their explicit expression (2.139), then we obtain

$$M_n(\phi,\phi';\theta_0,\theta_n)=A(\phi,\phi';\theta_0,\theta_n)\exp\left[B(\phi,\phi';\theta_0,\theta_n)\right], \tag{3.134}$$

$$A(\phi,\phi';\theta_0,\theta_n)=\frac{\cos\left[\frac{1}{2n}\left(\frac{\pi}{2}-\phi'+\theta_n\right)\right]\sin\left[\frac{1}{2n}\left(\frac{\pi}{2}+\phi+\theta_0\right)\right]\sin\left[\frac{1}{2n}\left(\frac{\pi}{2}+\phi'+\theta_0\right)\right]\mathrm{c}}{\cos\left[\frac{1}{2n}\left(\frac{\pi}{2}-\phi'-\theta_n\right)\right]\sin\left[\frac{1}{2n}\left(\frac{\pi}{2}+\phi-\theta_0\right)\right]\sin\left[\frac{1}{2n}\left(\frac{\pi}{2}+\phi'-\theta_0\right)\right]}$$

$$\frac{\cos\left[\frac{1}{2n}\left(\frac{\pi}{2}-\phi+\theta_n\right)\right]}{\cos\left[\frac{1}{2n}\left(\frac{\pi}{2}-\phi-\theta_n\right)\right]}, \tag{3.135}$$

$$B(\phi,\phi';\theta_0,\theta_n)=\frac{2}{n}\int_0^1 \mathrm{Log}\left\{\left[\left(\frac{1+w^2\cot^2\left[\frac{1}{2n}\left(\frac{\pi}{2}+\phi+\theta_0\right)\right]}{1+w^2\cot^2\left[\frac{1}{2n}\left(\frac{\pi}{2}+\phi-\theta_0\right)\right]}\right)\left(\frac{1+w^2\cot^2\left[\frac{1}{2n}\left(\frac{\pi}{2}+\phi'+\theta_0\right)\right]}{1+w^2\cot^2\left[\frac{1}{2n}\left(\frac{\pi}{2}+\phi'-\theta_0\right)\right]}\right)\right]C\right\}$$
$$\times\frac{dw}{\left(1+w^2\right)ch\left(\frac{2}{n}Argth\,w\right)},\tag{3.136}$$

$$C=\left(\frac{1+w^2tg^2\left[\frac{1}{2n}\left(\frac{\pi}{2}-\phi+\theta_n\right)\right]}{1+w^2tg^2\left[\frac{1}{2n}\left(\frac{\pi}{2}-\phi-\theta_n\right)\right]}\right)\left(\frac{1+w^2tg^2\left[\frac{1}{2n}\left(\frac{\pi}{2}-\phi'+\theta_n\right)\right]}{1+w^2tg^2\left[\frac{1}{2n}\left(\frac{\pi}{2}-\phi'-\theta_n\right)\right]}\right).$$

Inserting (E.4) and (E.9) into (3.133), we obtain the final expression for the uniform diffraction coefficient:

$$D(\phi,\phi';\theta_0,\theta_n,kr)=M_n(\phi,\phi';\theta_0,\theta_n)$$
$$\times\left\{L_n^{(1)}(\phi,\phi';\theta_0,\theta_n)\left[\bar{D}\left(\bar{\beta},K\right)+\Gamma_n(\phi',\theta_0)\bar{D}\left(\overset{+}{\beta},K\right)\right]\right.\tag{3.137}$$
$$\left.+L_n^{(2)}(\phi,\phi';\theta_0,\theta_n)\left[\overset{+}{D}\left(\bar{\beta},K\right)+\Gamma_n(\phi',\theta_n)\overset{+}{D}\left(\overset{+}{\beta},K\right)\right]\right\},$$

where M_n is given by (3.134) to (3.136) and where

$$L_n^{(1)}(\phi,\phi';\theta_0,\theta_n)=\frac{\sin\frac{\phi'}{n}+\sin\frac{\theta_0}{n}}{\sin\left(\frac{\phi-\pi}{n}\right)+\sin\frac{\theta_0}{n}}\quad C_n^{(1)}(\phi,\phi';\theta_0,\theta_n),\tag{3.138}$$

$$L_n^{(2)}(\phi,\phi';\theta_0,\theta_n)=\frac{\sin\frac{\phi'}{n}+\sin\frac{\theta_n}{n}}{\sin\left(\frac{\phi+\pi}{n}\right)+\sin\frac{\theta_n}{n}}\quad C_n^{2}(\phi,\phi';\theta_0,\theta_n).\tag{3.139}$$

Additionally, the coefficients $C_n^{(1)}$ and $C_n^{(2)}$ are given by (E.5) and (E.10), respectively. Equations (2.134) to (2.139) are the desired form of the uniform diffraction coefficient for numerical applications.

3.2.2.1.3. Uniform Asymptotic Solution Valid through the Shadow Boundaries of the Surface Waves

We have seen in chapter 2, section 2.2.2 that the surface waves are given by the residues at the poles of the function $\psi(\alpha + \varphi)$. According to (2.160), these poles are defined by

$$\begin{aligned} \varphi_p^+ &= +\left(\Phi + \alpha_p^+\right) - \varphi, \\ \varphi_p^- &= -\left(\Phi + \alpha_p^-\right) - \varphi, \end{aligned} \tag{3.140}$$

where the angles are referred to the bissectrix of the wedge angle, as shown in figure 2.6, and where $\alpha_p^\pm$ are the poles of the reflection coefficients that are the solutions of the characteristic equation

$$\sin \alpha_p^\pm = -Z^\pm = -\sin \theta^\pm. \tag{3.141}$$

We have also seen that, for a passive surface $\left(R_e Z^\pm > 0\right)$, the poles $\alpha_p^\pm$ are located outside the strip (0, π). Therefore if we consider the waves emanating from the edge with positive phase velocity, the corresponding poles are given by

$$\alpha_p^\pm = \pi + \theta^\pm, \quad R_e \theta^\pm > 0. \tag{3.142}$$

Hence only two poles of $\psi(\alpha + \varphi)$ may be located inside the closed contour $\gamma - \Gamma - \Gamma'$ (see fig. 2.15). When $\varphi = \Phi$, these poles are inside that contour if $\left|I_m \theta^\pm\right| > R_e \theta^\pm$.

When a pole crosses the SDP, the surface wave disappears. The discontinuity produced in the amplitude of the surface wave is compensated by a discontinuity with an opposite sign in the integral along the SDP that has been crossed by the pole. The asymptotic evaluation of that integral must therefore have the same discontinuity as the exact integral.

Two types of situations will be treated in this section. The first one concerns a pole crossing the SDP at a point located far from the saddle point. The second one concerns a pole crossing the SDP in the vicinity of the saddle point.

In a first step, we decompose the integrand into a sum of terms, each with a single pole. We start with the expression of $\psi(\alpha)$, given in (2.137), which we reproduce here:

$$\psi(\alpha) = \psi_\Phi\left(\alpha + \Phi + \frac{\pi}{2} - \theta^+\right)\psi_\Phi\left(\alpha + \Phi - \frac{\pi}{2} + \theta^+\right)\psi_\Phi\left(\alpha - \Phi - \frac{\pi}{2} + \theta^-\right)\psi_\Phi\left(\alpha - \Phi + \frac{\pi}{2} - \theta^-\right). \tag{3.143}$$

We also know that the Maliuzhinets special function $\psi_\Phi(u)$ is a meromorphic function having poles located on the real axis given by

$$u = \pm\left(\frac{3\pi}{2} + 2\Phi + 2N\pi\right), \quad N = 0,\ 1,\ 2,\ \ldots . \tag{3.144}$$

We see that the poles $\pm\left(\Phi + \alpha_p^{\pm}\right) = \pm\left(\Phi + \pi + \theta^{\pm}\right)$ of $\psi(\alpha)$ are also poles of the first and third function $\psi_\Phi(u)$ on the right-hand side of (3.143). Indeed, for $\alpha = \Phi + \pi + \theta^+$, the argument of the first function $\psi_\Phi(u)$ on the right-hand side of (3.143) is equal to $\frac{3\pi}{2} + 2\Phi$, which corresponds to a pole of this function. Likewise, $\alpha = -\Phi - \pi - \theta^-$ is a pole of $\psi_\Phi\left(\alpha - \Phi - \frac{\pi}{2} + \theta^-\right)$. The two other functions $\psi_\Phi(u)$, however, have no poles.

So that the poles appear explicitly, we set

$$\begin{aligned} u &= \alpha + \varphi - \left(\pi + \Phi + \theta^+\right), \\ \mathrm{v} &= \alpha + \varphi + \pi + \Phi + \theta^-. \end{aligned} \tag{3.145}$$

When α tends to φ_p^+, then, according to (3.145), u tends to 0. Likewise, when $\alpha \to \varphi_p^-$, then $\mathrm{v} \to 0$.

The change of variables (3.145) leads to

$$\psi_\Phi\left(\alpha + \Phi + \frac{\pi}{2} - \theta^+\right) = \psi_\Phi\left[\alpha + \varphi - \left(\pi + \Phi + \theta^+\right) + \frac{3\pi}{2} + 2\Phi\right] = \psi_\Phi\left(u + \frac{3\pi}{2} + 2\Phi\right),$$

$$\psi_\Phi\left(\alpha - \Phi - \frac{\pi}{2} + \theta^-\right) = \psi_\Phi\left(\alpha + \varphi + \pi + \Phi + \theta^+ - \frac{3\pi}{2} - 2\Phi\right) = \psi_\Phi\left(\mathrm{v} - \frac{3\pi}{2} - 2\Phi\right).$$

Now according to the second equation (2.140), we obtain

$$\psi_\Phi\left(u + \frac{3\pi}{2} + 2\Phi\right)\psi_\Phi\left(u + 2\Phi + \frac{\pi}{2}\right) = \psi_\Phi^2\left(\frac{\pi}{2}\right)\cos\left[\frac{\pi}{4\Phi}\left(u + \pi + 2\Phi\right)\right],$$

$$\psi_\Phi\left(u + 2\Phi + \frac{\pi}{2}\right)\psi_\Phi\left(u + 2\Phi - \frac{\pi}{2}\right) = \psi_\Phi^2\left(\frac{\pi}{2}\right)\cos\left[\frac{\pi}{4\Phi}\left(u + 2\Phi\right)\right].$$

Hence

$$\psi_\Phi\left(u + \frac{3\pi}{2} + 2\Phi\right) = \psi_\Phi\left(u + 2\Phi - \frac{\pi}{2}\right)\frac{\sin\left[\frac{\pi}{4\Phi}\left(\pi + u\right)\right]}{\sin\left(\frac{\pi}{4\Phi}u\right)}, \tag{3.146}$$

and since $\psi_\Phi(\alpha)$ is an even function of α, we also have

$$\psi_\Phi\left[\mathrm{v}-\left(\frac{3\pi}{2}+2\Phi\right)\right]=\psi_\Phi\left(-\mathrm{v}+2\Phi-\frac{\pi}{2}\right)\frac{\sin\left[\frac{\pi}{4\Phi}(\mathrm{v}-\pi)\right]}{\sin\left(\frac{\pi}{4\Phi}\mathrm{v}\right)}. \tag{3.147}$$

Inserting (3.146) and (3.147) into (3.143) yields

$$\begin{aligned}\psi(\alpha+\varphi)=&\psi_\Phi\left(\alpha+\varphi+\Phi-\frac{\pi}{2}+\theta^+\right)\psi_\Phi\left(\alpha+\varphi-\Phi+\frac{\pi}{2}-\theta^-\right)\psi_\Phi\left(\alpha+\varphi+\Phi-\frac{3\pi}{2}-\theta^+\right)\\&\times\psi_\Phi\left(-\alpha-\varphi+\Phi-\frac{3\pi}{2}-\theta^-\right)\frac{\sin\left[\frac{\pi}{4\Phi}(\alpha+\varphi-\Phi-\theta^+)\right]\sin\left[\frac{\pi}{4\Phi}(\alpha+\varphi+\Phi+\theta^-)\right]}{\sin\left[\frac{\pi}{4\Phi}(\alpha+\varphi-\pi-\Phi-\theta^+)\right]\sin\left[\frac{\pi}{4\Phi}(\alpha+\varphi+\pi+\Phi+\theta^-)\right]}\end{aligned} \tag{3.148}$$

We now refer the angles of incidence and observation to one of the faces of the wedge by using the relations (2.192). With these notations, along with (3.148), the function $\psi\left(\alpha+\frac{n\pi}{2}-\phi\right)$ appearing in (3.125) can be written in the form

$$\psi\left(\alpha+\frac{n\pi}{R}-\phi\right)=P_n(\alpha,\phi,\phi';\theta_0,\theta_n)S_n(\alpha,\phi,\phi';\theta_0,\theta_n), \tag{3.149}$$

where

$$\begin{aligned}P_n(\alpha,\phi,\phi';\theta_0,\theta_n)=&\psi_n\left(\alpha-\phi+n\pi-\frac{\pi}{2}+\theta_0\right)\psi_n\left(\alpha-\phi+\frac{\pi}{2}-\theta_n\right)\\&\times\psi_n\left(\alpha-\phi+n\pi-\frac{3\pi}{2}-\theta_0\right)\psi_n\left(-\alpha+\phi-\frac{3\pi}{2}-\theta_n\right),\end{aligned} \tag{3.150}$$

$$S_n(\alpha,\phi,\phi';\theta_0,\theta_n)=\frac{\sin\left[\frac{1}{2n}(\alpha-\phi-\theta_0)\right]\sin\left[\frac{1}{2n}(\alpha-\phi+n\pi+\theta_n)\right]}{\sin\left[\frac{1}{2n}(\alpha-\phi-\pi-\theta_0)\right]\sin\left[\frac{1}{2n}(\alpha-\phi+n\pi+\pi+\theta_n)\right]}. \tag{3.151}$$

Transforming (3.151) into a sum of terms, each with a single pole, we get

$$S_n\left(\alpha,\phi,\phi';\theta_0,\theta_n\right)=\mathrm{T}_n\left(\alpha,\phi,\phi';\theta_0,\theta_n\right)\left\{\frac{1}{\sin\left(\dfrac{\alpha-\alpha_0}{2n}\right)}+\frac{1}{\sin\left(\dfrac{\alpha-\alpha_n}{2n}\right)}\right\}, \tag{3.152}$$

where

$$\alpha_0=\phi+\pi+\theta_0\,,\quad \alpha_n=\phi-n\pi-\pi-\theta_n \tag{3.153}$$

are the poles of $\psi\left(\alpha+\dfrac{n\pi}{2}-\phi\right)$ and where

$$\mathrm{T}_n\left(\alpha,\phi,\phi';\theta_0,\theta_n\right)=\frac{\sin\left[\dfrac{1}{2n}\left(\alpha-\phi-\theta_0\right)\right]\sin\left[\dfrac{1}{2n}\left(\alpha-\phi+n\pi+\theta_n\right)\right]}{\sin\left[\dfrac{1}{2n}\left(\alpha-\phi+n\pi+\pi+\theta_n\right)\right]+\sin\left[\dfrac{1}{2n}\left(\alpha-\phi-\pi-\theta_0\right)\right]}. \tag{3.154}$$

It can be readily verified that T_n has no singularity located inside the closed contour $\gamma-\Gamma-\Gamma'$.

Next we insert (3.149) into (3.125) and take into account (3.150), (3.152), and the following identities:

$$\frac{1}{\sin\left[\dfrac{1}{2n}\left(\alpha-\alpha_{0_n}\right)\right]}\frac{1}{\sin\left[\dfrac{1}{2n}\left(\alpha-\overset{\pm}{\beta}\right)\right]}=\left\{\frac{1}{\sin\left(\dfrac{\alpha-\alpha_{0_n}}{2n}\right)}+\frac{1}{\sin\left(\dfrac{\alpha-\overset{\pm}{\beta}}{2n}\right)}\right\}\times\left(\sin\left[\frac{\alpha-\alpha_0}{2n}\right]+\sin\left[\frac{\alpha-\overset{\pm}{\beta}}{2n}\right]\right)^{-1}. \tag{3.155}$$

Then we obtain the required result for the diffracted field, composed of a sum of terms—each of which having a single pole,

$$P^d(M) = -\frac{P_0^i}{4\pi n i}\int_\Gamma e^{-ikr\cos\alpha}\frac{P_n \mathrm{T}_n}{\psi\left(\frac{n\pi}{2}-\phi'\right)}\frac{\sin\frac{\phi'}{n}+\sin\frac{\theta_0}{n}}{\sin\left(\frac{\phi-\alpha}{n}\right)+\sin\frac{\theta_0}{n}}$$

$$\left\{ h\left(\alpha,\alpha_0,\bar{\beta}\right)\cot\left(\frac{\alpha-\alpha_0}{2n}\right)+h\left(\alpha,\alpha_n,\bar{\beta}\right)\cot\left(\frac{\alpha-\alpha_n}{2n}\right)+\ell\left(\alpha,\bar{\beta}\right)\cot\left(\frac{\alpha-\bar{\beta}}{2n}\right)\right.$$

$$\left.+\Gamma_n(\phi',\theta_0)\left[h\left(\alpha,\alpha_0,\overset{+}{\beta}\right)\cot\left(\frac{\alpha-\alpha_0}{2n}\right)+h\left(\alpha,\alpha_n,\overset{+}{\beta}\right)\cot\left(\frac{\alpha-\alpha_n}{2n}\right)+\ell\left(\alpha,\overset{+}{\beta}\right)\cot\left(\frac{\alpha-\overset{+}{\beta}}{2n}\right)\right]\right\}d\alpha$$

$$-\frac{P_0^i}{4\pi n i}\int_{\Gamma'} e^{-ikr\cos\alpha}\frac{P_n \mathrm{T}_n}{\psi\left(\frac{n\pi}{2}-\phi'\right)}\frac{\sin\frac{\phi'}{n}+\sin\frac{\theta_n}{n}}{\sin\left(\frac{\phi-\alpha}{n}\right)+\sin\frac{\theta_n}{n}}$$

$$\left\{ h\left(\alpha,\alpha_0,\bar{\beta}\right)\cot\left(\frac{\alpha-\alpha_0}{2n}\right)+h\left(\alpha,\alpha_n,\bar{\beta}\right)\cot\left(\frac{\alpha-\alpha_n}{2n}\right)+\ell\left(\alpha,\bar{\beta}\right)\cot\left(\frac{\alpha-\bar{\beta}}{2n}\right)\right.$$

$$\left.+\Gamma_n(\phi',\theta_n)\left[h\left(\alpha,\alpha_0,\overset{+}{\beta}\right)\cot\left(\frac{\alpha-\alpha_0}{2n}\right)+h\left(\alpha,\alpha_n,\overset{+}{\beta}\right)\cot\left(\frac{\alpha-\alpha_n}{2n}\right)+\ell\left(\alpha,\overset{+}{\beta}\right)\cot\left(\frac{\alpha-\overset{+}{\beta}}{2n}\right)\right]\right\}d\alpha,$$

(3.156)

where

$$h\left(\alpha,\alpha_{0_n},\overset{\mp}{\beta}\right)=\frac{\cos\left(\frac{\alpha-\overset{\mp}{\beta}}{2n}\right)}{\cos\left(\frac{\alpha-\alpha_{0_n}}{2n}\right)}\frac{1}{\sin\left(\frac{\alpha-\alpha_{0_n}}{2n}\right)+\sin\left(\frac{\alpha-\overset{\mp}{\beta}}{2n}\right)},\qquad(3.157)$$

$$\ell\left(\alpha,\overset{\mp}{\beta}\right)=\frac{1}{\sin\left(\frac{\alpha-\alpha_0}{2n}\right)+\sin\left(\frac{\alpha-\overset{\mp}{\beta}}{2n}\right)}+\frac{1}{\sin\left(\frac{\alpha-\alpha_n}{2n}\right)+\sin\left(\frac{\alpha-\overset{\mp}{\beta}}{2n}\right)},\qquad(3.158)$$

and where P_n and T_n are given by (3.150) and (3.154). According to (3.153), we can also write (3.154) in the form

$$\mathrm{T}_n = \frac{\sin\left[\frac{\alpha-(\alpha_0-\pi)}{2n}\right]\sin\left[\frac{\alpha-(\alpha_n+\pi)}{2n}\right]}{\sin\left(\frac{\alpha-\alpha_0}{2n}\right)+\sin\left(\frac{\alpha-\alpha_n}{2n}\right)}. \tag{3.159}$$

In the case where the poles α_0 and α_n are far from the saddle points $\pm\pi$, the uniform asymptotic evaluation of (3.156) by the method of Pauli-Clemmow is similar to that of (3.125). By setting

$$\overset{\pm}{d}(x,n) = \frac{-e^{\frac{i\pi}{4}}}{n\sqrt{2\pi k}}\,\frac{1}{2}\cot\left(\frac{\pi\pm x}{2n}\right), \tag{3.160}$$

we obtain

$$P^d(M) = P^i(0)\,D(\phi,\phi';\theta_0,\theta_n;kr)\frac{e^{ikr}}{\sqrt{r}}, \tag{3.161}$$

where D is the diffraction coefficient given by

$$D(\phi,\phi';\theta_0,\theta_n;kr) = \frac{P_n(\pi,\phi,\phi';\theta_0,\theta_n)}{\psi\left(\frac{n\pi}{2}-\phi'\right)}\,\frac{\sin\frac{\phi'}{n}+\sin\frac{\theta_0}{n}}{\sin\left(\frac{\phi-\pi}{n}\right)+\sin\frac{\theta_0}{n}}\mathrm{T}_n(\pi,\phi,\phi';\theta_0,\theta_n)$$

$$\Big\{h\left(\pi,\alpha_0,\overset{-}{\beta}\right)\overset{-}{d}(\alpha_0,n)F\left[Ka(\alpha_0)\right]+h\left(\pi,\alpha_n,\overset{-}{\beta}\right)\overset{-}{d}(\alpha_n,n)F\left[Ka(\alpha_n)\right]$$

$$+\ell\left(\pi,\overset{-}{\beta}\right)\overset{-}{d}\left(\overset{-}{\beta},n\right)F\left[K\overset{-}{a}\left(\overset{-}{\beta}\right)\right]$$

$$+\Gamma_n(\phi',\theta_0)\left(h\left(\pi,\alpha_0,\overset{+}{\beta}\right)\overset{-}{d}(\alpha_0,n)F\left[Ka(\alpha_0)\right]+h\left(\pi,\alpha_n,\overset{+}{\beta}\right)\overset{-}{d}(\alpha_n,n)F\left[Ka(\alpha_n)\right]\right.$$

$$\left.+\ell\left(\pi,\overset{+}{\beta}\right)\overset{-}{d}\left(\overset{+}{\beta},n\right)F\left[K\overset{-}{a}\left(\overset{+}{\beta}\right)\right]\right)\Big\}$$

$$+\frac{P_n(-\pi,\phi,\phi';\theta_0,\theta_n)}{\psi\left(\frac{n\pi}{2}-\phi'\right)}\,\frac{\sin\frac{\phi'}{n}+\sin\frac{\theta_n}{n}}{\sin\left(\frac{\phi+\pi}{n}\right)+\sin\frac{\theta_n}{n}}\mathrm{T}_n(-\pi,\phi,\phi';\theta_0,\theta_n)$$

$$\begin{aligned}
&\Bigg\{ b\left(-\pi,\alpha_0,\overset{-}{\beta}\right)\overset{+}{d}\left(\alpha_0,n\right)F\left[Ka\left(\alpha_0\right)\right]+b\left(-\pi,\alpha_n,\overset{-}{\beta}\right)\overset{+}{d}\left(\alpha_n,n\right)F\left[Ka\left(\alpha_n\right)\right]\\
&+\ell\left(-\pi,\overset{-}{\beta}\right)\overset{+}{d}\left(\overset{-}{\beta},n\right)F\left[K\overset{+}{a}\left(\overset{-}{\beta}\right)\right]\\
&+\Gamma_n\left(\phi',\theta_n\right)\Bigg(b\left(-\pi,\alpha_0,\overset{+}{\beta}\right)\overset{+}{d}\left(\alpha_0,n\right)F\left[Ka\left(\alpha_0\right)\right]+b\left(-\pi,\alpha_n,\overset{+}{\beta}\right)\overset{+}{d}\left(\alpha_n,n\right)F\left[Ka\left(\alpha_n\right)\right]\\
&\qquad+\ell\left(-\pi,\overset{+}{\beta}\right)\overset{+}{d}\left(\overset{+}{\beta},n\right)F\left[K\overset{+}{a}\left(\overset{+}{\beta}\right)\right]\Bigg)\Bigg\},
\end{aligned} \tag{3.162}$$

where the argument of the transition function $F(x)$ is defined by $K = kr$ and

$$a\left(\alpha_p\right)=i\left[q\left(\alpha_p\right)-q\left(\alpha_s\right)\right]=1+i\,q\left(\alpha_p\right), \tag{3.163}$$

where $q(\alpha) = -i\cos\alpha$. For real poles corresponding to the incident and reflected waves, $a(\alpha_p)$ is defined by (3.86). For the poles α_0 and α_n corresponding to the surface waves, equation (3.163) yields

$$a\left(\alpha_p\right)=1+\cos\alpha_p. \tag{3.164}$$

We know from our analysis in section 3.2.1.1.1 that when a pole α_p crosses one of the SDP Γ or Γ', $R_e\left[a\left(\alpha_p\right)\right]$ passes through zero and the function $F\left[Ka\left(\alpha_p\right)\right]$ has a discontinuity that compensates the disappearance of the pole from the closed contour $\gamma-\Gamma-\Gamma'$. In the case of the surface wave corresponding to the pole α_0, this happens when $\phi+\theta'_0 = \text{Arc}\cos\left(1/ch\theta''_0\right)$, where $\theta_0=\theta'_0+i\theta''_0$.

The diffraction coefficient (3.162) is uniformly valid through the shadow boundaries of the incident, reflected, and surface waves, provided that the poles α_0 and α_n are not close to the saddle points $\alpha_s=\pi$ and $\alpha_s=-\pi$, respectively. According to (3.153), when the observation point is located on one of the faces of the wedge and when the impedance relative to this face tends to zero, the pole corresponding to the surface wave supported by that face tends to the saddle point.

In this case, the function $P_n(\pi,\phi,\phi';\theta_0,\theta_n)$ tends to zero. Consequently, the first term of the Pauli-Clemmow method given by (3.162) tends to zero. In order to obtain a uniform solution in the transition region of the surface wave, we must calculate the second term of this method.

Coming back to equation (3.12) of the Pauli-Clemmow method presented in section 3.2.1.1.1, we see that in the asymptotic expansion of an integral of the general form (3.7), limited to the order $0\left(k^{-1/2}\right)$, the second term of the Pauli-Clemmow method is given by

$$I(\Omega) = A_1 J_1 = \sqrt{\frac{\pi}{\Omega}}\left[1 - F\left(-i\Omega s_1^2\right)\right] A_1. \tag{3.165}$$

In the present case, $\Omega = K = kr$ and $-i\, s_1^2 = i\left[q\left(\alpha_p\right) - q\left(\alpha_s\right)\right] = a\left(\alpha_p\right)$. Hence

$$I(\Omega) = \sqrt{\frac{\pi}{\Omega}}\left[1 - F\left[Ka\left(\alpha_p\right)\right]\right] A_1. \tag{3.166}$$

The determination of the constant A_1 is straightforward. According to (3.11) and (3.8), we have $A_1 = T'(0)$ where

$$\mathrm{T}(s) = \left(s - s_1\right) G(s) = \left(s - s_1\right) g(z) \frac{dz}{ds} e^{\Omega q(z_s)}, \tag{3.167}$$

s_1 being a pole of $G(s)$ and z_s being the saddle point. Denoting by z_1 the pole of $g(z)$ corresponding to s_1, we set

$$g(z) = \frac{g_1(z)}{z - z_1}. \tag{3.168}$$

Hence

$$\mathrm{T}(s) = \frac{s - s_1}{z - z_1} g_1(z) \frac{dz}{ds} e^{\Omega q(z_s)} \tag{3.169}$$

and

$$\mathrm{T}'(s) = g_1(z) \frac{dz}{ds} e^{\Omega q(z_s)} \frac{d}{ds}\left(\frac{s - s_1}{z - z_1}\right) + \frac{s - s_1}{z - z_1}\left(\frac{dz}{ds}\right)^2 \frac{d\, g_1(z)}{dz} e^{\Omega q(z)}$$
$$+ \frac{s - s_1}{z - z_1} g_1(z) \frac{d^2 z}{ds^2} e^{\Omega q(z_s)}.$$

Since by hypothesis $g_1(z_s) = 0$, we obtain

$$\mathrm{T}'(0) = \frac{-s_1}{z_s - z_1} e^{\Omega q(z_s)} \left[\frac{d\, g_1(z)}{dz}\left(\frac{dz}{ds}\right)^2\right]_{z = z_s}$$

and, according to (3.52) and (3.54),

$$A_1 = T'(0) = \left(\frac{d\,g_1(z)}{dz}\right)_{z-z_s} \sqrt{\frac{-2}{q''(z_s)}}\,e^{\Omega q(z_s)}. \tag{3.170}$$

Moreover, since by definition

$$g_1(z) = (z - z_1)\,g(z),$$

it follows that

$$\left(\frac{d\,g_1(z)}{dz}\right)_{z-z_s} = (z_s - z_1)\left(\frac{dg}{dz}\right)_{z-z_s}. \tag{3.171}$$

Inserting (3.171) into (3.170) and comparing the expression for A_1 to that of $G(0)$, given by

$$G(0) = g(z_s)\sqrt{\frac{-2}{q''(z_s)}}\,e^{\Omega q(z_s)}, \tag{3.172}$$

we observe that A_1 is obtained by substituting $(z_s - z_1)\left(\frac{dg}{dz}\right)_{z-z_s}$ for $g(z_s)$ in (3.172). Hence the diffraction coefficient corresponding to the second term of the Pauli-Clemmow method is obtained by performing the following modifications in the formula (3.162):

- $P_n(\alpha_s, \phi, \phi'; \theta_0, \theta_n)$ needs to be changed in $\left(\frac{\partial P_n}{\partial \alpha}\right)_{\alpha=\alpha_s}$.
- $d(\alpha_p, n)$ is replaced by $d(\alpha_p, n)(\alpha_s - \alpha_p)$.
- $1 - F[Ka(\alpha_p)]$ is substituted for $F[Ka(\alpha_p)]$.

Finally the complete asymptotic diffraction coefficient, valid whatever the position of the poles may be, is given by the sum of $D(\phi, \phi'; \theta_0, \theta_n, kr)$ and $D^{(m)}(\phi, \phi'; \theta_0, \theta_n, kr)$, where $D^{(m)}$ is the modified diffraction coefficient corresponding to the second term of the Pauli-Clemmow method.

3.2.2.1.4. Oblique Incidence

The generalization to oblique incidence of the uniform diffraction coefficients derived in sections 3.2.2.1.1, 3.2.2.1.2, and 3.2.2.1.3 follows the same procedure as in chapter 2, section 2.2.2.4. The solution of the problem is given by the same formulas as for normal incidence, provided that the surface impedance is replaced by a modified surface impedance given by (2.201). Also, k has to be replaced by $k \sin \beta_0$, where β_0 is the angle between the incident direction and the edge, and φ and φ_0 are now the projected angles defined in figure 2.19(b).

3.2.2.2. Uniform Asymptotic Solution for a General Three-Dimensional Impedance Wedge

The uniform asymptotic solution valid through the shadow boundaries of the incident and reflected fields, established in section 3.2.2.1 for an impedance wedge with planar faces, can be extended to a three-dimensional impedance wedge with curved faces and a curved edge. The procedure is similar to that presented in section 3.2.1.2. It consists of modifying the distance parameter in the transition function $F(Ka)$. Since the asymptotic evaluation of the solution is founded on the Pauli-Clemmow method, the result will also be labeled as a UTD solution.

For a plane wave hitting a wedge with planar faces at oblique incidence, the distance parameter is contained in the parameter $K = kr \sin \beta_0$ and appears only in the transition functions $F\left[K\overset{-}{a}\left(\overset{-}{\beta}\right)\right]$, $F\left[K\overset{-}{a}\left(\overset{+}{\beta}\right)\right]$, $F\left[K\overset{+}{a}\left(\overset{-}{\beta}\right)\right]$, $F\left[K\overset{+}{a}\left(\overset{+}{\beta}\right)\right]$ of the diffraction coefficient. These functions have a jump discontinuity (see section 3.2.1.1.) at the shadow boundaries. This creates a jump in the diffracted field compensating exactly for the jump in the direct or reflected field at the shadow boundaries. The technique described in section 3.2.2.1 consists of introducing a distance parameter L at each shadow boundary by replacing K by $K = kL$ with $L = L^i$ at the shadow boundary of the incident field and $L = L^{ro}$ or $L = L^{rn}$ at the shadow boundaries of the field reflected by the faces *o* or *n*, respectively. With these modifications, the diffraction coefficient (3.133), extended to oblique incidence, reads

$$
\begin{aligned}
D\left(\phi,\phi';\theta_0,\theta_n;L^i,L^{ro},L^{rn}\right) = {} & \frac{\psi\left(\pi+\dfrac{n\pi}{2}-\phi\right)}{\psi\left(\dfrac{n\pi}{2}-\phi'\right)} \frac{\sin\dfrac{\phi'}{n}+\sin\dfrac{\theta_0}{n}}{\sin\left(\dfrac{\phi-\pi}{n}\right)+\sin\dfrac{\theta_0}{n}} \\
& \frac{1}{\sin\beta_0}\left(\overset{-}{d}\left(\overset{-}{\beta},n\right)F\left[K L^i \overset{-}{a}\left(\overset{-}{\beta}\right)\right]+\Gamma_n(\phi',\theta_0)\overset{-}{d}\left(\overset{+}{\beta},n\right)F\left[K L^{ro}\overset{-}{a}\left(\overset{+}{\beta}\right)\right]\right) \\
& +\frac{\psi\left(-\pi+\dfrac{n\pi}{2}-\phi\right)}{\psi\left(\dfrac{n\pi}{2}-\phi'\right)} \frac{\sin\dfrac{\phi'}{n}+\sin\dfrac{\theta_n}{n}}{\sin\left(\dfrac{\phi+\pi}{n}\right)+\sin\dfrac{\theta_n}{n}} \\
& \frac{1}{\sin\beta_0}\left(\overset{+}{d}\left(\overset{-}{\beta},n\right)F\left[K L^i \overset{+}{a}\left(\overset{-}{\beta}\right)\right]+\Gamma_n(\phi',\theta_0)\overset{+}{d}\left(\overset{+}{\beta},n\right)F\left[K L^{rn}\overset{+}{a}\left(\overset{+}{\beta}\right)\right]\right),
\end{aligned}
\tag{3.173}
$$

where θ is defined by $\sin\theta = Z^{(m)} = \dfrac{Z}{\sin\beta_0}$.

Furthermore, according to chapter 1, section 1.8, the field diffracted by a curved impenetrable wedge is given by

$$P^d(M) = P^i(Q)D\sqrt{\frac{\rho_e}{s(\rho_e + s)}}\,e^{iks}, \tag{3.174}$$

where D is the diffraction coefficient and where ρ_e is the radius of curvature of the diffracted wave in the plane formed by the diffracted ray reaching the observation point M and the tangent to the edge at the diffraction point Q.

As in section 3.2.1.2, in order to determine L^i, L^{ro}, L^{rn} we identify the jump discontinuity of the diffracted field $P^d(M)$ at each shadow boundary with the jump of the incident and reflected fields.

Using the same notations as in section 3.2.1.2, the jump of the incident field is given by

$$\delta P^i(M) = -P^i(Q)\sqrt{\frac{\rho_1^i\rho_2^i}{(\rho_1^i + s)(\rho_2^i + s)}}. \tag{3.175}$$

Likewise, the jump of the reflected field is given by

$$\delta P^R(M) = -P^i(Q)\Gamma_1\sqrt{\frac{\rho_1^r\rho_2^r}{(\rho_1^r + s)(\rho_2^r + s)}}, \tag{3.176}$$

where Γ_1 is the reflection coefficient defined in (3.126) for each face of the wedge.

When the face n is in the shadow of the incident field, we have $\phi = \phi' + \pi$ at the shadow boundary. Hence

$$\frac{\psi\left(\pi + \frac{n\pi}{2} - \phi\right)}{\psi\left(\frac{n\pi}{2} - \phi'\right)} = 1, \quad \frac{\sin\frac{\phi'}{n} + \sin\frac{\theta_0}{n}}{\sin\left(\frac{\phi - \pi}{n}\right) + \sin\frac{\theta_0}{n}} = 1$$

in (3.173) when the observation point is located on the shadow boundary.

The jump of the diffraction coefficient D is therefore equal to the jump of the expression

$$\frac{1}{\sin\beta_0}\bar{d}\left(\bar{\beta}, n\right)F\left[kL^i\bar{a}\left(\bar{\beta}\right)\right]$$

for $\overset{-}{\beta} = \phi - \phi' = \pi$. Since $\overset{-}{a}\left(\overset{-}{\beta}\right) = 1 + \cos\overset{-}{\beta}$, we obtain, according to (3.131) and (3.24),

$$\frac{1}{\sin\beta_0}\overset{-}{d}\left(\overset{-}{\beta},n\right)\delta F\left[kL^i\,\overset{-}{a}\left(\overset{-}{\beta}\right)\right] = -\frac{1}{\sin\beta_0}\sqrt{L^i}. \tag{3.177}$$

Likewise, when the face o is in the shadow of the incident field, we have $\phi = \phi' - \pi$ at the shadow boundary, and the jump of the diffraction coefficient is given by

$$\frac{1}{\sin\beta_0}\overset{+}{d}\left(\overset{-}{\beta},n\right)\delta F\left[kL^i\,\overset{+}{a}\left(\overset{-}{\beta}\right)\right] = -\frac{1}{\sin\beta_0}\sqrt{L^i}. \tag{3.178}$$

The reflection boundary of the field reflected on the face o occurs when $\phi = \pi - \phi'$. Then according to (3.127), we obtain

$$\frac{\psi\left(\pi + \dfrac{n\pi}{2} - \phi\right)}{\psi\left(\dfrac{n\pi}{2} - \phi'\right)}\;\frac{\sin\dfrac{\phi'}{n} + \sin\dfrac{\theta_0}{n}}{\sin\left(\dfrac{\phi - \pi}{n}\right) + \sin\dfrac{\theta_0}{n}} = -\frac{\psi\left(\dfrac{n\pi}{2} + \phi'\right)}{\psi\left(\dfrac{n\pi}{2} - \phi'\right)}\frac{1}{\Gamma_n\left(\phi',\theta_0\right)} = \frac{\Gamma_1\left(\phi',\theta_0\right)}{\Gamma_n\left(\phi',\theta_0\right)}.$$

Consequently, the jump of the diffraction coefficient is given by

$$\frac{\Gamma_n\left(\phi',\theta_0\right)}{\sin\beta_0}\overset{-}{d}\left(\overset{+}{\beta},n\right)\delta F\left[kL^{ro}\,\overset{-}{a}\left(\overset{+}{\beta}\right)\right] = -\frac{\Gamma_1\left(\phi',\theta_0\right)}{\sin\beta_0}\sqrt{L^{ro}}. \tag{3.179}$$

Likewise, at the reflection boundary $\phi = 2n\pi - \pi - \phi'$ on face n, the jump of the diffraction coefficient is given by

$$\frac{\Gamma_n\left(\phi',\theta_n\right)}{\sin\beta_0}\overset{+}{d}\left(\overset{+}{\beta},n\right)\delta F\left[kL^{rn}\,\overset{+}{a}\left(\overset{+}{\beta}\right)\right] = -\frac{\Gamma_1\left(\phi',\theta_n\right)}{\sin\beta_0}\sqrt{L^{rn}}. \tag{3.180}$$

Finally, equating the jump discontinuity of the diffracted field with the jump discontinuity of the incident or the reflected fields yields

$$L^i = \frac{s\left(\rho_e^i + s\right)\rho_1^i\,\rho_2^i\sin^2\beta_0}{\rho_e^i\left(\rho_1^i + s\right)\left(\rho_2^i + s\right)}, \tag{3.181}$$

$$L^r = \frac{s\left(\rho_e^r + s\right)\rho_1^r \rho_2^r \sin^2\beta_0}{\rho_e^r\left(\rho_1^r + s\right)\left(\rho_2^r + s\right)}, \tag{3.182}$$

where ρ_e^i and ρ_e^r are the values of ρ_e in the plane of incidence and reflection, respectively. We observe that L^i and L^r are the same as for a hard or soft wedge. This result should be expected since L^i and L^r depend only on geometrical parameters and not on the boundary conditions.

According to (1.212), the value of ρ_e in an arbitrary plane of diffraction is given by

$$\frac{1}{\rho_e} = \frac{1}{\rho^i} + \frac{\mathbf{n}\cdot\left(\mathbf{s}^i - \mathbf{s}\right)}{a\sin^2\beta_0}, \tag{3.183}$$

where $\mathbf{n}$ is the unit vector of the principal normal to the edge at the diffraction point Q and where $\mathbf{s}^i$ and $\mathbf{s}$ are, respectively, the unit vectors of the incident and diffracted rays at Q. Also, ρ^i is the radius of curvature of the incident wave front in the plane of incidence, and a is the radius of curvature of the edge at Q on the shadow boundary of the incident field $\mathbf{s}^i = \mathbf{s}$, and consequently

$$\frac{1}{\rho_e^i} = \frac{1}{\rho^i}. \tag{3.184}$$

Similarly, on the shadow boundary of the reflected field, $\mathbf{n}\cdot(\mathbf{s}^i - \mathbf{s}) = 2\mathbf{N}\cos\theta^i$, where $\mathbf{N}$ is the unit vector of the exterior normal to the face and θ^i is the angle of incidence. Then according to (3.183),

$$\frac{1}{\rho_e^r} = \frac{1}{\rho^i} - \frac{2\mathbf{n}\cdot\mathbf{N}\cos\theta^i}{a\sin^2\beta_0}. \tag{3.185}$$

The diffraction coefficient (3.162), established for a straight impedance wedge that is uniform through the transition region at the shadow boundaries of the surface waves, cannot be extended to a curved impedance wedge. The reason is that the surface waves on a curved surface change in nature. On a convex impedance surface they can be obtained by extending the exact asymptotic solution of the impedance cylinder, which is a canonical problem for such waves and is treated in chapter 2, section 2.3. A uniform excitation coefficient for a surface wave excited on a convex impedance surface, which remains valid through the shadow boundary of the surface wave, is derived in section 3.4.1.2. This solution can be extended to a surface wave excited at the edge of an impedance wedge with convex faces by applying the technique presented in chapter 5.

3.2.3. UNIFORM ASYMPTOTIC SOLUTIONS FOR THE ACOUSTIC FIELD SCATTERED BY A CONVEX WEDGE-SHAPED ELASTIC SHELL

We consider a convex wedge-shaped shell having elastic walls of constant thickness, separating a fluid from vacuum (fig. 3.9). In chapter 1, section 1.7, we have shown that the dominant term of the asymptotic expansion of the field reflected by a curved interface between a fluid and an elastic medium is obtained by multiplying the incident field by a reflection coefficient that is independent of the curvature of the interface. It can therefore be calculated by solving the problem of reflection of a sound wave by a planar interface. In the case of an elastic wall of constant thickness d separating two fluids, the same rule applies if the thickness is small compared to the wavelength. We suppose that this condition is satisfied. Thus the reflection coefficient $R^{\pm}$ on the face + or – of the wedge is given by

$$R^{\pm}(\alpha) = \frac{\sin\alpha - Z^{\pm}(\alpha)}{\sin\alpha + Z^{\pm}(\alpha)}, \tag{3.186}$$

where $Z^{\pm}(\alpha)$ is a quotient of two 4×4 determinants, the expression of which is given in appendix B.

When the medium inside the wedge is vacuum, no internal reflection between the two faces of the wedge can exist. The wedge-shaped shell behaves therefore like an impenetrable wedge. This behavior is also approximately verified when the interior of the wedge is filled with air and the outer medium is a more dense fluid like water.

The asymptotic solution of a wedge-shaped shell with convex faces separating a fluid from vacuum is therefore similar to that of a convex impedance wedge. In particular, away from the shadow boundaries of the direct, reflected, and surface waves, the total pressure field in the fluid has the same asymptotic form as (3.117). Moreover, in order to construct a uniform solution

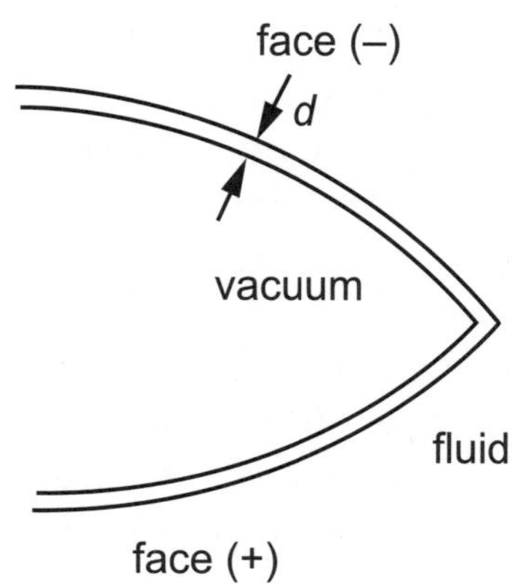

Figure 3.9. Wedge-shaped shell with convex faces.

valid in the transition regions and reducing to (3.117) outside these regions, we proceed as in section 3.2.2 by establishing first a uniform solution for the canonical problem represented by a wedge-shaped shell with planar faces separating the same media. Then a uniform solution for a wedge-shaped shell with convex faces is constructed by adapting the geometrical parameters of the solution of this canonical problem so that, outside the transition regions of the direct and reflected fields, it yields the nonuniform solution. As in the case of a convex impedance wedge, the edge-excited surface waves are treated in chapter 5.

3.2.3.1. Uniform Asymptotic Solution for a Wedge-Shaped Shell with Planar Faces

We assume that the surface impedance on each face of the wedge-shaped shell is the same as that of an infinite flat plate having the same thickness and separating the same media. As mentioned in the introduction of chapter 2, section 2.2.3, this property is only approximately verified since a coupling between both faces may take place close to the edge. However, this effect is negligibly small when the thickness of the walls is small compared to the wavelength in the fluid. When this condition is satisfied, the exact solution of the problem is the one derived in chapter 2, section 2.2.3. Only the asymptotic expansion of the solution will be reconsidered here.

According to (2.234), the integral form of the solution can be written in the general from (3.118). Applying the residue theorem, we obtain the exact expression of the diffracted field, which is again given by (3.119) and (3.120). Now if we set

$$\sin\theta(\alpha) = Z(\alpha), \tag{3.187}$$

then all the formulas from (3.121) to (3.129) remain unchanged, except that θ_0 and θ_n depend on the integration variable α. Consequently, (3.130) and (3.133) also remain unchanged, but θ_0 is now replaced with $\theta_0(\pi)$ and θ_n is replaced with $\theta_n(-\pi) = \theta_n(\pi)$ since $Z(\alpha)$ is an even function of α (see appendix B). In addition to these modifications, we must also keep in mind that $\psi(\alpha)$ is given by (2.218) and (2.222) and verifies the functional equation

$$\psi\left[\pm\left(\frac{n\pi}{2}+\alpha\right)\right] = -R_{\pm}(\alpha)\psi\left[\pm\left(\frac{n\pi}{2}-\alpha\right)\right], \tag{3.188}$$

where $R(\alpha)$ is the reflection coefficient. Equation (3.188) is essential for transposing the argument of $\psi(\alpha)$ in the fundamental domain $\left(-\frac{n\pi}{2},\frac{n\pi}{2}\right)$, where (2.222) is applicable. With these modifications, the uniform diffraction coefficient for a wedge-shaped shell with planar faces is given by

$$
\begin{aligned}
D(\phi,\phi';kr) = {} & \frac{\psi\left(\pi+\dfrac{n\pi}{2}-\phi\right)}{\psi\left(\dfrac{n\pi}{2}-\phi'\right)} \frac{\sin\dfrac{\phi'}{n}+\sin\dfrac{\theta_0(\pi)}{n}}{\sin\left(\dfrac{\phi-\pi}{n}\right)+\sin\dfrac{\theta_0(\pi)}{n}} \\
& \left(\overset{-}{d}\left(\overset{-}{\beta},n\right)F\left[K\overset{-}{a}\left(\overset{+}{\beta}\right)\right]+\Gamma_n\left(\phi',\theta_0(\pi)\right)\overset{-}{d}\left(\overset{+}{\beta},n\right)F\left[K\overset{-}{a}\left(\overset{+}{\beta}\right)\right]\right) \\
& +\frac{\psi\left(-\pi+\dfrac{n\pi}{2}-\phi\right)}{\psi\left(\dfrac{n\pi}{2}-\phi'\right)} \frac{\sin\dfrac{\phi'}{n}+\sin\dfrac{\theta_n(\pi)}{n}}{\sin\left(\dfrac{\phi+\pi}{n}\right)+\sin\dfrac{\theta_n(\pi)}{n}} \\
& \left(\overset{+}{d}\left(\overset{-}{\beta},n\right)F\left[K\overset{+}{a}\left(\overset{-}{\beta}\right)\right]+\Gamma_n\left(\phi',\theta_n(\pi)\right)\overset{+}{d}\left(\overset{+}{\beta},n\right)F\left[K\overset{+}{a}\left(\overset{+}{\beta}\right)\right]\right),
\end{aligned}
\tag{3.189}
$$

where $d(\beta, n)$ is given by (3.131) and where $F(x)$ is the transition function with the same arguments as in (3.133). Although the terms in (3.189) look identical to those of (3.133), their explicit expressions are much more complicated. Especially, $\theta_0(\alpha)$ and $\theta_n(\alpha)$ are complicated functions of α (see appendix B), and the numerical calculation of $\psi(\alpha)$ is more involved.

The generalization of (3.189) to oblique incidence follows the same procedure as in chapter 2, section 2.2.3.3. The diffraction coefficient is deduced from (3.189) by replacing k with $k \sin \beta_0$ and replacing the surface impedance with the modified surface impedance given by (2.201).

3.2.3.2. Uniform Asymptotic Solution for a Three-Dimensional Wedge-Shaped Shell

The procedure for establishing a uniform asymptotic solution, valid through the shadow boundaries of the incident and reflected fields of a three-dimensional wedge-shaped elastic shell separating a fluid from vacuum, is similar to that presented in sections 3.2.2.1 and 3.2.2.2. It consists of modifying the distance parameter in the transition function $F(Ka)$ in such a way that the jump of the direct or reflected fields through the shadow boundaries is exactly compensated by a jump with an opposite sign of the diffracted field. Since the term $d(\beta, n)F(Ka)$, responsible for the jump in the diffracted field, is independent of the boundary conditions, it can be readily shown that (3.177) and (3.178) still hold. Moreover, according to (3.127) and (3.128), which are derived from (3.188), equations (3.179) and (3.180) are still verified. Hence the distance parameters are independent of the boundary conditions and are given by (3.181) and (3.182). Consequently, the uniform diffraction coefficient is given by (3.173), in which θ_0 and θ_n are replaced by $\theta_0(\pi)$ and $\theta_n(\pi)$, respectively.

3.3. THE THREE-DIMENSIONAL SMOOTH CONVEX SURFACE

In chapter 1, section 1.9 and chapter 2, section 2.3.1.7, we derived a nonuniform solution for the acoustic field scattered by a three-dimensional soft, hard, or impedance surface. This solution is valid in certain subdomains of space when the source and the observation point are both far from the surface. A graphical representation of the different subdomains is shown in figure 3.10.

GA is valid in region I, whereas region III corresponds to the deep shadow beyond the surface boundary layer V. Region II is the transition zone between I and III and is divided into two parts by the shadow boundary. Regions IV and VI are the intersections of the transition zone II with the boundary layer in the vicinity of the surface. The nonuniform solutions derived in chapter 1, section 1.9 and chapter 2, section 2.3.1.7 are valid in regions I and III when the observation point is located far from the surface boundary layers VI and V, respectively.

Our objective is to extend these solutions to the transition region II. The method used is based on the solution of the canonical problem of a circular cylinder subjected to the sound field of a line source or a plane wave (treated in chapter 2, section 2.3). It consists of seeking a uniform solution of the circular cylinder problem by first transforming the exact integral form of the solution derived in chapter 2, section 2.3.1.1 in a manner such that the special functions appear in the description of the field in the transition zone. To extend this solution to the case of an arbitrary cylinder, one modifies the arguments of these functions in a way such that one reproduces the known asymptotic solutions in regions I and III, while simultaneously imposing the continuity of the solution in the shadow boundary. A further extension to a three-dimensional convex surface is then performed by the same constraints after introducing into the solution the divergence factors of the creeping rays on the surface and the divergence factors of the spatial waves that leave the surface tangentially.

3.3.1. UNIFORM ASYMPTOTIC SOLUTION THROUGH THE SHADOW BOUNDARY OF A SMOOTH CONVEX TWO-DIMENSIONAL SURFACE

A uniform asymptotic solution for this problem has been derived in reference [3.17] for a perfectly conducting cylinder, illuminated by an electromagnetic plane wave. This solution has been

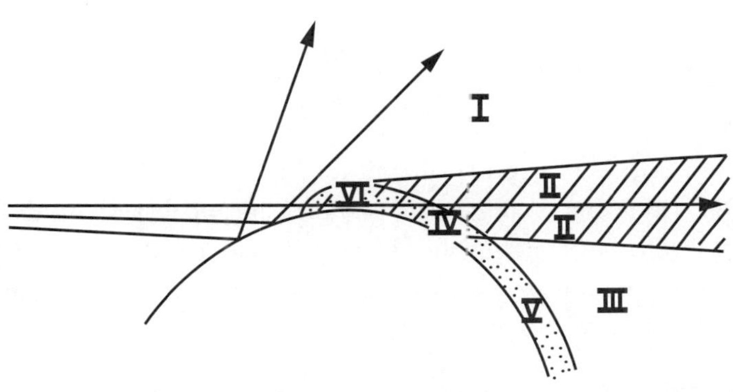

Figure 3.10. Boundary layers in the diffraction by a smooth convex impenetrable surface.

extended in reference [3.1] to an imperfectly conducting two-dimensional and three-dimensional smooth convex surface. The method used in these references and the formulas obtained can be readily transposed to acoustics. However, for completeness, we present again in this section the details of the solution, directly adapted to an acoustic impedance surface that contains, as special cases, the soft and hard boundary conditions.

3.3.1.1. Uniform Asymptotic Solution for a Circular Impedance Cylinder

We consider a normally incident acoustic plane wave striking an infinite circular cylinder with an acoustic surface impedance Z. Following the same procedure as in chapter 2, section 2.3.1, the total pressure field outside the cylinder is given by the solution of the Helmholtz equation, submitted to the boundary conditions on the surface and the radiation condition at infinity. As in the case of a line source, this problem can be solved by the method of separation of the variables, which leads to an infinite series involving Bessel and Hankel functions. By applying Watson's transformation followed by the Imai transform (2.279) and proceeding as in chapter 2, section 2.3.1.2, the series is transformed into the sum of two integrals $p_c + p$, where p_c gives the part of the contribution of the creeping waves, which is continuous through the shadow boundary, and where p is given by

$$p(\rho,\theta) = p_0 \int_D e^{i\nu\psi} \left[J_\nu(k\rho) - \frac{\Omega J_\nu(ka)}{\Omega H_\nu^{(1)}(ka)} H_\nu^{(1)}(k\rho) \right] d\nu, \tag{3.190}$$

where D is the contour $-\infty + i\varepsilon,\ \infty + i\varepsilon$, shown in figure 2.29; p_0 denotes the complex amplitude of the incident plane wave; and $\psi = |\theta| - \frac{\pi}{2}$ with $|\theta| < \pi$ or $|\psi| < \frac{\pi}{2}$, ψ being negative in the region directly submitted to the incident sound wave. The operator Ω is defined by (2.257).

Replacing $J_\nu(z)$ with a sum of Hankel functions (see equation (2.262)), the solution (3.190) can be written as

$$p(\rho,\theta) = \frac{1}{2} p_0 \int_D e^{i\nu\psi} \left[H_\nu^{(2)}(k\rho) - \frac{\Omega H_\nu^{(2)}(k\rho)}{\Omega H_\nu^{(1)}(k\rho)} H_\nu^{(1)}(k\rho) \right] d\nu. \tag{3.191}$$

In the transition zone II of figure 3.10, and for large ka, the contribution of the integrals (3.190) and (3.191) is more important when ν is of the same order as ka. As in chapter 2, section 2.3.1.2, this can be justified by replacing the Hankel functions occurring in (3.191) with the first term of their Debye asymptotic expansion. When this is done, a stationary phase point in the integrand (3.191) can be derived from

$$\psi + 2\operatorname{Arc}\cos\frac{\nu}{ka} - 2\operatorname{Arc}\cos\frac{\nu}{k\rho} = 0. \tag{3.192}$$

Equation (3.192) is equivalent to (2.305) for $r \to \infty$. Assuming that $\rho >> a$, we have $\operatorname{Arc\,cos}\frac{\nu}{k\rho} \cong \frac{\pi}{2}$. Hence when the observation point tends to the shadow boundary $\psi \to \frac{\pi}{2}$, the solution ν_s of (3.192) tends to *ka*. For *ka* >> 1, the Debye expansions of $H_\nu^{(1,2)}(ka)$ remain valid very close to ν = *ka*, and consequently the most important contribution to the integral comes from the neighborhood of ν = *ka*, where the saddle point is located. The same argument applies to the integral (3.190). Inside the transition region, this leads to the following change of variables:

$$\nu = ka + m\tau, \quad m = \left(\frac{ka}{2}\right)^{1/3}. \tag{3.193}$$

Now we replace the Bessel and Hankel functions by their expansions of the Olver type. For the Hankel function $H_\nu^{(1)}(ka)$, for instance, we have (see equation (2.328))

$$H_\nu^{(1)}(ka) \approx \frac{2}{m} e^{-i\frac{\pi}{3}} A_i(y), \tag{3.194}$$

and since

$$A_i(\sigma) = \frac{1}{2\pi i} \int_{-i\infty}^{+i\infty} e^{\sigma t - \frac{t^3}{3}} dt, \tag{3.195}$$

we obtain

$$A_i\left(e^{2i\frac{\pi}{3}}\tau\right) = \frac{e^{-2i\frac{\pi}{3}}}{2\pi i} \int_{\infty e^{i\frac{\pi}{6}}}^{\infty e^{-5\frac{\pi}{6}}} e^{\eta\tau - \frac{\eta^3}{3}} d\eta. \tag{3.196}$$

In (3.195) and (3.196), the contour of integration can be deformed upon recognizing that the behavior of the integrand as $|\xi| \to \infty$, where ξ states for *t* or η, is determined by $\exp\left(-\xi^2/3\right)$. Hence the integrals converge when $R_e(\xi^3) > 0$. This is verified in the following sectors of the complex σ-plane:

$$|\arg\sigma| < \frac{\pi}{6}, \quad \frac{\pi}{2} < \arg\sigma < \frac{5\pi}{6}, \quad -\frac{\pi}{2} > \arg\sigma > -\frac{5\pi}{6}.$$

We can therefore deform the contour of integration of (3.196) into the path $(-\Gamma_1)$ of figure 3.11.

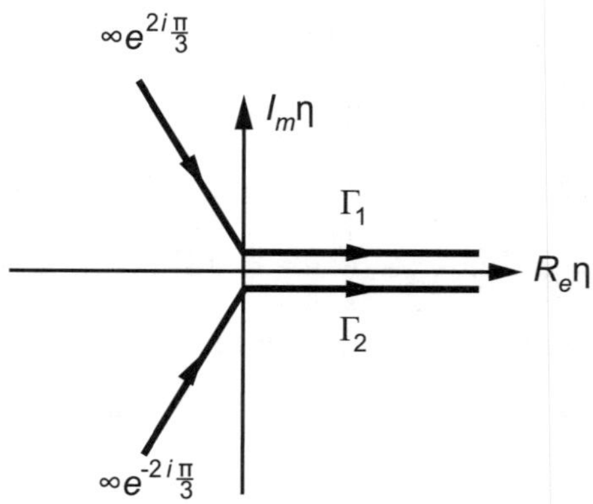

Figure 3.11. Integration paths for the Airy functions.

Inserting (3.196) into (3.194) we finally obtain, after deformation of the contour of integration,

$$H^{(1)}_{\nu(\tau)}(ka) = -\frac{i}{m\pi}\int_{\Gamma_1} e^{\eta\tau - \frac{\eta^3}{3}}\, d\eta = -\frac{i}{m\sqrt{\pi}} w_1(\tau), \tag{3.197}$$

where $w_1(\tau)$ is the Miller-type Airy function defined by

$$\mathrm{w}_1(\tau) = \frac{1}{\sqrt{\pi}}\int_{\Gamma_1} e^{\eta\tau - \frac{\eta^3}{3}}\, d\eta. \tag{3.198}$$

Similarly, starting with the Olver type asymptotic expansion of $H^{(2)}_{\nu}(ka)$, we obtain

$$H^{(2)}_{\nu(\tau)}(ka) = \frac{i}{m\sqrt{\pi}}\mathrm{w}_2(\tau), \tag{3.199}$$

where $w_2(\tau)$ is the Miller-type Airy function defined by

$$w_2(\tau) = \frac{1}{\sqrt{\pi}}\int_{\Gamma_2} e^{\eta\tau - \frac{\eta^3}{3}}\, d\eta, \tag{3.200}$$

where the contour Γ_2 is represented in figure 3.11.

Furthermore, replacing $J_\nu(x)$ with

$$J_\nu(x) = \frac{1}{2}\left[H^{(1)}_\nu(x) + H^{(2)}_\nu(x)\right], \tag{3.201}$$

we obtain

$$J_{\nu(\tau)}(ka) = \frac{1}{m\sqrt{\pi}} \frac{w_1(\tau) - w_2(\tau)}{2i} = \frac{1}{m\sqrt{\pi}} V(\tau). \tag{3.202}$$

We now express the integrals (3.190) and (3.191) in terms of the Miller-type Airy functions. Inserting (3.197), (3.199), and (3.202) into (3.190) and (3.191), we get

$$p(\rho,\theta) = m\, p_0 \int_{-\infty+i\varepsilon}^{+\infty+i\varepsilon} \left[J_{\nu(\tau)}(k\rho) - i \frac{\tilde{\Omega} V(\tau)}{\tilde{\Omega}\, w_1(\tau)} H^{(1)}_{\nu(\tau)}(k\rho) \right] e^{i(ka+m\tau)\psi} d\tau, \tag{3.203}$$

$$p(\rho,\theta) = \frac{m}{2} p_0 \int_{-\infty+i\varepsilon}^{+\infty+i\varepsilon} \left[H^{(1)}_{\nu(\tau)}(k\rho) + \frac{\tilde{\Omega}\, w_2(\tau)}{\tilde{\Omega}\, w_1(\tau)} H^{(1)}_{\nu(\tau)}(k\rho) \right] e^{i(ka+m\tau)\psi} d\tau, \tag{3.204}$$

where $\psi = |\theta| - \frac{\pi}{2}$ and

$$\tilde{\Omega} = -\frac{1}{m}\frac{\partial}{\partial \tau} + iZ. \tag{3.205}$$

It is now possible to express $p(r, \theta)$ by combining (3.203) and (3.204). Observing that the transformation that enables us to go from (3.203) to (3.204) is only applied to the integrand and not on the contour of integration, it follows that the integrals in (3.203) and (3.204) are identical along an arbitrary path of integration. In particular, the integral in (3.203) along the path C_1 starting at $-\infty + i\varepsilon$ and ending at $O + i\varepsilon$ is identical to the integral in (3.204) along the same path. Likewise, the two preceding integrals are identical along the path C_2, starting at $O + i\varepsilon$ and ending at $\infty + i\varepsilon$. Thus the field $p(r, \theta)$ may be expressed as the sum of the integral in (3.203) over the contour C_2 and the integral in (3.204) over the contour C_1.

After rearrangement of terms and use of (3.201), we obtain

$$p = I_1 + I_2, \tag{3.206}$$

where

$$I_1 = m\, p_0 \int_{C_1+C_2} J_{\nu(\tau)}(k\rho)\, e^{i\nu(\tau)\psi} d\tau - \frac{m\, p_0}{2} \int_{C_1} H^{(1)}_{\nu(\tau)}(k\rho)\, e^{i\nu(\tau)\psi} d\tau, \tag{3.207}$$

$$\begin{aligned} I_2 = {} & -im\, p_0 \int_{C_2} \frac{\tilde{\Omega} V(\tau)}{\tilde{\Omega}\, w_2(\tau)} H^{(1)}_{\nu(\tau)}(k\rho)\, e^{i\nu(\tau)\psi} d\tau \\ & + \frac{m\, p_0}{2} \int_{C_1} \frac{\tilde{\Omega}\, w_2(\tau)}{\tilde{\Omega}\, w_1(\tau)} H^{(1)}_{\nu(\tau)}(k\rho)\, e^{i\nu(\tau)\psi} d\tau, \end{aligned} \tag{3.208}$$

and $v(\tau) = ka + m\tau$.

We observe that I_1 is completely independent of the boundary conditions on the cylinder. To asymptotically evaluate the first integral in (3.207), we proceed as in chapter 2, section 2.3.1.2. Using (3.201), replacing the Hankel functions with their Debye approximations (2.285), and applying the steepest descent method, we obtain from the $H_\nu^{(2)}(k\rho)$ term the same expression for the saddle point condition as (2.289) with $\alpha_2 = \frac{\pi}{2}$. Hence the saddle point is at $\alpha_1 = \frac{\pi}{2} - |\theta|$. Similarly, for the $H_\nu^{(1)}(k\rho)$ term, the saddle point is found at $\alpha = |\theta| - \frac{\pi}{2}$. Evaluating the integral with both terms, it is seen that the result constitutes the incident field

$$m\, p_0 \int_{C_1+C_2} J_{\nu(\tau)}(k\rho)\, e^{i\nu(\tau)\psi} d\tau \cong p_0\, e^{-ik\rho\cos\theta}, \quad |\theta| < \pi. \tag{3.209}$$

The remaining integrals in (3.207) and (3.208) are treated separately in the shadowed portion of the transition region II of figure 3.10 and in the portion of region II situated in the region reached by the incident pressure field.

3.3.1.1.1. Part of the Transition Region Located in the Shadow

We consider again the term I_1. According to (3.209), the contribution of the first integral in (3.207) is also valid in the shadow region. We may therefore incorporate (3.209) in (3.207). The contribution of the second integral in (3.207) could also be calculated by the steepest descent method after replacing $H_\nu^{(2)}(k\rho)$ with its Debye approximation. However, since the contour C_1 has an end point at $O + i\varepsilon$, we also have to take into account the end point contribution in addition to the saddle point contribution.

Another method to asymptotically evaluate the second integral in (3.207) consists of replacing $H_\nu^{(1)}(k\rho)$ with its integral representation

$$H_{\nu(\tau)}^{(1)}(k\rho) = \frac{1}{\pi} \int_C e^{-ik\rho\sin\beta + i\nu(\tau)\beta}\, d\beta, \tag{3.210}$$

where the contour C is shown in figure 3.12. The resulting integral is evaluated by the method of steepest descent. This method is more accurate since an exact expression for the Hankel function is used. Inserting (3.210) into the second integral (3.207), we obtain the following after integrating over τ and adding the result (3.309):

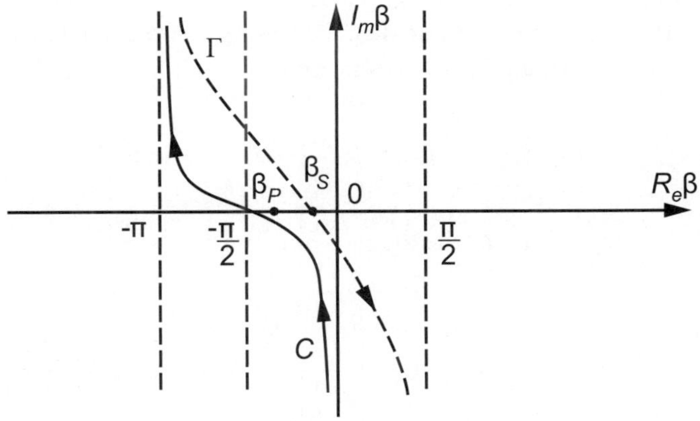

Figure 3.12. Contour of integration *C* and SDP *Γ*.

$$I_1 \approx p_0 e^{-ik\rho\cos\theta} - p_0 \frac{e^{ika\psi}}{2\pi i} \int_C \frac{e^{-ik\rho\left(\sin\beta - \frac{a}{\rho}\beta\right)}}{\psi + \beta} d\beta. \tag{3.211}$$

The integrand in (3.211) possesses a saddle point at $\beta = \beta_s = -\mathrm{Arc}\cos\frac{a}{\rho}$ and a pole at $\beta = \beta_p = -\psi$. On the shadow boundary, we have $\beta_s = -\psi_s$ or $\beta_s = \frac{\pi}{2} - |\theta_s|$, where $|\theta_s|$ is the value of $|\theta|$ at the shadow boundary. It follows that when $|\theta|$ tends to $|\theta_s|$, the pole and the stationary point coalesce. Furthermore, since $\psi > \psi_s$ in the shadow region, we have $\beta_p < \beta_s$ and $\phi < 0$ in the shadow region, where

$$\phi = \beta_p - \beta_s = \theta_s - |\theta| = -(\psi + \beta_s). \tag{3.212}$$

The domains in which $I_m\left(\sin\beta - \frac{a}{\rho}\beta\right) < 0$ when $I_m\beta \to +\infty$ are $-\frac{3\pi}{2} < R_e\beta < \frac{\pi}{2}$ and $\frac{\pi}{2} < R_e\beta < \pi$, whereas when $I_m\beta \to -\infty$, they are $-\frac{\pi}{2} < R_e\beta < \frac{\pi}{2}$. Thus one may deform the path C in (3.211) into the SDP Γ through β_s, as shown in figure 3.12. Since $\beta_p < \beta_s$ in the shadow region, the pole at β_p is crossed in this contour deformation. Therefore, we need to add to the integral along Γ the residue $-p_0 \exp(ik\rho\sin\beta)\, U(-\phi) = -p_0 \exp(-ik\cos\theta)\, U(|\theta| - \theta_s)$, where $U(x)$ is a step function whose value is unity for $x > 0$ and zero for $x < 0$.

The asymptotic evaluation of the integral along the SDP, with a pole close to the saddle point, follows the same method as the one described in section 3.2.1.1.1. Choosing the Pauli-Clemmow method and performing the transformation

$$q(\beta) = -i\left(\sin\beta - \frac{a}{\rho}\beta\right) = q(\beta_s) - \sigma^2, \quad \sigma \text{ real}, \tag{3.213}$$

we obtain the following expression for the the zero-order term:

$$\int_{\Gamma} \frac{-ik\rho\left(\sin\beta - \frac{a}{\rho}\beta\right)}{\psi + \beta} d\beta \cong \frac{-ik\rho\left(\sin\beta_s - \frac{a}{\rho}\beta_s\right)}{\beta_s - \beta_p} F\left(-ik\rho\, \sigma_1^2\right)\left(\frac{d\beta}{d\sigma}\right)_{\sigma=0}, \tag{3.214}$$

where $F(x)$ is the Kouyoumjian transition function defined by (3.17) and where

$$-i\sigma_1^2 \cong -i\left(q(\beta_p) - q(\beta_s)\right) \cong -i\,\frac{\left(\beta_p - \beta_s\right)^2}{2} q''(\beta_s) = \frac{\phi^2}{2\rho}\sqrt{\rho^2 - a^2}. \tag{3.215}$$

Since

$$q(\beta) - q(\beta_s) = -\sigma^2 \cong \frac{\left(\beta - \beta_s\right)^2}{2} q''(\beta_s),$$

we have

$$\beta - \beta_s = \sigma\sqrt{-\frac{2}{q''(\beta_s)}}$$

and

$$\left(\frac{d\beta}{d\sigma}\right)_{\sigma=0} = \sqrt{\frac{2}{\left|q''(\beta_s)\right|}}\; e^{i\alpha}, \tag{3.216}$$

where α is the phase of $d\beta / d\sigma$, which is identical to the phase of $d\beta$ since $d\sigma$ is positive along the path of integration. Thus $\alpha = -\frac{\pi}{4}$.

Inserting (3.215) and (3.216) into (3.214) and taking into account (3.212), after adding to (3.214) the residue at the pole, we obtain the following asymptotic approximation for the integral I_1:

$$I_1 \cong p_0 \frac{e^{ika|\phi|}}{\sqrt{2\pi k\,|\phi|}} e^{i\frac{\pi}{4}} F\left(kL\tilde{a}\right) \frac{e^{iks}}{\sqrt{s}}, \tag{3.217}$$

where ϕ is given by (3.212) and

$$L = \sqrt{\rho^2 - a^2} = s, \tilde{a} = \frac{\phi^2}{2}. \tag{3.218}$$

The geometrical parameters appearing in (3.217) are represented in figure 3.13.

Using (3.202), we replace $w_2(\tau)$ in the second integral in (3.208) to obtain

$$\begin{aligned} I_2 = -im\, p_0 \int_{C_1+C_2} \frac{\tilde{\Omega} v(\tau)}{\tilde{\Omega} w_1(\tau)} H^{(1)}_{v(\tau)}(k\rho)\, e^{iv(\tau)\psi} d\tau \\ + \frac{m\, p_0}{2} \int_{C_1} H^{(1)}_{v(\tau)}(k\rho)\, e^{iv(\tau)\psi} d\tau. \end{aligned} \tag{3.219}$$

Approximating the Hankel function $H^{(1)}_{v}(k\rho)$ by the dominant term of its Debye asymptotic expansion given by (2.285) and noting that in the transition region the major contribution to the integrals is for τ small, thus permitting to approximate sin γ in (2.285) by s/ρ, we obtain

$$I_2 = -p_0 \frac{e^{i\frac{\pi}{4}}}{\sqrt{2\pi k}} \frac{e^{ika\phi}}{\phi} \frac{e^{iks}}{\sqrt{s}} - m\, p_0 \sqrt{\frac{2}{k}}\, e^{ika\phi}\, P(\xi) \frac{e^{iks}}{\sqrt{s}}, \tag{3.220}$$

where $P(\xi)$ are generalized Pekeris functions defined by

$$P(\xi) = \frac{e^{i\frac{\pi}{4}}}{\sqrt{\pi}} \int_{-\infty}^{+\infty} \frac{\tilde{\Omega} v(\tau)}{\tilde{\Omega}\, w_1(\tau)} e^{i\xi\tau}\, d\tau \tag{3.221}$$

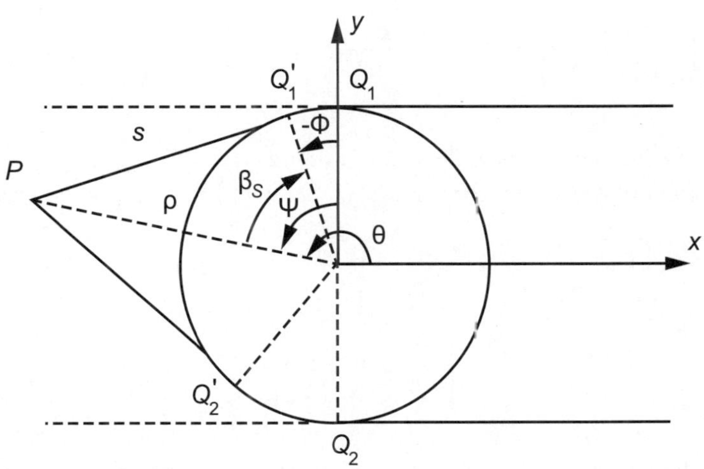

Figure 3.13. Geometrical parameters in the shadowed part of the transition region.

with

$$\xi = m\phi \tag{3.222}$$

in the shadow region.

For soft or hard boundary conditions, $P_{s,h}(\xi)$ are Pekeris caret functions tabulated by Logan [3.18] (see also reference [3.19]). Combining the results of (3.217) and (3.220) yields the following for an observation point in the shadow region:

$$p(\rho,\theta) = -p_0\, m \sqrt{\frac{2}{k}}\, e^{ikt} \left[\frac{e^{i\frac{\pi}{4}}}{2\xi\sqrt{\pi}} \left(1 - F\left[kL\tilde{a}\right]\right) + P(\xi) \right] \frac{e^{iks}}{\sqrt{s}}, \tag{3.223}$$

where

$$t = a\phi,\ \xi = m\phi. \tag{3.224}$$

This result may be interpreted geometrically as a field that propagates to the observation point P along the path $Q_1Q_1'P$ after being launched by the incident wave at Q_1. Noting that p_0 is the complex amplitude of the incident field with a phase reference at the origin, we see that p_0 is also equal to the incident field at Q_1. Thus p_0 in (3.223) must be interpreted locally to be $p^i(Q_1)$.

In the deep shadow region far from the shadow boundary, $kL\tilde{a}$ becomes large and $F\left(kL\tilde{a}\right) \to 1$. Hence only the $P(\xi)$ term in (3.223) is significant.

Using (3.202) in (3.221), we obtain the following in the shadow region:

$$P(\xi) = \frac{e^{i\frac{\pi}{4}}}{\sqrt{\pi}} \int_{-\infty+i\varepsilon}^{\infty+i\varepsilon} \frac{\tilde{\Omega} V(\tau)}{\tilde{\Omega}\, \mathrm{w}_1(\tau)} e^{i\xi\tau}\, d\tau = -\frac{e^{-i\frac{\pi}{4}}}{2\sqrt{\pi}} \int_{-\infty+i\varepsilon}^{\infty+i\varepsilon} \frac{\tilde{\Omega}\mathrm{w}_2(\tau)}{\tilde{\Omega}\, \mathrm{w}_1(\tau)} e^{i\xi\tau}\, d\tau,$$

where $0 << \varepsilon << 1$. We may now deform the contour of integration into a contour C surrounding the poles of the integrand, which are the zeros τ_n of $\tilde{\Omega}\, \mathrm{w}_1(\tau)$, and apply the residue theorem to get

$$P(\xi) = -\sqrt{\pi}\, e^{i\frac{\pi}{4}} \sum_{n=1}^{\infty} \frac{\tilde{\Omega}\mathrm{w}_2(\tau_n)}{\left[\dfrac{\partial}{\partial\tau} \tilde{\Omega}\, \mathrm{w}_1(\tau)\right]_{\tau=\tau_n}} e^{i\xi\tau_n}. \tag{3.225}$$

According to the general formula for the diffracted field in the shadow region given by (2.321), we have the following for a plane wave at normal incidence:

$$D_s D_n^2 = m\sqrt{\frac{2\pi}{k}}\, e^{i\frac{\pi}{4}} R_n,$$

where

$$D_s = \frac{e^{i\frac{\pi}{4}}}{\sqrt{8\pi k}}, \quad R_n = \frac{\tilde{\Omega}\, \mathrm{w}_2(\tau_n)}{\left[\frac{\partial}{\partial \tau}\tilde{\Omega}\, \mathrm{w}_1(\tau)\right]_{\tau=\tau_n}}. \tag{3.226}$$

Hence

$$D_n^2 = 4\pi m\ R_n. \tag{3.227}$$

It can be shown that D_n^2 is identical to the diffraction coefficient given by (2.330). The demonstration of this result is as follows.

Since τ_n is a solution of $\tilde{\Omega}\, \mathrm{w}_1(\tau) = 0$, we have

$$iZ = \frac{1}{m}\frac{\mathrm{w}_1'(\tau_n)}{\mathrm{w}_1(\tau_n)}, \tag{3.228}$$

and by using the Wronskian $W(w_1, w_2) = 2i$, we obtain

$$\tilde{\Omega}\, \mathrm{w}_2(\tau_n) = -\frac{1}{m\mathrm{w}_1(\tau_n)}\left[\mathrm{w}_1(\tau_n)\mathrm{w}_2'(\tau_n) - \mathrm{w}_1'(\tau_n)\mathrm{w}_2(\tau_n)\right] = -\frac{2i}{m\mathrm{w}_1(\tau_n)}. \tag{3.229}$$

Inserting (3.229) and (3.228) into (3.226) yields

$$R_n = -\frac{2i}{\mathrm{w}_1'^2(\tau_n) - \mathrm{w}_1(\tau_n)\,\mathrm{w}_1'(\tau_n)}.$$

According to (3.197) and (2.328), we have

$$\mathrm{w}_1(\tau) = 2\sqrt{\pi}\, e^{i\frac{\pi}{6}} A_i(y),$$

where

$$y = \frac{e^{-i\frac{\pi}{3}}}{m}(ka - \nu) = e^{2i\frac{\pi}{3}}\tau.$$

Since $A''(y) = yA(y)$, we obtain

$$w_1'^2(\tau_n) - w_1(\tau_n)\, w_1'(\tau_n) = 4\pi e^{i\frac{\pi}{3}} \left(\left[A_i'(y_n)\right]^2 - y_n\left[A_i(y_n)\right]^2 \right) \left(\frac{dy}{d\tau}\right)^2$$

and finally

$$D_n^2 = 4\pi m\ R_n = \frac{2^{2/3}\, e^{-i\pi/6} (ka)^{1/3}}{\left[A_i'(y_n)\right]^2 - y_n\left[A_i(y_n)\right]^2}, \quad y_n = e^{2i\frac{\pi}{3}}\tau_n. \tag{3.230}$$

Consequently, in the deep shadow region, the solution obtained in the shadow zone of the transition region (given by equation (3.223)) tends to the nonuniform creeping wave solution in the deep shadow region derived in chapter 2, sections 2.3.1.2 and 2.3.1.3 and limited to creeping waves reaching the observation point by propagating less than one tour around the cylinder. It is important to observe that the field of creeping waves traveling one tour or more around the cylinder is continuous through the shadow boundary and has no transition region.

3.3.1.1.2. Part of the Transition Region Reached by the Incident Field

The results given for I_1 and I_2 in (3.217) and (3.220), respectively, can be extended in the part of the transition region reached by the incident field. This is accomplished by extending the orthogonal coordinates σ and $\sigma + s$ with $\sigma > 0$ to negative values of σ and by adding the term $p_0 \exp(-ik\rho\cos\theta) = p_i(\rho, \theta)$. We can do this because $\beta_p > \beta_s$, and consequently the pole at β_p is not crossed by the contour deformation $C \to \Gamma$ (see fig. 3.12). We may therefore write the following at the observation point $P(\rho, \theta)$:

$$p(\rho,\theta) \approx p^i(P) - p_0 m \sqrt{\frac{2}{k}}\, e^{ikt} \left[\frac{e^{i\frac{\pi}{4}}}{2\xi\sqrt{\pi}} \left(1 - F\left[kL\tilde{a}\right]\right) + P(\xi) \right] \frac{e^{iks}}{\sqrt{s}}, \tag{3.231}$$

where the term $p_i(P)$ is now added to (3.223) and where t and ξ are now negative. The second term in (3.231), which supplements $p_i(Q)$, may be interpreted as a scattered field that, after being launched at Q, propagates backward from Q to Q' before being shed tangentially from Q' to P, as illustrated in figure 3.14.

We observe that this "pseudo-ray" path is different from the GA reflected ray path, which satisfies the generalized Fermat principle. It can be shown from reference [3.20] that the function

$$Q(\xi) = P(\xi) + \frac{e^{i\frac{\pi}{4}}}{2\xi\sqrt{\pi}}$$

is continuous at $\xi = 0$.

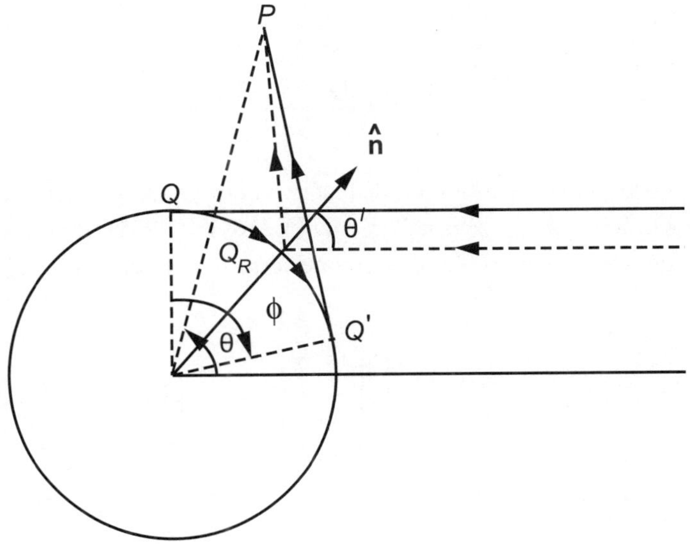

Figure 3.14. Reflected and pseudo-ray paths.

Hence the solution (3.231) is continuous through the shadow boundary. However, (3.231) does not reduce to the GA ray field, far from the shadow boundary. A way to match both solutions consists of modifying the parameter $\xi = -m\phi$ in (3.231). According to (3.212), we have

$$\phi = \beta_p - \beta_s = -\psi - \beta_s, \tag{3.232}$$

where $\beta_s = -\text{Arc}\cos\dfrac{a}{\rho}$.

In the far zone $\rho >> a$, $\beta_s \cong \dfrac{\pi}{2}$, and $\varphi \cong \dfrac{\pi}{2} - \psi$. Furthermore, if θ^i is the angle of reflection, we have for ρ large, $\theta = 2\theta^i$. Hence $\phi = 2\left(\dfrac{\pi}{2} - \theta^i\right)$ and ξ reduces to ξ', given by

$$\xi' = -2m\,\cos\theta^i. \tag{3.233}$$

In order to introduce in (3.231) the parameters of the GA reflected ray, we define ξ' by (3.233) and expand ϕ in a Taylor series with respect to ξ'. Denoting $\overline{\gamma} = \dfrac{\pi}{2} - \theta^i$, we have

$$\phi = 2\overline{\gamma}, \tag{3.234}$$

$$\xi' = -2m\sin\overline{\gamma} = -2m\cos\theta^i, \tag{3.235}$$

and for $|\xi|/2m<1$,

$$-\bar{\gamma}=\operatorname{Arc}\sin\frac{\xi'}{2m}=\frac{1}{2m}\xi'+\frac{1}{6}\left(\frac{1}{2m}\right)^{3}\xi'^{3}+\dots\,. \tag{3.236}$$

We limit the Taylor series expansion to the first two nonzero terms and introduce the expression of ϕ, given by (3.234), where $\bar{\gamma}$ is replaced by (3.236) in (3.231). By keeping two terms of (3.236) in the exponential exp (ikt) and only one term in the remainder of the expression, we obtain

$$p(\rho,\theta)\approx p^{i}(P)-p_{0}\,m\sqrt{\frac{2}{k}}\,e^{-2ika\cos\theta^{i}}\,e^{i\frac{\xi'^{3}}{12}}\left[\frac{e^{i\frac{\pi}{4}}}{2\xi'\sqrt{\pi}}\left(1-F\left[kL\tilde{a}'\right]\right)+P(\xi')\right]\frac{e^{iks}}{\sqrt{s}}, \tag{3.237}$$

where

$$\tilde{a}'=\frac{\phi^{2}}{2}=\frac{\xi'^{2}}{2m^{2}}=2\cos^{2}\theta^{i}, \tag{3.238}$$

$L=s$, and $s^{2}=\rho^{2}-a^{2}=\ell^{2}+2a\ell\cos\theta^{i}$, where ℓ is the distance Q_RP from the reflection point Q_R to the observation point P.

In the far zone ($\rho>>a$), $s\cong\ell$, and

$$\frac{e^{iks}}{\sqrt{s}}\cong\frac{e^{ik\left(\ell+a\cos\theta^{i}\right)}}{\sqrt{\ell}}. \tag{3.239}$$

Inserting (2.239) into (2.237) and replacing p_0 with $p(Q_R)e^{-ika\cos\theta^{i}}$, we obtain

$$p(P)=p^{i}(P)+p^{i}(Q_{R})\left(-m\sqrt{\frac{2}{k}}\,e^{i\frac{\xi'^{3}}{12}}\left[\frac{e^{i\frac{\pi}{4}}}{2\xi'\sqrt{\pi}}\left(1-F\left[kL\tilde{a}'\right]\right)+P(\xi')\right]\right)\frac{e^{ik\ell}}{\sqrt{\ell}}. \tag{3.240}$$

The divergence factor of the reflected wave, given by the GA, is

$$A=\sqrt{\frac{\rho_{c}}{\rho_{c}+\ell}}, \tag{3.241}$$

where $\rho_c=\dfrac{a\cos\theta^{i}}{2}$ is the reflected ray caustic distance. For $\ell>>a$, (3.241) reduces to

$$A \cong \sqrt{\frac{\rho_c}{\ell}} = \frac{1}{\sqrt{\ell}}\sqrt{\frac{a\cos\theta^i}{2}}. \tag{3.242}$$

By replacing $\frac{1}{\sqrt{\ell}}$ in (3.240) with $\sqrt{\frac{\rho_c}{\ell}}\sqrt{\frac{2}{a\cos\theta^i}}$, the solution (3.240) can also be written in the following form:

$$p(P) = p^i(P) + p^i(Q_R)\left(-\sqrt{-\frac{4}{\xi'}}\, e^{i\frac{\xi'^3}{12}}\left[\frac{e^{i\frac{\pi}{4}}}{2\xi'\sqrt{\pi}}\left(1 - F\left[kL\tilde{a}'\right]\right) + P(\xi')\right]\right)\sqrt{\frac{\rho_c}{\ell}}\, e^{ik\ell}. \tag{3.243}$$

When the observation point crosses the shadow boundary, the term T, given by

$$\mathrm{T} = \frac{e^{i\frac{\pi}{4}}}{2\xi'\sqrt{\pi}} F\left[kL\tilde{a}'\right],$$

has a jump that can be calculated using (3.24), where s_1 states for σ_1 and where, according to (3.215), we have $kL\tilde{a}' = -ik\rho\sigma_1^2 = k\frac{\phi^2}{2}\sqrt{\rho^2 - a^2}$. At the shadow boundary, $\theta^i = \frac{\pi}{2}$ and consequently $\phi = 0$. It follows that σ_1 and $\xi' = -2m\cos\theta^i = m\phi$ both pass through zero at the shadow boundary. Hence when ξ' varies from 0^- to 0^+, we obtain the following result for the jump $\Delta\mathrm{T}$ by taking the limit when $\phi \to 0$:

$$\Delta\mathrm{T} = \frac{e^{i\frac{\pi}{4}}}{2\xi'\sqrt{\pi}}\,\delta F\left[kL\tilde{a}'\right] = \frac{1}{m}\sqrt{\frac{k}{2}}\left(\rho^2 - a^2\right)^{1/4},$$

where we have replaced ξ' with $m\phi$, θ' with $\frac{\pi}{2}$, and used (3.238). Finally, by taking into account (3.241), the jump of the diffracted field is given by

$$p_i(Q_R)\sqrt{-\frac{4}{\xi'}}\sqrt{\frac{\rho_c}{\ell}}\,\Delta\mathrm{T}\, e^{ik\ell} = p_i(Q_R)\, e^{ik\sqrt{\rho^2 - a^2}} = p^i(P).$$

Consequently, the pressure field is continuous through the shadow boundary. It can also be shown that, far from the shadow boundary, (3.243) reduces to the GA field. As a matter of fact, for $|\xi'| >> 1$, $F\left[k L\tilde{a}'\right] \to 1$. Furthermore, $P(\xi')$ is asymptotically approximated for $\xi' < 0$ and $|\xi'| >> 1$, as [3.19]

$$P\left(\xi\right) \cong -R\sqrt{-\frac{\xi'}{4}}\, e^{-i\frac{\xi'^3}{12}},$$

where R is the reflection coefficient. Consequently, by replacing $A \cong \sqrt{\frac{\rho_c}{\ell}}$ by its expression (3.241), we recover the well-known formula of GA:

$$p(P) \cong p^i(P) + p^i(Q_R) R \sqrt{\frac{\rho_c}{\rho_c + \ell}}\, e^{ik\ell}. \tag{3.244}$$

3.3.1.2. Uniform Asymptotic Solution for a Smooth Convex Impedance Cylinder

The generalization of the uniform solution obtained in section 3.3.1.1 for a circular cylinder to a convex cylinder of slowly varying curvature, at normal incidence, is accomplished by treating each point on the convex cylinder as if it were locally on a circular cylinder of the same radius of curvature. Thus the radius a in the solution of the previous section is replaced by the local radius of curvature $a(t)$ of the convex cylinder for normal incidence $a(t) = \rho_g(t)$.

In the shadow region, the various parameters occurring in (3.223) are generalized in the following way:

$$\xi = \int_Q^{Q'} \frac{m(t')}{\rho_g(t')} dt', \quad t = \int_Q^{Q'} dt', \quad m(t') = \left[\frac{k\rho_g(t')}{2}\right]^{1/3}, \tag{3.245}$$

and

$$L = s, \quad \tilde{a} = \frac{\xi^2}{2m(Q)m(Q')}, \tag{3.246}$$

where Q is the launching point of the creeping wave and where Q' is the point of tangential shedding of the diffracted field from the surface toward the observation point P. Moreover, the Fock parameter m in (3.223) is replaced by $\sqrt{m(Q)m(Q')}$ to be consistent with reciprocity. This symmetrical splitting of m is also essential for preserving the uniform property of the result so that in the deep shadow region it reduces to the GA result. With these modifications, the uniform solution in the shadow region for a smooth convex impedance cylinder is given by

$$p(P) \cong p^i(Q)\left(-\sqrt{m(Q)m(Q')}\, e^{ikt}\sqrt{\frac{2}{k}}\left[\frac{e^{i\frac{\pi}{4}}}{2\xi\sqrt{\pi}}\left(1-F\left[kL\tilde{a}'\right]\right)+P(\xi)\right]\right)\frac{e^{iks}}{\sqrt{s}}. \tag{3.247}$$

This result is valid in the shadowed part of region II, excluding region IV, and also in region III (see fig. 3.10).

In the region reached by the incident wave, the parameters in (3.243) take the generalized form

$$\xi' = -2m(Q_R)\cos\theta^i, \quad m(Q_R) = \left[\frac{k\rho_g(Q_R)}{2}\right]^{1/3} \tag{3.248}$$

and

$$L = \ell, \quad \tilde{a}' = 2\cos^2\theta^i, \tag{3.249}$$

where ℓ is the distance from the reflection point Q_R to the observation point P. With these modifications, the uniform solution in the region reached by the incident wave, generalized to a convex impedance cylinder, is given by

$$p(P) \cong p^i(Q) + p^i(Q_R)\left(-\sqrt{-\frac{4}{\xi'}}\, e^{i\frac{\xi'^3}{12}}\left[\frac{e^{i\frac{\pi}{4}}}{2\xi'\sqrt{\pi}}\left(1-F\left[kLa'\right]\right)+P(\xi')\right]\right)\sqrt{\frac{\rho_c}{\rho_c+\ell}}\, e^{ik\ell}, \tag{3.250}$$

where the reflected ray caustic distance is now given, more generally, by

$$\rho_c = \frac{\rho_g(Q_R)\cos\theta^i}{2}. \tag{3.251}$$

This result is valid in region I and also in the portion of region II reached by the incident wave, which excludes region VI.

Expressions (3.247) and (3.250) remain valid at oblique incidence, provided that the impedance Z in the Pekeris function $P(\xi)$ be replaced by $Z' = Z/\sin\beta$. Also, $\xi' = -2m(Q_R)\cos\vartheta^i$, where θ^i is the angle of reflection of the oblique ray, the reflection angle θ_t^i of the projected ray on a plane perpendicular to the generatrices being related to θ^i by $\cos\theta_t^i = \cos\theta^i / \sin\beta$.

3.3.1.3. Field in the Boundary Layer of a Smooth Convex Impedance Cylinder

The region of validity of (3.247) and (3.250) excludes the boundary layer and its vicinity in the shadow region. When the field point is very close to the surface of a circular cylinder, such that $\rho - a << a$, the asymptotic approximation (3.197) and (3.199) take the forms

$$H_\nu^{(1)}(k\rho) \cong -\frac{i}{m\sqrt{\pi}} \mathrm{w}_1(\tau'),$$

$$H_\nu^{(2)}(k\rho) \cong \frac{i}{m\sqrt{\pi}} \mathrm{w}_2(\tau'),$$

where $\nu = k\rho + m\tau'$. Keeping the transformation $\nu = ka + m\tau$, we see that $\tau' = \tau - \frac{kn}{m}$, where $n = \rho - a$ is the normal distance of the observation point to the surface. We may therefore approximate $J_\nu(k\rho)$ and $H_\nu^{(1)}(k\rho)$ in (3.203) by

$$H_\nu^{(1)}(k\rho) \cong -\frac{i}{m\sqrt{\pi}} \mathrm{w}_1(\tau - h),$$
$$J_\nu(k\rho) \cong \frac{i}{m\sqrt{\pi}} V(\tau - h), \tag{3.252}$$

where

$$h = \frac{kn}{m}.$$

Inserting these results into (3.203) yields the following expression for the field in the boundary layer:

$$p(\rho,\theta) \cong \frac{e^{ika}}{\sqrt{\pi}} p_0 \int_{-\infty+i\varepsilon}^{+\infty+i\varepsilon} \left[V(\tau - h) - \frac{\tilde{\Omega} V(\tau)}{\tilde{\Omega}\, \mathrm{w}_1(\tau)} w_1(\tau - h) \right] e^{i\, m\psi\tau}\, d\tau, \tag{3.253}$$

where $\psi = |\theta| - \frac{\pi}{2}$, as before.

The circular cylinder result in (3.253) may now be generalized to an arbitrary convex cylinder case by replacing $\psi = \frac{m}{a}$ with ξ and replacing ka with t, where ξ and t are defined in (2.245), with Q' replaced by the projection P_N of the observation point P on the surface. We observe also that the arc length integrals for ξ and t are evaluated along the surface geodesic from Q to P_N in the shadow region. Moreover, since p_0 in the circular cylinder case is also the field at the launching point Q of the creeping wave, it has to be replaced by $p^i(Q)$ in (3.253). With these modifications, the field in the boundary layer close to the shadow boundary of a smooth convex impedance surface is given by

$$p(P) = p^i(Q)\frac{e^{ikt}}{\sqrt{\pi}} \int_{-\infty+i\varepsilon}^{+\infty+i\varepsilon} \left[V(\tau - b) - \frac{\tilde{\Omega} V(\tau)}{\tilde{\Omega}\, w_1(\tau)} w_1(\tau - b) \right] e^{i\,\xi\tau}\, d\tau, \tag{3.254}$$

where $\tilde{\Omega}$ is given by (3.205).

3.3.2. *UNIFORM ASYMPTOTIC SOLUTION THROUGH THE SHADOW BOUNDARY OF A SMOOTH CONVEX THREE-DIMENSIONAL SURFACE*

A uniform asymptotic solution for a perfectly conducting three-dimensional convex surface, illuminated by an electromagnetic plane wave, has been reported in reference [3.21]. An extension of this solution to an imperfectly conducting three-dimensional convex surface is presented in reference [3.1].

These solutions can be readily adapted to the diffraction of an acoustic plane wave by a general convex, hard, soft, or impedance surface. In this section, the formulas for the acoustic three-dimensional case are derived by generalizing those obtained in section 3.3.1 for a smooth convex two-dimensional surface.

Following the method described in the introduction of section 3.3, we heuristically extend the uniform solution for a circular cylinder (derived in section 3.3.1) to a general three-dimensional convex surface. We do this by modifying the arguments of the Pekeris function $P(\xi)$ and the transition function $F(x)$ in such a way that the total field is continuous through the shadow boundary and tends to the GA field outside the transition region, in the domain reached by the incident wave. In this operation, we must take into account the divergence factors of the creeping rays along the surface and the divergence factors of the space rays away from the surface, as in chapter 2, section 2.3.2.

We consider first the case of a torsionless geodesic. Starting from the solution obtained in section 3.3.1.2, we write the following for the three-dimensional case:

$$p(P) = p^i(Q) \left(-\sqrt{m(Q)m(Q')}\, e^{ikt} \sqrt{\frac{2}{k}} \left[\frac{e^{i\frac{\pi}{4}}}{2\xi\sqrt{\pi}} \left(1 - F\left[k L^d \tilde{a}\right]\right) + P(\xi) \right] \right) \sqrt{\frac{d\Sigma(Q)}{d\Sigma(Q')}} \sqrt{\frac{\rho_c}{s(\rho_c + s)}}\, e^{iks}, \tag{3.255}$$

where P is an observation point located in the shadowed part of the transition region and where Q is the launching point of the creeping ray that follows the geodesic QQ' and sheds energy tangentially at Q', along the space ray $Q'P$ of length $s = Q'P$. It is important to observe that the trajectory of the ray and the divergence factors are determined independently by applying the Fermat principle (see chapter 1). The convex surface is a caustic for the diffracted rays. Hence the radius of curvature of the diffracted wave front at Q' is zero in the plane perpendicular to the plane formed by the binormal to the geodesic at Q' and the diffracted ray $Q'P$. However, the radius of curvature at Q' of this wave front in the plane formed by the binormal and $Q'P$ is equal to the distance ρ_c from Q' to the other branch of the caustic of the diffracted rays. The distance ρ_c to the caustic is also the radius of curvature of the wave front at Q' of the creeping wave launched by the incident plane wave at Q. The divergence factor of the creeping rays at an arbitrary point M on the geodesic QQ' is defined by the square root of the quotient of the cross sections $d\Sigma(Q)$ and $d\Sigma(Q')$ of the tube of surface rays at Q and Q', respectively.

The other parameters occurring in (3.255) are generalized, as in (3.245). The Pekeris and transition functions $P(\xi)$ and $F(x)$ are identical to those introduced in the two-dimensional case. However, the parameter L^d in the argument of F will be modified. As shown later on in this chapter, its value will be determined by imposing the continuity condition of the total field across the shadow boundary.

In the region where the incident wave exists, the total field is given by (3.250), in which the divergence factor $\sqrt{\rho_c / \rho_c + \ell}$ must be replaced by

$$A = \sqrt{\frac{\rho_1^r \, \rho_2^r}{\left(\rho_1^r + s_1\right)\left(\rho_2^r + s_1\right)}},$$

where $s_1 = Q_R P$ and ρ_1^r, ρ_2^r are the principal radii of curvature of the reflected wave at the reflection point Q_R. This modification is necessary since the field in the region reached by the incident field must tend to the GA field when the observation point is far from the shadow boundary. Hence (3.250) is modified as follows:

$$p(P) = p^i(P) + p^i(Q_R)\left(-\sqrt{-\frac{4}{\xi'}}\, e^{\frac{i\xi'^3}{12}} \left[\frac{e^{i\frac{\pi}{4}}}{2\xi'\sqrt{\pi}}\left(1 - F\left[kL^r\tilde{a}'\right]\right) + P(\xi')\right] A e^{iks_1}\right), \quad (3.256)$$

where s_1 is the distance from the reflection point Q_R to the observation point P and where ξ' is given by (3.248).

When the observation point leaves the transition region, we have $F(x) \to 1$ and, as shown in section 3.3.1.1, the Pekeris function behaves like

$$P(\xi) \cong -R\sqrt{-\frac{\xi'}{4}}\, e^{\frac{-i\,\xi'^3}{12}},$$

where R is the reflection coefficient. Consequently, (3.256) tends to the GA field as it should. Hence it remains only to impose the continuity of the field across the shadow boundary. Since the term

$$P(x) + \frac{e^{i\frac{\pi}{4}}}{2x\sqrt{\pi}}$$

is well defined at $x = 0$, owing to the fact that the singular part of $P(x)$ compensates exactly the singularity of the remainder of this term (see reference [3.20]), only the term

$$\mathrm{T} = \frac{e^{i\frac{\pi}{4}}}{2\xi'\sqrt{\pi}}\, F\left(k L^r \tilde{a}'\right)$$

has a jump when the observation point crosses the shadow boundary. Taking (3.24) into account and noting that $\tilde{a}' = 2\cos^2\theta^i$, we obtain

$$\Delta\mathrm{T} = \frac{1}{m(Q_R)}\sqrt{\frac{k L^r}{2}}\, e^{ik L^r \tilde{a}'}.$$

On the shadow boundary, we have

$$\rho_1^r = \frac{\rho_g(Q')\cos\theta^i}{2}, \quad \rho_2^r = \rho_2^i.$$

Consequently,

$$\sqrt{\rho_1^r} = \sqrt{\frac{\rho_g}{m(Q_R)}}\sqrt{-\frac{\xi'}{4}} = m(Q_R)\sqrt{\frac{2}{k}}\sqrt{-\frac{\xi'}{4}}$$

since at the shadow boundary Q_R and Q' coalesce. Finally, by inserting these relations into (3.256), we see that the jump $\delta p(P)$ of the pressure field is equal to

$$\delta_p(P) = -p^i(Q_R)\sqrt{L^r}\sqrt{\frac{\rho_2^r}{(\rho_2^r + s_1)s_1}}.$$

The jump of the incident field is given by

$$\delta p_i = -p^i(Q_R)\sqrt{\frac{\rho_1^i \rho_2^i}{(\rho_1^i + s_1)(\rho_2^i + s_1)}}.$$

Hence the solution (3.256) is continuous if we choose

$$L^r = \frac{\rho_1^i \rho_2^i}{(\rho_1^i + s_1)(\rho_2^i + s_1)} \frac{s_1(\rho_2^r + s_1)}{\rho_2^r}.$$

Similarly, by applying the same procedure to (3.255), we see that the jump of $p(P)$, when P crosses the shadow boundary on the side of the shadow zone, is equal to

$$\delta p(P) = -p^i(Q)\sqrt{L^d}\sqrt{\frac{\rho}{s(\rho + s)}}.$$

In order for this jump to compensate for the jump of the incident field, we must have

$$L^d = \frac{\rho_1^i \rho_2^i}{(\rho_1^i + s)(\rho_2^i + s)} \frac{s_1(\rho + s)}{\rho}.$$

For an incident plane wave, we obtain $L^r = L^d = s$.

Inserting the expressions derived for L^d and L^r into (3.255) and (3.256), respectively, we obtain two asymptotic solutions that match on the shadow boundary. The solution so defined is continuous on the shadow boundary, but its derivative with respect to the observation angle is generally discontinuous.

We consider now a creeping wave propogating along a geodesic with torsion. Since the uniform solutions (3.247) and (3.250) for oblique incidence on a smooth convex cylindrical impedance surface do not depend explicitly on the angle of incidence β, the same property will hold on a three-dimensional surface approximated locally by a circular cylindrical surface of radius $a(t)$ defined by

$$\frac{\sin^2 \beta(t)}{a(t)} = \frac{1}{\rho_g(t)},$$

where t is the abscissa of a point Q on the geodesic and where $\beta(t)$ is the angle between the geodesic and the principal direction $\mathbf{t}_1$, corresponding to the largest principal radii of curvature $a_1(t)$ of the surface at Q.

It is important to observe that $a(t) < a_2(t)$, where $a_2(t)$ is the radius of curvature in the direction orthogonal to $\mathbf{t}_1$, since

$$\frac{1}{\rho_g(t)} = \frac{\cos^2\beta(t)}{a_1(t)} + \frac{\sin^2\beta(t)}{a_2(t)}.$$

However, $a(t)$ is close to $a_2(t)$ when $a_1(t)$ is large or when $\beta(t)$ approaches $\frac{\pi}{2}$.

In the frame of the local cylindrical approximation described previously, the uniform solutions (3.255) and (3.256) remain valid for the general case of creeping waves, propagating along a geodesic with torsion on a three-dimensional impedance surface. Particularly, the parameters L^d and L^r are given by the same formulas, and the integrand of the Pekeris function is unchanged and given by

$$P(x) = -\frac{e^{-i\frac{\pi}{4}}}{2\sqrt{\pi}} \int_{-\infty+i\varepsilon}^{\infty+i\varepsilon} \frac{v'(\tau) - imZv(\tau)}{w_1'(\tau) - imZw_1(\tau)} e^{ix\tau}\, d\tau,$$

where $m = m(t) = \left[k\rho_g(t)/2\right]^{1/3}$ is a function of the position of the point $Q(t)$ on the geodesic.

In the transition region, $a(t)$ can be replaced by its value $a(0)$ at Q. The same approximation holds for $\beta(t) \cong \beta(0)$. Hence ρ_g and consequently m are constant when the observation point is located in the transition region. Outside the transition region, the dependence of m on the position on the geodesic needs to be taken into account in the residue series expression of $P(x)$, in order to recover the nonuniform solution established in section 3.3.1 for the deep shadow region. In the region reached by the incident field, the parameter m must be calculated at the reflection point Q_R.

3.4. THE THREE-DIMENSIONAL SMOOTH CONVEX SHELL

In chapter 2, section 2.3.3, we have derived a nonuniform solution for a hollow elastic circular cylinder. We have also given some extensions of the solution to a noncircular cylindrical smooth convex shell and a three-dimensional smooth convex shell. All solutions presented thus far are nonuniform through the shadow boundaries of the incident and reflected waves and also nonuniform through the shadow boundaries of the elastic surface waves. In this chapter, we first derive a uniform solution through the forementioned shadow boundaries for the canonical problem of a circular hollow elastic cylinder excited by a plane wave at normal incidence, treated in chapter 2, section 2.3.3. Then we extend this solution to oblique incidence and to a general three-dimensional smooth convex shell.

3.4.1. UNIFORM SOLUTION FOR A HOLLOW ELASTIC CYLINDER

As in chapter 2, section 2.3.3, we consider a hollow elastic circular cylinder embedded in a fluid. The cylinder is closed and the medium inside the cylinder is vacuum. A plane wave is incident on the cylinder in a direction normal to its axis. Three types of waves are considered (fig. 3.15):

- the reflected wave
- the elastic surface waves launched at Q_1, Q_2, … , Q_1', Q_2', where the direction of propagation of the incident field makes a critical angle with the normal to the surface and circumvents the cylinder in both directions
- creeping waves launched at T_1 and T_1'

The asymptotic solutions derived in chapter 2, section 2.3.3 are nonuniform in narrow zones that lie in the vicinity of the shadow boundaries T_1 and T_1' of the direct and reflected waves and are

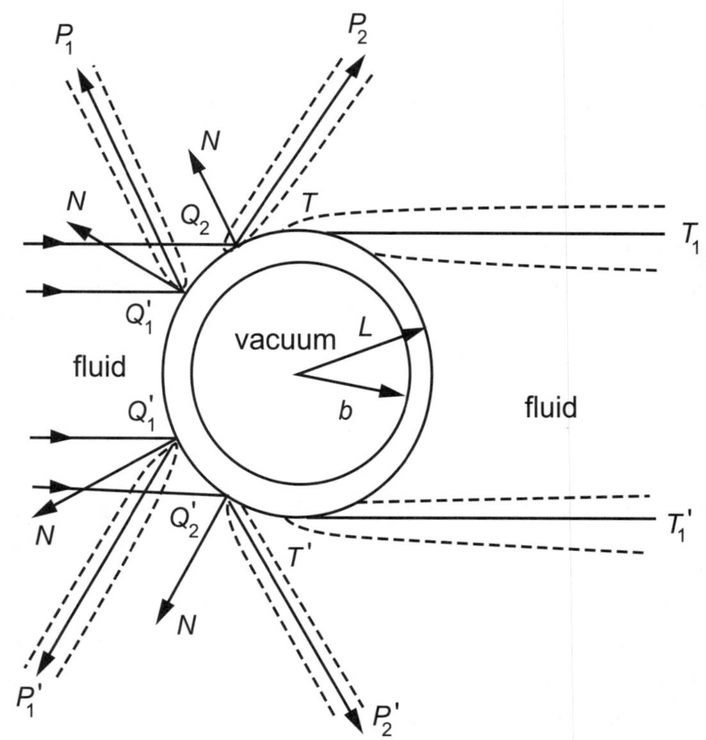

Figure 3.15. Shadow boundaries on a circular hollow cylinder excited by a plane wave.

also nonuniform in the vicinity of the shadow boundaries $P_1, P_2, \ldots$, and $P_1', P_2', \ldots$, of the elastic surface waves (see fig. 3.15).

3.4.1.1. Uniform Solution for the GA Field Associated with the Creeping Wave Field

According to (2.35), (2.36), and (2.37), the contributions of the geometric direct and reflected waves are contained in

$$p_2 = \frac{1}{2} p_0 \int_{C_0} e^{+i\nu\left(\frac{\pi}{2}-\theta\right)} \left[H_\nu^{(2)}(k_1 r) - \frac{\Omega_\nu H_\nu^{(2)}(k_1 a)}{\Omega_\nu H_\nu^{(1)}(k_1 a)} H_\nu^{(1)}(k_1 r) \right] d\nu, \tag{3.257}$$

whereas the contribution of the creeping waves is partially contained in (3.257), augmented by the series

$$p_c = \pi p_0 \sum_{n=1}^{\infty} e^{i\nu_n \frac{\pi}{2}} \frac{\cos\nu_n(\pi-\theta)}{\sin\nu_n \pi} \frac{\Omega_{\nu_n} H_{\nu_n}^{(2)}(k_1 a)}{\left[\frac{\partial}{\partial\nu}\Omega_\nu H_\nu^{(1)}(k_1 a)\right]_{\nu=\nu_n}} H_{\nu_n}^{(1)}(k_1 r). \tag{3.258}$$

We have shown in chapter 2, section 2.3.2.4 that the series (3.258) gives the creeping wave contribution for an observation point located outside the shadow region. It also gives the creeping wave contribution in the shadow region, except for the creeping wave having the shortest path length, which is absent in (3.258) but contained in (3.257). As a consequence, p_c is continuous through the shadow boundaries. In order to get a uniform solution for the direct and reflected field associated with the creeping waves, we must therefore construct a uniform asymptotic expansion of the integral (3.257). Apart from the fact that the equivalent impedance is now a function of ν, the integral (3.257) is identical to (3.190). Hence by introducing into the integrand the same asymptotic expansions of the Hankel functions as in section 3.3.1 and by following the same procedure as in that section, we obtain similar results.

In the shadowed part of the transition region close to the shadow boundary, a uniform solution for the creeping wave field is given by the same formula as (3.223):

$$p_2(\rho,\theta) = -p_0 m \sqrt{\frac{2}{k}}\, e^{ikt} \left[\frac{e^{i\frac{\pi}{4}}}{2\xi\sqrt{\pi}} \left(1 - F\left[k_1 L\tilde{a}\right]\right) - P(\xi) \right] \frac{e^{iks}}{\sqrt{s}}, \tag{3.259}$$

where

$$t = a\phi, \quad \tilde{a} = m\varphi, \quad L = \sqrt{\rho^2 - a^2} = s \tag{3.260}$$

and where the geometrical parameters s, ϕ are defined in figure 3.16.

The functions $F(X)$ and $P(X)$ are, respectively, the Kouyoumjian transition function defined by (3.17) and the generalized Pekeris function given by

$$\tilde{P}(X) = -\frac{e^{i\frac{\pi}{4}}}{2\sqrt{\pi}} \int_{-\infty+i\varepsilon}^{+\infty+i\varepsilon} \frac{\tilde{\Omega}_\tau V(\tau)}{\tilde{\Omega}_\tau \mathrm{w}_1(\tau)} e^{x\tau}\, d\tau, \tag{3.261}$$

where

$$\tilde{\Omega}_\tau = -\frac{1}{m} \frac{\partial}{\partial \tau} + i\eta(\tau). \tag{3.262}$$

In the deep shadow region, only the term $P(\xi)$ in (3.259) is significant, since $k L\tilde{a}$ becomes large and $F\left(kL\tilde{a}\right) \to 1$. Moreover, by using (3.202) we obtain the following in the shadow region:

$$\tilde{P}(\xi) = -\frac{e^{-i\frac{\pi}{4}}}{2\xi\sqrt{\pi}} \int_{-\infty+i\varepsilon}^{\infty+i\varepsilon} \frac{\tilde{\Omega}_\tau w_2(\tau)}{\tilde{\Omega}_\tau w_1(\tau)} e^{i\xi\tau}\, d\tau. \tag{3.263}$$

Finally, by applying the residue theorem, it follows that

$$\tilde{P}(\xi) = -\sqrt{\pi}\, e^{i\frac{\pi}{4}} \sum_{n=1}^{\infty} \frac{\tilde{\Omega}_n \mathrm{w}_2(\tau_n)}{\left[\frac{\partial}{\partial \tau}\tilde{\Omega}_n \mathrm{w}_1(\tau)\right]_{\tau=\tau_n}} e^{i\xi\tau_n} = -\sqrt{\pi}\, e^{i\frac{\pi}{4}} \left(\sum_{n=1}^{\infty} R_n e^{i\xi\tau_n}\right)$$

and

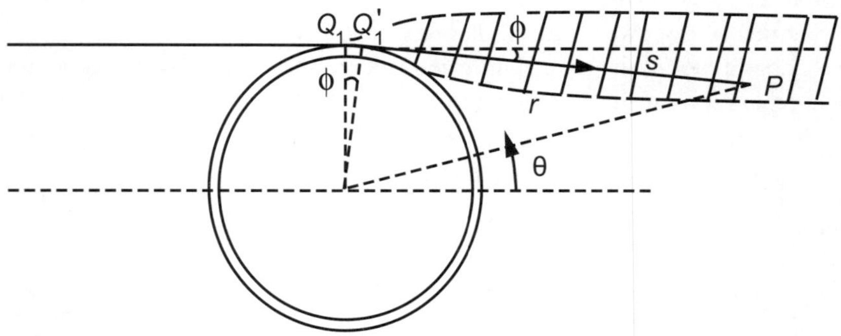

Figure 3.16. Geometrical parameters appearing in the formula for the field diffracted in the transition region by the creeping wave $Q_1Q_1'Q$.

$$p_2(\rho,\theta) \cong p_0\, m \sqrt{\frac{2\pi}{k_1}}\, e^{i\frac{\pi}{4}} \left(\sum_{n=1}^{\infty} R_n\, e^{i\xi\tau_n} \right) e^{ik_1 t}\, \frac{e^{ik_1 s}}{\sqrt{s}}. \tag{3.264}$$

Let $D_n^2 = 4\pi\, m\, R_n$. By noting that the different steps from (3.227) to (3.230) remain unchanged, we obtain

$$p_2(P) = p^i(Q_1) e^{ik_1 t} \left(\sum_{n=1}^{\infty} D_n^2\, e^{i\xi_n \tau_n} \right) \frac{e^{iks}}{\sqrt{s}}, \tag{3.265}$$

where D_n^2 is given by (3.230).

Formula (3.265) corresponds to the creeping wave that was missing in (3.258). Hence by adding (3.255) to (3.258), we obtain the complete creeping wave contribution derived earlier in the shadow region outside the transition region.

In the part of the transition region reached by the incident field we obtain, by following the same procedure as in 3.3.1.1,

$$p_2(\rho,\theta) \cong p^i(\rho,\theta) - p_0\, m \sqrt{\frac{2}{k_1}}\, e^{ik_1 t} \left[\frac{e^{i\frac{\pi}{4}}}{2\xi\sqrt{\pi}} \left(1 - F\left[k_1 L\tilde{a}\right]\right) + \tilde{P}(\xi) \right] \frac{e^{iks}}{\sqrt{s}}, \tag{3.266}$$

where t and ξ are now negative and where $\tilde{a}$ and L are still given by (3.260). The diffracted ray follows a pseudo-ray path, as shown in figure 3.17.

As in section 3.3.1, it can be shown [3.20] that the function

$$Q(\xi) = \tilde{P}(\xi) + \frac{e^{i\frac{\pi}{4}}}{2\xi\sqrt{\pi}} \tag{3.267}$$

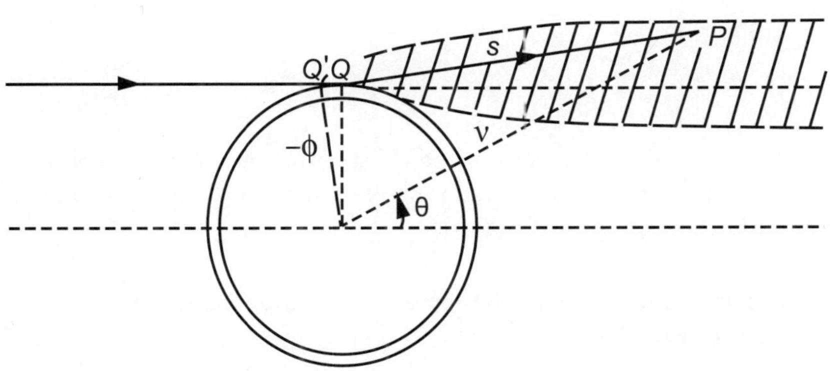

Figure 3.17. Pseudo-ray path *QQ′P*.

is continuous at $\xi = 0$. Furthermore, as in section 3.3.1.1, equation (3.231), the transition function $F\left[kL\tilde{a}\right]$ divided by ξ has a jump discontinuity when ξ passes through zero. This discontinuity multiplied by the rest of the coefficients in (3.266) is equal to $p^i(Q)$ when the observation point moves into the shadow region. Consequently, $p_2(\rho, \theta)$ is continuous through the shadow boundary. However, in section 3.3.1.1, the formula (3.266) does not reduce to the GA ray field, far from the shadow boundary, but with a slight modification of the parameters ξ and $\tilde{a}$ in the region reached by the incident field (see section 3.3.1.1), a uniform solution is obtained that tends to the GA field outside the transition region. The derivation of this solution follows the same steps as in section 3.3.1.1 and leads to

$$p_2(P) = p^i(P) + p^i(Q_R) - \left(\sqrt{-\frac{4}{\xi'}} e^{\frac{\xi'^3}{12}} \left[\frac{e^{i\frac{\pi}{4}}}{2\xi'\sqrt{\pi}} \left(1 - F\left[k_1 L\tilde{a}\right]\right) + \tilde{P}(\xi') \right] \right) \sqrt{\frac{\rho_c}{\ell}}\, e^{ik\ell}, \tag{3.268}$$

where

$$\xi' = -2m\cos\theta^i, \quad \tilde{a} = \frac{\phi^2}{2} \cong \frac{\xi'^2}{2m^2} = 2\cos^2\theta^i. \tag{2.269}$$

By considering the triangle QQ_RP in figure 3.14, where Q_R is the reflection point, the following relation follows:

$$s^2 = \rho^2 - a^2 = \ell^2 + 2a\,\ell\cos\theta.$$

Hence for $a/\ell << 1$ the distance s to the observation point P along the pseudo-ray can be replaced by $s \cong \ell$ in the amplitude term and $s \cong \ell + a\cos\theta^i$ in the phase term, where ℓ is the distance Q_RP along the reflected ray. Also, ρ_c is the reflected ray caustic distance given by

$$\rho_c = \frac{a\cos\theta^i}{2}. \tag{2.270}$$

As in section 3.3.1.1, it can be shown that $p_2(P)$ is continuous through the shadow boundary and that (3.268) tends to the GA field when the observation moves away from the shadow boundary in the region reached by the incident field.

3.4.1.2. Uniform Solution for the Reflected Field Associated with the Elastic Surface Wave Field

The solution presented in chapter 2, section 2.3.3.7 for the elastic surface waves is discontinuous at the launching points Q_ℓ and Q'_ℓ of these waves. It is therefore nonuniform in a neighborhood of the direction of reflection at those points called the transition region (see fig. 3.15). We suppose that the transition regions of the N surface waves, launched in both directions on a hollow cylinder, do not overlap.

Consider the contour C_s of figure 2.3 that passes through the saddle points ν_R and ν_D. In the neighborhood of ν_R and ν_D, C_s is supposed deformed into the SDP passing through these points (fig. 3.18).

Let ν_1 be one of the poles of the reflection coefficient. When the observation angle θ augments from 0 to π, the SDP moves from right to left and crosses the pole ν_1. When ν_1 is outside the contour C_s, an elastic surface wave exists, and its field is given by the residue of the integral p_1 (see equation (2.401)) at the pole ν_1. When ν_1 is inside the contour C_s, its contribution to the integral p_1 is zero, and the surface wave does not exist. Also, when ν_1 is close to the SDP, the asymptotic evaluation of the integral along $C_{s,}$ which gives the reflected field, is affected by the presence of the pole ν_1 of its integrand. Consequently, the expression of the reflected field is modified.

In order to construct a uniform asymptotic expansion of an integral having a pole close to the SDP passing through an isolated saddle point of its phase, two different methods have been presented in section 3.2.1.1.1—namely, the Pauli-Clemmow method and the Van der Waerden method. The method of Pauli-Clemmow is the easiest to apply when the residue function $T(\nu) = (\nu - \nu_1)\, R(\nu)$, where $R(\nu)$ is the reflection coefficient, varies slowly in the neighborhood of ν_1.

On elastic metallic surfaces, it has been found that $R(\nu)$ has a zero very close to its pole. Consequently, this condition is generally not satisfied, and the Van der Waerden method is the most appropriate one.

A detailed description of the Van der Waerden method is given in section 3.2.1.1.1. The main steps of this method are summarized in the following section.

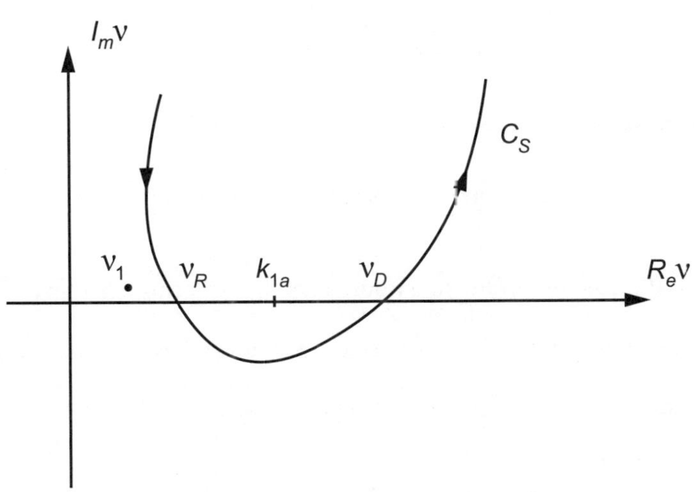

Figure 3.18. Deformation of C_s into the SDP in the vicinity of ν_R and ν_D.

3.4.1.2.1. Brief Description of Van der Waerden's Method

We consider an integral of the type

$$I(\Omega) + \int_{SDP} g(z,\alpha)\, e^{\Omega q(z)}\, dz,$$

where $\Omega > 0$ is a real number, large compared to unity, α is a complex valued parameter, and $g(z, \alpha)$, $q(z)$ are holomorphic functions of z. We assume that $q(z)$ has a stationary point of order 1 at z_s, which is a solution of $q'(z) = 0$ with $q''(z) \neq 0$. We also assume that the function

$$g(z,\alpha) = \frac{h(z,\alpha)}{z - z_0(\alpha)}$$

has a simple pole at $z = z_0(\alpha)$.

Next, we apply the following transformation:

$$q(z) = q(z_s) - s^2, \tag{3.271}$$

which transforms the SDP into the real axis of the complex s-plane, the stationary phase point z_s into $s = 0$, and the pole into $s = s_1(\alpha)$. The integral reads

$$I(\Omega,\alpha) = \int_{-\infty}^{+\infty} G(s,\alpha)\, e^{-\Omega s^2}\, ds,$$

where

$$G(s,\alpha) = g(z,\alpha)\frac{dz}{d\alpha}\, e^{\Omega q(z_s)}.$$

In the Van der Waerden method, we write

$$G(s) = \frac{A}{s - s_1} + H(s), \quad T(s) = (s - s_1)\, G(s),$$

where $H(s)$ and $T(s)$ are holomorphic functions. Then by replacing $H(s)$ by its Taylor series and inserting the corresponding expression for $G(s)$ into the integral, we obtain an asymptotic expansion of $I(\Omega,\alpha)$ that, at the order $0\left(\Omega^{-1/2}\right)$, takes the following form:

$$\overline{I}^a(\Omega) = \sqrt{\frac{\pi}{\Omega}}\,\frac{1}{s_1}\left\{T(s_1)\left[1 - F\left(-i\Omega s_1^2\right)\right] - T(o)\right\},$$

where $F(x)$ is the Kouyoumjian transition function given by (3.17).

It has been shown in section 3.2.1.1.1 that $\bar{I}^a(\Omega)$ has a jump when the pole crosses the SDP that compensates exactly the disappearance of the residue at that pole. A uniform asymptotic expansion $\bar{S}_u$ as a function of the location of the pole is obtained by adding $\bar{I}^a(\Omega)$ to the residue at the pole in the region where it exists:

$$\bar{S}_u = 2\pi i R_1 U\left[-R_e\left(-i\Omega s_1^2\right)\right] + \bar{I}^a(\Omega), \tag{3.272}$$

where

$$R_1 = \mathrm{T}(s_1)\, e^{-\Omega s_1^2}$$

is the residue of the integral at the pole and where $U(\cdot)$ is the Heaviside function. The line above the letter I and S indicates that the asymptotic expansion is limited to $0\left(\Omega^{-1/2}\right)$.

An alternative form of $\bar{I}^a(\Omega)$ is obtained by defining

$$\begin{aligned} V(s_1) &= -\frac{1}{s_1}\frac{1}{2\pi i}\sqrt{\frac{\pi}{\Omega}}\, F\left(-i\Omega s_1^2\right) e^{\Omega s_1^2}, \\ \hat{V}(s_1) &= -\frac{1}{s_1}\frac{1}{2\pi i}\sqrt{\frac{\pi}{\Omega}}\, e^{\Omega s_1^2}, \end{aligned} \tag{3.273}$$

where $\hat{V}(s_1)$ is the first term of the asymptotic expansion of $V(s_1)$.

Then the uniform solution takes the form

$$\bar{S}_u = 2\pi i R_1\left[V(s_1) - \hat{V}(s_1)\right] + I_{nu}(\Omega) + 2\pi i R_1 U\left[-R_e\left(-i\Omega s_1^2\right)\right], \tag{3.274}$$

where $I_{nu}(\Omega)$ is the asymptotic expansion of $I(\Omega)$, when the pole has no influence on the result given by

$$I_{nu}(\Omega) = \sqrt{\frac{\pi}{\Omega}}\, G(o) = -\sqrt{\frac{\pi}{\Omega}}\,\frac{1}{s_1}\,\mathrm{T}(o). \tag{3.275}$$

3.4.1.2.2. Application to the Hollow Cylinder

The part of the integral (2.36), giving the reflected field, reads

$$p_R = -\frac{p_0}{2} \int_{C_s} e^{+i\nu\left(\frac{\pi}{2}-\theta\right)} \frac{\Omega_\nu H_\nu^{(2)}(x_1)}{\Omega_\nu H_\nu^{(1)}(x_1)} H_\nu^{(1)}(k_1 r)\, d\nu, \tag{3.276}$$

where C_s is the contour shown in figure 2.3, passing through the saddle points ν_R and ν_D, in the vicinity of which it is identical to the SDP passing through these points. We have seen in chapter 2, section 2.3.2.3 that, for $R_e\nu < ka$, by replacing the Hankel functions with the dominant term of their asymptotic expansion, equation (3.276) can also be written as

$$p_R = \frac{p_0}{2}\sqrt{\frac{2}{\pi}} \int_{C_s} \frac{\psi'(x_1,\nu)-\eta_\nu}{\psi'(x_1,\nu)+\eta_\nu}\left[(k_1 r)^2 - \upsilon^2\right]^{-1/4} e^{iq(\nu)}\, d\nu, \tag{3.277}$$

where ψ is defined in (2.302) and where

$$\begin{aligned} q(\nu) &= \psi(k_1 r,\nu) - 2\psi(x_1,\nu) + \nu\left(\frac{\pi}{2}-\theta\right) \\ &= -\nu\theta + 2\nu \operatorname{Arc\,cos}\left(\frac{\nu}{x_1}\right) - 2\sqrt{x_1^2-\nu^2} + \left(\frac{\pi}{2} - \operatorname{Arc\,cos}\frac{\nu}{k_1 r}\right)\nu \\ &+\sqrt{(k_1 r)^2-\nu^2} + \frac{\pi}{4} = k_1 f(\nu) + \frac{\pi}{4}. \end{aligned} \tag{3.278}$$

Also, according to (2.443),

$$\frac{\psi'(x_1,\nu)-\eta_\nu}{\psi'(x_1,\nu)+\eta_\nu} = \frac{\sqrt{1-\left(\frac{\nu}{x_1}\right)^2} - \eta_\nu}{\sqrt{1-\left(\frac{\nu}{x_1}\right)^2} + \eta_\nu} = R(\nu). \tag{3.279}$$

In order to apply (3.274), we need to calculate the residue R_1 of (3.277) for $\nu = \nu_1$ and the parameter s_1. The residue of the integral (3.277) at the pole ν_1 is given by

$$R_1 = -\frac{p_0}{\sqrt{2\pi}} e^{i\frac{\pi}{4}}\, r_1\, e^{ik_1 f(\nu_1)}, \tag{3.280}$$

where

$$r_1 = \frac{2\sqrt{1-\left(\frac{\nu_1}{x_1}\right)^2}}{\left[\frac{\partial}{\partial\nu}\left(\sqrt{1-\left(\frac{\nu_1}{x_1}\right)^2} - \eta_\nu\right)\right]_{\nu=\nu_1}} \left[(k_1 r)^2 - \upsilon^2\right]^{-1/4}. \tag{3.281}$$

The expression of the parameter s_1 follows from (2.271), which according to the notations used in (3.278), gives

$$s_1^2 = i\left[f\left(v_R\right) - f\left(v_1\right)\right]. \tag{3.282}$$

Keeping the notations of section 3.2.1.1.2, we state

$$s_1^2 = i\,\bar{a}. \tag{3.283}$$

Hence

$$\begin{aligned} V\left(s_1\right) &= -\frac{1}{\sqrt{i\,\bar{a}}}\,\frac{1}{2\pi i}\sqrt{\frac{\pi}{k_1}}\;F\left(k_1\,\bar{a}\right)e^{ik_1\left[f\left(v_R\right)-f\left(v_1\right)\right]}, \\ \hat{V}\left(s_1\right) &= -\frac{1}{\sqrt{i\,\bar{a}}}\,\frac{1}{2\pi i}\sqrt{\frac{\pi}{k_1}}\;e^{ik_1\left[f\left(v_R\right)-f\left(v_1\right)\right]}. \end{aligned} \tag{3.284}$$

Inserting (3.280) and (3.284) into (3.274) yields the following:

$$p^u = \frac{p_0}{\sqrt{2\pi}}\,e^{i\frac{\pi}{4}}\sqrt{\frac{\pi}{k_1}}\;e^{i k_1 f\left(v_R\right)}\,\frac{r_1}{\sqrt{i\,\bar{a}}}\left[1 - F\left(k_1\,\bar{a}\right)\right] + I_{nu} + 2\pi i\,R_1 U\left(-R_e\bar{a}\right). \tag{3.285}$$

We have already shown in chapter 2, section 2.3.3.3 that the solution I_{nu} is just the GA field p_G. Consequently, at an observation point M in the transition region, the uniform solution for the reflected field associated with the elastic surface waves field is given by

$$p^u\left(M\right) = p_R^{(1)}\left(M\right) + p_G\left(M\right) + P_s\left(M\right)U\left(-R_e\bar{a}\right), \tag{3.286}$$

where $p_R^{(1)}$ is given by the first term on the right-hand side of (3.285), and P_s is the elastic surface wave field. The term $p_R^{(1)}$ enforces the continuity of the field p^u, as will be shown briefly now.

In the neighborhood of the saddle point, we have

$$f\left(v\right) = f\left(v_R\right) + \frac{1}{2}\left(v - v_R\right)^2 f''\left(v_R\right).$$

Hence by introducing the real and imaginary parts of v_1 denoted by v_{1r} and v_{1i}, it follows that

$$\bar{a} = -\frac{1}{2}\left(v - v_R\right)^2 f''\left(v_R\right) = -\frac{1}{2}f''\left(v_R\right)\left[\left(v_{1r} - v_R\right)^2 - v_{1i}^2 + 2i\,v_{1i}\left(v_{1r} - v_R\right)\right].$$

When the pole ν_1 is located outside the contour C_s as in figure 3.18, the elastic surface wave exists, and since the SDP makes the angle $-\pi/4$ with the real ν-axis, we have

$$\nu_{1r} < \nu_R, \quad \left(\nu_{1r} - \nu_R\right)^2 - \nu_{1i}^2 > 0.$$

Consequently, since $f''(\nu_R)$ is positive, we obtain

$$R_e\left(\bar{a}\right) > 0, \quad I_m\left(\bar{a}\right) < 0. \tag{3.287}$$

Recalling that $F(X)$, defined by (3.17), is defined on the upper Riemann sheet of the complex x-plane as

$$-\frac{\pi}{2} < \arg\left(X\right) < \frac{3\pi}{2}$$

with a cut along the negative imaginary axis, we see that the surface wave exists in the domain where the inequalities (3.287) are verified. When the pole crosses the SDP, the corresponding point in the complex $X = k_1\bar{a}$ plane crosses the cut in the direction from $R_e X < 0$ to $R_e X > 0$. Then $F\left(k_1\,\bar{a}\right)$ has a jump equal to (see equation (3.24))

$$-2\pi i \sqrt{\frac{k_1}{\pi}}\,\sqrt{i\,\bar{a}}\; e^{-ik_1\bar{a}}.$$

Consequently, $p_R^{(1)}$ has a jump $\delta\, p_R^{(1)}$ equal to

$$\delta\, p_R^{(1)}\left(M\right) = -p_0\,\sqrt{2\pi}\; e^{-i\frac{\pi}{4}}\, r_1\, e^{ik_1 f_1(\nu_1)}. \tag{3.288}$$

This result is equal to the residue at the pole in (2.401), in which only the zero-order circumferential mode corresponding to an elastic surface wave having encircled the cylinder less than one tour has been conserved. Since it enforces the continuity of the surface wave field, it invites us to write $p_R^{(1)}\left(M\right)$ in another form in which the expression of the elastic surface wave appears explicitly. Such an expression is given by (3.273), in which R_1 is replaced by (3.280), which leads to

$$p_R^{(1)}\left(M\right) = p_0\,\sqrt{2\pi}\; e^{i\frac{\pi}{4}}\, r_1\, e^{ik_1 f_1(\nu_1)}\left[V\left(s_1\right) - \hat{V}\left(s_1\right)\right] \tag{3.289}$$

where $V(s_1)$ and $\hat{V}\left(s_1\right)$ are defined by (3.284) with

$$s_1 = \sqrt{i\bar{a}} = e^{i\frac{\pi}{4}} \sqrt{\bar{a}}.$$

The residue at the pole ν_1 in (2.401), limited to the zero-order mode having encircled the cylinder less than one tour, may be written in the form

$$p_s(M) = p_0 \sqrt{2\pi}\, e^{3i\frac{\pi}{4}} r_1 e^{ik_1 f_1(\nu_1)}\, U(-R_e \bar{a}), \tag{3.290}$$

where $U(\cdot)$ is the Heaviside function. By using these results, it can be easily verified that the jump of $V(s_1) - \hat{V}(s_1)$ when the pole crosses the SDP is equal to 1. Hence the field

$$p(M) = p_R^{(1)}(M) + p_S(M) \tag{3.291}$$

is continuous at the observation point.

3.4.1.3.3. Geometrical Interpretation of the Solution

In chapter 2, section 2.3.3.7, we have given a geometrical interpretation of $p_s(M)$ in terms of rays, which is essential for the generalization of the nonuniform solution of an elastic hollow circular cylinder to a three-dimensional convex elastic shell. A geometrical interpretation in terms of rays is also necessary for the extension of the uniform solution (3.286) to a three-dimensional convex elastic shell. For this purpose, let $\nu_1 = k_1 a \sin\theta_1$, where $\theta_1 = \theta_1' + i\theta_1''$ is a complex angle. We assume, as in (2.446), that $\theta_1'' << \theta_1'$. Then by neglecting terms of order $\theta_1''^2$ in the phase and θ_1'' in the amplitude, we obtain, as in section 2.3.3.7,

$$r_1 \cong -2\sqrt{k_1}\, \frac{a\cos^2\theta_1'}{\sin\theta_1' - \left(\dfrac{\partial\eta}{\partial\theta}\right)_{\theta=\theta_1'}}\, \frac{1}{\sqrt{s + a\cos\theta_1'}}, \tag{3.292}$$

$$k_1 f(\nu_1) = k_1 s - k_1 a\cos\theta_1' + \nu_1 \phi_1, \tag{3.293}$$

and

$$p_R^{(1)} \cong p^i(Q_1) D_S^{-1}\, \frac{a\cos^2\theta_1'}{\sin\theta_1' - \left(\dfrac{\partial\eta}{\partial\theta}\right)_{\theta=\theta_1'}}\, e^{i\nu_1\phi_1}\, \frac{e^{ik_1\ell_1}}{\sqrt{s + a\cos\theta_1'}} \left[V(s_1) - \hat{V}(s_1)\right], \tag{3.294}$$

where $\ell_1 = Q_1'M$ and $\phi_1 = \text{arc}\ Q_1Q_1'$ (fig. 3.19). When the observation point is located in the part of the transition region where the elastic surface wave exists, the arc ϕ_1 is positive, whereas it is negative in the part of the transition region where the elastic surface wave is absent. Hence when

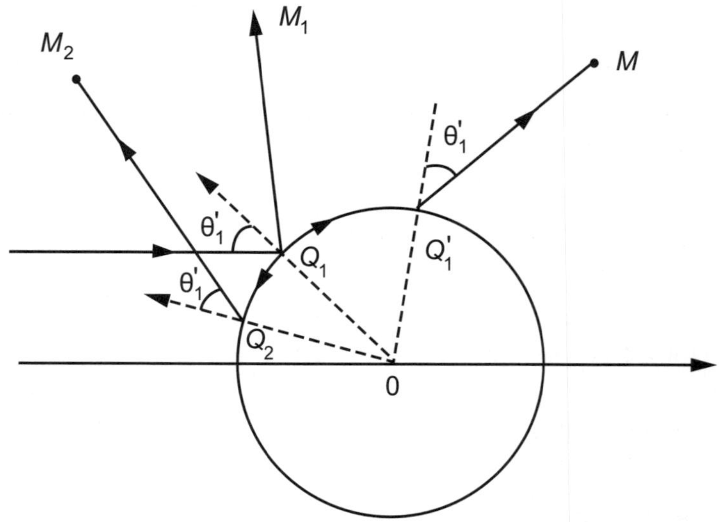

Figure 3.19. Elastic surface ray $Q_1Q_1'M_1$ and pseudo-elastic surface ray $Q_1Q_2'M_2$.

the surface wave exists, the field $p_R^{(1)}$ can be interpreted as a wave of the same nature as the elastic surface wave excited at Q_1. It propagates along the surface with the same propagation constant $k_{1s} = v_1/a$. But $p_R^{(1)}$ takes nonzero values in the part of the transition region where the elastic surface wave does not exist. In this case, it is possible to extend the geometrical interpretation of the solution by introducing the concept of a pseudo surface ray. Such a ray is a mathematical extension of the true surface ray in the region where the latter does not exist.

A pseudo surface ray can be represented geometrically, as shown in figure 3.19. It remains to express $V(s_1) - \hat{V}(s_1)$ in terms of geometrical parameters. Since $f(v_R)$ and $f(v_l)$ depend only on the trajectories of the reflected ray and the surface or pseudo surface ray, respectively, the only parameter that remains to be interpreted geometrically is $\bar{a}$. The parameter $\bar{a}$ is defined by

$$\bar{a} = f(v_R) - f(v_1). \tag{3.295}$$

We have seen that $R_e\,\bar{a} = 0$ when the pole v_1 crosses the SDP corresponding to the saddle point v_R. However, $I_m\,\bar{a} \neq 0$. The phase of the reflected field is known and given by

$$f(v_R) = -a\cos\theta_R + s_R, \tag{3.296}$$

where $s_R = Q_RM$ is the distance from the reflection point Q_R to the observation point M. An approximation to the phase $f(v_1)$ is given by (3.293). Hence if we insert (3.296) and (3.293) into (3.295), then we obtain the following approximation for $\bar{a}$:

$$\bar{a} \cong -a\cos\theta_R + s_R - s + a\cos\theta_1' - \frac{v_1}{k_1}\phi_1. \tag{3.297}$$

When the observation point crosses the shadow boundary of the elastic surface wave field, we have $\theta_R = \theta_1'$, $s_R = s$, and $\phi_1 = 0$. Hence $R_e\bar{a} = I_m\bar{a} \neq 0$. The approximation (3.297) for $\bar{a}$ can be used for evaluating $R_e\bar{a}$, since $R_e\bar{a} = 0$ on the shadow boundary. But $I_m\bar{a}$ cannot be approximated by (3.397) because $\hat{V}(s_1)$ would be infinite on the shadow boundary, according to (3.284). As a matter of fact, $I_m\bar{a}$ has a small imaginary part when θ_1'' is small. Therefore, $\hat{V}(s_1)$ takes a large value on the shadow boundary, which is compensated by the reflected field $p_G(M)$. The latter is calculated at the saddle point v_R on the real v-axis and also takes a large value, since v_R is close to the pole v_1 of the reflected coefficient. Therefore, we need a better approximation for $\bar{a}$ than the one given in (3.397). Since $|v_1 - v_R|$ is small close to the shadow boundary for $\theta'' << \theta'$, we can approximate $f(v_R) - f(v_1)$ by

$$f(v_R) - f(v_1) = -\frac{1}{2}(v_1 - v_R)^2 f''(v_R) \tag{3.298}$$

and consequently

$$I_m\bar{a} \cong -f''(v_R)\, v_{1i}(v_{1r} - v_R). \tag{3.299}$$

Since v_1 is a solution of the characteristic equation, v_{1r} and v_{1i} are known as well as $v_R = ka\sin\theta_R$, where θ_R is angle of reflection. Hence it remains to find a geometrical interpretation of $f''(v_R)$. According to (3.278), we have

$$q''(v_R) = \frac{1}{k_1}\left[\frac{1}{\sqrt{r^2 - a^2\sin^2\theta_R}} - \frac{2}{a\cos\theta_R}\right] = k\, f''(v_R), \tag{3.300}$$

and since

$$\sqrt{r^2 - a^2\sin^2\theta_R} = s_R + a\cos\theta_R, \tag{3.301}$$

we obtain

$$f''(v_R) = -\frac{1}{k_1^2}\,\frac{1}{s_R + a\cos\theta_R}\left(\frac{s_R + \rho_{cR}}{\rho_{cR}}\right), \tag{3.302}$$

where ρ_{cR} is the radius of curvature of the reflected wave front at the reflection point Q_R, given by

$$\rho_{cR} = \frac{a\cos\theta_R}{2}. \tag{3.303}$$

ρ_{cR} is also the distance from Q_R to the caustic of the reflected waves. Consequently,

$$I_m(\bar{a}) = \frac{1}{k_1^2} v_{1i}\left(v_{1r} - v_{1i}\right)\frac{s_R + \rho_{cR}}{\rho_{cR}}\left(\ell + a\cos\theta_R\right)^{-1}. \tag{3.304}$$

It would also be possible to obtain an expression for $R_e\bar{a}$ from (3.298). However, this result would be only valid in a narrow zone close to the shadow boundary. If the transition zone is large, a better approximation for $R_e(\bar{a})$ is given by (3.297)

$$R_e\,\bar{a} = a\cos\theta_R + s_R - s + a\cos\theta_1' - \frac{v_{1r}}{k_1}\phi_1. \tag{3.305}$$

The geometrical interpretation given in chapter 2, section 2.3.3.7, for the diffraction coefficients and the divergence factor of an elastic surface wave can also be extended to (3.294), which reads

$$p_R^{(1)} \cong p^i(Q_1)D_{sr,su}^{(1)}(Q_1)e^{ik_sQ_1Q_1'}D_{su,sr}^{(1)}(Q_1')\frac{e^{ik_1\ell_1}}{\sqrt{s+\rho_1}}\left[V(s_1) - \hat{V}(s_1)\right], \tag{3.306}$$

where $k_s = v_1/a$ is the propagation constant of the electric surface wave and where $\rho_1 = a\cos\theta_1'$. Also, the excitation and radiation coefficients at Q_1 and Q_2 are given by (2.485).

3.4.1.3. Oblique Incidence

The uniform solution for the reflected field associated with the elastic surface wave field obtained for normal incidence on a circular cylindrical shell can be extended to oblique incidence by applying, as in chapter 2, section 2.3.3.8, the translation invariance, which reads

$$p_R^{(1)}(\rho,\theta,z) = p_R^{(1)}(\rho,\theta;k_{1t})\,e^{ik_zz}, \tag{3.307}$$

where $p_R^{(1)}(\rho,\theta;k_t)$ verifies the Helmholtz equation for normal incidence and where $k_t = k_1 \sin\beta$, $k_z = k_1\cos\beta$, β being the angle of incidence.

As in chapter 2, section 2.3.3.8, the expression of $p_R^{(1)}(\rho_t,\theta;k_{1t})$ can be deduced from the solution of the diffraction problem for normal incidence by replacing the equivalent impedance η_n by $\bar{\eta}_n$, given by (2.468).

Observing that the uniform solution (3.294) at normal incidence differs from the corresponding nonuniform solution only by the factor $\left[V(s_1) - \hat{V}(s_1)\right]$, the geometrical interpretation of the common part, for oblique incidence, will be the same. Hence according to (2.481) we obtain

$$p_R^{(1)}(\rho,\theta,z) = p_i(Q_1)\, D_1^2\, e^{ik_s'\ell}\, e^{-\gamma\ell}\, e^{ik_1 s} \sqrt{\frac{\rho_c}{s+\rho_c}} \left[V(s_1) - \hat{V}(s_1)\right], \tag{3.308}$$

where γ and D_1^2 are, respectively, given by (2.482) and (2.483) and where $k_s = k_1 \sin\theta'$. Also, $\ell = arc\, Q_1\, Q_1'$, where Q_1' is defined by the position of the observation point in the transition region and where ρ_c is the distance of Q_1' to the caustic of the radiated waves. For normal incidence, the functions $V(s_1)$ and $\hat{V}(s_1)$ are defined by (3.284). According to (3.282), they depend only on $k_1\bar{a}$. An expression for $\bar{a}$ in terms of geometrical parameters is given by (3.304) and (3.305). For oblique incidence, $k_1\bar{a}$ needs to be replaced by $k_t\bar{a}$, where $k_t = k_1 \sin\beta$ and all other geometrical parameters in (3.304) and (3.305) have to be considered as projected geometrical parameters on a plane perpendicular to the cylinder axis. Introducing the nonprojected parameters, we obtain

$$k_t\bar{a} = k_1\Big[(s_R - s)\sin^2\beta + \rho_g \sin^2\alpha(\cos\theta_1' - \cos\theta_R) - \ell\sin\theta_1'\sin\alpha\sin\beta + i\rho_g\sin\theta_1'\sin^3\alpha\, sh\theta_{1t}''\left(\frac{s_R+\rho_{cR}}{s_R+2\rho_{cR}}\right)\Big], \tag{3.309}$$

where α is given by (2.475).

3.4.1.4. Noncircular Cylindrical Shell

The extension of (3.308) to a noncircular cylindrical shell is obtained by invoking the fact that at high frequencies, the field depends only on the local geometrical properties of the surface. The term $p_R^{(1)}(\rho,\theta,z)$ can therefore be directly derived from (2.487) and (3.308). It is given by

$$p_R^{(1)}(\rho,\theta,z) = p_i(Q_1)\, D_{sr,su}(Q_1)\, D_{su,sr}(Q_2)\, e^{i\int_{Q_1}^{Q_2} k_s'(\ell)\,d\ell}\, e^{-\int_{Q_1}^{Q_2}\gamma(\ell)\,d\ell}\, e^{ik_1 s} \sqrt{\frac{\rho_c(Q_2)}{s+\rho_c(Q_2)}} \left[V(s_1) - \hat{V}(s_1)\right], \tag{3.310}$$

where k_s' and γ are given by (2.488) and (2.494), respectively, and where $V(s_1)$ and $\hat{V}(s_1)$ are defined by (3.284) with $k_1\bar{a}$, given (3.309), which here takes the form

$$\begin{aligned} k_t\bar{a} = k_1\Big\{&\left[s(Q_R) - s(Q_2)\right]\sin^2\beta + \rho_g(Q_2)\sin^2\alpha(Q_2)\left[\cos\theta_1'(Q_2) - \cos(\theta_R)\right] \\ &- \ell(Q_2)\sin\theta_1'(Q_2)\sin\alpha(Q_2)\sin\beta(Q_2) + i\rho_g(Q_2)\sin\theta_1'(Q_2)\sin^3\alpha(Q_2)\, sh\theta_{1t}''(Q_2) \\ &\left(\frac{s_R(Q_R)+\rho_{cR}(Q_R)}{s_R(Q_R)+2\rho_{cR}(Q_R)}\right)\Big\}. \end{aligned} \tag{3.311}$$

The radiation and diffraction coefficients in (3.310) are identical to their homologues, given by (2.485), with η replaced by $\bar{\eta}$.

3.4.2. UNIFORM SOLUTION FOR A THREE-DIMENSIONAL CONVEX SHELL

The transition region of the creeping waves and the elastic surface waves are always separated, as shown in figure 3.20. A uniform solution for a convex shell is therefore composed of two solutions, one around the transition region 1 of the creeping waves and another corresponding to the transition regions 2 of the elastic surface waves. These two solutions match in region 3, with the GA solution augmented by the nonuniform solution for the elastic surface waves.

3.4.2.1. Uniform Solution for Creeping Waves

The general form of the uniform solution for the creeping waves propagating on a three-dimensional convex shell is identical to the one presented in section 3.3.2 for a convex impedance surface since the boundary conditions modify only the integrand of the Pekeris function.

By limiting our analysis to the case where the creeping wave follows a torsionless geodesic, we obtain the following for the field in the shadowed part of the transition region (as in section 3.3.2):

$$p(P) = p^i(Q)\left(-\sqrt{m(Q)m(Q')}\, e^{ik_1 t}\sqrt{\frac{2}{k_1}}\left[\frac{e^{i\frac{\pi}{4}}}{2\xi\sqrt{\pi}}\left(1 - F\left[k_1 L^d \tilde{a}\right]\right) + \tilde{P}(\xi)\right]\right) \tag{3.312}$$

$$\sqrt{\frac{d\Sigma(Q)}{d\Sigma(Q')}}\sqrt{\frac{\rho}{s(\rho+s)}}\, e^{ik_1 s},$$

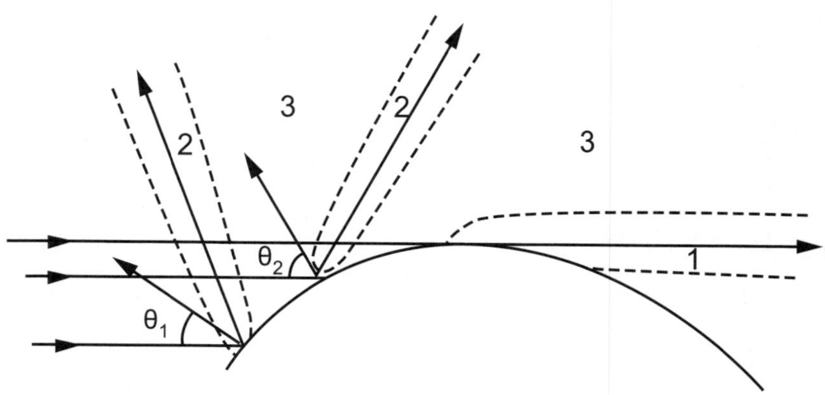

Figure 3.20. Positions of the transition regions: (1) creeping rays, (2) elastic surface rays, (3) GA rays and elastic surface rays.

where P is the observation point and where Q is the launching point of the creeping ray following the geodesic QQ' and shedding energy tangentially at Q' to the observation point P. Also, $s = Q'P$; ρ is the distance to the caustic of the diffracted rays in the plane defined by the binormal to the geodesic at Q' and the ray $Q'P$; $d\Sigma(Q)$ and $d\Sigma(Q')$ are the cross sections of a tube of surface rays at Q and Q', respectively; and

$$\xi = \int_{Q}^{Q'} \frac{m(t')}{\rho_g(t)} dt', \quad t = \int_{Q'}^{Q} dt', \quad m(t') = \left[\frac{k_1 \rho_g(t')}{2}\right]^{1/3}, \quad \tilde{a} = \frac{\xi^2}{2m(Q)m(Q')}.$$

The Pekeris function $\tilde{P}(X)$ is given by (3.261). In the deep shadow region, $F(k_1 L\tilde{a}) \to 1$, and the solution (3.312) reduces to

$$p(P) = -p^i(Q)\left(\sqrt{m(Q)m(Q')}\right) e^{ik_1 t} \frac{e^{i\frac{\pi}{4}}}{2\xi\sqrt{\pi}} \tilde{P}(\xi) \sqrt{\frac{d\Sigma(Q)}{d\Sigma(Q')}} \sqrt{\frac{\rho}{s(\rho+s)}} e^{ik_1 s}. \quad (3.313)$$

In (3.313), $\tilde{P}(X)$ can be calculated by applying the residue theorem, which gives

$$\tilde{P}(\xi) = -\sqrt{\pi}\, e^{i\frac{\pi}{4}} \sum_{n=1}^{\infty} R_n e^{i\xi\tau_n}, \quad (3.314)$$

where

$$R_n = \frac{\tilde{\Omega}_\tau \mathrm{w}_2(\tau_n)}{\left[\frac{\partial}{\partial\tau} \tilde{\Omega}_\tau \mathrm{w}_1(\tau)\right]_{\tau=\tau_n}}$$

and where τ_n are the zeros of $\tilde{\Omega}_\tau \mathrm{w}_1(\tau)$. It can be shown, as in (3.264) to (3.265), that by adding (3.313) to (3.258) we obtain the complete creeping wave contribution, derived earlier in chapter 2, section 2.3.4.1, in the shadow region outside the transition region.

Similarly, the solution in the part of the transition region reached by the incident field is given by

$$p(P) = p^i(P) + p^i(Q_R)\left(-\sqrt{-\frac{4}{\xi'}}\, e^{\frac{\xi'^3}{12}} \left[\frac{e^{i\frac{\pi}{4}}}{2\xi'\sqrt{\pi}}\left(1 - F\left[k_1 L^r \tilde{a}'\right]\right) + \tilde{P}(\xi')\right]\right) A\, e^{ik_1 s_1}, \quad (3.315)$$

where the divergence factor A is given by

$$A = \sqrt{\frac{\rho_1^r \rho_2^r}{(\rho_1^r + s_1)(\rho_2^r + s_1)}}$$

and

$$s_1 = Q_R P, \quad \tilde{a}' = 2\cos^2\theta^i, \quad \xi' = -2m(Q_R)\cos\theta^i.$$

In the region reached by the incident field, away from the transition region, $F\left(k_1 L^r \tilde{a}\right) \to 1$, and $\tilde{P}(\xi)$ behaves like

$$\tilde{P}(\xi) = -R\sqrt{-\frac{\xi'}{4}}\, e^{-\frac{\xi'^3}{12}}. \tag{3.316}$$

Hence the second term on the right-hand side of (3.315) tends to the GA reflected field. Finally, by imposing the continuity of the field at the shadow boundary, we obtain (see section 3.3.2)

$$L^r = \frac{\rho_1^i\,\rho_2^i}{\left(\rho_1^i + s_1\right)\left(\rho_2^i + s_1\right)}\,\frac{s_1\left(\rho + s_1\right)}{\rho}, \quad L^d = \frac{\rho_1^i\,\rho_2^i}{\left(\rho_1^i + s\right)\left(\rho_2^i + s\right)}\,\frac{s_1\left(\rho + s\right)}{\rho}. \tag{3.317}$$

On the shadow boundary, $s_1 = s$, but in the transition region away from this boundary, $s_1 \neq s$. For plane wave incidence, $L^r = s_1$ and $L^d = s$.

3.4.2.2. Uniform Solution for Elastic Surface Waves

We limit our analysis to the case where the elastic surface wave propagates along a geodesic without torsion on a three-dimensional convex shell. Under this condition, the uniform solution for the elastic surface waves can be readily derived from (3.310) by adapting the divergence factors of the surface waves along the geodesic followed by these waves and the divergence factors of the space waves radiated in the fluid. The final result is given by

$$\begin{aligned} p_R^{(1)}(\rho,\theta,z) = {} & p_i(Q_1) D_{sr,su}(Q_1) D_{su,sr}(Q_2)\, e^{i\int_{Q_1}^{Q_2} k_s'(\ell)\,d\ell}\, e^{-\int_{Q_1}^{Q_2} \gamma(\ell)\,d\ell}\, e^{ik_1 s} \\ & \sqrt{\frac{d\Sigma(Q_1)}{d\Sigma(Q_2)}}\sqrt{\frac{\rho_1\,\rho_2}{(\rho_1 + s)(\rho_2 + s)}}\left[V(s_1) - \hat{V}(s_1)\right], \end{aligned} \tag{3.318}$$

where $V(s_1)$ and $\hat{V}(s_1)$ are functions of $k_1\bar{a}$ defined by (3.284). According to (3.304) and (3.305), the expression of $k_1\bar{a}$ is given by

$$\begin{aligned} k_1\bar{a} = k_1\Big[& s_R(Q_R) - s(Q_2) + \rho_g(Q_2)\left(\cos\theta_1'(Q_2) - \cos\theta_R\right) \\ & -\ell(Q_2)\sin\theta_1'(Q_2) + i\rho_g \sin\theta_1'(Q_2)\, sh\theta_{1't}''(Q_2)\frac{s_R(Q_R) + \rho_{cR}(Q_R)}{s_R(Q_R) + 2\rho_{cR}(Q_R)}\Big]. \end{aligned} \tag{3.319}$$

The diffraction coefficients $D_{sr,\,su}(Q_1)$ and $D_{su,\,sr}(Q_2)$ are given by (2.485).

3.5. NUMERICAL RESULTS

In this section, we present some numercal results obtained by the author on a straight wedge and on a circular cylinder.

3.5.1. RESULTS CONCERNING THE WEDGE

Figures 3.21 and 3.22 concern the diffraction of a plane wave at normal incidence by a hard and soft wedge, respectively. The wedge has an angle of 90° and the incident wave makes an angle of 30° with the reference face.

The shadow boundaries of the reflected and direct fields are therefore at 150° and 210°, respectively. The observation direction varies from 0 to 270°, and $kr = 40$ where k is the wave number in the fluid and where r is the distance of the observation point to the edge. The solid curve, the dotted curve, and the dashed curve give the modulus of the direct field, the diffracted

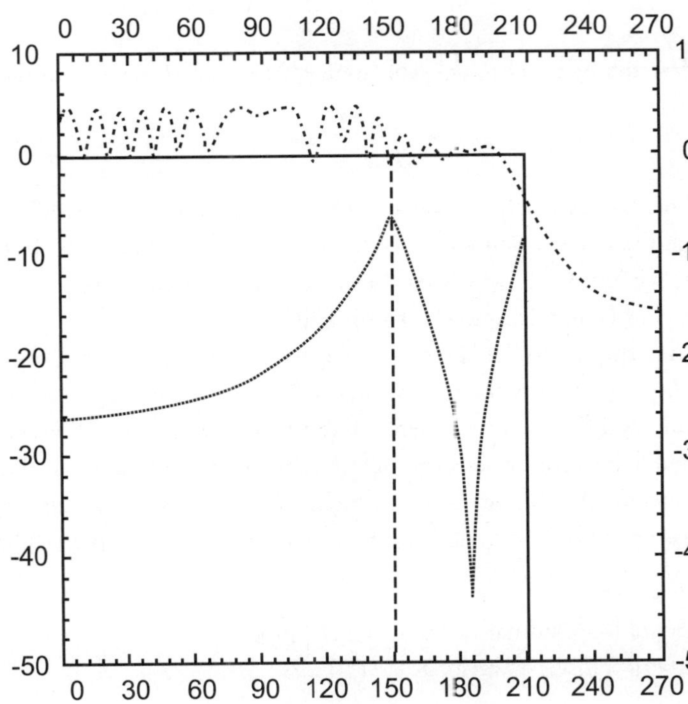

Figure 3.21. Diffraction by a right-angle hard wedge.

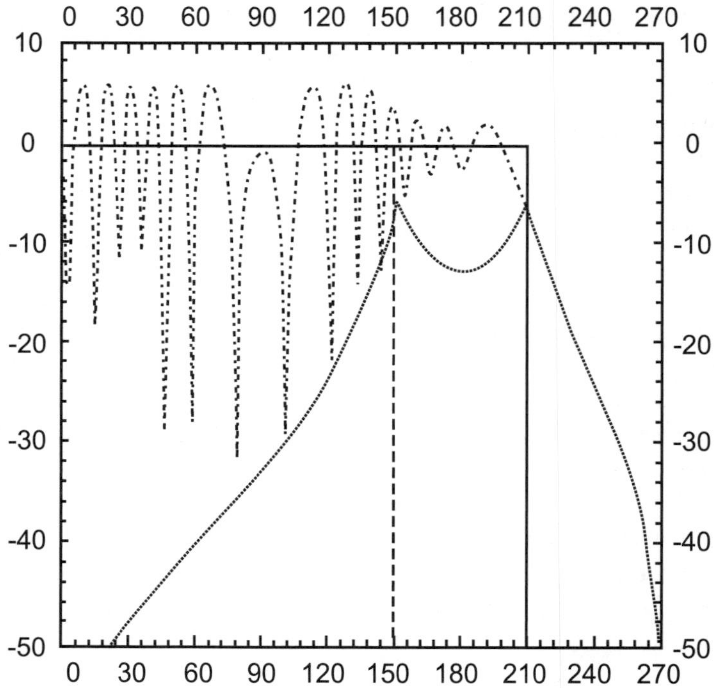

Figure 3.22. Diffraction by a right-angle soft wedge. The dashed curve gives the total field.

field, and the total field, respectively, in decibels. These results show that the diffracted field is discontinuous through the shadow boundaries of the reflected and direct fields. However, the total field established in section 3.2.1, which has been calculated by applying the uniform solution based on the method of Pauli-Clemmow, is continuous.

Figure 3.23 gives the results for the same wedge with an impedance Z on both faces, and figure 3.24 gives the same results for an impedance $1/Z$. In these cases, the uniform solution involving Maliuzhinets's diffraction coefficient, established in section 3.2.2, has been applied. We observe that the total field is continuous through the shadow boundaries, as it should be.

Figure 3.25 concerns the diffraction of an incident plane wave by an elastic wedge-shaped shell. The wedge has an angle of 90° and is made of brass having the following characteristics:

- density = 8.6 kg/dm^3
- sound velocity of longitudinal waves: c_L = 4410 m/s
- sound velocity of transverse waves: c_T = 2150 m/s

The thickness of the wall is d = 8 mm. The wedge-shaped shell is immerged in water, the sound velocity of which has been taken equal to c = 1480 m/s. The medium inside the wedge is vacuum.

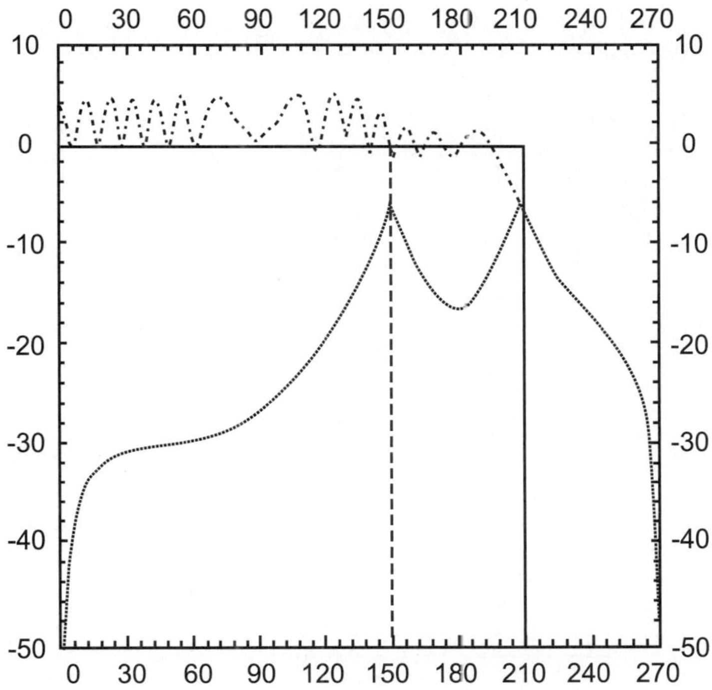

Figure 3.23. Diffraction by a right-angle impedance wedge with a surface impedance $Z = 1/\sqrt{2}$.

In practice, such conditions exist when the wedge belongs to a rectangular waveguide or box. The medium inside the box can also be replaced by air without appreciable differences in the results.

The incident wave makes an angle of 180° with the reference face. Therefore, the reflected field has a shadow boundary at 180°. The numerical results presented in figure 3.26 correspond to a frequency of 0.34 MHz and a distance of observation $r = 0.02$ m. The dotted curve corresponds to the diffracted field, and the solid curve gives the sum of the reflected and diffracted fields. These results, which were obtained with the uniform solution established in section 3.2.3, show that the sum of the reflected and edge-diffracted fields is continuous through the transition region at 180°, as expected. The effect of the Lamb waves on the total field is not represented. It slightly modifies the sum of the reflected and diffracted fields between 0° and 40°.

3.5.2. RESULTS CONCERNING THE CIRCULAR CYLINDER

Numerical results are presented for the diffraction of a plane wave at normal incidence by a circular cylindrical shell immerged in water. The medium inside the cylinder is vacuum. The cylindrical shell is made of stainless steel having the following characteristics:

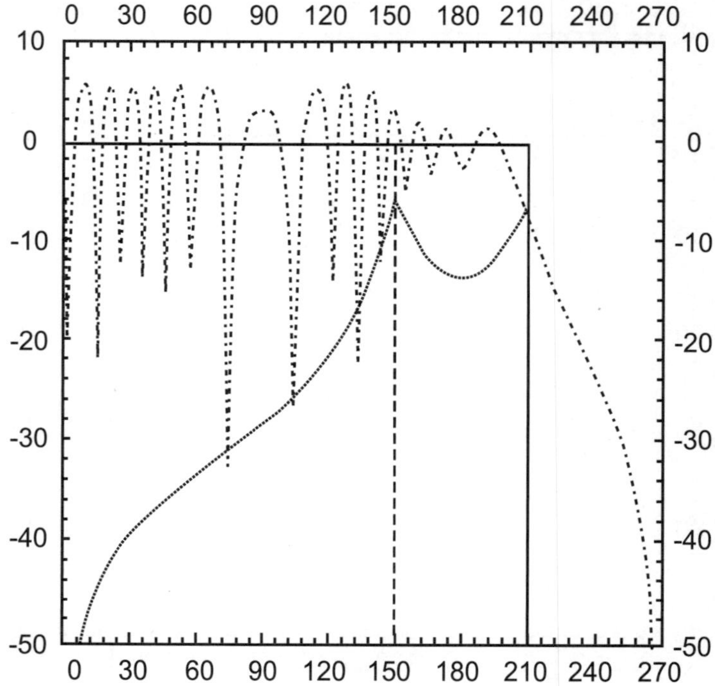

Figure 3.24. Diffraction by a right-angle impedance wedge (surface impedance $Z = \sqrt{2}$).

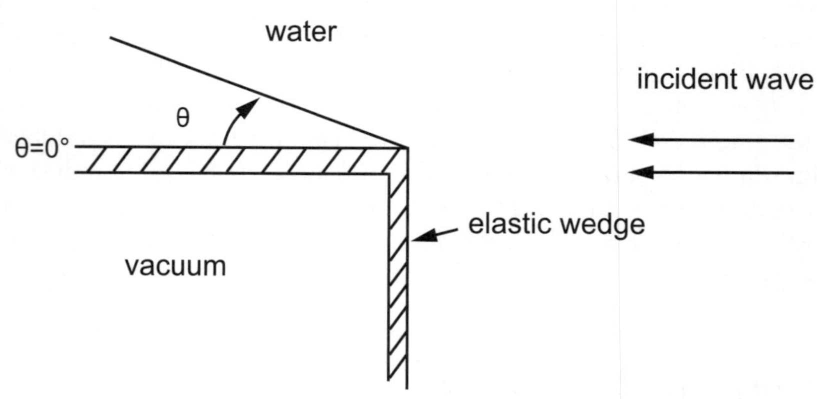

Figure 3.25. Elastic right-angle wedge–shaped shell: Definition of the incident and observation angles.

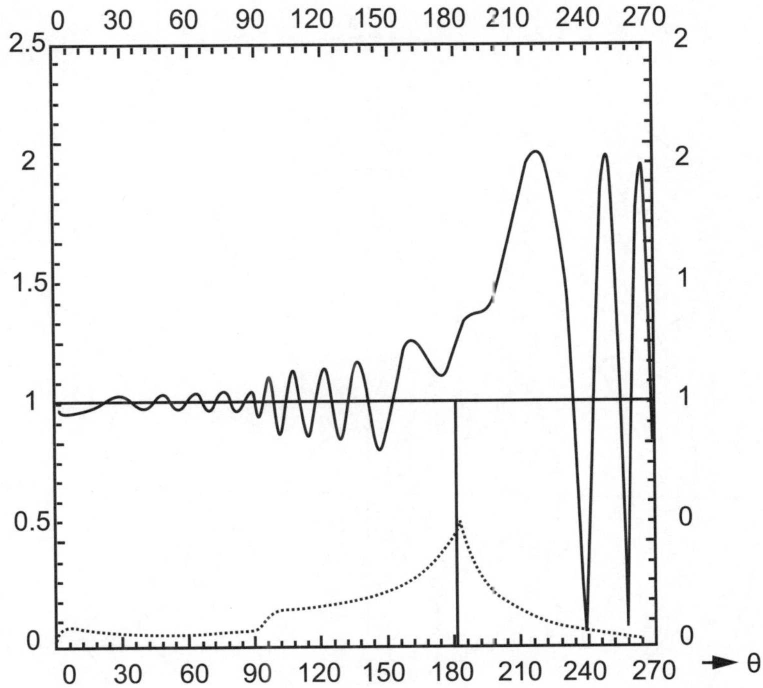

Figure 3.26. Diffraction by an elastic right-angle wedge–shaped shell. The solid curve corresponds to the sum of the reflected and diffracted fields.

- Density: 7.8 kg/dm^3
- Sound velocity of longitudinal waves: $c_L = 5790$ m/s
- Sound velocity of transverse waves: $c_T = 3100$ m/s

The geometry of the cylindrical shell is defined by its exterior radius a and its interior radius $b = 0.97a$. The plane wave is incident in the direction $\theta = 0$, and the scattered field comprising the reflected waves, the surface waves, and the creeping waves has been calculated for observation angles varying from 0° to 360°.

Figures 3.27 and 3.28 give the angular diagram in decibels (each graduation corresponds to 10 dB) of the relative amplitude of the total field for $a = 1$ cm, $ka = 8.48$, and a distance of observation $r = 10$ cm. The diagram of figure 3.27 corresponds to the nonuniform solution at the shadow boundaries of the surface waves, whereas the diagram of figure 3.28 corresponds to the uniform solution based on the Van der Waerden method, established in section 3.4.1. In this last diagram, we observe the continuity of the field through the shadow boundaries of the elastic surface waves. It can be shown that for the frequency and characteristics of the cylindrical shell chosen in our application, the reflection coefficient has only one pole. Therefore, the discontinuity

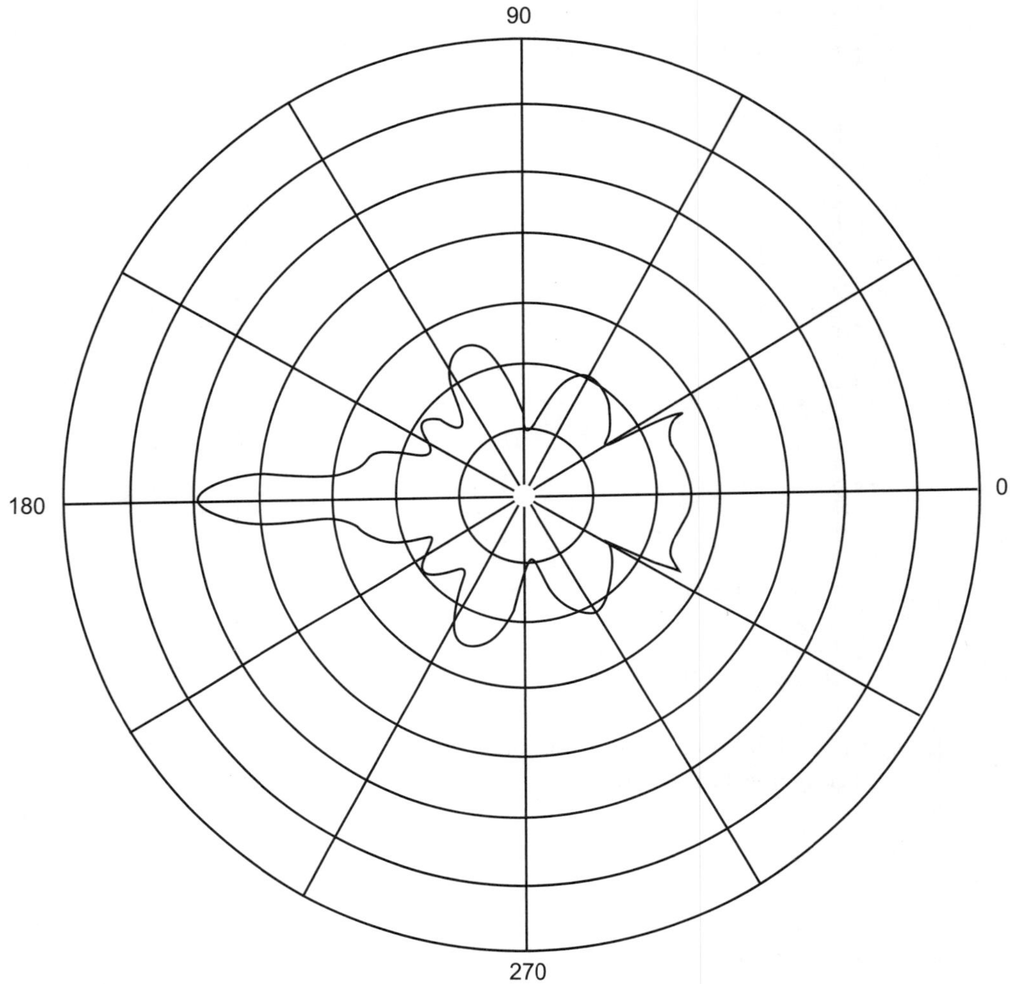

Figure 3.27. Diffraction of a plane wave by a circular cylindrical shell made of stainless steel, immerged in water. This is the nonuniform solution for the amplitude of the sum of the contributions of the creeping waves, elastic surface waves, and reflected waves. Each graduation corresponds to 10 dB: a = 1 cm, ka = 8.48, b/a = 0.97, and r = 10 a.

that appears in figure 3.27 corresponds to the shadow boundaries of two surface waves surrounding the cylinder in opposite directions.

Figures 3.29 and 3.30 give the diagram obtained with the uniform solution established in section 3.4.1 for a circular cylindrical shell made of stainless steel, defined by a = 1.5 cm, b/a = 0.97. These results have been obtained at a distance $r = 10a$ for ka = 30 and ka = 40, respectively. They confirm the validity of our uniform formulation of the elastic surface waves.

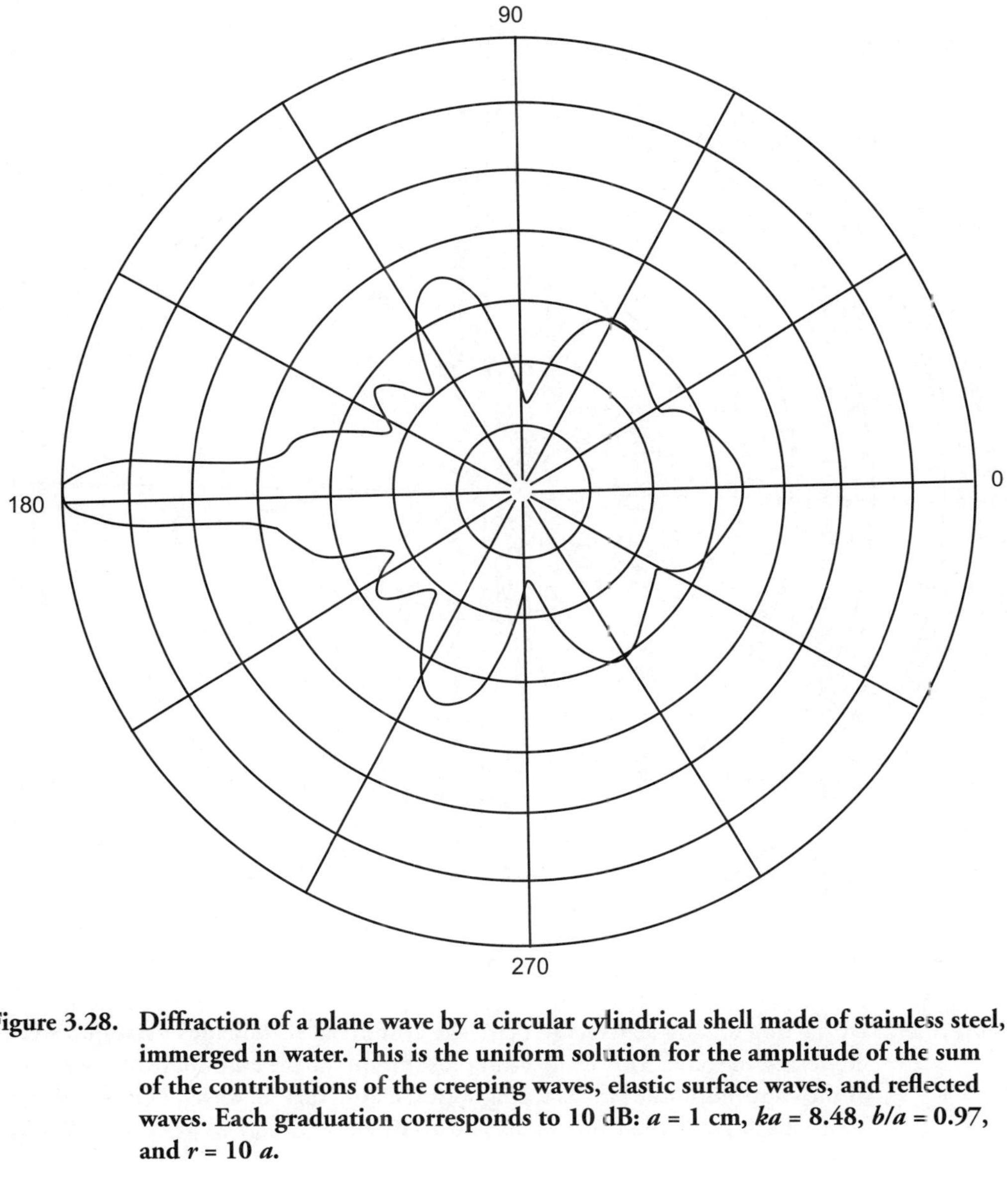

Figure 3.28. Diffraction of a plane wave by a circular cylindrical shell made of stainless steel, immerged in water. This is the uniform solution for the amplitude of the sum of the contributions of the creeping waves, elastic surface waves, and reflected waves. Each graduation corresponds to 10 dB: a = 1 cm, ka = 8.48, b/a = 0.97, and r = 10 a.

REFERENCES

[3.1] Bouche, D., F. Molinet, and R. Mittra. 1997. *Asymptotic Methods in Electromagnetics*. Berlin: Springer-Verlag.

[3.2] Buchal, R. N., and J. B. Keller. 1960. "Boundary Layer Problem in Diffraction Theory." *Communications on Pure and Applied Mathematics* 13: 85–114.

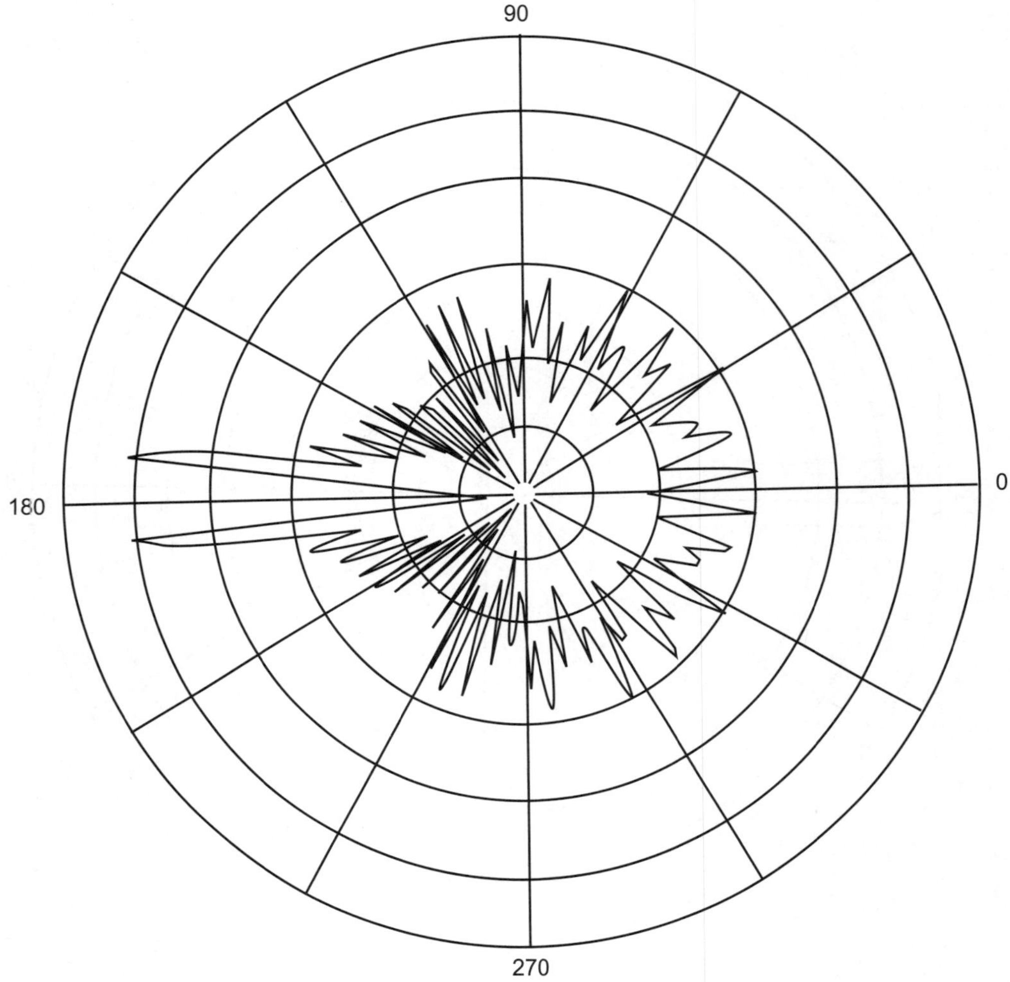

Figure 3.29. Diffraction of a plane wave by a circular cylindrical shell made of stainless steel, immerged in water. This is the uniform solution for the amplitude of the sum of the contributions of the creeping waves, elastic surface waves, and reflected waves. Each graduation corresponds to 10 dB: a = 1.5 cm, ka = 30, b/a = 0.97, and r = 10 a.

[3.3] Babich, V. M., and N. Y. Kirpicnikova. 1979. *The Boundary Layer Method in Diffraction Problems.* Berlin: Springer-Verlag.

[3.4] Pauli, W. 1938. "On the Asymptotic Series for Functions in the Theory of the Diffraction of Light." *Physical Review* 54: 924–31.

[3.5] Clemmow, P. C. 1950. "Some Extensions of the Method of Integration by Steepest Descents." *Quarterly Journal of Mechanics and Applied Mathematics* 3: 241–56.

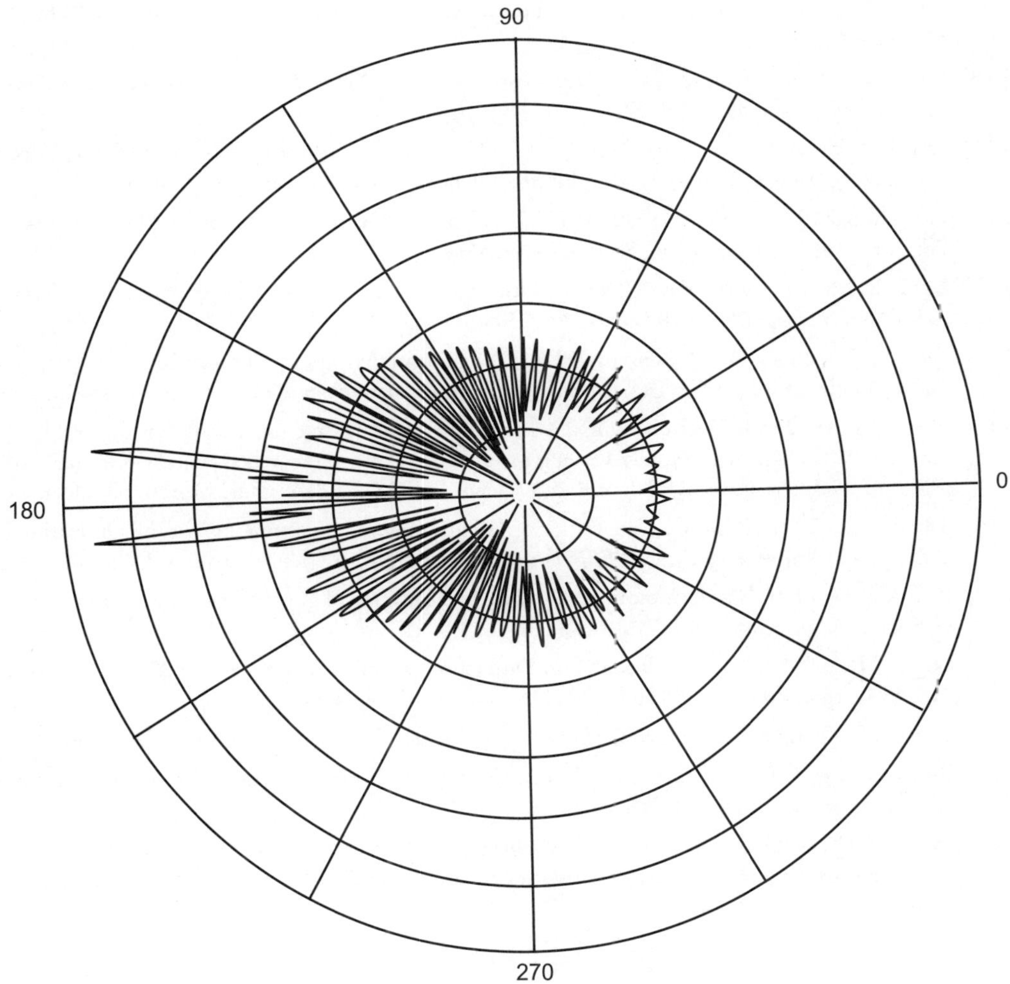

Figure 3.30. Diffraction of a plane wave by a circular cylindrical shell made of stainless steel, immerged in water. This is the uniform solution for the amplitude of the sum of the contributions of the creeping waves, elastic surface waves, and reflected waves. Each graduation corresponds to 10 dB: a = 1.5 cm, ka = 40, b/a = 0.97, and r = 10 a.

[3.6] Van der Waerden. 1951. "On the Method of Saddle Points." *Applied Scientific Research* B2: 33–45.

[3.7] Hutchins, D. L., and R. G. Kouyoumjian. 1969. "Asymptotic Series Describing the Diffraction of a Plane Wave by a Wedge." Report 2183-3, Electro-Science Laboratory, Ohio State University.

[3.8] Bleistein, N. 1967. "Uniform Asymptotic Expansion of Integrals with Many Nearby Stationary Points and Algebraic Singularities." *Journal of Mathematics and Mechanics* 17: 533–59.

[3.9] Felsen, L. B., and N. Marcuvitz. 1973. *Radiation and Scattering of Waves.* Englewood Cliffs, NJ: Prentice-Hall.

[3.10] Rojas, R. G. 1987. "Comparison between Two Asymptotic Methods." *IEEE Transactions on Antennas and Propagation* 35: 1489–92.

[3.11] Pathak, P. H., and R. G. Koujoumjian. 1973. "The Dyadic Diffraction Coefficient for a Perfectly Conducting Wedge." Report 2183-4, Electro-Science Laboratory, Ohio State University.

[3.12] Kouyoumjian, R. G., and P. H. Pathak. 1974. "A Uniform Geometrical Theory of Diffraction for an Edge in a Perfectly Conducting Surface." *Proceedings of the IEEE* 62: 1448–61.

[3.13] Lewis, R. M., and J. Boersma. 1969. "Uniform Asymptotic Theory of Edge Diffraction." *AIP Journal of Mathematical Physics* 10 (12): 2291–2305.

[3.14] Lee, S. N., and G. A. Deschamps. 1976. "A Uniform Asymptotic Theory of Electromagnetic Diffraction by a Curved Wedge." *IEEE Transactions on Antennas and Propagation* 24: 25–34.

[3.15] Hwang, Y. M., and R. G. Kouyoumjian. 1974. "A Dyadic Diffraction Coefficient for an Electromagnetic Wave Which Is Rapidly Varying at an Edge." Paper presented at the US National Committee of the International Union of Radio Science (USNC/URSI) Annual Meeting, Boulder, CO.

[3.16] Tiberio, R., G. Pelosi, and G. Manara. 1985. "A Uniform GTD Formulation for the Diffraction by a Wedge with Impedance Faces." *IEEE Transactions on Antennas and Propagation* 33 (8): 867–73.

[3.17] Pathak, P. H. 1979. "An Asymptotic Analysis of the Scattering of Plane Waves by a Smooth Convex Cylinder." *Radio Science* 14 (3): 419–35.

[3.18] Logan, N. A. 1959. "General Research in Diffraction Theory," Report LMSD-288087, Lockheed Aircraft Corporation, Missiles and Space Division, Sunnyvale, CA.

[3.19] Logan, N. A., and K. S. Lee. 1962. *Electromagnetic Waves.* Madison: University of Wisconsin Press.

[3.20] Bowman, J. J., T. B. A. Senior, and P. L. E. Uslenghi. 1969. *Electromagnetic and Acoustic Scattering of Simple Shapes.* Amsterdam: Hemisphere.

[3.21] Pathak, P. H., W. D. Burnside, and R. J. Marhefka. 1980. "A Uniform GTD Analysis of the Diffraction of Electromagnetic Waves by a Smooth Convex Surface." *IEEE Transactions on Antennas and Propagation* 28 (5): 631–42.

Chapter 4

WAVE FIELD NEAR A CAUSTIC

4.1. INTRODUCTION

The rays reflected by a smooth regular surface, the rays diffracted by a sharp edge, the creeping rays leaving a smooth convex surface tangentially, and all classes of space rays all have a common property: they are the characteristics of a first-order partial differential equation (the eikonal equation) and therefore constitute a family of curves depending on two parameters. Such a family of curves always has an envelope that in ray theory is called a caustic.

We limit our analysis to a homogeneous medium. In this case, the family of rays forms a congruence of straight lines. Each class of rays has a caustic that, in general, is composed of two separated sheets. A caustic may be real or virtual. For instance, the caustic corresponding to the reflection of a plane wave that is normally incident on a convex cylinder is virtual (fig. 4.1). It is formed by two regular surfaces located inside the cylinder, the cross section of which has a cusp.

Another example, shown in figure 4.2, is the caustic of the rays diffracted by the edge of a truncated circular cylinder submitted to a sound wave emanating from a point source. We observe that, even for this simple problem, the caustic is rather complicated. It is composed of regular surfaces joined along cuspid edges or at focal points.

Other examples are shown in figures 4.3 and 4.4. Figure 4.3 shows the cross section of the caustic of the rays reflected by a convex-concave surface generated by a curve having an inflection point and rotating around an axis Δ, which is also the direction of propagation of the incident rays. We observe that the caustic has a virtual sheet corresponding to the envelope of the rays reflected by the convex part of the surface and a real sheet corresponding to the rays reflected by the concave part. Both sheets are tangent to a cone at infinity. Figure 4.4 shows some examples of degenerated caustics. Figures 4.4(a) and 4.4(b) show a caustic degenerated into a straight line, and figure 4.4(c) shows a caustic degenerated into a focal point at infinity.

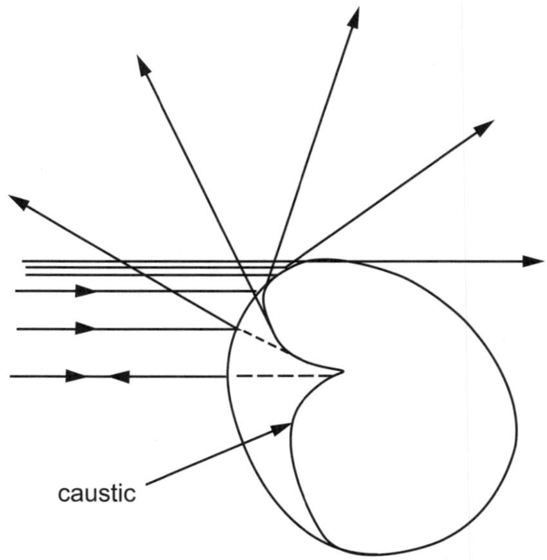

Figure 4.1. Virtual caustic inside a cylindrical body struck by parallel rays.

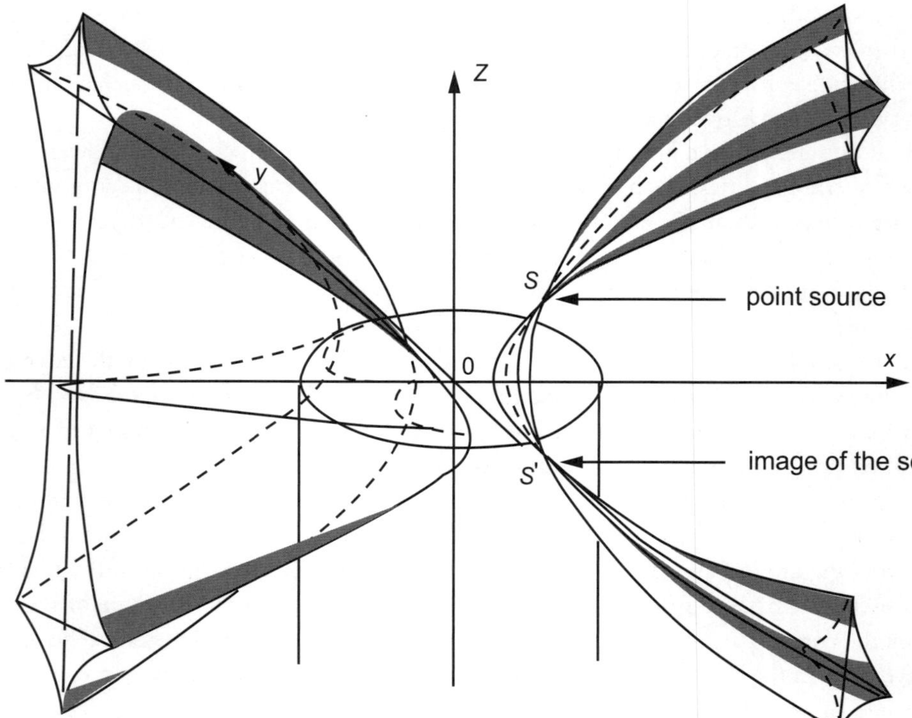

Figure 4.2. Caustic of rays emanating from a point source *S* and diffracted by a circular edge.

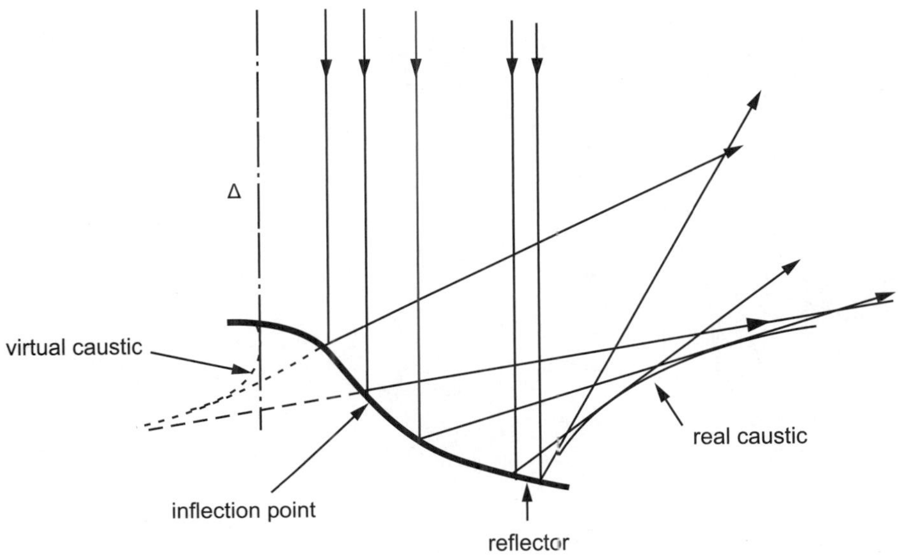

Figure 4.3. Real and virtual caustic of parallel rays reflected by a convex-concave rotational symmetric reflector.

4.2. TECHNIQUES FOR CALCULATING THE FIELD ON A CAUSTIC AND IN ITS NEIGHBORHOOD

The laws of geometrical optics (GO) or geometrical acoustics (GA) are not valid in the layer around the caustic, which we call the boundary layer. The thickness of this layer depends on the wavelength and will be specified for each type of caustic. The techniques for calculating the field in the vicinity of a caustic depend on the geometrical characteristics of the caustic. The most common cases are a regular caustic, a cusped caustic, and a caustic degenerated into a line or a point. Techniques for degenerated caustics are well known and will not be presented here. They comprise the Kirchhoff approximation of the scattered field on flat surfaces and the equivalent line source technique on circular edges. We limit our analysis to the calculation of the field near a regular caustic. Some comments and references are also given on the techniques that permit us to calculate the field near a cusp line.

4.2.1. OBSERVATION POINT LOCATED CLOSE TO A REGULAR CAUSTIC

At an observation point located on the lit side of a regular caustic, the field passes two rays belonging to the congruence of rays having this caustic as their envelope (fig. 4.5).

Each of these rays constitutes a stationary path having a point on the body that may be a point of reflection or a point of diffraction. When the observation point approaches the caustic

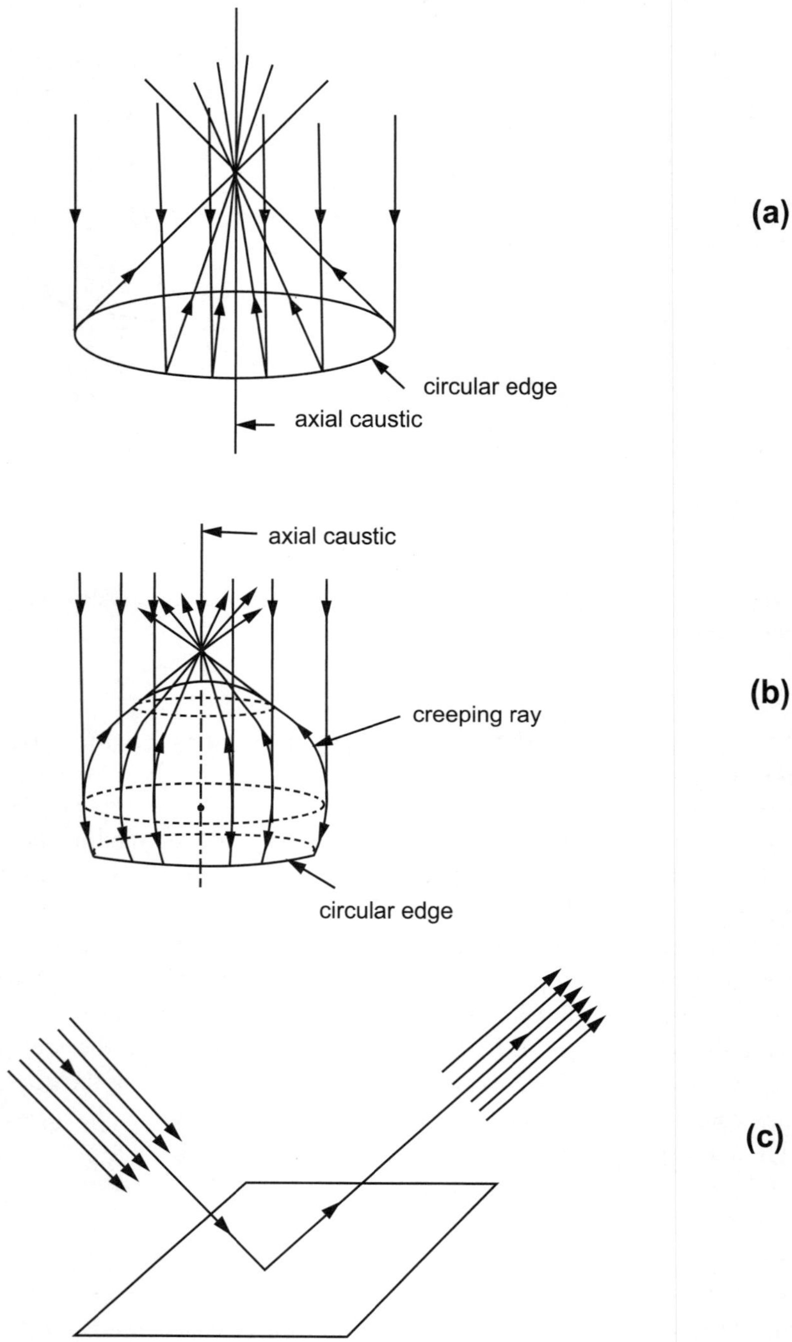

Figure 4.4. Degenerated caustics: (a) and (b) illustrate an axial caustic and (c) is a point caustic at infinity.

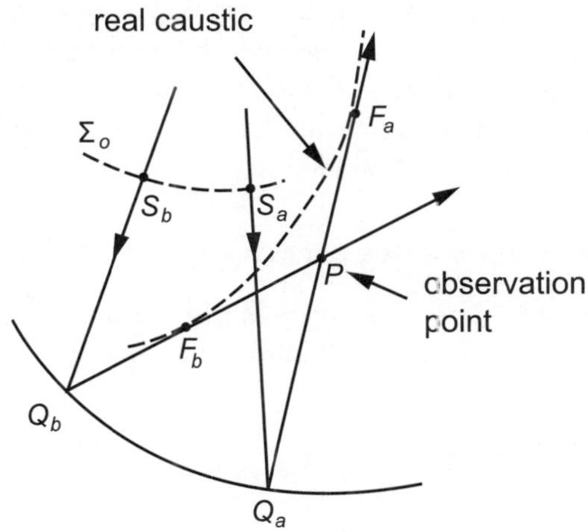

Figure 4.5. At an observation point close to a real caustic, the field passes two rays tangent to the caustic.

on the lit side, the two points of interaction with the body (reflection points or diffraction points) meet. This indicates that a correspondence must exist between the relative position of the reflection or diffraction points on the body and the distance of their point of intersection with the caustic.

Let Σ be the wave front of the incident wave. Σ is orthogonal to the congruence of incident rays, which in the most general case of multiple interactions with the body are the reflected or diffracted rays emanating from the previous interaction. On the surface Σ, the phase is constant and supposed known along each ray.

Let Σ_0 be a reference wave front and let S_aQ_aP and S_bQ_bP be the two rays passing through the observation point P and intercepting the surface of the body at the interaction points Q_a and Q_b (see fig. 4.5). We set

$$s_a^i = S_a Q_a, \quad s_b^i = S_b Q_b, \quad s_a^r = Q_a P, \quad s_b^r = Q_b P \tag{4.1}$$

and

$$\rho_{1a}^r = -Q_a F_a, \quad \rho_{1b}^r = -Q_b F_b, \tag{4.2}$$

where F_a and F_b are the points where the rays are tangent to the caustic.

According to the general expression (1.129) of the GA or GO field, this solution is not valid when the observation point P tends to the caustic, since F_a and F_b are focal points for the wave front. Our objective is to find a uniform asymptotic solution that is valid close to the caustic and on the caustic and reduces to the GO field when the observation point is far from the caustic.

As mentioned in the introduction of chapter 3, a direct solution that avoids a matching procedure consists of first choosing an ansatz for the uniform solution that has the same structure as the uniform solution of the corresponding canonical problem. Additionally, we modify the arguments of the special functions contained in it in such a way that, outside the boundary layer of the caustic, the solution tends to the GO solution.

4.2.1.1. Canonical Problem for a Regular Caustic

A simple canonical solution of the wave equation such that the rays corresponding to this solution are tangent to a caustic is provided in chapter 3 of reference [4.1]:

$$u(r,\theta) = J_{ka}(kr)e^{ika\theta}, \tag{4.3}$$

where (r, θ) are polar coordinates and $J_{ka}(kr)$ is a Bessel function, while a = constant > 0.

If we replace the Bessel function for $r > a$ by its asymptotic expansion of Debye (see page 419 in [4.2]), then we find

$$u = \frac{1}{2}\sqrt{\frac{2}{\pi ka}}\left(\frac{a^2}{r^2-a^2}\right)^{1/4}\left\{e^{ik\left(\sqrt{r^2-a^2}-a\arccos\left(\frac{a}{r}\right)+a\theta\right)}e^{-\frac{i\pi}{4}}\right.$$
$$\left.+e^{ik\left(-\sqrt{r^2-a^2}+a\arccos\left(\frac{a}{r}\right)+a\,\theta\right)}e^{\frac{i\pi}{4}}\right\}\cdot\left(1+O\left(k^{-1}\right)\right), \tag{4.4}$$

where $O(k^{-1})$ are asymptotic expansions in integral powers of k^{-1}. Expression (4.4) is in the form of a ray expansion. The two terms in this expansion correspond to two rays passing through a point P with coordinates (r, ϕ)—the first term behaving like a ray that leaves the circle, which has a radius a, and the second corresponding to a ray going toward the circle. The two rays are tangent to the circle, and the wave fronts described by the equations

$$\tau_{\pm}(r,\theta) = a\theta \mp \left(\sqrt{r^2-a^2} - a\arccos\frac{a}{r}\right) = \text{constant} \tag{4.5}$$

are involutes of the circle $r = a$. Consequently, the circle $r = a$ coincides with the envelope of the family of normals to the wave fronts defined by (4.5) and is therefore the caustic. We note that there is $a - \pi/2$ jump of the phase at the crossing of the caustic.

The region $r < a$ corresponds to the shadow zone. In this region, the Debye asymptotic form of the Bessel function [4.2] gives

$$u = \frac{1}{2}\sqrt{\frac{2}{\pi ka}}\left(\frac{a^2}{r^2-a^2}\right)^{1/4} e^{-k\left[a\,\mathrm{Log}\left(\frac{a}{r}+\sqrt{\left(\frac{a}{r}\right)^2-1}\right)+\sqrt{a^2-r^2}\right]}e^{ika\theta}\times\left(1+O\left(k^{-1}\right)\right). \tag{4.6}$$

Thus in the shadow zone the wave field decays exponentially with increasing k. But the expansion (4.6), like (4.4), is not valid for $r \cong a$ (i.e., close to the caustic).

A uniform asymptotic expansion of the solution (4.3) can be derived by using uniform asymptotic formulas for the Bessel function [4.3]. Then we get for $r \geq a$

$$u \cong (kr)^{-1/3}\left(\left[\sqrt{2}\,\frac{r^{1/3}\zeta_e^{1/4}}{\sqrt[4]{r^2-a^2}}+\sum_{n=1}^{\infty}\frac{b_n\left(\frac{a}{r}\right)}{(kr)^{2n}}\right]A_i(X)+\frac{1}{(kr)^{1/3}}\sum_{n=0}^{\infty}\frac{d_n\left(\frac{a}{r}\right)}{(kr)^{2n+1}}A_i'(X)\right)e^{ika\theta}, \tag{4.7}$$

where

$$X=-k^{2/3}\zeta_e,\qquad \zeta_e^{3/2}=\frac{3}{4}(\tau_- - \tau_+)=\frac{3}{4}(2\ell - s) \tag{4.8}$$

and where $A_i(X)$ and $A_i'(X)$ are, respectively, the Airy function and its derivation with respect to X. According to simple trigonometric formulas, we also have $(\tau_- - \tau_+) = 2\ell - s$ (fig. 4.6).

For $r \leq a$, we obtain

$$u \cong (kr)^{-1/3}\left(\left[\sqrt{2}\,\frac{r^{1/3}\zeta_0^{1/4}}{\sqrt[4]{a^2-r^2}}+\sum_{n=1}^{\infty}\frac{b_n\left(\frac{a}{r}\right)}{(kr)^{2n}}\right]A_i(X)+\frac{1}{(kr)^{1/3}}\sum_{n=0}^{\infty}\frac{d_n\left(\frac{a}{r}\right)}{(kr)^{2n+1}}A_i'(X)\right)e^{ika\theta}, \tag{4.9}$$

where

$$X=-k^{2/3}\zeta_0,\qquad \zeta_0^{3/2}=\frac{3}{2i}\left[a\,\mathrm{Log}\left(\frac{a}{r}+\sqrt{\left(\frac{a}{r}\right)^2-1}\right)+\sqrt{a^2-r^2}\right]. \tag{4.10}$$

An integral form of $A_i(X)$ is given by

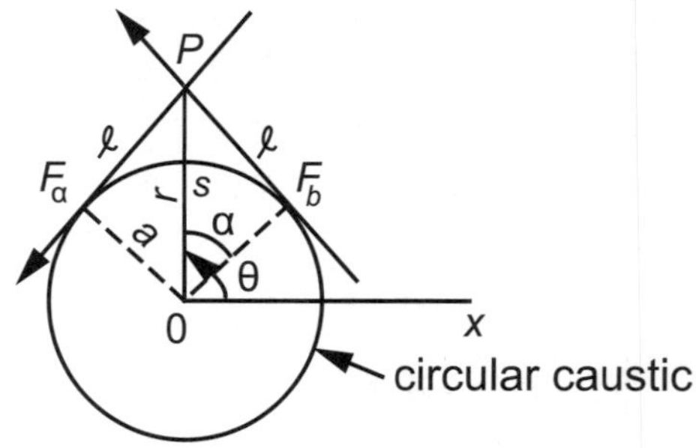

Figure 4.6. Rays tangent to a circular caustic.

$$A_i(X) = \frac{1}{2\pi}\int_{-\infty}^{+\infty} e^{i\left(\frac{t^3}{3}+xt\right)}\, dt. \tag{4.11}$$

It is important to observe that the integrand in (4.11) has two stationary points of order 1 at $t^2 + X = 0$. When $r \geq a$, we have $X \leq 0$. Hence the stationary points are given by $t_s = \pm\sqrt{|X|}$. When the observation point tends to the caustic, $X \to 0$. Consequently, the two stationary points coalesce. This property shows that the Airy function is the special function characterizing the behavior of the field in the vicinity of the caustic.

By replacing $A_i(X)$ by its asymptotic expansion for $X < 0$, it can be readily shown that, up to the order k^{-1}, (4.7) reduces to (4.4). Similarly, by replacing $A_i(X)$ with its asymptotic expansion for $X > 0$, the expression (4.9) reduces to (4.6) at the order k^{-1}. Indeed, coming back to the asymptotic expansion of $A_i(X)$ for $(X \leq 0)$, we see that when $|X|$ is large, the saddle points are widely separated, and when $|X| \to \infty$, one may obtain an asymptotic expansion of $A_i(X)$ by separately treating the contribution from each relevant saddle point. Furthermore, the contour of integration in (4.11) can be deformed away from the real axis upon recognizing that the behavior of the integrand as $|t| \to \infty$ is determined by $\exp(it^3 / 3)$. The integral (4.11) converges when $R_e(it^3) <$ 0 or $I_m t^3 > 0$. Now $I_m t^3$ is positive in the following sectors of the complex t-plane: $0 < |\arg t| < \frac{\pi}{3}$, $-\frac{2\pi}{3} < |\arg t| < -\frac{\pi}{3}$, and $\frac{2\pi}{3} < |\arg t| < \pi$ (fig. 4.7).

Thus by Cauchy's theorem the contour of integration can be deformed in any contour in the t-plane, which begins at $|t| \to \infty$ in sector 2 and at $|t| \to \infty$ in sector 3. Especially, the contour of integration of (4.11), which is the real axis, can be deformed into the contour L_{23}, which itself can be replaced by the difference of the contours L_{21} and L_{31}. Hence (4.11) can be written as

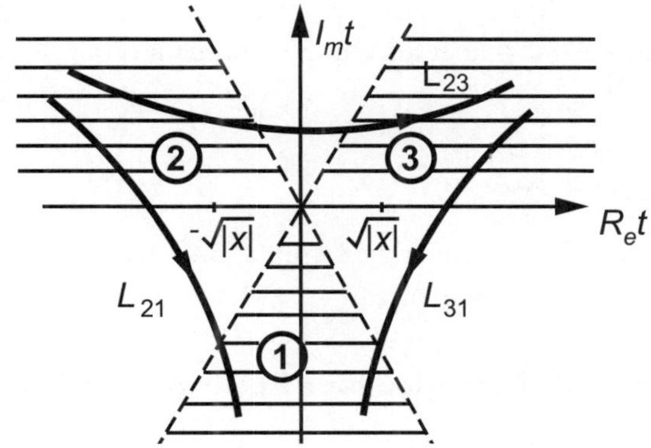

Figure 4.7. Airy integral paths; $I_m t^3 > 0$ in the shadowed regions.

$$A_i(X) = \frac{1}{2\pi} \int_{L_{21}} e^{i\left(\frac{t^3}{3}+Xt\right)} dt - \frac{1}{2\pi} \int_{L_{31}} e^{i\left(\frac{t^3}{3}+Xt\right)} dt. \tag{4.12}$$

The contours L_{31} and L_{21} can be deformed into the steepest descent paths (SDP) passing through the saddle points $\pm\sqrt{\leq|X|}$, and the derivation of the asymptotic expansions of the two integrals in (4.12) follows the classical technique for an isolated first-order saddle point presented in chapter 2, section 2.2.1.3. The result for the dominant order is

$$A_i(X) = \frac{1}{\sqrt{\pi}} \frac{1}{(-X)^{1/4}} \sin\left[\frac{2}{3}(-X)^{3/2} + \frac{\pi}{4}\right] \tag{4.13}$$

or

$$A_i\left(-k^{2/3}\zeta_e\right) \cong \frac{1}{\sqrt{\pi}} k^{-1/6} \zeta_e^{-1/4} \sin\left[\frac{2}{3} k \zeta_e^{3/2} + \frac{\pi}{4}\right]. \tag{4.14}$$

For $r < a$ we obtain the following expression for ζ_0 from (4.10):

$$\zeta_0 = -\left(\frac{3}{2}\right)^{2/3} \left[a \operatorname{Log}\left(\frac{a}{r} + \sqrt{\frac{a^2}{r^2} - 1}\right) + \sqrt{a^2 - r^2}\right]^{2/3}. \tag{4.15}$$

Hence $\zeta_0 < 0$ and $X > 0$. In this case, the integrand of (4.11) has two stationary points of order 1 at $t_s = \pm i\sqrt{X}$. These points are on the imaginary axis. When X is large they are widely separated, and the contour L_{23} of figure 4.7 can be deformed into the SDP passing through the saddle point

$t_s = i\sqrt{X}$. The formal asymptotic expansion can now be written directly by applying the standard method for an isolated saddle point. The result for the dominant order is

$$A_i(X) = \frac{1}{2\sqrt{\pi}} \frac{1}{X^{1/4}} e^{-2/3 X^{3/2}}. \tag{4.16}$$

When (4.16) is inserted into (4.9) we recover the formula (4.6), which shows that the field in the shadow region of the caustic decays exponentially.

4.2.1.2. Uniform Asymptotic Expansion for a Regular Three-Dimensional Caustic

The objective in this section is to construct an asymptotic expansion of the field near a three-dimensional regular caustic, which formally satisfies the Helmholtz equation in three dimensions:

$$(\Delta + k)\, u = 0 \tag{4.17}$$

as $k \to \infty$. A formal solution of (4.17), in a form suggested by the solution of the canonical problem of section 4.2.1.1, is

$$u = A(\mathbf{r}, k)\, A_i(X) + k^{-1/3}\, B(\mathbf{r}, k)\, A_i'(X), \tag{4.18}$$

where A and B depend on the coordinates of the observation point and are asymptotic expansions of the Luneburg-Kline type,

$$A(\mathbf{r}, k) = k^{\gamma} \sum_{n=0}^{N} \frac{A_n}{(ik)^n} + o\left(\frac{1}{k^N}\right),\ B(\mathbf{r}, k) = k^{\gamma} \sum_{n=0}^{N} \frac{B_n}{(ik)^n} + o\left(\frac{1}{k^N}\right), \tag{4.19}$$

where γ is a constant.

It should be noted that in the canonical problem of section 4.2.1.1, the coefficients A_n for odd values of n and the coefficients B_n for even values of n were all zero. This is due to the symmetry of the canonical problem. In the general case, the introduction of these coefficients is necessary. On the side of the caustic where the rays exist, we make the assumption that the argument X of the Airy function is still given by (4.8), where τ_+ is the eikonal of the ray going toward the caustic and where τ_- is the eikonal of the ray leaving the caustic. According to the notations introduced in 4.2.1, we have

$$\tau_+ = s_a^i + s_a^r, \quad \tau_- = s_b^i + s_b^r, \tag{4.20}$$

$$\tau_- - \tau_+ = s_b^i + s_b^r - (s_a^i + s_a^r). \tag{4.21}$$

The ansatz (4.18) was first proposed by Kravtsov [4.4] and Ludwig [4.5], who derived a uniform solution near a general two-dimensional regular caustic.

In order to justify our assumption for X, we consider first a regular two-dimensional, noncircular caustic (C). Let P be the observation point, F_bP be the ray leaving the caustic and passing through P, and PF_a be a ray going toward the caustic (fig. 4.8). Let (C_1) be the involute of (C) passing through F_a and intersecting the ray F_bP at Q. If we denote by s the arc of F_bF_a along the caustic, we have $s = F_bQ$. Hence we obtain

$$\tau_- - \tau_+ = F_bP - \left(s - PF_a\right) = \ell_1 + \ell_2 - s. \tag{4.22}$$

Now let $n = P_1P$ be the normal distance of P to the caustic and a be the radius of curvature of the caustic at P_1. It is evident that if a is large compared to n, we can replace the caustic with its osculating circle in the neighborhood of P_1. Hence the formulas established for a circular caustic apply also locally to a noncircular, two-dimensional regular caustic.

Consider now a three-dimensional caustic and again let P be the observation point, F_bP be the ray leaving the caustic, and PF_a be the ray going toward the caustic. We denote by (G) the geodesic curve of the caustic, tangent at F_b to the ray F_bQ. If (G) is a planar curve, then it passes through F_a. Therefore, on the plane F_bPF_a, we find the same geometrical figure as in figure 4.9. Hence the geometrical relation (4.22) still holds, and when a is large compared to n, the geodesic (G) can be replaced in the neighborhood of P_1 by its osculating circle and the formulas established for a circular caustic apply again locally.

If (G) is not planar, then we are confronted with a torsion geodesic. However, if the observation point P is close to the caustic, then the torsion of the geodesic leaving the contact point F_b can be neglected, and again we find that when P approaches the caustic, the difference of the eikonals of the two rays passing through P tends to that obtained for a circular caustic.

In the boundary layer, close to the caustic, the argument X of the Airy function can also be expressed as a function of the distance n of P to the caustic. Replacing the caustic by its osculating circle, we obtain (see fig. 4.6)

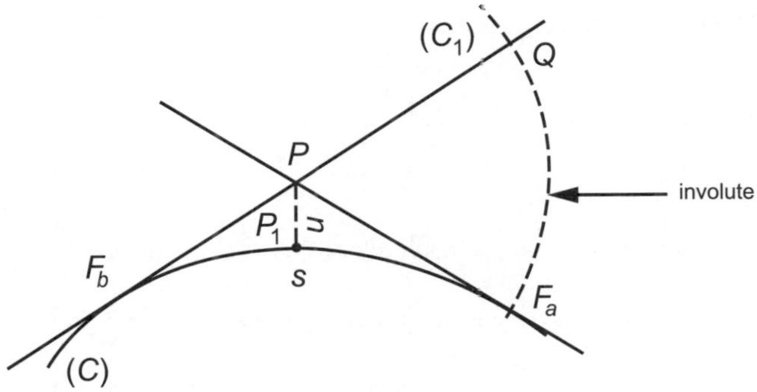

Figure 4.8. Involute C_1 of C.

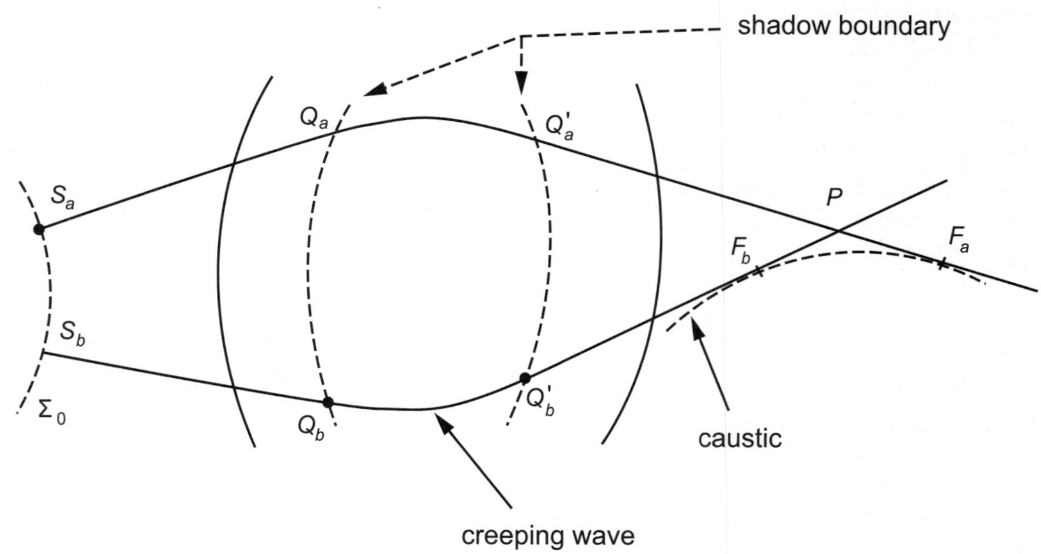

Figure 4.9. Paths followed by two creeping rays passing through a point near a real caustic.

$$\ell - \frac{s}{2} = a(\tan\alpha - \alpha) = a\frac{\alpha^3}{3} + O(a\alpha^5). \tag{4.23}$$

Using $\alpha = \frac{\ell}{a} + O\left(\left(\frac{\ell}{a}\right)^3\right)$ and $n = \frac{\ell^2}{2a} + O\left(\left(\frac{\ell}{a}\right)^4\right)$, we get

$$\ell - \frac{s}{2} = \frac{\ell^3}{3a^2} = \frac{(2an)^{3/2}}{3a^2}. \tag{4.24}$$

Finally, by taking (4.8) into account, we find

$$X = -k^{2/3}\zeta_e = -k^{2/3}\left(\frac{2}{a}\right)^{1/3} n = -\frac{kn}{m}, \tag{4.25}$$

where $m = \left(\frac{ka}{2}\right)^{1/3}$ is the Fock parameter. Hence, in a neighborhood $n \approx \frac{m}{k} = k^{-2/3}\left(\frac{a}{2}\right)^{1/3}$ of the caustic, the argument X of the Airy function is of order -1. It is equal to the opposite of the stretched variable ν in the normal direction, occurring in the boundary layer method. Since the solution obtained by this method is given by

$$u \approx e^{iks} A(s)\, A_i(-\nu), \tag{4.26}$$

where $A(s)$ is an amplitude factor, (s,n) are the geodesic coordinates of the observation point P in the boundary layer, and

$$\nu = \frac{kn}{m}, \tag{4.27}$$

we see that the special function $A_i(X)$ in (4.18) is also justified by the boundary layer method. It remains to determine the coefficients $A(\mathbf{r},k)$ and $B(\mathbf{r},k)$. These coefficients are obtained by identifying the asymptotic expansion of (4.18) for $|X|$ large, with the field obtained by the GO (GA) or the geometrical theory of edge and surface diffraction (GTD), according to reflected, edge-diffracted, or surface-diffracted rays.

4.2.1.3. Caustic of Rays Reflected by a Smooth Surface

The procedure for determining the coefficients A and B of (4.18) for a caustic of the reflected rays consists of identifying, far from the caustic, the sum of the field of the two reflected rays passing through the observation point P with the field given by (4.18) for $|X|$ large. The GA field, corresponding to the two reflected rays passing through P, is given by

$$\begin{aligned} p(P) = {} & p^i(Q_a)\, R(Q_a) \sqrt{\frac{\rho_{1a}^r\, \rho_{2a}^r}{(\rho_{1a}^r + s_a^r)(\rho_{2a}^r + s_a^r)}}\; e^{ik s_a^r} \\ & + p^i(Q_b)\, R(Q_b) \sqrt{\frac{\rho_{1b}^r\, \rho_{2b}^r}{(\rho_{1b}^r + s_b^r)(\rho_{2b}^r + s_b^r)}}\; e^{ik s_b^r}, \end{aligned} \tag{4.28}$$

where $p^i(Q)$ is the incident field at Q, $R(Q)$ is the reflection coefficient at Q, $\rho_1^r(Q)$ and $\rho_2^r(Q)$ are the radii of curvature of the reflected wave front at Q, and s_a^r and s_b^r are the phases of the wave front at Q_a and Q_b, respectively.

Formula (4.28) corresponds to the dominant term of the asymptotic expansion of the reflected field. This term must be identified with the dominant term of (4.18), which according to (4.19), reads

$$u = k^{\gamma}\left[A_0(\mathbf{r})\, A_i(X) + k^{-1/3} B_0(\vec{r})\, A_i'(X)\right]. \tag{4.29}$$

The asymptotic expansion of $A_i(X)$ for $|X|$ large is given by (4.13). Since

$$A_i'(X) = \frac{1}{2\pi} \int_{L_{21}-L_{31}} (it)\, e^{i\left(\frac{t^3}{3} + Xt\right)} dt, \tag{4.30}$$

the dominant term of its asymptotic expansion for $|X|$ large is obtained by multiplying, respectively, the dominant terms of the asymptotic expansions of the integrals (4.12) along the contours L_{21} and L_{31} by $i\sqrt{-X}$ and $-i\sqrt{-X}$, which gives

$$A_i'(X) = -\frac{1}{\sqrt{\pi}}(-X)^{1/4}\cos\left[\frac{2}{3}(-X)^{3/2} + \frac{\pi}{4}\right]. \tag{4.31}$$

In (4.29), substituting for $A_i(X)$ and $A_i'(X)$ the dominant term of their asymptotic expansion and identifying the result with (4.28), leads to $\gamma = 1/6$ and

$$\begin{Bmatrix} A_0 \\ B_0 \end{Bmatrix} = -2\pi i\ e^{ik\delta_e} \begin{Bmatrix} P_e \\ -i\zeta_e^{-1/2}\ Q_e \end{Bmatrix}, \tag{4.32}$$

where

$$\delta_e = \frac{1}{2}\left(s_b^i + s_b^r + s_a^i + s_a^r\right) \tag{4.33}$$

and

$$\begin{Bmatrix} P_e \\ Q_e \end{Bmatrix} = \frac{\zeta_e^{\frac{1}{4}}}{2\sqrt{\pi}} \left[p^i(Q_a)e^{-iks_a^i}\ D(Q_a)\sqrt{\frac{\rho_{1a}^r\rho_{2a}^r}{\left(\rho_{1a}^r + s_a^r\right)\left(\rho_{2a}^r + s_a^r\right)}}\ e^{i\frac{\pi}{4}} \right. $$
$$\left. \pm p^i(Q_b)\ e^{-iks_b^i}\ R(Q_b)\sqrt{\frac{\rho_{1b}^r\rho_{2b}^r}{\left(\rho_{1b}^r + s_b^r\right)\left(\rho_{2b}^r + s_b^r\right)}}\ e^{-i\frac{\pi}{4}} \right]. \tag{4.34}$$

When the observation point tends toward the caustic, one of the radii of curvature of the wave front (ρ^1, for instance) tends to $-s^r$, and the denominators $\rho_{1a}^r + s_a^r$ and $\rho_{1b}^r + s_b^r$ in (4.34) tend to zero. However, by performing a Taylor expansion of ζ_e and $\left(\rho_1^r + s^r\right)$, it can be readily shown that the product $\zeta_e^{\frac{1}{4}}\left(\rho_1^r + s^r\right)$ tends to a finite limit.

4.2.1.4. Caustic of Rays Diffracted by an Edge

We start again with the ansatz (4.18) for the representation of the uniform asymptotic solution of the diffracted field in the neighborhood of a caustic of rays diffracted by an edge. Away from the caustic, this solution must match with the edge-diffracted field given by the GTD.

The dominant term of the asymptotic expansion of the diffracted field is given by (see chapter 1, section 1.6)

$$p(P) = u^i(Q_a)\,D(Q_a)\sqrt{\frac{\rho_a^r}{\rho_a^r + s_a^r}}\,e^{iks_a^r} + u^i(Q_b)\,D(Q_b)\sqrt{\frac{\rho_b^r}{\rho_b^r + s_b^r}}\,e^{iks_b^r}, \tag{4.35}$$

where Q_a and Q_b are the diffraction points on the edge of the two rays passing through P, $\rho^r_{a,b}$ is the radius of curvature of the wave fronts at Q_a and Q_b in the plane of diffraction, and D is the diffraction coefficient. Identifying (4.35) with the dominant term (4.29) or (4.18) for $|X|$ large leads to $\gamma = -\frac{1}{3}$ and

$$\begin{Bmatrix} A_0 \\ B_0 \end{Bmatrix} = -2\pi i\, e^{ik\delta_e} \begin{Bmatrix} P_e^d \\ -i\zeta_e^{-1/2}\, Q_e^d \end{Bmatrix}, \tag{4.36}$$

where

$$\begin{Bmatrix} P_e^d \\ Q_e^d \end{Bmatrix} = \frac{\zeta^{1/4}}{2}\sqrt{\frac{k}{\pi}}\left[p^i(Q_a)e^{-iks_a^i}\,D(Q_a)\sqrt{\frac{\rho_a^r}{\rho_a^r + s_a^r}}\,e^{i\frac{\pi}{4}} \mp p^i(Q_b)\,e^{-iks_b^i}\,D(Q_b)\sqrt{\frac{\rho_b^r}{\rho_b^r + s_b^r}}\,e^{-i\frac{\pi}{4}}\right],$$

where D is of order $k^{-1/2}$.

4.2.1.5. Caustic of Rays Diffracted by a Smooth Surface (Creeping Rays)

We consider a space wave exciting creeping waves on a smooth convex surface separating a fluid from an impenetrable media. As shown in chapter 1, section 1.7, the diffraction process can be described by space rays exciting creeping rays at the shadow boundary. The creeping rays follow geodesics of the surface and shed energy in the fluid along space rays that are tangent to the surface. As mentioned in the introduction of this chapter, such rays always have an envelope or caustic composed of two sheets. One of them is the surface of diffraction, while the other is a real or virtual caustic, depending on whether the rays leaving the surface diverge or converge.

Let P be a point of observation close to a real caustic of surface diffracted rays. In the case of a regular caustic, there are two rays passing through P, as shown in figure 4.9. Let Σ_0 be the wave front of the incident wave. The rays passing through P are orthogonal to Σ_0 and tangent to the caustic. Let S_a and S_b be the points on Σ_0 where the rays passing through P emanate, and let Q_a and Q_b be the points on the shadow boundary of the surface where the space rays S_aQ_a and S_bQ_b excite the creeping rays $Q_aQ'_a$ and $Q_bQ'_b$, respectively. Furthermore, let $Q_aQ'_a$ and $Q_bQ'_b$ be the arc length of the geodesics followed by the creeping rays, and let Q'_aP and Q'_bP be the space rays launched at Q'_a and Q'_b, respectively. We introduce the notations

$$\begin{aligned} s_a^i &= S_a Q_a, \quad s_b^i = S_b Q_b, \\ s_a^{\prime r} &= \widehat{Q_a Q_a'} + s_a^r, \quad s_b^{\prime r} = \widehat{Q_b Q_b'} + s_b^r, \\ s_a^r &= Q_a' P, \quad s_b^r = Q_b' P. \end{aligned} \tag{4.37}$$

The difference between the eikonals of the two rays at P is given by

$$\tau_- - \tau_+ = s_b^i + s_b^{\prime r} - \left(s_a^i + s_a^{\prime r}\right) = q, \tag{4.38}$$

and the half sum of the eikonals is denoted by

$$\delta_e = \frac{1}{2}\left(s_a^i + s_a^{\prime r} + s_b^i + s_b^{\prime r}\right). \tag{4.39}$$

We consider the case where the observation point is close to a regular part of the caustic. According to the general expression for the uniform asymptotic expansion of the field near a regular caustic, introduced in section 4.2.1.2, we write the dominant term in the form (4.29), where

$$X = -k^{2/3}\zeta_e, \quad \zeta_e^{3/2} = \frac{3}{4}q. \tag{4.40}$$

The coefficients A_0 and B_0 and the constant γ are determined by identifying the dominant term (4.29) for $|X|$ large with the dominant term of the sum of the GTD fields associated to each creeping ray, given by (see chapter 1, section 1.7)

$$\begin{aligned} p(P) = {} & p^i(Q_a)\sum_{p=1}^{\infty} D_p(Q_a) D_p(Q_a') e^{-\int_{Q_a}^{Q_a'} \alpha_p(t)\,dt} \sqrt{\frac{d\eta(Q_a)}{d\eta(Q_a')}} \sqrt{\frac{\rho_a^r}{s_a^r\left(\rho_a^r + s_a^r\right)}}\, e^{iks_a^{\prime r}} \\ & + p^i(Q_b)\sum_{p=1}^{\infty} D_p(Q_b) D_p(Q_b') e^{-\int_{Q_b}^{Q_b'} \alpha_p(t)\,dt} \sqrt{\frac{d\eta(Q_b)}{d\eta(Q_b')}} \sqrt{\frac{\rho_b^r}{s_b^r\left(\rho_b^r + s_b^r\right)}}\, e^{iks_b^{\prime r}}, \end{aligned} \tag{4.41}$$

where $D_p(Q)$ and $D_p(Q')$ are, respectively, the excitation and the launching coefficients of the creeping ray $\widehat{QQ'}$; α_p is the attenuation coefficient along the geodesic followed by the ray; and ρ^r is the radius of curvature of surface diffracted rays given by $\rho^r = -Q'F$, where Q' is the launching point of the creeping ray and F is the contact point with the caustic. Explicit expressions for $D_p(Q)$ and α_p are given in chapter 2, section 2.3 for different boundary conditions.

Proceeding as in section 4.2.1.3 by substituting for $A_i(X)$ and $A_i'(X)$ the dominant term of their asymptotic expansion, the identification of (4.29) with (4.41) leads readily to $\gamma = 1$ and

$$\begin{Bmatrix} A_0 \\ B_0 \end{Bmatrix} = -2\pi i \; e^{ik\delta_e} \begin{Bmatrix} P_e^{sd} \\ -i\zeta_e^{-1/2} \; Q_e^{sd} \end{Bmatrix}, \tag{4.42}$$

where

$$\begin{Bmatrix} P_e^{sd} \\ Q_e^{sd} \end{Bmatrix} = \frac{\zeta_e^{1/4} k^{1/6}}{2\sqrt{\pi}} \left[ip(Q_a) e^{-ik\left(s_a^i + \widehat{Q}_a \widehat{Q}_a'\right)} D(Q_a, Q_a') \sqrt{\frac{\rho_a^r}{s_a^r\left(\rho_a^r + s_a^r\right)}} \; e^{i\frac{\pi}{4}} \right. \\ \left. \mp p^i(Q_b) \; e^{-ik\left(s_b^i + \widehat{Q}_b \widehat{Q}_b'\right)} D(Q_b, Q_b') \sqrt{\frac{\rho_b^r}{s_b^r\left(\rho_b^r + s_b^r\right)}} \; e^{-i\frac{\pi}{4}} \right], \tag{4.43}$$

where

$$D(Q, Q') = \left(\sum_{p=1}^{\infty} D_p(Q) D_p(Q') e^{-\int_Q^{Q'} \alpha_p(t)\,dt} \right) \sqrt{\frac{d\eta(Q)}{d\eta(Q')}} \; e^{ik\widehat{Q}\widehat{Q}'}, \tag{4.44}$$

and where, according to the results of chapter 2.3, $D_p(Q)D_p(Q')$ is of order $k^{-1/6}$.

4.2.1.6. Comments on the Solutions for a Regular Caustic

The uniform solutions established in sections 4.2.1.3 and 4.2.1.5 for a regular caustic cover all the situations usually encountered in practical applications. In each case, the diffracted field is expressed in terms of geometrical parameters, depending on the paths of the rays and on the characteristics of the wave front. Since these components are always calculated in GA and GTD applications, they can be used for testing the distance of the observation point to the caustic, as well as for determining the field near and at the caustic, on the side where the rays exist.

In the shadow region of the caustic, it is in principle possible to calculate the field by an analytical prolongation of the formulas that we have established for the lit region. However, this operation, which needs to introduce complex rays, is very complicated for a two-dimensional geometry and almost impossible to achieve for a general three-dimensional geometry. An approximate method that gives satisfactory results consists of calculating the argument X of the Airy function at a point P in the shadow region of the caustic by relating this argument to the argument X' of the Airy function at the point P' in the lit region, symmetric of P with respect to the caustic, by stating $X = -X' > 0$, where

$$X' = -k^{2/3}\zeta_e(P'). \tag{4.45}$$

Moreover, the constant γ and the coefficients A_0 and B_0 are replaced by their values at P'. Thus the uniform asymptotic solution in the shadow region of the caustic keeps the same form as in (4.29) but with $X = k^{2/3}\zeta_e(P') > 0$. This solution matches with the solution in the lit region when $X \to 0$ and decreases exponentially in the shadow region when X becomes large, as it should. Indeed, it can be readily shown that the dominant terms of the asymptotic expansion of $A_i(X)$ and $A_i'(X)$ for X positive and large are given by

$$A_i(X) \cong \frac{1}{2\sqrt{\pi}}(X)^{-1/4}\, e^{-2/3\, X^{3/2}}, \qquad A_i'(X) \cong -\frac{1}{2\sqrt{\pi}}(X)^{1/4}\, e^{-2/3\, X^{3/2}}. \tag{4.46}$$

4.2.2. OBSERVATION POINT LOCATED CLOSE TO A LINE CUSP OF A CAUSTIC

At an observation point located on the lit side of the caustic, close to a line cusp, three rays pass that belong to the congruence of rays having the caustic as their envelope—two of them being tangent to one regular branch of the caustic and another being tangent to the other branch (fig. 4.10).

As in section 4.2.1, we have to find the special function characterizing the behavior of the field in the vicinity of the cusp. Since each of the three rays constitutes a stationary path, the special function must be an integral of the form

$$I(k) = \int f(z)e^{ikq(z)}\, dz,$$

where $q(z) = \tau(s)$ has three saddle points. If we employ

$$kq(z) = \tau(s) = Ys + Xs^2 + s^4$$

then we obtain a reference integral, which has been studied by Pearcey [4.6] and is known as the Pearcey function:

$$I(X,Y) = \int_{-\infty}^{+\infty} e^{i\left(s^4 + Xs^2 + Ys\right)}\, ds. \tag{4.47}$$

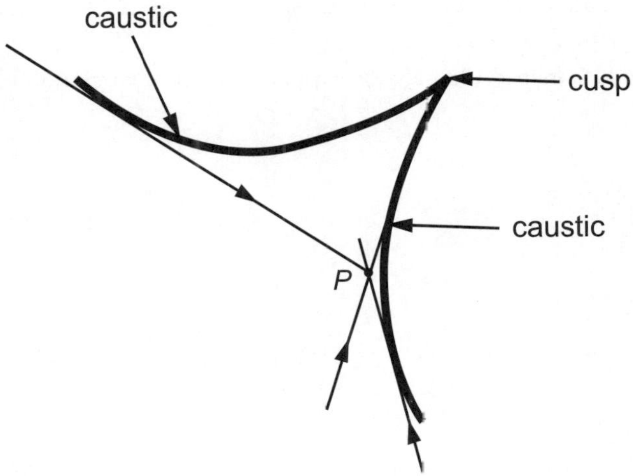

Figure 4.10. Geometry of the rays near a cusp of a caustic.

A summary of the results of Pearcey and certain numerical data can be found in chapter 6 of reference [4.7]. Another approach using the boundary layer method is given in chapter 6, section 6.2.2.2 of reference [4.8] and leads to the same special function.

The relations between X, Y, and the geometrical parameters of the problem are rather complicated. In practical applications, the previously described technique for calculating the field close to a cusp of a caustic is seldom applied. An alternative approach, based on an asymptotic evaluation of the displacement on the boundary and subsequent use of a representation integral to compute the scattered field, is generally preferred.

•

REFERENCES

[4.1] Babich, V. M., and V. S. Buldyrev 1991. *Short Wavelength Diffraction Theory*. Berlin: Springer-Verlag.

[4.2] Felsen, L. B., and N. Marcuvitz. 1972. *Radiation and Scattering of Waves*. Englewood Cliffs, NJ: Prentice-Hall.

[4.3] Olver, F. W. J. 1974. *Asymptotic and Special Functions*. New York: Academic.

[4.4] Kravtsov, Yu. A. 1964. "Asymptotic Solution of Maxwell's Equations Near a Caustic." *Radiofizika* 7: 1049–56.

[4.5] Ludwig, D. 1966. "Uniform Asymptotic Expansions at a Caustic." *Communications on Pure and Applied Mathematics* 19: 215–50.

[4.6] Pearcey, T. 1946. "The Structure of an Electromagnetic Field in the Neighbourhood of a Cusp of a Caustic." *Philosophical Magazine* 37: 311–17.

[4.7] Brekhovskikh, L. M., and O. A. Godin. 1992. *Acoustics of Layered Media*. 2nd ed. Berlin: Springer-Verlag.

[4.8] Bouche, D., and F. Molinet. 1987. *Asymptotic Methods in Electromagnetics*. Berlin: Springer-Verlag.

Chapter 5

HYBRID DIFFRACTION COEFFICIENTS

5.1. INTRODUCTION

Formulas for single fundamental diffraction phenomena that are described by elementary diffraction coefficients (e.g., reflection on general doubly curved surfaces; edge diffraction by a straight or curved wedge; and smooth convex and concave surface diffraction involving creeping waves, whispering gallery waves, and elastic surface waves) have been established in chapters 2 and 3.

In this chapter, we present diffraction coefficients corresponding to two fundamental diffraction phenomena (i.e., edge-diffracted or edge-excited creeping waves on a convex surface). The corresponding diffraction coefficient, which is a combination of the diffraction coefficients of the two elementary diffraction phenomena, is called a "hybrid diffraction coefficient" according to Albersten and Christiansen [5.1], who introduced this term. Our investigation concerns two main topics:

- edge-excited and edge-diffracted waves propagating on a convex surface
- edge-excited and edge-diffracted waves propagating on a concave surface

In each topic, all types of surface waves propagating on a rigid, soft, elastic, or impedance surface will be considered.

5.2. EDGE-EXCITED AND EDGE-DIFFRACTED CREEPING WAVES

We consider a convex surface having a line of discontinuity of the tangent plane, forming a convex three-dimensional wedge. Hybrid diffraction phenomena involving creeping waves are shown in figure 5.1.

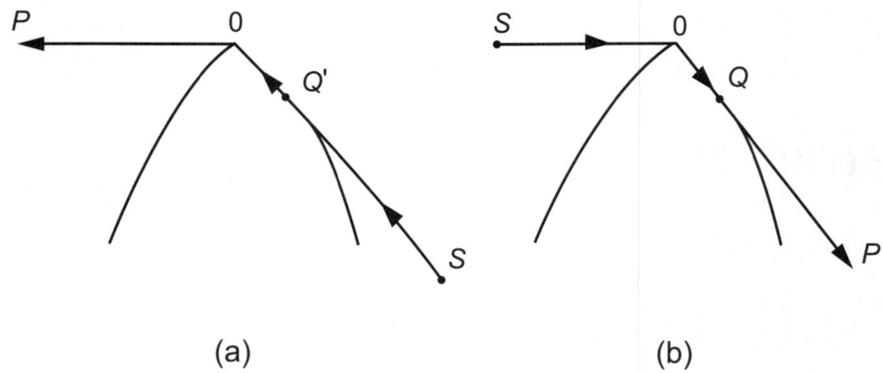

Figure 5.1. Hybrid diffraction phenomena: (a) an edge-diffracted creeping wave and (b) an edge-excited creeping wave.

In figure 5.1(a), a space wave emanating from a source S excites a creeping wave at Q', which is diffracted by the edge of the wedge at O into a space wave reaching the observation point P. In figure 5.1(b), a space wave emanating from a source S is diffracted by the edge of the wedge at O and excites a creeping wave traveling along the surface of the wedge to the point Q, where it is tangentially shed from the surface to the observation point P. We will first treat the diffraction phenomenon shown in figure 5.1(a) by following a procedure from Michaeli [5.2]. In this case, the edge diffraction follows the surface diffraction. The surface field close to the edge is decomposed into a spectrum of local inhomogeneous plane waves, and the diffracted field is obtained by applying to each component of this spectrum the formulas established for the diffraction of a plane wave by the edge of the wedge. This procedure is also known as the spectral theory of diffraction (STD; see reference [5.3]). The solution of the diffraction phenomenon shown in figure 5.1(b) follows figure 5.1(a) by applying the reciprocity theorem.

A critical situation appears at grazing incidence or grazing observation when SO in figure 5.1(b) or OP in figure 5.1(a) are close to the plane tangent at O to the face of the wedge supporting the creeping wave. In this situation, the transition regions of the direct, reflected, and diffracted waves overlap. In this case, the special functions describing the field contain incomplete Airy functions. In [5.4], Michaeli shows that this behavior can be deduced from the solution of the canonical problem of a curved sheet established by Idemen and Felsen [5.5]. Some comments on this subject are also given in reference [5.6].

In what follows, we first consider the spectral decomposition of the Fock field on a smooth, convex, two-dimensional surface. Next we apply the STD to a two-dimensional wedge with convex faces. Then we extend the solution to a three-dimensional wedge.

5.2.1. SPECTRAL REPRESENTATION OF THE FOCK FIELD ON A SMOOTH CONVEX CYLINDRICAL SURFACE

We consider first a two-dimensional smooth convex surface with hard, soft, or impedance boundary conditions. Let P' be the location of a line source and P be an observation point close to the surface at a distance n from its orthogonal projection P_N on the surface (fig. 5.2).

We denote by t the arc length from the launching point Q' of the creeping wave to P_N and suppose that P is in the vicinity of a fixed point O on the creeping ray, which will be the origin of our spectral representation of the field. The lengths ℓ and ℓ'_0 of the curvilinear arcs along the creeping ray are positive on the right of the origin O and negative on the left, so that $t = \ell'_0 - \ell$. According to the formulas established in chapter 3, section 3.3.1.3, the zero-order term of the asymptotic expansion of the total field for hard, soft, or impedance boundary conditions is given by

$$p(P) = p(\xi,n) = p^i(Q')\frac{e^{ikt}}{\sqrt{\pi}}\int_{-\infty+i\varepsilon}^{+\infty+i\varepsilon}\left[V(\tau-h) - \frac{\tilde{\Omega}V(\tau)}{\tilde{\Omega}w_1(\tau)}w_1(\tau-h)\right]e^{i\xi\tau}d, \tag{5.1}$$

where

$$\tilde{\Omega} = -\frac{1}{m}\frac{\partial}{\partial\tau} + iZ, \tag{5.2}$$

$$h = \frac{kn}{m}, \quad m = \left(\frac{k\rho}{2}\right)^{1/3}, \quad \xi = \frac{mt}{a} \tag{5.3}$$

and where $\rho = \alpha$ is the radius of curvature of the surface at the origin O. Formula (5.1) is valid in the penumbra region, where the surface around O can be replaced by its circular cylinder approximation.

Proceeding as in reference [5.6], we expand the functions $v(\tau - h)$ and $w_1(\tau - h)$ in a Taylor series in the parameter h, which is assumed to be small. Then by taking into account $w_1''(\tau) = \tau w_1(\tau)$ and $V''(\tau) = \tau V(\tau)$ in (5.1), we obtain at an observation point close to the origin O:

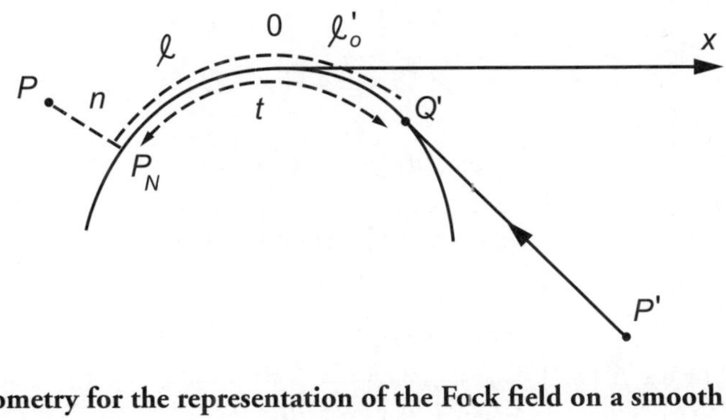

Figure 5.2. Geometry for the representation of the Fock field on a smooth convex surface.

$$p(\xi,n) = p^i(Q') \int_{-\infty+i\varepsilon}^{+\infty+i\varepsilon} e^{i(\ell_0'-\ell)\left(k+\tau\frac{m}{a}\right)} \frac{\left[1+h^2\frac{\tau}{2} - imZ\left(h+\frac{h^3\tau}{6}\right)+O\left(h^4\right)\right]}{\sqrt{\pi}\left[w_1'(\tau)-imZw_1(\tau)\right]}\,d\tau, \qquad (5.4)$$

where t has been replaced by $\ell_0' - \ell$ and where ρ has been replaced by the radius of curvature a of the surface at the origin.

The functions $w_1(\tau)$ and $w_1'(\tau)$ in the denominator of (5.4) increase as $\exp\left(\frac{2}{3}|\tau|^{3/2}\right)$ on the contour $\left(\infty e^{2i\pi/3}, +\infty\right)$, into which the path of integration can be deformed. Hence the dominant contribution to the integral (5.4) comes from $|\tau| \le 1$ and, since h is supposed to be small, we have $\left|h\sqrt{\tau}\right| << 1$. We can therefore write

$$\begin{aligned} h + h^3\frac{\tau}{6} &\cong \frac{1}{i\sqrt{\tau}}\sin\left(ih\sqrt{\tau}\right),\\ 1+h^2\frac{\tau}{2} &\cong \cos\left(ih\sqrt{\tau}\right). \end{aligned} \qquad (5.5)$$

By replacing the sine and cosine functions by exponentials, we finally obtain

$$p(\xi,n) = \int_{\Gamma_2} \left\{A(\tau)\left[C_\tau^i(\ell,n)+C_\tau^r(\ell,n)\right]+B(\tau)\left[C_\tau^i(\ell,n)-C_\tau^r(\ell,n)\right]\right\}d\tau, \qquad (5.6)$$

where Γ_2 is defined in figure 3.11.

The coefficients $C_\tau^{i,r}(\ell,n)$ are given by

$$C_\tau^{i,r}(\ell,n) = \exp\left[-ik\ell\left(1+\frac{\tau}{2m^2}\right) \mp ikn\left(\frac{i\sqrt{\tau}}{m}\right)\right] \qquad (5.7)$$

and

$$\begin{aligned} A(\tau) &= p^i(Q')\,\frac{\exp\left[i\,\ell_0'\left(k+\tau\frac{m}{a}\right)\right]}{2\sqrt{\pi}\left[w_1'(\tau)-imZw_1(\tau)\right]},\\ B(\tau) &= p^i(Q')\,\frac{-imZ\exp\left[i\,\ell_0'\left(k+\tau\frac{m}{a}\right)\right]}{2\sqrt{\pi\xi}\left[w_1'(\tau)-imZw_1(\tau)\right]}. \end{aligned} \qquad (5.8)$$

We see that (5.7) represents quasi-plane waves since if we state $\theta = i\sqrt{\tau}\, m^{-1}$, then for $|\tau| \leq 1$ we have

$$\frac{i\sqrt{\tau}}{m} \cong \sin\theta, \quad 1 + \frac{\tau\, m^{-2}}{2} \cong \cos\theta \tag{5.9}$$

and

$$C_{\tau}^{i,r}(\ell, n) = \exp\left[-ik\left(\ell\cos\theta \pm n\sin\theta\right)\right]. \tag{5.10}$$

In the tangent plane approximation $\ell \cong x\,, n \cong y$, (5.10) represents an inhomogeneous incident and reflected plane wave. Hence (5.6) is locally a spectrum of inhomogeneous plane waves. The amplitude of the spectral component of the incident wave is given by the coefficient of $C^{i}\tau$, which we write in the form

$$A(\tau) + B(\tau) = \frac{p^{i}(Q')\exp\left[i\,\ell'_0\left(k + \tau\dfrac{m}{a}\right)\right]}{2\sqrt{\pi}\,\left[w'_1(\tau) - im\,Z\,w_1(\tau)\right]}\,\frac{2}{1 + R(\theta)}, \tag{5.11}$$

where $R(\theta)$ is the reflection coefficient given by

$$R(\theta) = \frac{\sin\theta - Z}{\sin\theta + Z}. \tag{5.12}$$

We will now apply these results to the construction of hybrid diffraction coefficients for a creeping wave diffracted by the edge of a curved wedge.

5.2.2. HYBRID DIFFRACTION COEFFICIENTS FOR CREEPING WAVES ON A CURVED WEDGE

5.2.2.1. Two-Dimensional Wedge

The notations are defined in figure 5.3. The line source S is in the deep shadow or in the penumbra region corresponding to the reciprocal problem, and the observation point P is supposed far away from the origin O, which is taken at the edge in the direction φ.

According to (5.6) and (5.11), the incident field at a point $M(\ell, n)$ in the vicinity of O is given by the spectrum of inhomogeneous plane waves:

$$p(\ell, n) = \frac{1}{2}\,p^{i}(Q')\,e^{ik\,\ell'_0}\,\frac{1}{\sqrt{\pi}}\int_{\Gamma_2}\frac{e^{i\,\xi'_0\,\tau}}{w'_1(\tau) - im\,Z\,w_1(\tau)}\;\frac{2}{1 + R(\theta)}\,e^{-ik\left(\ell\cos\theta\, + n\sin\theta\right)}\,d\tau, \tag{5.13}$$

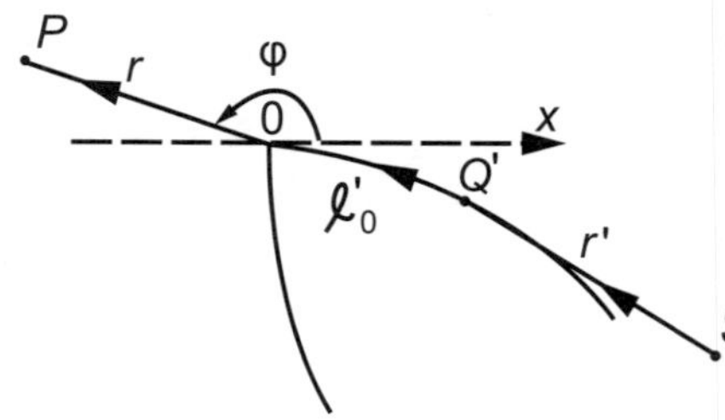

Figure 5.3. Creeping wave diffracted into a space wave.

where $\xi_0' = \dfrac{m\,\ell_0'}{a}$ and where θ is a function of τ defined by (5.9). Applying the STD, we directly obtain the edge-diffracted field at P:

$$p(P) = \frac{1}{2} p^i(Q')\, e^{ik\ell_0'} \left(\frac{1}{\sqrt{\pi}} \int_{\Gamma_2} e^{i\xi_0'\tau} \frac{D(\varphi,\theta;Z)}{w_1'(\tau) - imZ\,w_1(\tau)} \frac{2}{1+R(\theta)} d\theta \right) \frac{e^{ikr}}{\sqrt{r}}, \tag{5.14}$$

where $D(\varphi,\theta;Z)$ is the analytical continuation of the diffraction coefficient of a space wave by the edge of a wedge.

For soft $(Z = \infty)$ or hard $(Z = 0)$ boundary conditions, $D(\varphi,\theta)$ is the analytical continuation of Keller's edge-diffraction coefficient for a wedge given by (2.64). Since these coefficients depend only on trigonometric functions, they vary slowly with $\theta(\tau)$ for φ not close to π. This occurs in the reciprocal problem away from the situation where the transition regions of the direct, reflected, and diffracted rays overlap. Outside this region, which can be defined by $\pi - \varphi \geq O(m^{-1})$ since the angular extent of the transition region is defined by $\xi_0' \cong 1$, Keller's diffraction coefficients may be replaced by the first term of their Taylor series expansion, with respect to θ:

$$\begin{aligned} D_s(\varphi,\theta) &\cong \theta \left[\frac{\partial}{\partial\theta} D_s(\varphi,\theta) \right]_{\theta=0} + O(m^{-3}), \\ D_h(\varphi,\theta) &\cong D_h(\varphi,0) + O(m^{-2}), \end{aligned} \tag{5.15}$$

where $\theta = \dfrac{i\sqrt{\tau}}{m}$.

Inserting (5.15) into (5.14) yields

$$p_s(P) = \frac{1}{2} p^i(Q') \frac{i}{m} f(\xi_0')\, e^{ik\ell_0'} \left[\frac{\partial}{\partial\theta} D_s(\varphi,\theta) \right] \frac{e^{ikr}}{\sqrt{r}}, \tag{5.16}$$

$$p_b(P) = \frac{1}{2} p^i(Q')\, g(\xi_0')\, e^{ik\ell_0'}\, D_b(\varphi,0)\, \frac{e^{ikr}}{\sqrt{r}}, \tag{5.17}$$

where $f(x)$ and $g(x)$ are the Fock functions defined by

$$f(x) = \frac{1}{\sqrt{\pi}} \int_{\Gamma_2} \frac{e^{ix\tau}}{w_1(\tau)}\, d\tau, \quad g(x) = \frac{1}{\sqrt{\pi}} \int_{\Gamma_2} \frac{e^{ix\tau}}{w_1'(\tau)}\, d\tau. \tag{5.18}$$

In (5.16) and (5.17), the parameter ξ_0' is given by $\xi_0' = \dfrac{m\,\ell_0'}{a}$ where $\ell_0' = \text{arc}\, Q'O$, a is the radius of curvature of the surface at O, and $m = \left(\dfrac{ka}{2}\right)^{1/3}$. Formulas (5.16) and (5.17) are in agreement with the result of Pathak and Kouyoumjian [5.7].

By reciprocity, equations (5.16) and (5.17) also describe the diffracted field at S for a line source at P. Interchanging the positions of S and P and, accordingly, the primed parameters with the unprimed ones and using the symmetry properties of $D_{s,h}(\varphi, \theta)$, we obtain (see fig. 5.1[b])

$$p_s(P) = \frac{1}{2} p^i(O) \frac{i}{m} f(\xi_0)\, e^{ik\ell_0} \left[\frac{\partial}{\partial\theta'} D_s(\theta',\varphi)\right]_{\theta'=0} \frac{e^{ikr'}}{\sqrt{r'}}, \tag{5.19}$$

$$p_h(P) = \frac{1}{2} p^i(O)\, g(\xi_0) e^{ik\ell_0} D_h(0,\varphi) \frac{e^{ikr'}}{\sqrt{r'}}, \tag{5.20}$$

where

$$r' = QP, \quad \xi_0 = \frac{m\,\ell_0}{a}, \quad \ell_0 = \text{arc}\, OQ. \tag{5.21}$$

For an impedance boundary condition, $D(\varphi,\theta;Z)$ is the analytical continuation of the Maliuzhinets diffraction coefficient derived in chapter 2, section 2.2.2 and given by (2.193). Since the integrand in (5.14) is rapidly decreasing with $|\tau|$ along the integration path, $|\theta| << 1$ is in the domain of τ, where the integrand is not negligibly small. Hence

$$\left(\frac{\sin\theta + z}{\sin\theta}\right) D(\varphi,\theta;Z) \cong Z \left[\frac{\partial}{\partial\theta} D(\varphi,\theta;Z)\right]_{\theta=0}. \tag{5.22}$$

Consequently, (5.14) may also be written in the form

$$p(P) = \frac{1}{2} p^i(Q') \frac{i}{m} \bar{f}(\xi_0')\, e^{ik\ell_0'} \left[\frac{\partial}{\partial\theta} D(\varphi,\theta;Z)\right]_{\theta=0} \frac{e^{ikr}}{\sqrt{r}}, \tag{5.23}$$

where

$$\bar{f}(x) = \frac{1}{\sqrt{\pi}} \int_{\Gamma_2} e^{ix\tau} \frac{-imZ}{w_1'(\tau) - imZ\,w_1(\tau)}\,d\tau. \tag{5.24}$$

When $|Z| \to \infty$, (5.24) shows that $\bar{f}(x) \to f(x)$. Hence formula (5.23) tends to the result (5.16), obtained for the soft boundary condition. However, when $Z \to 0$, it follows from (5.24) that $\bar{f}(x) \to 0$. Formula (5.23) is therefore not valid for $Z = 0$.

If $|Z|$ is small compared to $|\sin\theta|$, we have the following instead of (5.22):

$$\left(\frac{\sin\theta + Z}{\sin\theta}\right) D(\varphi,\theta;Z) \cong D(\varphi,\theta;Z), \tag{5.25}$$

and when $Z \to 0$, the Maliuzhinets diffraction coefficient tends to zero. Hence we do not recover the result for the hard boundary condition given by (5.17). In fact, when $|\theta|$ is small and $Z \to 0$, Maliuzhinets's solution has to be replaced by a uniform solution with respect to Z since we must also add the contribution of the surface wave to the contribution of the space wave diffracted by the edge. Formula (5.23) is therefore valid for $|Z|$ different from zero and not too small $\left(|Z| >> m^{-1}\right)$.

By reciprocity, the field diffracted at S by a line source at P is given by

$$p(S) = \frac{1}{2} p^i(O) \frac{i}{m} \bar{f}(\xi_0)\, e^{ik\ell_0} \left[\frac{\partial}{\partial\theta'} D(\theta',\varphi;Z)\right]_{\theta'=0} \frac{e^{ikr'}}{\sqrt{r'}}, \tag{5.26}$$

where r', ξ_0, and ℓ_0 are defined in (5.21).

If, in figure 5.3, S lies in the deep shadow region of the reciprocal ray, then $\xi_0' >> 1$ and $f(\xi_0')$ and $g(\xi_0')$ in (5.16) and (5.17), as well as $\bar{f}(\xi_0')$ in (5.23), can be replaced by their residue series corresponding to the creeping modes. However, the radius of curvature of the geodesic followed by the creeping ray is no longer constant. As shown in chapter 3, section 3.3.1.2., they can be replaced by

$$\xi_0' = \int_{Q'}^{O} \frac{m(t)\,dt}{\rho(t)}, \quad m(t) = \left(\frac{k\rho(t)}{2}\right)^{1/3}. \tag{5.27}$$

In addition, the diffracted field has to be multiplied by $\left(\frac{\rho(t)}{a}\right)^{1/6}$ (see chapter 1, section 1.9).

If the source S in figure 5.3 lies above the *ox* axis in the transition region of the diffracted ray in the reciprocal problem, then the ray coordinate system for creeping waves can be extended by introducing the pseudo-ray path $SQ'O$ defined in figure 5.4. (see also the comment after equation (3.231)).

The face of the wedge that is reached by the incident pressure wave is extended continuously behind the edge by a fictitious surface having the same curvature as the real face at O. The incident ray is tangent at Q' to this surface. The field in the vicinity of O is still given by the

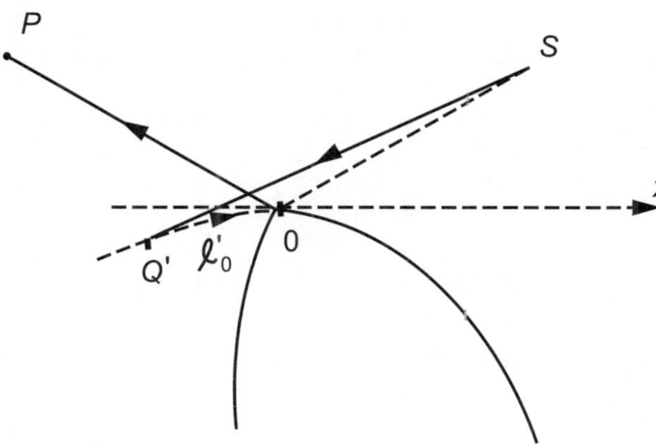

Figure 5.4. Pseudo-ray path.

inhomogeneous plane wave spectrum (5.13), but now $\ell'_0 < 0$ and consequently $\xi'_0 < 0$. Formulas (5.16), (5.17), and (5.23) remain valid, and the edge-diffracted field is continuous when S crosses OX. We will show that, by a proper choice of the parameter ξ'_0, it is possible to match this solution with the geometrical theory of diffraction (GTD) solution when S moves away from the transition region into the region where the incident field is present. In a first step, since the GTD solution is usually written with the phase reference of the incident field on the edge, we express $p^i(Q')$ in (5.14) with respect to $p^i(O)$.

At the periphery of the transition region, the radius of curvature of the fictitious surface at Q' is approximately equal to the radius of curvature a of this surface at O. By using single trigonometric relations and the conditions $|\ell'_0|/a \ll 1$, we get

$$SQ' \cong SO + a\sin\left(\frac{|\ell'_0|}{a}\right) \cong SO + |\ell'_0| - \frac{|\ell'_0|^3}{5a^2}.$$

Hence for $\ell'_0 < 0$ we have

$$p^i(Q')\, e^{ik\ell'_0} = p^i(O)\; e^{i\frac{\xi'^3_0}{3}}. \tag{5.28}$$

At a point M close to O on the face reached by the incident wave, the phase difference is $\phi(M) - \phi(O) \cong -k\ell\cos\alpha$, $\ell > 0$, where α is the angle between OS and the tangent at M to the surface. Moreover, if M is at a distance n from the surface, then we have $\phi(M) - \phi(O) \cong -k(\ell\cos\alpha + n\sin\alpha)$. Hence (5.10) has to be replaced by

$$C_\tau^{i,r}(\ell,n) = \exp\left[-ik\left(\ell\cos(\alpha+\theta) \pm n\sin(\alpha+\theta)\right)\right]. \tag{5.29}$$

For $\alpha = 0$ this equation is identical to (5.10). Therefore, the continuity of the field through the shadow boundary of the diffracted rays (in the reciprocal problem) is still satisfied. With these modifications and the fact that α tends to the angle ψ between OP' and OX when M tends to O, the solution in the part of the transition region reached by the incident field is given by

$$p_s(P) = \frac{1}{2} p^i(O) \frac{i}{m} f(\xi_0') e^{i\frac{\xi_0'^3}{3}} \frac{D_s(\varphi,\psi)}{\psi} \frac{e^{ikr}}{\sqrt{r}}, \tag{5.30}$$

$$p_h(P) = \frac{1}{2} p^i(O) g(\xi_0') e^{i\frac{\xi_0'^3}{3}} D_h(\phi,\psi) \frac{e^{ikr}}{\sqrt{r}} \tag{5.31}$$

for the soft and hard boundary conditions, where we have used the approximation

$$\left[\frac{\partial}{\partial\theta} D_s(\varphi,\psi+\theta)\right]_{\theta=0} = \frac{\partial}{\partial\psi} D_s(\varphi,\psi) \cong \frac{D_s(\varphi,\psi)}{\psi}, \tag{5.32}$$

since for S located in the transition region, ψ is small $\left(\psi \cong m^{-1}\right)$.

Similarly, for an impedance boundary condition, we obtain the following for $|Z| >> m^{-1}$:

$$p(P) = \frac{1}{2} p^i(O) \frac{i}{m} \bar{f}(\xi_0') e^{i\frac{\xi_0'^3}{3}} \frac{D(\varphi,\psi;Z)}{\psi} \frac{e^{ikr}}{\sqrt{r}}. \tag{5.33}$$

When the source S leaves the transition region on the side of the transition region reached by the incident wave, we have $-\xi_0' >> 1$. Thus the Fock functions can be calculated by the steepest descent method by replacing the Airy functions in the integrand by their asymptotic expressions $w_1(\tau) \approx \tau^{-1/4} \exp\left(\frac{2}{3}\tau^{3/2}\right)$ and $w_1'(\tau) \approx \tau^{1/4} \exp\left(\frac{2}{3}\tau^{3/2}\right)$.

For $x < 0$, we obtain

$$f(x) \cong 2ix\, e^{-i\frac{x^3}{3}}, \quad g(x) \cong 2e^{-i\frac{x^3}{3}}, \quad \bar{f}(x) \cong 2i\left(\frac{-imZ}{x - imZ}\right) x e^{-i\frac{x^3}{3}}. \tag{5.34}$$

Inserting (5.34) into (5.30) and (5.31), we obtain the GTD results for soft and hard boundary conditions if we state

$$\xi_0' = -m\psi. \tag{5.35}$$

For an impedance boundary condition with $|Z| >> m^{-1}$, we have $|x| << mZ$ at the periphery of the transition region and consequently, if we state (5.35), the solution (5.33) matches with the GTD solution. Since $\psi = 0$ on the shadow boundary, the continuity of (5.30), (5.31), and (5.33) with ξ_0' replaced by (5.35), with the corresponding edge-diffracted creeping wave solution given by (5.16), (5.17), and (5.23), respectively, is verified.

5.2.2.2. Three-Dimensional Wedge

The generalization of the previously described results to a three-dimensional wedge comprises two steps. In a first step, the solution for a two-dimensional wedge at normal incidence is extended to oblique incidence. In a second step, this last solution is then further extended, somewhat heuristically, to a general three-dimensional wedge by invoking the local nature of the diffraction phenomena at high frequencies.

Consider the creeping wave excited by a planar wave obliquely incident on a two-dimensional wedge (fig. 5.5). Let β be the acute angle between the incident direction and the *oz* axis, which is directed along the edge. A traveling wave dependence allows one to deduce the diffracted field from the solution corresponding to normal incidence by multiplying the latter by $e^{ik_z z}$ and replacing k with $k_t = k_t \sin\beta$. The results for soft or hard boundary conditions deduced from (5.19) and (5.20) are therefore given at the observation point P by

$$p_s(P) = \frac{1}{2} p^i(O) \frac{i}{m_t} f(\xi_{0t}) e^{ik_t \ell_t} \left[\frac{\partial}{\partial \theta'_t} D_s(\theta'_t, \varphi; k_t) \right]_{\theta'_t = 0} \frac{e^{ik_t \rho}}{\sqrt{\rho}} e^{ik_z z}, \tag{5.36}$$

$$p_h(P) = \frac{1}{2} p^i(O)\, g(\xi_{0t}) e^{ik_t \ell_t} D_h(O.\varphi; k_t) \frac{e^{ik_t \rho}}{\sqrt{\rho}} e^{ik_z z}, \tag{5.37}$$

where

$$m_t = \left(\frac{k_t a}{2} \right)^{1/3}, \quad \xi_{0t} = \frac{m_t \ell_t}{a}, \quad \rho = Q_1 P_1, \quad \theta'_t = i\sqrt{\xi}\, m_t^{-1}. \tag{5.38}$$

Since

$$\ell_t = \ell \sin\beta, \; \rho_g = \frac{a}{\sin^2\beta}, \; \rho = s\sin\beta, \; k_t = k\sin\beta, \; k_z = k\cos\beta,$$

where ℓ is the arc of geodesic OQ and $\rho_g = \rho_g(O)$ is its radius of curvature at O, equations (5.36) and (5.37) may also be written as follows:

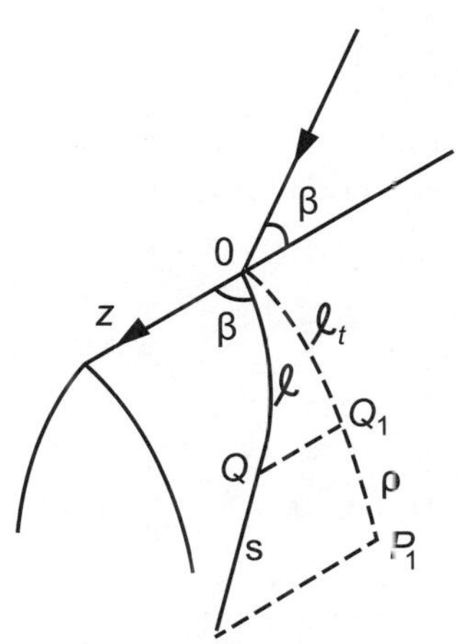

Figure 5.5. Oblique incidence of a plane wave on a two-dimensional wedge.

$$p_s(P) = \frac{1}{2} p^i(O) \frac{i}{m \sin\beta} f(\xi_0) e^{ik\ell} \left[\frac{\partial}{\partial \theta'_t} \frac{1}{\sin\beta} D_s(\theta'_t, \varphi; k_t) \right]_{\theta'_t=0} \frac{e^{iks}}{\sqrt{s}}, \tag{5.39}$$

$$p_h(P) = \frac{1}{2} p^i(O)\, g(\xi_0)\, e^{ik\ell} \left(\frac{1}{\sin\beta} D_h(O, \varphi; k_t) \right) \frac{e^{iks}}{\sqrt{s}}, \tag{5.40}$$

where $D_{sh}(\theta'_t, \varphi; k_t)$ is given by (2.64) and where

$$m = \left(\frac{k \rho_g(O)}{2} \right)^{1/3}, \quad \xi_0 = \frac{m\ell}{\rho_g(O)}. \tag{5.41}$$

For an impedance boundary condition, we obtain the following for $|Z| > m^{-1}$ by using the same procedure:

$$p(P) = \frac{1}{2} p^i(O) \frac{i}{m \sin\beta} \bar{f}(\xi_0) e^{ik\ell} \left[\frac{\partial}{\partial \theta'_t} \left(\frac{1}{\sin\beta} D\left(\theta'_t, \varphi; Z^{(m)}; k \right) \right) \right]_{\theta'_t=0} \frac{e^{iks}}{\sqrt{s}}, \tag{5.42}$$

where, according to (2.203), $Z^{(m)} = \dfrac{Z}{\sin\beta}$ and where $D_{sh}(\theta'_t, \varphi; k_t)$ is given by (2.193). When the observation point P is in the deep shadow zone, we have to adapt (5.36), (5.37), and (5.42) to the variation of the curvature by replacing ξ_0 with

$$\xi_0 = \int_0^{\ell} \frac{m(\ell)\, d\ell}{\rho_g(\ell)} \tag{5.43}$$

and by introducing the factor $\left[\dfrac{\rho_g(O)}{\rho_g(Q)} \right]^{1/6}$.

Similarly, the two-dimensional solution for oblique incidence, in the part of the transition region reached by the incident wave, is given for soft and hard boundary conditions by

$$p_s(P) = \frac{1}{2} p^i(O) \frac{i}{m} f(\xi'_0)\, e^{i\frac{\xi'^3_0}{3}} \frac{1}{\sin\beta} \frac{D_s(\varphi, \psi_t; k)}{\hat{n}\cdot\hat{s}} \frac{e^{iks}}{\sqrt{s}}, \tag{5.44}$$

$$p_h(P) = \frac{1}{2} p^i(O)\, g(\xi'_0)\, e^{i\frac{\xi'^3_0}{3}} \frac{1}{\sin\beta} D_h(\varphi, O; k) \frac{e^{iks}}{\sqrt{s}}, \tag{5.45}$$

where $\hat{n}$ is the unit vector, along the outward normal at O, to the face of the wedge supporting the creeping wave and where $\hat{s}$ is the unit vector along the observation direction.

In (5.44) and (5.45), we have also used the following approximation, valid for ψ_t small:

$$\psi_t \approx \sin\psi_t \approx \frac{\sin\psi}{\sin\beta} = \frac{\cos\gamma^i}{\sin\beta} = \frac{\hat{n}\cdot\hat{s}}{\sin\beta}, \quad \gamma^i = \frac{\pi}{2} - \psi \tag{5.46}$$

and

$$\xi_0' = -m_t\,\psi_t \cong -m\sin\psi = -m\cos\gamma^i . \tag{5.47}$$

For an impedance boundary condition, we obtain, similarly for $|Z| > m^{-1}$,

$$p(P) = \frac{1}{2}p^i(O)\frac{i}{m}\bar{f}(\xi_0')\,e^{i\frac{\xi_0'^3}{3}}\,\frac{1}{\sin\beta}\,\frac{D\left(\varphi,\psi_t;Z^{(m)};k\right)}{\hat{n}\cdot\hat{s}}\,\frac{e^{iks}}{\sqrt{s}}. \tag{5.48}$$

By using (5.34), it can readily be verified that in the region reached by the incident pressure field, away from the transition zone, the formulas (5.44), (5.45), and (5.48) tend to the GTD formulas for the two-dimensional edge-diffracted field.

The generalization of these formulas to an arbitrary incident field diffracted by a three-dimensional wedge is straightforward. Invoking the local nature of the diffraction phenomena, we replace, locally in the vicinity of the point of diffraction Q_E determined by applying Fermat's principle, the actual surface with a cylindrical surface whose generatrices are parallel to the tangent to the edge at Q_E. In the transition region, the formulas obtained for a two-dimensional wedge are then applied locally and extended away from the transition region on the shadow side by replacing ξ_0 in (5.39), (5.40), and (5.42) with

$$\xi_0 = \int_0^{\ell} \frac{m\,d\ell}{\rho_g}, \quad m = \left(\frac{k\rho_g}{2}\right)^{1/3}, \tag{5.49}$$

where $\rho_g(\ell)$ is the local radius of curvature of the geodesic followed by the creeping ray on the three-dimensional surface. Likewise, in the region reached by the incident field, ξ_0' in (5.44), (5.45), and (5.48) is replaced by

$$\xi_0' = -m(\hat{n}\cdot\hat{s}), \quad m = \left(\frac{k\rho_g(O)}{2}\right)^{1/3}, \tag{5.50}$$

where $\hat{n}$ is the unit normal to the reference face at Q_E, $\rho_g(O)$ is the radius of curvature of the geodesic at Q_E, and $\hat{s}$ is the unit vector along the observation direction. It is important to observe that the cylindrical approximation of the actual surface at Q_E is defined by the direction of the tangent to the edge at Q_E, the angles of obliqueness β and $\rho_g(O)$, and not by the actual radius of curvature of the normal cross section of the reference face at Q_E.

In addition to this modification, it is necessary to introduce in (5.39), (5.40), and (5.41) the divergence factor of the creeping rays $\left[\rho_g(Q_E)/\rho_g(Q)\right]^{1/2}$. This is because the geodesics on a three-dimensional surface are no longer parallel and also because the factor $\left[\rho_g(Q_E)/\rho_g(Q)\right]^{1/6}$ depends on the variation of the thickness $h = kn / m$ of the boundary layer. Furthermore, we have to adapt the divergence factor of the space rays emitted at Q in the shadow region by replacing $1/\sqrt{s}$ in the previously stated formulas with

$$A = \sqrt{\frac{\rho^d}{s\left(\rho^d + s\right)}}, \tag{5.51}$$

where ρ^d is the distance of Q to the caustic of the space rays emitted at Q, in the plane tangent to the three-dimensional surface at Q. Likewise, the divergence factor $1/\sqrt{s}$ in (5.44), (5.45), and (5.48) must be replaced by

$$A' = \sqrt{\frac{\rho^e}{s\left(\rho^e + s\right)}}, \tag{5.52}$$

where ρ^e is the radius of curvature of the edge-diffracted wave in the plane of diffraction defined by the tangent to the edge at Q_E and the direction of observation. The expressions (5.51) and (5.52) of the divergence factors have been derived in chapter 1, sections 1.9 and 1.8, respectively.

The three-dimensional solutions for soft and hard boundary conditions, obtained with the previously stated modifications to the two-dimensional solution, are uniform through the transition region of the edge-diffracted rays. They were presented for the first time in reference [5.8]. For an impedance boundary condition, the three-dimensional solution, like the two-dimensional solution, matches with the GTD solution at the periphery of the transition region.

5.2.2.3. Solution Valid at Grazing Incidence and Grazing Observation

Grazing incidence and grazing observation happen when the source is located in the transition region I and when the observation point is located in the transition region II of figure 5.6. In this case, the transition regions corresponding to the shadow boundaries of the edge-diffracted rays overlap with the transition regions corresponding to the shadow boundaries of the reflected and direct rays. When this happens, the diffracted coefficient $D(\varphi, \varphi'; Z)$ becomes very large since it has a singularity when $\varphi' = 0$ and $\varphi = \pi$. Hence it is no longer possible to consider $D(\varphi, \varphi'; Z)$ as a slowly verifying function in (5.14). In addition, the reflected field has a caustic close to the edge and consequently the tangent plane approximation used in the spectral method is no longer valid.

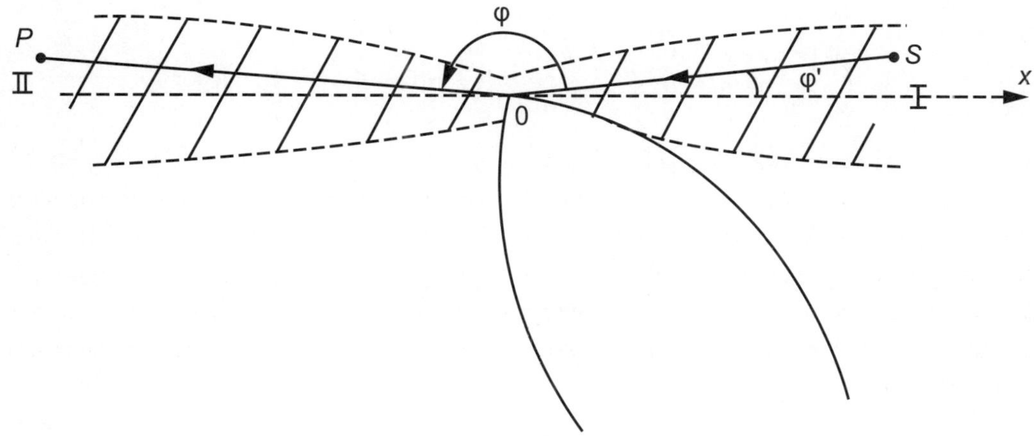

Figure 5.6. Grazing incidence and observation on a wedge.

Another method to treat the problem needs to be found. In 1980, Idemen and Felsen [5.5] solved a new canonical problem: the diffraction of a whispering gallery mode by a truncated cylindrical shell. Their approach, based on the solution of a Hilbert (or Wiener-Hopf) problem, also applies to the creeping waves excited on the convex face of the cylindrical shell. The solution of this canonical problem was the starting point for a refined approach to the edge-diffraction problem in the transition regions of the diffracted rays, developed several years later by Michaeli [5.4]. We do not give the details of his approach; it is described in chapter 5 of reference [5.9]. The main result is that the edge-diffracted field by a two-dimensional wedge for soft, hard, and impedance boundary conditions for a source S in the shadow part of region I and the observation P in the region II (see fig. 5.6) is given by

$$p(P) = \frac{1}{2} p^i(Q') e^{ik\ell'_0}\, \tilde{S}(\xi'_0, \xi_0) \frac{e^{ikr}}{\sqrt{r}}, \tag{5.53}$$

where

$$\tilde{S}(\xi'_0, \xi_0) = 4m\sqrt{\pi}\, e^{i\frac{\xi_0'^3}{3}}\, \tilde{M}(\xi'_0 - \xi_0, \xi_0) \tag{5.54}$$

with ξ'_0 and ξ_0 defined by $\xi'_0 = m\ell'_0 a$ and $\xi_0 = m\psi$, $\psi = \pi - \varphi$, respectively, and where $\tilde{M}(x, y)$ is the transition function defined by

$$\tilde{M}(x, y) = \frac{1}{2\pi} \int_{\Gamma_2} e^{ix\tau} \frac{\frac{\partial I_1}{\partial \tau}(\tau, y) - imZ\, I_1(\tau, y)}{w'_1(\tau) - im\, Z\, w_1(\tau)}\, d\tau. \tag{5.55}$$

In (5.55), $I_1(\tau, y)$ is the incomplete Airy function given by

$$I_1(\tau, y) = \int_y^{\infty} e^{i\left(t\tau - \frac{t^3}{3}\right)} dt. \tag{5.56}$$

Formula (5.55) remains valid for $|Z| \to 0$ and $|Z| \to \infty$.

In order to get the total edge-diffracted field it is necessary to add to (5.53) a correction term, called Ufimtsev's correction, as shown in reference [5.9]. The solution (5.53) is nonuniform when the source or the observation points leave the overlapping transition regions I or II of figure 5.6. A uniform solution for a two-dimensional wedge is derived in reference [5.4]. This uniform solution has been extended in reference [5.10] to a three-dimensional wedge reached by a plane wave at grazing incidence.

5.3. EDGE-EXCITED AND EDGE-DIFFRACTED SURFACE WAVES

5.3.1. *IMPEDANCE WEDGE*

The field diffracted by the edge of a straight impedance wedge, which has been treated in chapter 2, section 2.2.2, behaves like the field radiated by an anisotropic line source located on an infinite impedance plane. In order to establish the equivalence between both problems, we consider first the case of an isotropic line source parallel to an impedance plane.

The problem is two-dimensional. Let $S(\boldsymbol{\rho}_0)$ and $M(\boldsymbol{\rho})$ be the source and the observation point, respectively, in a plane perpendicular to the line source. The field of a line source in an infinite homogeneous fluid is given by

$$p(\boldsymbol{\rho}) = \frac{i}{4} H_0^{(1)}\left(k|\boldsymbol{\rho} - \boldsymbol{\rho}_0|\right), \tag{5.57}$$

where k is the propagation constant in the fluid and where $\boldsymbol{\rho}_0$ and $\boldsymbol{\rho}$ are the position vectors of S and M in the *xy* plane. By replacing the Hankel function with its spectral representation in terms of inhomogeneous plane waves, we obtain, for a line source parallel to the *oz* axis in front of an impedance plane merging with the *xy* plane,

$$p(\boldsymbol{\rho}) = \frac{i}{4} H_0^{(1)}\left(k|\boldsymbol{\rho} - \boldsymbol{\rho}_0|\right) + \frac{i}{4\pi} \int_C R(\xi) e^{ik|\boldsymbol{\rho} - \boldsymbol{\rho}_0'|\cos(\xi - \alpha)} d\xi, \tag{5.58}$$

where $S'(\rho_0')$ is the image of S in the *xy* plane and where $R(\xi)$ is the reflection coefficient given by

$$R(\xi) = \frac{\sin\xi - Z}{\sin\xi + Z}, \tag{5.59}$$

where Z is the relative impedance of the plane. The angles ξ and α are defined in figure 5.7. The contour of integration C in the complex ξ-plane is shown in figure 5.8.

The contour C can be deformed into the steepest descent path (SDP) Γ crossing the $R_e\xi$ axis at the saddle point $\xi = \alpha$. Since the second term on the right-hand side of (5.58) corresponds to the scattered field p_s, we obtain

$$p_s(\boldsymbol{\rho}) = \frac{i}{4\pi}\left[\int_{\Gamma} R(\xi)\, e^{ik|\boldsymbol{\rho}-\boldsymbol{\rho}_0'|\cos(\xi-\alpha)} d\xi - 2\pi i \sum_n Q_n\right], \tag{5.60}$$

where Q_n are the residues at the poles of the reflection coefficient crossed by the deformation of the contour C into the contour Γ.

The asymptotic evaluation of the integral on the right-hand side of (5.60) at the saddle point $\xi = \alpha$ gives the reflected field at M,

$$p_R(\boldsymbol{\rho}) = \frac{e^{i\frac{\pi}{4}}}{\sqrt{8\pi k}} \frac{\sin\alpha - Z}{\sin\alpha + Z} \frac{e^{ikr}}{\sqrt{r}} = D_s \frac{\sin\alpha - Z}{\sin\alpha + Z} \frac{e^{ikr}}{\sqrt{r}}, \tag{5.61}$$

where $r = S'M$. The contribution of the poles gives the surface waves.

For a constant impedance, the poles are the roots of the characteristic equation

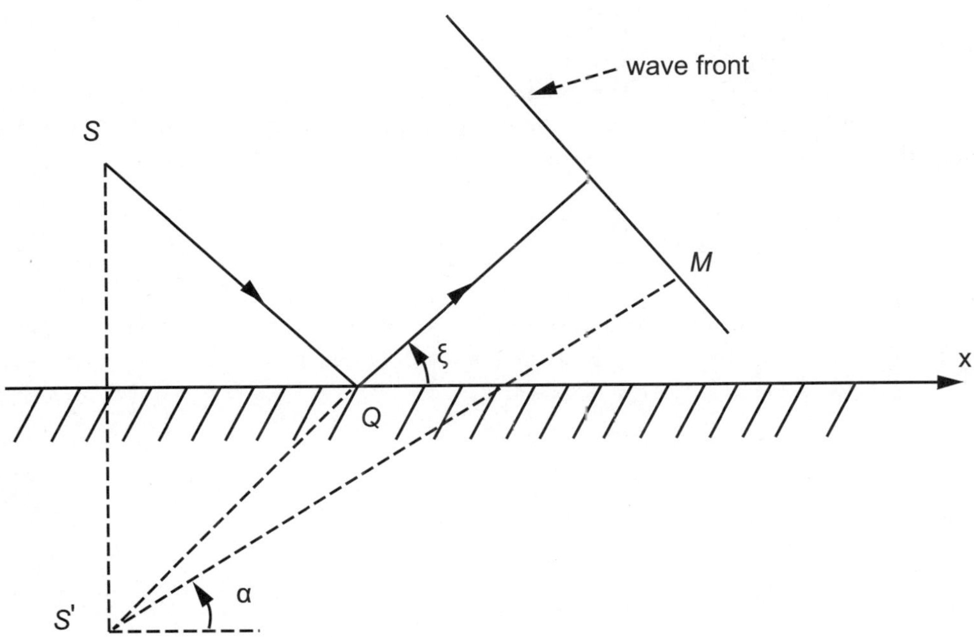

Figure 5.7. Schematic representation of the complex ray path and reflected wave front.

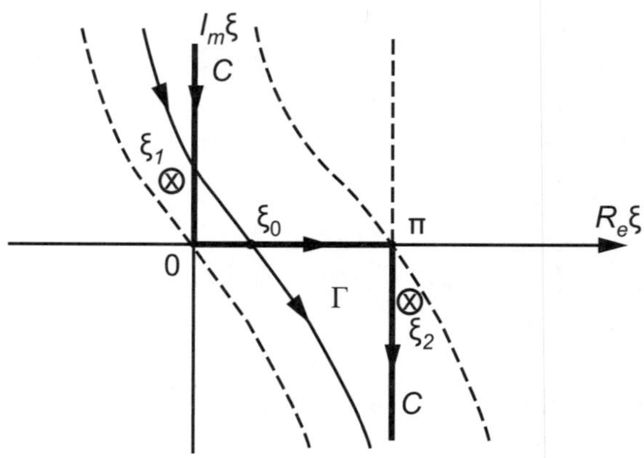

Figure 5.8. Contour *C* and steepest descent path (SDP) *Γ* in the complex *ξ*-plane.

$$\sin\alpha = -Z. \tag{5.62}$$

Let θ be a complex angle defined by $Z = \sin\,\theta$. Then the poles, which may be crossed by the deformation of C into Γ, are given by

$$\xi_1 = -\theta\,,\, \xi_2 = \pi + \theta.$$

For a passive surface ($R_e Z > 0$), these poles correspond to surface waves when ($R_e\theta > 0$) and ($I_m\theta < 0$). They are therefore located in the complex ξ-plane, as shown in figure 5.8. Consequently, the field of the surface wave propagating in the direction $x > 0$ is given by

$$P_{sw}(\boldsymbol{\rho}) = tg\,\xi_1\, e^{ik|\boldsymbol{\rho}-\boldsymbol{\rho}_0'|\cos(\xi_1-\alpha)}U(\alpha-\alpha_\ell), \tag{5.63}$$

where $U(\cdot)$ is the Heaviside function and where α_ℓ is the value of α corresponding to the limit of existence of the surface wave when ξ_1 is situated on the SDP. Since the equation of Γ is

$$\cos(R_e\xi-\alpha)ch(I_m\xi) = 1,$$

we obtain

$$\cos(R_e\xi_1-\alpha_\ell)ch(I_m\xi_1) = 1$$

and consequently

$$\alpha_\ell = R_e\xi_1 - \text{Arc}\cos\left[\frac{1}{ch\left(I_m\xi_1\right)}\right]. \tag{5.64}$$

The phase term in (5.63) is compatible with the Fermat principle extended to complex rays (see chapter 1, section 1.3). When the distance from the source S to the impedance plane tends to zero, (5.63) reduces to

$$P_{sw}(\boldsymbol{\rho}) = tg\,\xi_1\, e^{ik\rho\cos(\xi_1-\alpha)}\, U(\alpha-\alpha_\ell), \tag{5.65}$$

where $\rho = SM$.

We now consider the field radiated by an anisotropic line source parallel to an impedance plane. Suppose that its radiation in an unbounded fluid, at large distance $|\boldsymbol{\rho}-\boldsymbol{\rho}_0|$ from the source, is given by

$$p_i(\boldsymbol{\rho}) = \frac{e^{i\frac{\pi}{4}}}{\sqrt{8\pi k}}\, A(\alpha)\, \frac{e^{ik|\boldsymbol{\rho}-\boldsymbol{\rho}_0|}}{\sqrt{|\boldsymbol{\rho}-\boldsymbol{\rho}_0|}}, \tag{5.66}$$

where α is the angle between $|\boldsymbol{\rho}-\boldsymbol{\rho}_0|$ and the ox axis. Formula (5.66) is also the asymptotic evaluation of the integral

$$I = \frac{i}{4\pi}\int_C A(\xi)\, e^{ik|\boldsymbol{\rho}-\boldsymbol{\rho}_0|\cos(\xi-\alpha)}\, d\xi. \tag{5.67}$$

Formula (5.67) is a spectrum of inhomogeneous plane waves multiplied by the complex weight function A(ξ). In the presence of an impedance plane, (5.67) generates a surface wave given by

$$P_{sw}(\boldsymbol{\rho}) = tg\,\xi_1\, A(\xi_1)\, e^{ik|\boldsymbol{\rho}-\boldsymbol{\rho}_0'|\cos(\xi_1-\alpha)}\, U(\alpha-\alpha_\ell). \tag{5.68}$$

If we replace $A(\alpha)$ by the diffraction coefficient $D(\varphi_i, \alpha)$ of a straight impedance wedge, we obtain the following for a line source located in the impedance plane:

$$P_{sw}(\boldsymbol{\rho}) = tg\,\xi_1\, D(\varphi_i, \xi_1)\, e^{ik\rho\cos(\xi_1-\alpha)}\, U(\alpha-\alpha_\ell), \tag{5.69}$$

where $\rho = SM$, φ_i is the angle of incidence with the ox axis and where α is the polar angle of $\boldsymbol{\rho}$.

It can be shown that (5.69) is identical to the surface wave generated by the scattering of an incident plane wave of unit amplitude by an impedance wedge. Similarly, if we consider a surface wave propagating toward the edge of a straight wedge, the field diffracted by the edge of the wedge, at the distance r from the edge, in the direction making an angle φ_d with the face supporting the surface wave is given by

$$P_d(r,\varphi_d) = P_{sw}(Q_1)\, D(\pi - \xi_1, \varphi_d)\, D_s \,\frac{e^{ikr}}{\sqrt{r}}, \tag{5.70}$$

where

$$P_{sw}(Q_1) = tg\,\xi_1\; e^{ik\rho_0' \cos(\xi_1 - \alpha)} \tag{5.71}$$

and where $\boldsymbol{\rho}_0'$ is the distance of the image of the source that generates the surface wave to the edge and where α is the angle between $\boldsymbol{\rho}_0'$ and the ox axis. If the source is in the face of the wedge, which happens when the surface wave is excited by a space wave at a first edge located at Q_0 in the same plane (fig. 5.9), the incident surface wave is given by

$$P_{sw}(Q_1) = A(\xi_1)\, tg\,\xi_1\; e^{ikL\cos\xi_1}, \tag{5.72}$$

where L is the distance Q_0Q_1 between both edges, $A(\xi_1) = 1$ for an isotropic line source, $A(\xi_1) = D(\varphi_0, \xi_1)$ for an edge-excited surface wave, and the incident space wave makes an angle φ_0 with the face supporting the surface wave.

5.3.2. *ELASTIC WEDGE-SHAPED SHELL*

As in section 5.3.1, we consider first an isotropic line source parallel to an elastic infinite plate separating a fluid from vacuum. The only difference with the impedance surface is the expression of the reflection coefficient, which now reads

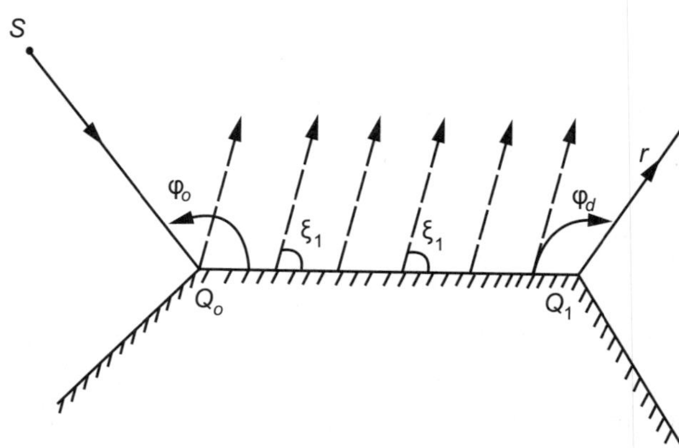

Figure 5.9. Edge-excited surface wave at Q_0 diffracted into a space wave at Q_1.

$$R(\xi)=\frac{\sin\xi-\eta(\xi)}{\sin\xi+\eta(\xi)}, \tag{5.73}$$

where $\eta(\xi)$ is the impedance of the elastic plate divided by the impedance of the fluid. The roots of the characteristic equation

$$\sin\xi+\eta(\xi)=0$$

give the different types of surface waves (see appendix C) that may be excited on the plate. The pressure field of each wave is given by the residue of the integral (5.58) at the pole corresponding to this wave multiplied by ($-2\pi i$). At the pole ξ_p it yields the following instead of (5.63):

$$P_{sw}=\frac{\sin\xi_p}{\cos\xi_p+\left(\frac{\partial\eta}{\partial\xi}\right)_{\xi=\xi_p}}e^{ik|\boldsymbol{\rho}-\boldsymbol{\rho}_0'|\cos(\xi_p-\alpha)}U\left(\alpha-\alpha_\ell^{(p)}\right), \tag{5.74}$$

where

$$\alpha_\ell^{(p)}=R_e\xi_p-Arc\cos\left[\frac{1}{ch\left(I_m\,\xi_p\right)}\right]. \tag{5.75}$$

Similarly, the diffraction of a surface wave propagating along a face of a straight elastic wedge-shaped shell, by its edge, is given by

$$P_d(r,\varphi_d)=P_{sw}(Q_1)\,D\left(\pi-\xi_p,\varphi_d\right)D_s\,\frac{e^{ikr}}{\sqrt{r}}, \tag{5.76}$$

where

$$P_{sw}(Q_1)=\frac{\sin\xi_p}{\cos\xi_p+\left(\frac{\partial\eta}{\partial\xi}\right)_{\xi=\xi_p}}e^{ik\rho_0'\cos(\xi_p-\alpha)}, \tag{5.77}$$

the significance of ρ_0' and α being the same as in (5.71).

These formulas can also be applied to multiple diffraction processes involving space and surface waves. In particular, for a shell having the shape shown in figure 5.9, the doubly diffracted field involving a surface wave is given by

$$P_d(r,\varphi_d)=D(\varphi_0,\xi_p)\frac{\sin\xi_p}{\cos\xi_p+\left(\frac{\partial\eta}{\partial\xi}\right)_{\xi=\xi_p}}e^{ikL\cos\xi_p}\,D\left(\pi-\xi_p,\varphi_d\right)D_s\,\frac{e^{ikr}}{\sqrt{r}}. \tag{5.78}$$

Formula (5.78) applies in particular to the excitation and diffraction of a Scholte-Stoneley wave or a Lamb wave by the edge of a wedge-shaped shell.

Formula (5.78) has been applied to a rectangular waveguide made of duralumin. It is immerged in water, and the medium inside the waveguide is vacuum. The dimensions of the cross section are specified in figure 5.10.

An incident plane wave of frequency 3.2 MHz and unit amplitude impinges the edge *A* at an angle of 38° with the face *AD*. A Scholte-Stoneley wave is excited at *A* and diffracted at *B* into a space wave. The modulus in decibels of the field of the diffracted wave at *B* is represented as a function of the observation angle γ in figure 5.11. The directions $\gamma = 0$ and $\gamma = -90°$ correspond to the directions defined by *AB* and *BC*, respectively. As expected, the amplitude of the diffracted field is maximum in the direction $\gamma = 0$. There are also Lamb waves excited at *A* and diffracted at *B*, but their contribution is not included in the results shown.

5.3.3. ELASTIC SURFACE WAVE ON A CURVED WEDGE

The formula giving the field of an elastic surface wave propagating on a curved elastic shell was established in chapter 2, section 2.3.3.7 for a circular cylindrical shell at normal incidence and in chapter 2, section 2.3.3.9 for a noncircular cylindrical shell at normal and oblique incidence. We limit our analysis to a circular cylindrical curved wedge-shaped shell at normal incidence, since its generalization to a noncircular cylindrical shell follows the same rules as those applied in chapter 2, section 2.3.3.9.

The field of an elastic surface wave of propagation constant $k_{s\ell} = k_1 \sin\theta_\ell$, excited at a point Q_ℓ on a circular cylindrical shell and observed at a point *M* on the surface of the shell (fig. 5.12), is obtained by letting the distance s_ℓ in (2.457) tend to zero.

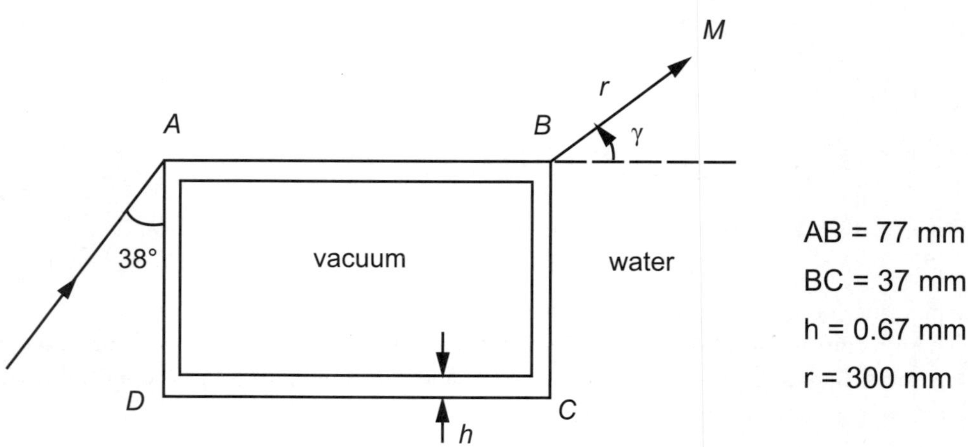

Figure 5.10. Rectangular waveguide immerged in water.

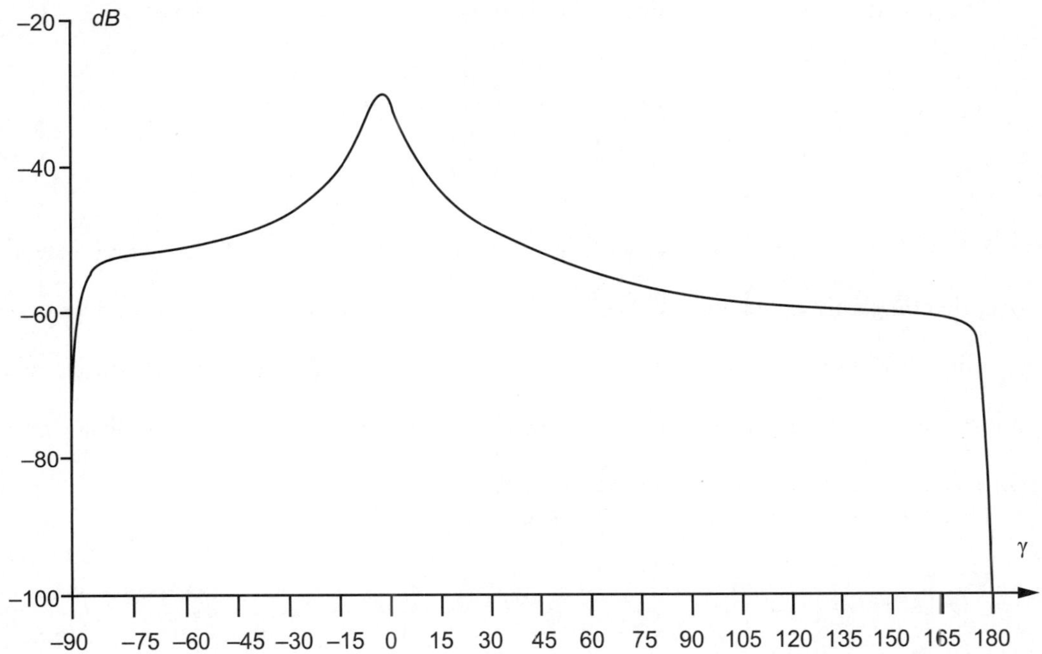

Figure 5.11. Diffraction at the edge *B* of a Scholte-Stoneley wave excited at *A*.

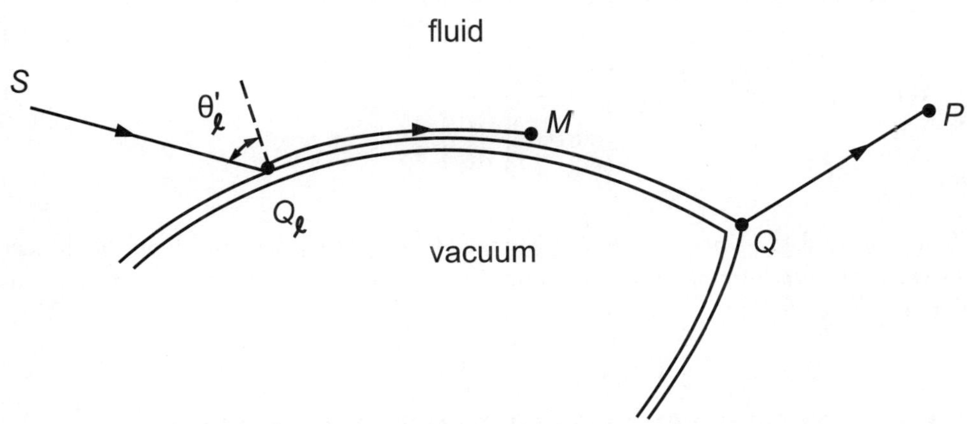

Figure 5.12. Circular cylindrical curved wedge-shaped shell.

The field at M on the surface, due to the elastic surface wave, is therefore given by

$$p(M) = p^i(Q_\ell)\, D_s^{-1}\, \frac{\sqrt{a\cos^3\theta'_\ell}}{\sin\theta'_\ell - \left(\dfrac{\partial\eta}{\partial\theta}\right)_{\theta=\theta'_\ell}}\, e^{ik_1 \sin\theta_\ell\, \sigma_\ell(M)}, \tag{5.79}$$

where $\sigma_\ell(M)$ is the curvilinear distance from Q_ℓ to M. Since $\theta_\ell = \theta'_\ell + i\,\theta''_\ell$ is a complex angle, we see that the elastic surface wave at M is an inhomogeneous plane wave. If M is at the edge Q of the wedge-shaped shell, the diffraction coefficient at Q is given by the diffraction coefficient of a complex incident plane wave, making a complex angle $\frac{\pi}{2} + \theta_\ell$ with the face of the straight elastic wedge-shaped shell tangent to the curved shell at Q.

Therefore, we obtain the following for the diffracted field at an observation point P:

$$p^d(P) = p^i(Q_\ell)\, \frac{\sqrt{a\cos^3\theta'_\ell}}{\sin\theta'_\ell - \left(\dfrac{\partial\eta}{\partial\theta}\right)_{\theta=\theta'_\ell}}\, e^{ik_1 \sin\theta_\ell\, \sigma_\ell(Q)}\, D\!\left(\frac{\pi}{2} + \theta_\ell, \varphi_d\right) \frac{e^{i k_1 QP}}{\sqrt{QP}}, \tag{5.80}$$

where $D(\varphi_i, \varphi_d)$ is a Maliuzhinets diffraction coefficient given by (2.193), in which $\psi(\alpha)$ is given by (2.218) and (2.222).

Proceeding as in sections 5.3.1 and 5.3.2, by introducing into the diffraction problem the concept of an anisotropic line source radiating in the presence of a curved elastic shell, the field on the surface of the edge-excited elastic surface waves is given by

$$p(M) = p_i(Q)\; D\!\left(\varphi_i, \frac{\pi}{2} - \theta_\ell\right) \frac{\sqrt{a\cos^3\theta'_\ell}}{\sin\theta'_\ell - \left(\dfrac{\partial\eta}{\partial\theta}\right)_{\theta=\theta'_\ell}}\, e^{ik_1 \sin\theta_\ell\, \sigma(M)}. \tag{5.81}$$

Since for an elastic surface wave, θ''_ℓ is small, θ_ℓ can be replaced by θ'_ℓ in the diffraction coefficient D in both formula (5.80) and formula (5.81). However, θ''_ℓ must be retained in the phase term.

5.4. EDGE-EXCITED AND EDGE-DIFFRACTED WHISPERING GALLERY WAVES

In chapter 2, section 2.4, we established the formulas giving the pressure field of a line source in front of a concave hard, soft, or impedance surface. These formulas depend on the curvilinear distance ℓ between the projections of the source and the observation points on the concave surface and on their heights n' and n above the surface.

For $\sigma = m\ell / a$ of small value, $h = kn / m \ll 1$ and $h' = kn' / m \ll 1$, where m is the Fock parameter and a is the radius of the cylinder, the field at the observation point is given by (2.560) to (2.563). These formulas can be readily extended to a concave circular shell with a fluid on the concave side and vacuum on the convex side by replacing Z with the equivalent impedance $\eta(\xi)$, given by (2.398).

We consider the case of a line source located on the concave surface. Here, $h' = 0$ and since the Wronskian between w_1 and w_2 is equal to $2i$, we obtain

$$F_\xi(0,Z) = \frac{2i}{\Omega w_2(\xi)}.$$

By expanding $F_\xi(h)$ in a Taylor series with respect to h, it is possible to represent $S^{(0)}(\sigma, h, 0)$ in the form of a spectrum of local inhomogeneous plane waves. Applying the STD to each component of this spectrum leads to the solution of the diffraction problem of a whispering gallery-type field excited by a line source, edge in a concave impedance surface, or a concave elastic shell (fig. 5.13).

For a rigid surface, the diffracted field at an observation point P in the fluid is given by

$$p^d(P) = -D_s \frac{e^{ik\ell}}{\sqrt{\ell}} \bar{S}_b(\sigma) D_b(0,\varphi_d) \frac{e^{ikr}}{\sqrt{r}}, \tag{5.82}$$

where $r = AP$ and where

$$\bar{S}_b(\sigma) = \frac{e^{-i\frac{\pi}{4}}}{2}\left(\frac{\sigma}{\pi}\right)^{1/2} \int_{C_\xi} \frac{e^{i\sigma\xi}}{V'(\xi)\,w_2'(\xi)} d\xi. \tag{5.83}$$

$D_b(0, \varphi_d)$ is the diffraction coefficient for a hard wedge given by (2.64).

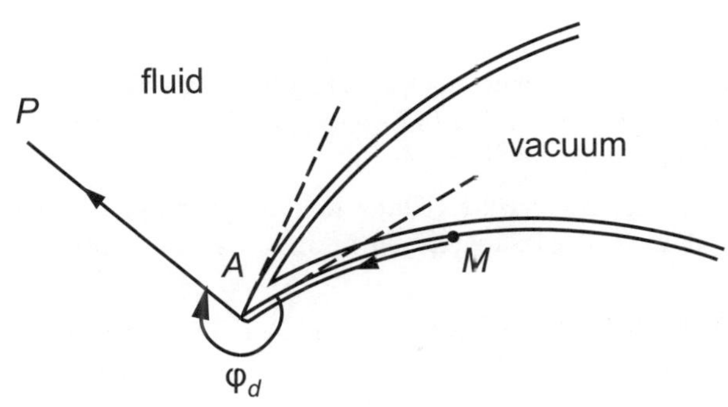

Figure 5.13. Wedge in a cylindrical concave shell.

By applying the reciprocity theorem to (5.82), we obtain the field at a point M on a concave rigid surface:

$$p(M) = -p^i(A) D_s \frac{e^{ik\ell}}{\sqrt{\ell}} \bar{S}_b(\sigma) D_b(\varphi_i, 0). \tag{5.84}$$

Similarly, for a soft surface we obtain

$$\frac{\partial p}{\partial n}(M) = p^i(A) D_s \frac{e^{ik\ell}}{(\ell)^{3/2}} \bar{S}_s(\sigma) \left[\frac{\partial}{\partial \theta} D_s(\varphi_i, \theta) \right]_{\theta=0}, \tag{5.85}$$

where

$$\bar{S}_s(\sigma) = \frac{e^{-3i\frac{\pi}{4}}}{2} \left(\frac{\sigma^3}{\pi} \right)^{1/2} \int_{C_\xi} \frac{e^{i\sigma\xi}}{V(\xi) \mathrm{w}_2(\xi)} d\xi \tag{5.86}$$

and where $D_s(\varphi_i, \theta)$ is the diffraction coefficient for a soft wedge given by (2.64), in which the minus sign has to be chosen. Similar formulas can be readily derived for a wedge in a concave impedance surface or a concave shell.

REFERENCES

[5.1] Albertsen, N. C., and P. L. Christiansen. 1978. "Hybrid Diffraction Coefficients for First- and Second-Order Discontinuities of Two-Dimensional Scatterers." *SIAM Journal of Applied Mathematics* 34: 398–414.

[5.2] Michaeli, A. 1989. "Transition Functions for High-Frequency Diffraction by a Curved Perfectly Conducting Wedge." *IEEE Transactions on Antennas and Propagation* 37: 1073–85.

[5.3] Mittra, R., Y. Rhamat-Samii, and W. L. Ko. 1976. "Spectral Theory of Diffraction." *Applied Physics* 10: 1–13.

[5.4] Michaeli, A. 1989. "Transition Functions for High-Frequency Diffraction by a Curved Perfectly Conducting Wedge, Part 3: Extension to Overlapping Transition Regions." *IEEE Transactions on Antennas and Propagation* 37: 1086–92.

[5.5] Idemen, M., and L. B. Felsen. 1981. "Diffraction of a Whispering Gallery Mode by the Edge of a Thin Cylindrically Curved Surface." *IEEE Transactions on Antennas and Propagation* 29 (4): 571–79.

[5.6] Molinet, F. A. 2005. "Edge-Excited Rays on Convex and Concave Structures: A Review." *IEEE Antennas and Propagation* 47 (5): 34–46.

[5.7] Pathak, P. H., and R. G. Koujoumjian. 1977. "On the Diffraction of Edge-Excited Surface Rays." Paper presented at the US National Committee of the International Union of Radio Science (USNC/URSI) Annual Meeting, Stanford, CA.

[5.8] Hill, K. C., and P. H. Pathak. 1987. "A UTD Analysis of the Excitation of Surface Rays by an Edge in an Otherwise Smooth Perfectly Conducting Convex Surface." Paper presented at the US National Committee of the International Union of Radio Science (USNC/URSI) Annual Meeting, Blacksburg, VA.

[5.9] Bouche, D., F. Molinet, and R. Mittra. 1997. *Asymptotic Methods in Electromagnetics*. Berlin: Springer-Verlag.

[5.10] Molinet, F. A. 1996. "Uniform Asymptotic Solution for the Diffraction of a Plane Wave by a Perfectly Conducting Three-Dimensional Curved Wedge." In *Electromagnetic Wave Interactions*, edited by A. Guran, R. Mittra, and P. J. Moser, 285–311. London: World Scientific.

Appendix A

A BRIEF PRESENTATION OF THE GOVERNING EQUATIONS FOR WAVE PROCESSES IN FLUIDS

A.1. WAVE PROPAGATION IN IDEAL FLUIDS

The fundamental equations of fluid dynamics are the equation of continuity, which expresses the conservation of matter, and the equation of motion of a volume element in the fluid called Euler's equation.

Let $\mathbf{v}$, p, and ρ be the particle velocity, pressure, and medium density, respectively. Since all thermodynamic quantities are determined by the value of two of them, together with the equation of state, the five quantities—namely, the three components of $\mathbf{v}$, the pressure p, and the density ρ—completely determine the state of the moving fluid and obey the hydrodynamic equations given in reference [A.1]:

$$\frac{\partial \rho}{\partial t} + div(\rho\,\mathbf{v}) = 0 \quad \text{continuity,} \tag{A.1}$$

$$\frac{\partial \mathbf{v}}{\partial t} + (\mathbf{v} \cdot \mathbf{grad})\mathbf{v} = -\frac{1}{\rho}\,\mathbf{grad}\,p \quad \text{Euler,} \tag{A.2}$$

where t is time and where $\frac{\partial}{\partial t} + \mathbf{v} \cdot \mathbf{grad} = \frac{d}{dt}$ is the convective time derivative (i.e., the rate of change of a physical quantity of a given fluid particle as it moves about in space). If the fluid is submitted to external forces (sources or gravity), then an additional force $\rho\mathbf{F}$, where $\mathbf{F}$ is the force per unit mass, acts on any unit volume and must be added to the right-hand side of (A.2).

The preceding equations do not account for any energy dissipation. Therefore they hold only for the motion of fluids in which thermal conductivity and viscosity are unimportant. Such fluids are said to be *ideal.* The absence of heat exchange between different parts of the fluid means that the motion is *adiabatic*. The entropy density of a chosen particle is then constant. Denoting by S the entropy per unit mass, we can express the condition of adiabatic motion as

$$\frac{dS}{dt} = 0. \tag{A.3}$$

Hence from the state equation

$$p = p(\rho, S), \tag{A.4}$$

we have

$$\left(\frac{\partial}{\partial t} + \mathbf{v} \cdot \mathbf{grad}\right) p = c^2 \left(\frac{\partial}{\partial t} + \mathbf{v} \cdot \mathbf{grad}\right) \rho, \tag{A.5}$$

where

$$c = c(\rho, S) = \sqrt{\left(\frac{\partial p}{\partial \rho}\right)_S}. \tag{A.6}$$

A.2. SOUND WAVES

We consider now small oscillations in a compressible fluid at rest, where ρ_o and p_o are constant equilibrium density and pressure. An oscillatory motion with small amplitude in a compressible fluid is called a sound wave. Since the oscillations are small and $\mathbf{v}$ is small also, the term $(\mathbf{v} \cdot \mathbf{grad})\mathbf{v}$ in Euler's equation may be neglected. For the same reason, the relative changes in the fluid density and pressure are small, and we can write

$$p = p_o + p', \quad \rho = \rho_o + \rho', \tag{A.7}$$

where p' and ρ' are disturbances of p_o and ρ_o caused by the sound wave. By neglecting small quantities of the second order, the equation of continuity (A.1) becomes

$$\frac{\partial \rho'}{\partial t} + \rho_o \operatorname{div} \mathbf{v} = 0, \tag{A.8}$$

and Euler's equation (A.2) reduces in the same approximation to

$$\frac{\partial \mathbf{v}}{\partial t} + \frac{1}{\rho_o} \mathbf{grad}\ p' = 0. \tag{A.9}$$

Equations (A.8) and (A.9) contain the unknown functions **v**, p', and ρ'. To eliminate one of these, we notice that a sound wave in an ideal fluid is also adiabatic. Hence (A.5) is verified and since according to (A.2) $\nabla p_o = 0$ for a fluid at rest, it follows from (A.4) that $\nabla \rho_o = 0$. Using these results in (A.5), the first-order terms in this equation give

$$\frac{\partial p'}{\partial t} = c_o^2 \frac{\partial \rho'}{\partial t}, \tag{A.10}$$

where c_o is the value of c not influenced by the wave. It is possible to show directly that

$$p' = c_o^2\ \rho'. \tag{A.11}$$

Indeed since the sound motion is adiabatic, the small change p' in the pressure is related to the small change ρ' in the density by

$$p' = \left(\frac{\partial p}{\partial \rho}\right)_{\substack{S=Cte \\ \rho'=o}} \rho' = c_o^2\ \rho'.$$

After applying the divergence operator to (A.9) and $\dfrac{\partial}{\partial t}$ to (A.8), then subtracting one from the other, and then using (A.10), we get

$$\frac{\partial^2 p'}{\partial t^2} - c_o^2\ \Delta p' = 0. \tag{A.12}$$

An equation having the form (A.12) is called a wave equation.

Applying the operator $\dfrac{\partial}{\partial t}$ to (A.9) and the gradient operator to (A.8), we find that each of the three components of the velocity **v** satisfies an equation having the same form. On substituting (A.11) into (A.12), we see that ρ' also satisfies the wave equation.

Two important solutions of (A.11) are worth mentioning:

$$p_1' = f\left(\mathbf{n} \cdot \frac{\mathbf{r}}{c_o} \pm t\right), \tag{A.13}$$

$$p_2' = r^{-1} g\left(\frac{r}{c_o} \pm t\right), \quad r \equiv |\mathbf{r}|, \tag{A.14}$$

where $\mathbf{n}$ is a constant unit vector and where f and g are arbitrary smooth functions. The unit vector $\mathbf{n}$ is normal to the planes where the argument of f is constant. The meaning of this solution is evident: in any plane $\mathbf{n} \cdot \mathbf{r}$ = constant, the pressure varies with time, and at any given time, it is different for a different $\mathbf{n} \cdot \mathbf{r}$ but is the same for positions $\mathbf{r}$ and times t, such that $\mathbf{n} \cdot \mathbf{r} + c_o t$ = constant or $\mathbf{n} \cdot \mathbf{r} - c_o t$ = constant. This means that if at some instant $t = 0$ and if at some point the pressure has a certain value, then after a time t the same value of the pressure is found at a distance $c_o t$ along the direction $\mathbf{n}$ from the original point. Thus the pressure and the other quantities $\mathbf{v}$ and ρ' are propagated through the medium in the $\mathbf{n}$ direction with a velocity c_o, which is called the velocity of sound. The waves described by (A.13) are called plane waves. They propagate along $\mathbf{n}$ without changing their shape or amplitude. The wave p_2' is called a spherical wave. It is spherically symmetric; that is, the sound pressure and other characteristics of the sound field are constant at spheres r = constant at fixed moments of time.

A.3. BOUNDARY CONDITIONS

The equations established thus far describe acoustic fields in unbounded, homogeneous liquids. If the liquids are bounded, or their parameters are discontinuous at some surfaces, boundary conditions must be specified for these equations. We consider three cases: the absolutely rigid boundary, the absolutely soft boundary, and the interface between two fluids at rest.

A.3.1. ABSOLUTELY RIGID BOUNDARY

In this case, a given boundary S_o is not deformed by the wave. Hence

$$\mathbf{v}_n(\mathbf{r}) = 0, \quad \mathbf{r} \in S_o, \tag{A.15}$$

where $\mathbf{v}_n$ is the component of the particle velocity vector $\mathbf{v}$, which is normal to S_o.

The undisturbed fluid being at rest, we have also $(\mathbf{v}_o)_n = 0$. According to (A.9), the condition (A.15) can also be written in terms of pressure field:

$$\frac{\partial p'}{\partial n} = 0, \quad \mathbf{r} \in S_o. \tag{A.16}$$

A.3.2. ABSOLUTELY SOFT BOUNDARY

For an absolutely soft boundary, the total pressure is zero. In other words,

$$p' + p_o = 0, \quad \mathbf{r} \in S, \tag{A.17}$$

where S is the position of the surface under the action of the sound waves. Since, for an undisturbed fluid, $\nabla p_o = 0$ and since p_o satisfies the boundary condition on S_o in the absence of the wave, we obtain the following for the linear approximation:

$$p' = 0, \quad \mathbf{r} \in S_o. \tag{A.18}$$

A.3.3. INTERFACE BETWEEN TWO FLUIDS AT REST

In this case, two boundary conditions need to be satisfied: first, the equality of the normal components of the particle velocities at both sides of the interface

$$\left[\mathbf{v}_n\right]_{S_o} = 0, \tag{A.19}$$

where the difference of values of $\mathbf{v}_n$ at both sides of S_o is denoted by $[\mathbf{v}_n]$ and, second, the equality of the forces acting upon each part of the interface from both sides,

$$\left[p\right]_{S_o} = 0. \tag{A.20}$$

According to (A.9), the first condition may be written

$$\left[\frac{1}{\rho_o}\frac{\partial p'}{\partial n}\right]_{S_o} = 0, \tag{A.21}$$

where the derivative of p' is taken along the normal to S_o.

A.4. HARMONIC WAVES

By using the spectral representation

$$p(\mathbf{r}, t) = \int_{-\infty}^{+\infty} p\,(\mathbf{r}, \omega)\exp(-i\omega t)\,d\omega, \tag{A.22}$$

where ω is the frequency, and keeping in mind that only the real part has physical meaning, the wave equation (A.12) is considerably simplified. Equation (A.12) now becomes

$$\Delta\, p(\mathbf{r},\omega) + k^2 p(\mathbf{r},\omega) = 0, \tag{A.23}$$

where $k = \dfrac{\omega}{c_o}$ is the wave number. Equation (A.23) is called the Helmholtz equation.

Waves with fixed ω are called monochromatic. Equation (A.22) can be interpreted as a continuous spectrum of elementary harmonic (periodic) waves, unlimited in time. For a plane wave, by taking the inverse Fourier transform of (A.22), unlimited in time, and replacing $p(\mathbf{r}, t)$ with its expression (A.13), we obtain

$$p(\mathbf{r},\omega) = A(\omega)\, e^{\pm ik\,\mathbf{n}\cdot\mathbf{r}}, \tag{A.24}$$

where $A(\omega)$ is the complex amplitude of the wave.

The functions $p(\mathbf{r},\omega)$ given by (A.24) are two solutions of the Helmholtz equation (A.23). They correspond to homogeneous plane waves traveling in opposite directions with respect to **n**.

A monochromatic traveling plane wave is a wave that propagates in the direction of the unit vector **n**. In such a wave, all quantities are functions of $\mathbf{k} \cdot \mathbf{r} - \omega t$ only, where $\mathbf{k} = k\mathbf{n}$ is the wave vector. The pressure is given by

$$p(\mathbf{r},\omega,\mathrm{t}) = A(\omega)\, e^{i\,(\mathbf{k}\cdot\mathbf{r}-\omega\mathbf{t})}. \tag{A.25}$$

A.5. REFLECTION AT A BOUNDARY

When a sound wave is incident on a boundary, it undergoes reflection. Let the plane $z = 0$ be the surface of the boundary that delimits the homogeneous fluid at rest, with sound velocity c and density ρ occupying the half-space $z > 0$. A monochromatic traveling wave is assumed to be incident from the half-space $z > 0$ upon this boundary. Let the xz plane coincide with the incident plane defined by the normal to the boundary surface and the wave vector of the incident wave (fig. A.1). The amplitude of the incident wave is assumed to be unity. Then omitting the time factor $\exp(-i\omega t)$, the incident and the reflected waves can be written accordingly:

$$p_i = \exp\left[\, ik\left(x \sin\theta_i - z\cos\theta_i\right) \right], \tag{A.26}$$

$$p_r = R\exp\left[ik\left(x\sin\theta_r + z\cos\theta_r \right) \right],\ k = \frac{\omega}{c}, \tag{A.27}$$

where θ_i is the incidence angle (i.e., the angle between the wave vector $\mathbf{k}_i$ of the incident wave and the z-axis), θ_r is the angle between the wave vector $\mathbf{k}_r$ of the reflected wave and the z-axis, and R is the reflection coefficient defined by the ratio between the complex amplitudes of the reflected and incident waves. The plane of reflection defined by $\mathbf{k}_r$ and the normal to the boundary and the incident plane are identical. This fact was already accounted for in (A.27). It is an immediate consequence of both the equation of motion and the fact that the boundary conditions do not depend on y.

On the surface $z = 0$, the total field $p = p_i + p_r$ must satisfy the boundary conditions. We consider three types of boundaries: absolutely rigid, absolutely soft, and impedance boundaries.

A.5.1. ABSOLUTELY RIGID BOUNDARY

According to (A.16), the total field p must satisfy $\dfrac{\partial p}{\partial z} = 0$ on the surface $z = 0$. It follows from (A.26) and (A.27) that the equation

$$-\cos\theta_i \exp\left(ikx\sin\theta_i \right) + R\cos\theta_r \exp\left(ikx\sin\theta_r \right) = 0$$

should be verified for all values of x, which implies that $\theta_i = \theta_r$ and that

$$R = 1. \tag{A.28}$$

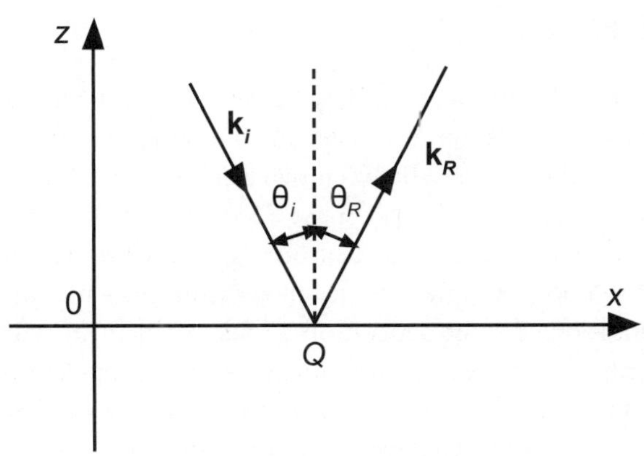

Figure A.1. Reflection of a sound wave at a boundary.

A.5.2. ABSOLUTELY SOFT BOUNDARY

According to (A.18), the total field must satisfy $p = 0$ at $z = 0$. By the same reasoning, it follows from (A.26) and (A.27) that $\theta_i = \theta_r$ and that

$$R = -1. \tag{A.29}$$

A.5.3. IMPEDANCE BOUNDARY

On an impedance boundary, the wave satisfies the impedance boundary condition

$$\frac{\partial p}{\partial z} + ik\gamma p = 0, \quad z = 0, \tag{A.30}$$

where $p = p_i + p_r$. According to (A.26) and (A.27), it follows from (A.30) that the equation

$$\left(-\cos\theta_i + \gamma\right)\exp\left(ikx\sin\theta_i\right) + R\left(-\cos\theta_r + \gamma\right)\exp\left(ikx\sin\theta_r\right) = 0$$

should be satisfied for all values of z, which implies that $\theta_i = \theta_r$ and that

$$R = \frac{\cos\theta_i - \gamma}{\cos\theta_i + \gamma}. \tag{A.31}$$

A.6. REFLECTION AND REFRACTION AT THE INTERFACE OF TWO HOMOGENEOUS FLUIDS

When a sound wave is incident on the boundary between two different fluid media, it undergoes reflection and refraction. As in section A.5, reflection and refraction planes are identical.

Let the plane $z = 0$ be the interface between two homogeneous fluids at rest, one with sound velocity c_1 and density ρ_1 occupying the half-space $z > 0$ and the other with sound velocity c_2 and density ρ_2 occupying the half-space $z < 0$. Consider a monochromatic traveling wave incident from the half-space $z > 0$ upon this interface. It follows from the equation of motion and the fact that the boundary conditions do not depend on y that the incident plane, reflected plane, and refracted plane are identical. Let the xz plane coincide with the incident plane (fig. A.2).

The amplitude of the incident wave is assumed to be unity. Then omitting the factor $\exp(-i\omega t)$, the incident and the reflected waves can be written as

$$p_i = \exp\left[ik_1\left(x\sin\theta_1 - z\cos\theta_1\right)\right], \tag{A.32}$$

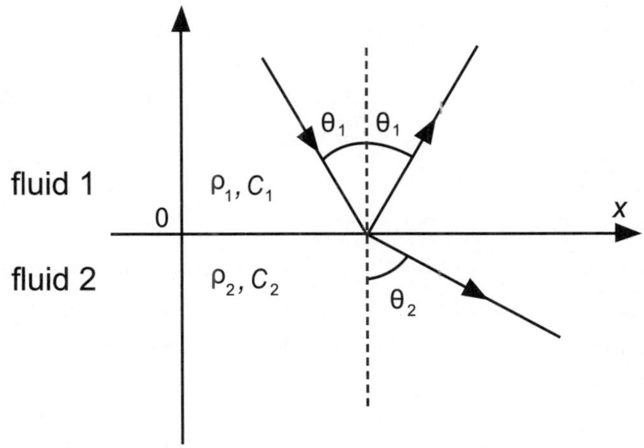

Figure A.2. Reflection and refraction of a plane sound wave at the interface between two media.

$$p_r = R\exp\left[ik_1\left(x\sin\theta_1' + z\cos\theta_1'\right)\right], \quad k_1 = \frac{\omega}{c_1}, \tag{A.33}$$

where θ_1 is the incident angle, θ'_1 is the reflection angle, and R is the reflection coefficient. The total sound pressure in the half-space $z > 0$ is

$$p_1 = p_i + p_r. \tag{A.34}$$

The refracted wave in the half-space $z < 0$ can be written as

$$p_2 = \mathrm{T}\exp\left[ik_2\left(x\sin\theta_2 - z\cos\theta_2\right)\right], \quad k_2 = \frac{\omega}{c_2}, \tag{A.35}$$

where θ_2 is the refraction angle and T is the transmission coefficient. Applying the boundary conditions (A.20), we obtain

$$\exp\left(ik_1 x\sin\theta_1\right) + \mathrm{R}\exp\left(ik_1 x\sin\theta_1'\right) = \mathrm{T}\exp\left(ik_2 x\sin\theta_2\right).$$

This equation must be satisfied for all values of x, which implies that

$$k_1\sin\theta_1 = k_1\sin\theta_1' = k_2\sin\theta_2 \tag{A.36}$$

and

$$1 + R = \mathrm{T}. \tag{A.37}$$

Equation (A.36) yields the law of reflection $\theta_1' = \theta_1$ and the Snell-Descartes law of refraction

$$k_1 \sin\theta_1 = k_2 \sin\theta_2. \tag{A.38}$$

Equation (A.37) is the relation between the reflection and refraction coefficients. A second equation involving R and T is obtained by enforcing the continuity of the normal derivative of the particle velocities at both sides of the interface, which can be written as (see A.21)

$$\frac{1}{\rho_1}\frac{\partial p_1}{\partial z} = \frac{1}{\rho_2}\frac{\partial p_2}{\partial z}. \tag{A.39}$$

Applied to (A.32), (A.33), and (A.35), equation (A.39) gives

$$\frac{1}{\rho_1} k_1 \cos\theta_1 (1 - R) = \frac{1}{\rho_2} k_2 \cos\theta_2 \mathrm{T}, \tag{A.40}$$

which combined with (A.37), immediately yields

$$R = \frac{\rho_2 \cos\theta_1 - n\rho_1 \cos\theta_2}{\rho_2 \cos\theta_1 + n\rho_1 \cos\theta_2} \qquad \mathrm{T} = \frac{2\rho_2 \cos\theta_1}{\rho_2 \cos\theta_1 + n\rho_1 \cos\theta_2}, \tag{A.41}$$

where $n = \dfrac{k_2}{k_1} = \dfrac{c_1}{c_2}$ is the refraction index.

According to Snell's law, we have $sin^2\theta_1 = n^2 sin^2\theta_2$. Hence we also have

$$R = \frac{\rho_2 \cos\theta_1 - \rho_1\sqrt{n^2 - \sin^2\theta_1}}{\rho_1 \cos\theta_1 + \rho_1\sqrt{n^2 - \sin^2\theta_1}}, \qquad \mathrm{T} = \frac{2\rho_2 \cos\theta_1}{\rho_2 \cos\theta_1 + \sqrt{n^2 - \sin^2\theta_1}}. \tag{A.42}$$

Expressions (A.42) are called the Fresnel coefficients. Additional information on wave propagation in fluids can be found in references [A.1], [A.2], and [A.3].

REFERENCES

[A.1] Landau, L. D., and E. M. Lifshitz. 2004. *Fluid Mechanics*. Vol. 6 of *Course of Theoretical Physics*. 2nd ed., revised. New York: Elsevier.

[A.2] Brekhowskikh, L. M., and O. A. Godin. 1990. *Acoustics of Layered Media I: Plane and Quasi-Plane Waves*. Berlin: Springer-Verlag.

[A.3] Bruneau, M. 1983. *Introduction aux Théories de L'acoustique*. Le Mans, France: Université du Maine.

Appendix B

A BRIEF PRESENTATION OF THE GOVERNING EQUATIONS OF LINEARIZED ELASTICITY

B.1. DEFORMATION

An elastic body has a natural undeformed state to which it returns when all external loads are removed. Under the action of applied forces, elastic bodies exhibit deformations to a certain extent (i.e., they change in shape and volume). The position of any point in the undeformed body is described mathematically by its position vector **x** with components $x_i = (1, 2, 3)$ in some fixed rectangular Cartesian coordinate system. Let the position vector of some particular point before the deformation be **x** and after the deformation have a different value **x**′. The displacement of this point due to the deformation is given by $\mathbf{x}' - \mathbf{x}$, which we shall denote by **u** with components $u_i = x_i' - x_i$. The vector **u** is called the displacement vector.

Let us consider two points that are very close together, and let dx_i and dx_i' be, respectively, the radius vectors joining these points before and after the deformation. The distance between the points is $d\ell$ before the deformation and $d\ell'$ after it. Using the general summation rule, we can write $d\ell^2 = dx_i^2$, $d\ell'^2 = \left(dx_i + du_i\right)^2$.

Substituting $du^i = (\partial u^i / \partial x_j)dx_j$, we can write

$$d\ell'^2 = d\ell^2 + 2\frac{\partial u_i}{\partial x_j} dx_i \, dx_j + \frac{\partial u_i}{\partial x_j} \frac{\partial u_i}{\partial x_\ell} dx_j \, dx_\ell. \tag{B.1}$$

The second term on the right can be put in the symmetrical form

$$\left(\frac{\partial u_i}{\partial x_j}+\frac{\partial u_j}{\partial x_i}\right)dx_i\,dx_j.$$

If we interchange the suffixes i and ℓ in the third term, then $d\ell'^2$ takes the final form

$$d\ell'^2 = d\ell^2 + 2u_{ij}\,dx_i\,dx_j, \tag{B.2}$$

where the tensor u_{ik} is given by

$$u_{ij} = \frac{1}{2}\left(\frac{\partial u_i}{\partial x_j}+\frac{\partial u_j}{\partial x_i}+\frac{\partial u_\ell}{\partial x_i}\frac{\partial u_\ell}{\partial x_j}\right). \tag{B.3}$$

The tensor u_{ij} is called the strain tensor. It is symmetrical ($u_{ij} = u_{ji}$) but nonlinear with respect to the displacement vector. It is however a matter of wide experience that many wave propagation effects in elastic solids can adequately be described by a linearized theory. Within the restrictions of the linearized theory, by neglecting the nonlinear term in (B.3), we see that the deformation is described in a very simple manner by the small-strain tensor ε with components

$$\varepsilon_{ij} = \frac{1}{2}\left(\frac{\partial u_i}{\partial x_j}+\frac{\partial u_j}{\partial x_i}\right). \tag{B.4}$$

The nine partial derivatives $\partial u_i / \partial x_j$ are usually written in the shorter notation $u_{i,j}$. With this notation, we write

$$\varepsilon_{ij} = \frac{1}{2}\left(u_{i,j}+u_{j,i}\right). \tag{B.5}$$

It is evident that $\varepsilon_{ij} = \varepsilon_{ji}$ (i.e., ε is a symmetric tensor of rank two).

B.2. LINEAR MOMENTUM AND STRESS TENSOR

In the theory of continuous media, a basic postulate is that the mechanical action of the material points situated on one side of an arbitrary material surface within a body upon those on the other side can be completely accounted for by prescribing a suitable surface traction on this surface. Let a surface element with unit outward normal **n** be submitted to a force **t** per unit area. The surface traction **t** depends generally on the orientation of **n**, as well as on the location **x** of the surface element.

Suppose we remove from a body a closed region V with boundary S. The surface S is subjected to a distribution of surface tractions $\mathbf{t}(\mathbf{x}, t)$, and each mass element of v may be subjected to a body force per unit mass $f(\mathbf{x}, t)$. According to the principle of balance of linear momentum, we have

$$\int_S \mathbf{t}\,ds + \int_V \rho \mathbf{f}\,dV = \int_V \rho \frac{\partial^2 \mathbf{u}}{\partial t^2}\,dV. \tag{B.6}$$

If we replace the vector $\mathbf{t}$ with a contracted product of the vector $\mathbf{n}$ with a tensor of rank two, then it is possible to transform the surface integral into a volume integral. Indeed, by stating

$$t_i = \sigma_{ij} n_j \tag{B.7}$$

and applying Gauss's theorem, we obtain

$$\int_S \sigma_{ij} n_j\,ds = \int_V \sigma_{ij,j}\,dV. \tag{B.8}$$

The tensor σ_{ij} is called the stress tensor, and (B.7) is the Cauchy stress formula. The component σ_{ij} is the projection of the force acting on the unit area perpendicular to the axis i on the axis j.

By introducing (B.8) into (B.6), it follows that

$$\int_V \left(\sigma_{ij,j} + \rho f_i - \rho \frac{\partial^2 u_i}{\partial t^2} \right) dV = 0. \tag{B.9}$$

Since V may be an arbitrary part of the body, it follows that whenever the integrant is continuous, we have

$$\sigma_{ij,j} + \rho f_i = \rho \frac{\partial^2 u_i}{\partial t^2}. \tag{B.10}$$

This is the Cauchy's first law of motion. It can be shown by applying the principle of balance of moment of momentum that the stress tensor is symmetric.

For the linearized theory, the principle of moment of momentum states

$$\int_S (\mathbf{x} \times \mathbf{t})\,ds + \int_V (\mathbf{x} \times \mathbf{f})\rho\,dV = \int_V \rho \frac{\partial}{\partial t}\left(\mathbf{x} \times \frac{\partial \mathbf{u}}{\partial t} \right) dV. \tag{B.11}$$

Using the special tensor of rank three called the alternating tensor, whose components are defined as

$$e_{ijk} \quad \begin{cases} +1 & \text{if } ijk \text{ represents an even permutation of } 1\ 2\ 3 \\ 0 & \text{if any two of the } ijk \text{ indices are equal} \\ -1 & \text{if } ijk \text{ represents an odd permutation of } 1\ 2\ 3, \end{cases}$$

and the summation convention, the components of the cross product $\mathbf{h} = \mathbf{u} \times \mathbf{v}$ may be expressed as

$$h_i = e_{ijk}\, u_j\, v_k. \tag{B.12}$$

Making use of (B.12) in (B.11), this equation can be written as

$$\int_S e_{ijk}\, x_j\, t_k\, ds + \int_V \rho e_{ijk}\, x_j\, f_k\, dV = \int_V \rho\, e_{ijk}\, x_j \frac{\partial^2 u_k}{\partial t^2} dV. \tag{B.13}$$

Substituting expression (B.7) in the surface integral and applying the Gauss theorem, we obtain

$$\int_S e_{ijk}\, x_j\, \sigma_{kj}\, n_j\, ds = \int_V \varepsilon_{ijk} \frac{\partial}{\partial x_j}\left(x_j\, \sigma_{kj}\right) ds = \int_V \varepsilon_{ijk}\left(\sigma_{kj} + x_j\, \sigma_{kj,j}\right) dV. \tag{B.14}$$

By virtue of (B.10), equation (B.13) then reduces to

$$\int_V \varepsilon_{ijk}\, \sigma_{kj}\, dV = 0, \tag{B.15}$$

and since V may be an arbitrary part of the body, it follows that

$$\varepsilon_{ijk}\ \sigma_{kj} = 0,$$

which implies that

$$\sigma_{kj} = \sigma_{jk} \tag{B.16}$$

(i.e., the stress tensor is symmetric).

B.3. HOOKE'S LAW

We saw in section B.2 that the state of stress in continuous media is completely determined by the stress tensor σ_{ij}, and we also saw that the state of deformation is determined by the strain tensor u_{ij}, which in linearized theory reduces to the small-strain tensor ε_{ij}. We shall now assume that when an elastic medium is maintained at a fixed temperature, there is a one-to-one analytical relation between σ_{ij} and u_{ij} and that σ_{ij} vanishes when the strains u_{ij} are all zero. In the linear theory of elasticity, this analytical relation reduces to

$$\sigma_{ij} = C_{ijk\ell}\,\varepsilon_{k\ell}\left(i,j,k,\ell = 1,2,3\right), \tag{B.17}$$

where the coefficient $C_{ijk\ell}$ will generally vary from point to point of the medium. We confine our attention to the media, called elastically homogeneous, in which $C_{ijk\ell}$ does not vary throughout the region under consideration.

The law (B.17) is a natural generalization of Hooke's law, published in 1676, which states that for a thin rod, the extension is proportional to the force, under the limit of elasticity. It is called the generalized Hooke's law or, for short, Hooke's law.

Inasmuch as the components σ_{ij} are symmetric, an interchange of the indices i and j in (B.17) does not alter these formulas, so that

$$C_{ijk\ell} = C_{jik\ell}.$$

Moreover, since $\varepsilon_{ij} = \varepsilon_{ji}$, we deduce from (B.17) that the coefficients $C_{ijk\ell}$ are also symmetric with respect to the last two indices. Hence there are at most 36 independent constants in the general stress-strain law (B.17).

Most structural materials are formed of crystalline substances. However, the dimensions of most crystals are so small in comparison to the dimensions of the body and so chaotically distributed that, globally, the substance behaves as though it were isotropic.

From the definition of the isotropic medium, it follows that its elastic properties are independent of the orientation of the coordinate axes. In particular, the coefficients $C_{ijk\ell}$ must remain invariant when we introduce new coordinate axes. In this case it can be shown, by rotating the x_1, x_2, x_3 system by a right angle about the x_1 axis and successively a right angle and an angle of 45° about the x_3 axis, that the number of essential elastic constants reduces to two and that the generalized Hooke's law for a homogeneous isotropic body can be written in the following form:

$$\sigma_{ij} = \lambda\,\delta_{ij}\left(\varepsilon_{11} + \varepsilon_{22} + \varepsilon_{33}\right) + 2\mu\,\varepsilon_{ij}, \tag{B.18}$$

where the Kronecker symbol δ_{ij} is 1 for $i = j$ and 0 for $i \neq j$.

The constants λ and μ were introduced by Lamé and are called the Lamé constants. Making use of the formula (B.5), we can write the stress-strain relations (B.18) in terms of displacements u_i in the following form:

$$\sigma_{ij} = \lambda \delta_{ij} u_{k,k} + \mu\left(u_{i,j} + u_{j,i}\right). \tag{B.19}$$

The equations (B.19) are the relation between stresses and deformations in the linear theory of elasticity and in the most simple case of locally isotropic solids.

B.4. WAVES IN AN ELASTIC MEDIUM

The differential equations of motion of an elastic solid can be obtained by substituting the values of the stress tensor components (B.19) in Cauchy's first law of motion (B.10). Differentiating (B.19) with respect to x_j, we get

$$\begin{aligned}\sigma_{ij,j} &= \lambda \delta_{ij} u_{k,kj} + \mu\left(u_{i,jj} + u_{j,ij}\right)\\ &= \lambda \delta_{ii} u_{k,ki} + \mu\left(u_{i,jj} + u_{j,ji}\right)\\ &= \left(\lambda + \mu\right) u_{j,ji} + \mu u_{i,jj}.\end{aligned}$$

Hence by inserting this expression into (B.10) we obtain

$$\mu u_{i,jj} + \left(\lambda + \mu\right) u_{j,ji} + \rho f_i = \rho \frac{\partial^2 u_i}{\partial t^2}. \tag{B.20}$$

Equation (B.20) is the dynamical equation in terms of the displacement.

When the body forces **f** are equal to zero, we obtain the equation of motion for the medium under the influence of its own elastic restoring forces. The vector form of (B.20) for $\mathbf{f} = 0$ is

$$\rho \frac{\partial^2 \mathbf{u}}{\partial t^2} = \left(\lambda + \mu\right) \mathbf{grad}\left(\operatorname{div} \mathbf{u}\right) + \mu \Delta \mathbf{u}. \tag{B.21}$$

Equation (B.21) describes the propagation of elastic waves in a homogeneous isotropic medium. For this equation, it is necessary to adjoin the initial and the boundary conditions. Thus at each point of the surface S of the nondeformed medium, the surface tractions **t** or the displacements **u** must be prescribed. The functions u_i prescribed on the surface S, in general, are functions of the space coordinates x_i and the time t. If the tractions **t** are prescribed as functions of x_i and t, then the components of stress must satisfy (B.7) on the surface S. Additionally, one must know the initial conditions on the displacement u_i and on their time derivatives (see reference [B.1]).

In many applications, it is important to consider mixed boundary conditions involving **u** and **t**. We will next consider some special cases in which the surface S is an interface between two different media.

B.5. BOUNDARY CONDITIONS AT INTERFACES

At interfaces between solids, the boundary conditions can be rather diverse and depend on the kind of contact between the solids. We are mainly interested in the boundary conditions between a solid and a nonviscous fluid. In that case, four conditions take place:

$$\left[u_n\right]_S = 0\,,\ \left[\sigma_{nn}\right]_S = 0\,,\quad \sigma_{nj} = 0,\quad j \neq n, \tag{B.22}$$

where the difference of values of a function f at both sides of S is denoted $[f]$. At the boundary of a solid and vacuum (free boundary), three conditions exist:

$$\sigma_{nj} = 0,\quad j = 1, 2, 3. \tag{B.23}$$

B.6. ELASTIC WAVES IN HOMOGENEOUS ISOTROPIC SOLIDS

By using the identity

$$\Delta\mathbf{a} = \mathbf{grad}\left(\operatorname{div}\mathbf{a}\right) - \mathbf{curl}\left(\operatorname{curl}\mathbf{a}\right),$$

equation (B.21) can be transformed into

$$\frac{\partial^2\mathbf{u}}{\partial t^2} = \frac{\lambda + 2\mu}{\rho}\mathbf{grad}\left(\operatorname{div}\mathbf{u}\right) - \frac{\mu}{\rho}\mathbf{curl}\left(\operatorname{curl}\mathbf{u}\right). \tag{B.24}$$

The form of the equation (B.24) suggests that part of the solution can be expressed in terms of the gradient of a scalar function φ and that another part can be expressed in terms of the **curl** of a vector **A**, since the equation for each part simplifies considerably.

If we state

$$\mathbf{u} = \mathbf{u}_\ell + \mathbf{u}_t\,,\quad \mathbf{u}_\ell = \mathbf{grad}\,\varphi\,,\quad \mathbf{u}_t = \operatorname{curl}\mathbf{A}, \tag{B.25}$$

then by introducing (B.25) into (B.24), we obtain

$$\frac{\partial^2\mathbf{u}_\ell}{\partial t^2} - \frac{\lambda + 2\mu}{\rho}\Delta\mathbf{u}_\ell + \frac{\partial^2\mathbf{u}_t}{\partial t^2} - -\frac{\mu}{\rho}\Delta\mathbf{u}_t = 0. \tag{B.26}$$

After successively applying the operators div and **curl** to (B.26), we find that

$$\operatorname{div}\left(\frac{\partial^2 \mathbf{u}_\ell}{\partial t^2} - \frac{\lambda + 2\mu}{\rho}\Delta \mathbf{u}_\ell\right) = 0\,,\ \mathbf{curl}\left(\frac{\partial^2 \mathbf{u}_\ell}{\partial t^2} - \frac{\lambda + 2\mu}{\rho}\Delta \mathbf{u}_\ell\right) = 0,$$

$$\operatorname{div}\left(\frac{\partial^2 \mathbf{u}_t}{\partial t^2} - \frac{\mu}{\rho}\Delta \mathbf{u}_t\right) = 0\,,\ \mathbf{curl}\left(\frac{\partial^2 \mathbf{u}_t}{\partial t^2} - \frac{\mu}{\rho}\Delta \mathbf{u}_t\right) = 0.$$

It follows from these relations that the quantities in brackets are vectors that depend only on t. Therefore, we have

$$\frac{\partial^2 \mathbf{u}_\ell}{\partial t^2} - \frac{\lambda + 2\mu}{\rho}\Delta \mathbf{u}_\ell = \mathbf{F}(t), \tag{B.27}$$

$$\frac{\partial^2 \mathbf{u}_t}{\partial t^2} - \frac{\mu}{\rho}\Delta u_t = \mathbf{G}(t). \tag{B.28}$$

The homogeneous equations obtained from (B.27) and (B.28) by taking $\mathbf{F} = \mathbf{G} = 0$ describe the propagation of elastic waves. By writing these equations in the form

$$\frac{\partial^2 \mathbf{u}_\ell}{\partial t^2} - C_\ell^2\,\Delta \mathbf{u}_\ell = 0, \quad C_\ell = \sqrt{(\lambda + 2\mu)/\rho}, \tag{B.29}$$

$$\frac{\partial^2 \mathbf{u}_t^2}{\partial t^2} - C_t^2\,\Delta \mathbf{u}_t = 0, \quad C_t = \sqrt{\mu / \rho}, \tag{B.30}$$

we see that the components $\mathbf{u}_\ell$ and $\mathbf{u}_t$ propagate independently in a homogeneous solid. Note that since $\operatorname{curl} \mathbf{u}_\ell = 0$, there is no twisting of the medium, only stretching and squeezing. Waves of this sort are called longitudinal or compressional waves. On the other hand, since the relative change in the volume of the particle during deformation is div**u**, the component $\mathbf{u}_t$, which verifies $\operatorname{div}\mathbf{u}_t = 0$, corresponds to a displacement without expansion or contraction, and therefore, the strain must be a type of shear. Waves of that sort are called transverse or shear waves.

From (B.29) and (B.30), we see that the wave velocity C_ℓ of compressional waves is always greater than the wave velocity C_t of the shear waves. For both types of waves, the wave velocity is greater for larger moduli λ and μ (i.e., a stiffer medium) and is smaller for larger ρ (i.e., a dense medium). When the shear modulus μ is zero, only compressional waves similar to those propagated in liquid and gaseous media remain.

According to the definition of the potentials, we obtain the following from (B.29) and (B.30) for a homogeneous media:

$$\mathbf{grad}\left(\frac{\partial^2\varphi}{\partial t^2} - C_\ell^2\,\Delta\varphi\right) = 0, \tag{B.31}$$

$$\mathbf{curl}\left(\frac{\partial^2\mathbf{A}}{\partial t^2} - C_t^2\,\Delta\mathbf{A}\right) = 0. \tag{B.32}$$

Hence, within an accuracy of an arbitrary function of t for φ and the gradient of an arbitrary scalar function of ($\mathbf{r}$, t), the equations for the potentials are

$$\frac{\partial^2\varphi}{\partial t^2} - C_\ell^2\,\Delta\varphi = 0, \tag{B.33}$$

$$\frac{\partial^2\mathbf{A}}{\partial t^2} - C_t^2\,\Delta\mathbf{A} = 0. \tag{B.34}$$

Equations (B.29), (B.30), (B.33), and (B.34) are called wave equations and are analogous to (A.12). Important solutions of these equations are plane wave solutions

$$\mathbf{u}_\ell = \mathbf{f}\left(\boldsymbol{\rho}\cdot\mathbf{r} - C_\ell t\right), \quad \varphi = \phi\left(\boldsymbol{\rho}\cdot\mathbf{r} - C_\ell t\right), \tag{B.35}$$

$$\mathbf{u}_t = \mathbf{g}\left(\boldsymbol{\rho}\cdot\mathbf{r} - C_t t\right), \quad \mathbf{A} = \boldsymbol{\psi}\left(\boldsymbol{\rho}\cdot\mathbf{r} - C_t t\right), \tag{B.36}$$

where $\boldsymbol{\rho}$ is a constant unit vector and where ϕ and the components of $\mathbf{f}$, $\mathbf{g}$, and $\boldsymbol{\psi}$ are arbitrary smooth functions. The vector $\mathbf{r}$ denotes the position vector and $\boldsymbol{\rho}\cdot\mathbf{r}$ = constant describes a plane normal to the unit vector $\boldsymbol{\rho}$, which defines the direction of propagation of the wave. Equations (B.35) and (B.36) thus represent plane waves whose plane of constant phase are normal to $\boldsymbol{\rho}$.

The plane wave (B.35) is longitudinal, since we have

$$\mathbf{u}_\ell = \mathbf{grad}\,\varphi = \boldsymbol{\rho}\dot{\phi}\left(\boldsymbol{\rho}\cdot\mathbf{r} - C_\ell t\right), \quad \dot{\phi} = \frac{d\phi}{dz}(z).$$

In other words, displacement occurs along the direction of propagation. The plane wave in (B.36) is transverse. The displacement is normal to the wave propagation. Indeed, we have the following for the displacement $\mathbf{u}_t$:

$$\boldsymbol{\rho}\cdot\mathbf{u}_t = \boldsymbol{\rho}\cdot\mathbf{curl}\;\boldsymbol{\psi} = \boldsymbol{\rho}\cdot\left[\boldsymbol{\rho}\times\dot{\boldsymbol{\psi}}\left(\boldsymbol{\rho}\cdot\mathbf{r} - C_t t\right)\right] = 0.$$

B.7. HARMONIC WAVES

By using the spectral representations

$$Q(\mathbf{r},t)=\int_{-\infty}^{+\infty} Q(\mathbf{r},\omega)\exp(-i\omega t)\,d\omega, \tag{B.37}$$

where ω is the frequency and where Q states for the potential φ or for any component of $\mathbf{u}_\ell$, $\mathbf{u}_t$, or $\mathbf{A}$. Keeping in mind that only the real part has physical meaning, the equations (B.29), (B.30), (B.33), and (B.34) become

$$\Delta\mathbf{u}_\ell(\mathbf{r},\omega)+k_\ell^2\,\mathbf{u}_\ell(\mathbf{r},\omega)=0, \tag{B.38}$$

$$\Delta\mathbf{u}_t(\mathbf{r},\omega)+k_t^2\,\mathbf{u}_t(\mathbf{r},\omega)=0, \tag{B.39}$$

$$\Delta\varphi(\mathbf{r},\omega)+k_\ell^2\,\varphi(\mathbf{r},\omega)=0, \tag{B.40}$$

$$\Delta\,\mathbf{A}(\mathbf{r},\omega)+k_t^2\,\mathbf{A}(\mathbf{r},\omega)=0, \tag{B.41}$$

where $k_\ell=\dfrac{\omega}{C_\ell}$ and $k_t=\dfrac{\omega}{C_t}$ are the wave numbers for the longitudinal and transverse elastic waves, respectively.

Waves with fixed ω are called monochromatic. Equation (B.37) can be interpreted as a continuous spectrum of elementary harmonic waves, unlimited in space. For a plane wave, by taking the inverse Fourier transform of (B.37), unlimited in time, and replacing $Q(\mathbf{r},t)$ with $Q(\boldsymbol{\rho}\cdot\mathbf{r}\pm ct)$, we obtain

$$Q(\mathbf{r},\omega)=A^{\pm}(\omega)\exp(\pm ik\,\boldsymbol{\rho}\cdot\mathbf{r}), \tag{B.42}$$

where $A^{\pm}(\omega)$ are the complex amplitudes of the waves and $k=\dfrac{\omega}{C}$. $Q(\mathbf{r},\omega)$, given by (B.42), has two expressions corresponding to two solutions of the Helmholtz equation:

$$\Delta Q(\mathbf{r},\omega)+k^2\,Q(\mathbf{r},\omega)=0. \tag{B.43}$$

It corresponds to homogeneous plane waves traveling in opposite directions with respect to $\boldsymbol{\rho}$.

Consider now a plane wave propagating in the direction of the unit vector $\boldsymbol{\rho}$. By inserting (B.42) with the positive sign in the exponent into (B.37), we obtain

$$Q(\mathbf{r},t) = \int_{-\infty}^{+\infty} A(\omega)\exp\left[\, i\left(k\boldsymbol{\rho}\cdot\mathbf{r} - \omega t\right)\right] d\omega. \tag{B.44}$$

Equation (B.44) is a continuous spectrum of elementary harmonic plane waves. For a monochromatic wave, the amplitude A is constant and the corresponding elementary wave is called a traveling wave. For the longitudinal displacement, such a wave is given by

$$\mathbf{u}_\ell(\mathbf{r},t) = \boldsymbol{\rho} A_\ell \exp\left[\, i\left(k_\ell \boldsymbol{\rho}\cdot\mathbf{r} - \omega t\right)\right] = \boldsymbol{\rho} A_\ell \exp\left[\, i k_\ell\left(\boldsymbol{\rho}\cdot\mathbf{r} - C_\ell t\right)\right]. \tag{B.45}$$

For the transverse displacement, we have

$$\mathbf{u}_t(\mathbf{r},t) = (\mathbf{b}\times\boldsymbol{\rho}) A_t \exp\left[\, i k_t\left(\boldsymbol{\rho}\cdot\mathbf{r} - C_t t\right)\right], \tag{B.46}$$

where $\mathbf{b}$ is a unit vector perpendicular to the unit vector $\boldsymbol{\rho}$, which determines the direction of the propagation of the wave. Similar expressions are obtained for the potentials φ and $\mathbf{A}$.

B.8. REFLECTION AND REFRACTION AT A PLANE INTERFACE

We limit our analysis to the important case of reflection at the boundary of fluid and elastic half-spaces. Consider a rectangular Cartesian coordinate system x_1, x_2, x_3, and assume that the half-space $x_2 > 0$ is filled by a fluid at rest in which only sound waves propagate. The fluid is supposed homogeneous, with density ρ_o and sound velocity C_o. The half-space $x_2 < 0$ is occupied by a homogeneous isotropic elastic material characterized by its mass density ρ_1 and the Lamé elastic constants λ and μ. A monochromatic traveling wave is assumed to be incident from the half-space $x_2 > 0$, on the interface between the fluid and the elastic material. Let the x_1, x_2 plane coincide with the incident plane defined by the normal to the interface and the wave vector $\mathbf{k}_o = k_o\boldsymbol{\rho}$ (with $k_o = \omega/C_o$) of the incident wave.

The question then is what combination of additional waves is required in order for the stresses and displacements to be continuous at the interface. These additional waves are called reflected and refracted waves. Moreover, since in an elastic medium, longitudinal and transverse waves exist and propagate independently, we have to consider refracted waves of both types. Now since the incident wave has displacement components only in the x_1, x_2 plane, it follows from the continuity requirement that the displacement components of the reflected and refracted waves are also in the x_1, x_2 plane.

Thus the displacement of the incident, reflected, and refracted waves can be written accordingly:

$$\begin{aligned}
\mathbf{u}^i &= A_o\,\mathbf{d}^{(o)} \exp\left[\, ik_o \left(x_1 \sin\theta_o - x_2 \cos\theta_o - C_o t\right)\right],\\
\mathbf{u}^r &= A_1\,\mathbf{d}^{(1)} \exp\left[\, ik_o \left(x_1 \sin\theta_1 - x_2 \cos\theta_1 - C_o t\right)\right],\\
\mathbf{u}^\ell &= A_2\,\mathbf{d}^{(2)} \exp\left[\, ik_2 \left(x_1 \sin\theta_2 - x_2 \cos\theta_2 - C_2 t\right)\right],\\
\mathbf{u}^t &= A_3\,\mathbf{d}^{(3)} \exp\left[\, ik_3 \left(x_1 \sin\theta_3 - x_2 \cos\theta_3 - C_3 t\right)\right],
\end{aligned} \tag{B.47}$$

where θ_o is the incident angle, θ_1 is the reflected angle, θ_2 is the reflected angle of the longitudinal elastic wave, and θ_3 is the refracted angle of the transverse elastic wave (fig. B.1). A_n, k_n, C_n ($n = 0, 1, 2, 3$) are, respectively, the amplitude, the wave number, and the velocity of the waves, and $\mathbf{d}^{(n)}$ ($n = 0, 1, 2, 3$) are unit vectors defining the direction of the displacement. If $\boldsymbol{\rho}^{(n)}$ ($n = 0, 1, 2, 3$) is the unit vector defining the direction of propagation of the waves, we have

$$\mathbf{d}^{(o)} = \boldsymbol{\rho}^{(o)}, \quad \mathbf{d}^{(1)} = \boldsymbol{\rho}^{(1)}, \quad \mathbf{d}^{(2)} = \boldsymbol{\rho}_2, \quad \mathbf{d}^{(3)} = \mathbf{i}_3 \times \boldsymbol{\rho}_3 \tag{B.48}$$

with

$$\begin{aligned}
\boldsymbol{\rho}^{(0)} &= \sin\theta_0\ \mathbf{i}_1 - \cos\theta_0\ \mathbf{i}_2,\\
\boldsymbol{\rho}^{(1)} &= \sin\theta_1\ \mathbf{i}_1 + \cos\theta_1\ \mathbf{i}_2,\\
\boldsymbol{\rho}^{(2)} &= \sin\theta_2\ \mathbf{i}_1 - \cos\theta_2\ \mathbf{i}_2,\\
\boldsymbol{\rho}^{(3)} &= \sin\theta_3\ \mathbf{i}_1 - \cos\theta_3\ \mathbf{i}_2,
\end{aligned} \tag{B.49}$$

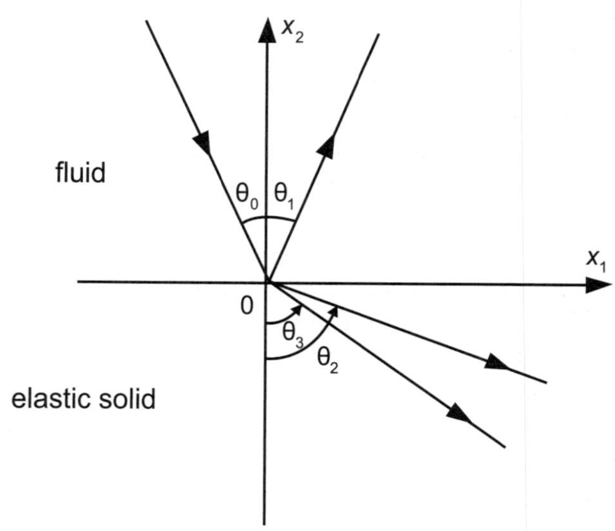

Figure B.1. Incident, reflected, and refracted waves.

where $\mathbf{i}_1$, $\mathbf{i}_2$, $\mathbf{i}_3$ are the unit vectors of the coordinate system x_1, x_2, x_3.

For a given incident wave, the amplitudes, the unit propagation vectors, and the wave numbers of the reflected and refracted waves must be computed from the conditions on the displacements and stresses at the interface between the two media. The boundary conditions between an elastic solid and a nonviscous fluid are given by (B.22). In that case, three conditions take place:

1. The normal displacement is continuous:

$$u_2^i + u_2^r = u_2^\ell + u_2^t. \tag{B.50}$$

2. The normal stress is continuous:

$$\sigma_{22}^i + \sigma_{22}^r = \sigma_{22}^\ell + \sigma_{22}^t. \tag{B.51}$$

3. The tangential stress vanishes:

$$\begin{aligned} \sigma_{21}^\ell + \sigma_{21}^t &= 0, \\ \sigma_{23}^\ell + \sigma_{23}^t &= 0. \end{aligned} \tag{B.52}$$

Since in a fluid the stress corresponds to the pressure, which is normal to the interface, and since $\mathbf{u}^\ell$ and $\mathbf{u}^t$ are in the plane x_1, x_2, the tangential components σ_{23}^ℓ and σ_{23}^t of the stress tensor are zero. Hence the equation

$$\sigma_{23}^\ell + \sigma_{23}^t = 0$$

is automatically satisfied.

Applying condition 1, equation (B.50), to (B.47) and taking into account (B.48) and (B.49), we obtain the following at the interface ($x_2 = 0$):

$$\begin{aligned} &-A_o \cos\theta_o \exp\left[ik_o \left(x_1 \sin\theta_o - C_o t\right) \right] + A_1 \cos\theta_1 \exp\left[ik_1 \left(x_1 \sin\theta_1 - C_0 t\right) \right] \\ &= -A_2 \cos\theta_2 \exp\left[ik_2 \left(x_1 \sin\theta_2 - C_2 t\right) \right] + A_3 \cos\theta_3 \exp\left[ik_3 \left(x_1 \sin\theta_3 - C_3 t\right) \right]. \end{aligned} \tag{B.53}$$

Since this equation must be satisfied for all values of x_1 and t, the exponentials must be equal, which requires

$$k_0 \sin\theta_0 = k_0 \sin\theta_1 = k_2 \sin\theta_2 = k_3 \sin\theta_3, \tag{B.54}$$

$$k_0\, C_0 = k_2\, C_2 = k_3\, C_3. \tag{B.55}$$

Equation (B.54) implies that the angle of reflection is equal to the angle of incidence:

$$\theta_0 = \theta_1. \tag{B.56}$$

Substituting these results into (B.53), it also follows that the amplitude must satisfy

$$\left(A_1 - A_0\right)\cos\theta_0 = -A_2\cos\theta_2 + A_3\sin\theta_3. \tag{B.57}$$

In order to apply condition 2, we first express the stresses on the interface, due to the incident and reflected waves, in terms of the corresponding pressures. According to (A.9), we have

$$\mathbf{u} = \frac{1}{\rho_0\,\omega^2}\,\mathbf{grad}\,P, \tag{B.58}$$

and since

$$P = P_0\, e^{ik_0\left(\boldsymbol{\rho}\cdot\mathbf{r} - C_0 t\right)}, \tag{B.59}$$

we obtain

$$\mathbf{u} = \frac{ik_0}{\rho_0\,\omega^2}\,\boldsymbol{\rho}\,P_0\, e^{ik_0\left(\boldsymbol{\rho}\cdot\mathbf{r} - C_0 t\right)}. \tag{B.60}$$

Substituting (B.60) into the two first equations, (B.47) gives

$$P_0 = -i\,\rho_0\,\omega C_0\,A_0 \qquad P_1 = -i\,\rho_0\,\omega C_0\,A_1. \tag{B.61}$$

The total pressure is normal to the interface and in the opposite direction of the normal **n** outside the solid that we have chosen in the definition of the stresses. Taking this into account, we have the following on the interface ($x_2 = 0$):

$$\sigma_{22}^i + \sigma_{22}^r = i\,\rho_0\,\omega C_0\left(A_0 + A_1\right)\exp\left[ik_0\left(x_1\sin\theta_0 - C_0 t\right)\right] \tag{B.62}$$

In order to calculate the stresses on the interface due to the transmitted longitudinal and transverse waves, we substitute the components of the corresponding displacements, given by the two last equations (B.47), into Hooke's law (B.19). With our notations, the stress tensor is obtained as

$$\sigma_{ij} = \left[\lambda \delta_{ij} \left(d_n \rho_k \right) + \mu \left(d_i p_j + d_j \rho_i \right) \right] ik\, e^{ik(\mathbf{x} \cdot \mathbf{p} - Ct)}. \tag{B.63}$$

Applying the condition (B.51) by taking into account (B.63), (B.48), and (B.49), we obtain

$$i \rho_0 \omega C_0 \left(A_0 + A_1 \right) = ik_2 A_2 \left(\lambda + 2\mu \cos^2 \theta_2 \right) - ik_3 A_3 \sin 2\theta_3. \tag{B.64}$$

Since the Lamé constants λ and μ are related to the velocities C_2 and C_3 by (see equations (B.29) and (B.30))

$$C_2^2 = \frac{\lambda + 2\mu}{\rho_1}, \quad C_3^2 = \frac{\mu}{\rho_1}, \tag{B.65}$$

(B.64) gives

$$\left(A_0 + A_1 \right) = \frac{\rho_1}{\rho_0 C_0} \left[A_2 \left(C_2 - \frac{2C_3^2}{C_2} \sin^2 \theta_2 \right) \right] - A_3 C_3 \sin 2\theta_3. \tag{B.66}$$

Expressing condition (3) from (B.52) with the help of (B.63) gives

$$k_2 A_2 \sin 2\theta_2 + k_3 A_3 \cos 2\theta_3 = 0$$

and consequently

$$A_3 = -\frac{C_3}{C_2} \frac{\sin 2\theta_2}{\cos 2\theta_3} A_2. \tag{B.67}$$

Inserting (B.67) into (B.64) and (B.57) gives

$$A_1 = A_0 - \frac{Q \dfrac{\rho_1}{\rho_0 C_0} \cos \theta_0 - 1}{Q \dfrac{\rho_1}{\rho_0 C_0} \cos \theta_0 + 1}, \tag{B.68}$$

where, after some elementary trigonometric calculus, Q can be written as

$$Q = \frac{C_2}{\cos\theta_2}\cos^2 2\theta_3 + \frac{C_3}{\cos\theta_3}\sin^2 2\theta_3,$$

and finally, the reflection coefficient R of the reflected wave is given by

$$R = \frac{A_1}{A_0} = \frac{Z_\ell \cos^2 2\theta_3 + Z_t \sin^2 2\theta_3 - Z_0}{Z_\ell \cos^2 2\theta_3 + Z_t \sin^2 2\theta_3 + Z_0}, \tag{B.69}$$

where we have introduced the following expressions:

$$Z_0 = \frac{\rho_0 C_0}{\cos\theta_0}, \quad Z_\ell = \frac{\rho_1 C_2}{\cos\theta_2}, \quad Z_t = \frac{\rho_1 C_3}{\cos\theta_3}. \tag{B.70}$$

The transmission coefficient for the refracted (or transmitted) longitudinal wave is obtained by inserting (B.67) into (B.57) or (B.66). The result is

$$\mathrm{T}_\ell = \frac{A_2}{A_0} = 2\frac{C_0}{C_2}\,\frac{\rho_1^{-1}\rho_0\, Z_\ell \cos 2\theta_3}{Z_\ell \cos^2 2\theta_3 + Z_t \sin^2 2\theta_3 + Z_0}. \tag{B.71}$$

Finally, from (B.67) and (B.71), we obtain

$$\mathrm{T}_t = \frac{A_3}{A_0} = -2\frac{C_0}{C_3}\,\frac{\rho_1^{-1}\rho_0\, Z_t \sin 2\theta_3}{Z_\ell \cos^2 2\theta_3 + Z_t \sin^2 2\theta_3 + Z_0}. \tag{B.72}$$

By analyzing these formulas, we see that for normal incidence $\theta_0 = \theta_2 = \theta_3 = 0$. $T_t = 0$, while

$$R = \frac{Z_\ell - Z_0}{Z_\ell + Z_0}, \quad \mathrm{T}_\ell = 2\frac{\rho_0}{\rho_1}\frac{C_0}{C_2}\frac{Z_\ell}{Z_\ell + Z_0}. \tag{B.73}$$

As could be expected, in this case the transverse (or shear) waves are not excited. On the other hand, for $\theta_3 = \frac{\pi}{4}$, which corresponds to $\theta_o = \mathrm{Arc}\sin\left(C_0 / \sqrt{2}\, C_3\right)$, we obtain

$$R = \frac{Z_t - Z_0}{Z_t + Z_0}, \quad \mathrm{T}_\ell = 0, \quad \mathrm{T}_t = 2\frac{\rho_0}{\rho_1}\frac{C_0}{C_3}\frac{Z_t}{Z_t + Z_0} \tag{B.74}$$

(i.e., only transverse waves are excited).

B.9. REFLECTION ON AN ELASTIC PLATE SEPARATING A FLUID FROM VACUUM

The derivation of the reflection coefficient of a plate immerged in a fluid is a classical problem, the solution of which may be found in textbooks (see, for instance, references [B.2] or [B.3]). In what follows, we give a short presentation of the solution of an elastic plate separating a fluid from vacuum. The geometrical and acoustic parameters are defined in figure B.2.

The incident pressure field is a plane wave in the fluid (medium O) given by

$$P_i = A \exp[\mathrm{ik}(x \sin \theta - z \cos \theta)]. \tag{B.75}$$

If we denote by R the reflection coefficient, the reflected field reads

$$P_r = AR \exp[\mathrm{ik}(x \sin \theta - z \cos \theta)]. \tag{B.76}$$

According to (A9), the displacement **u** in the fluid is given by

$$\mathbf{u} = \frac{1}{\rho_0 \omega^2} \mathrm{grad}(P_i + P_r), \tag{B.77}$$

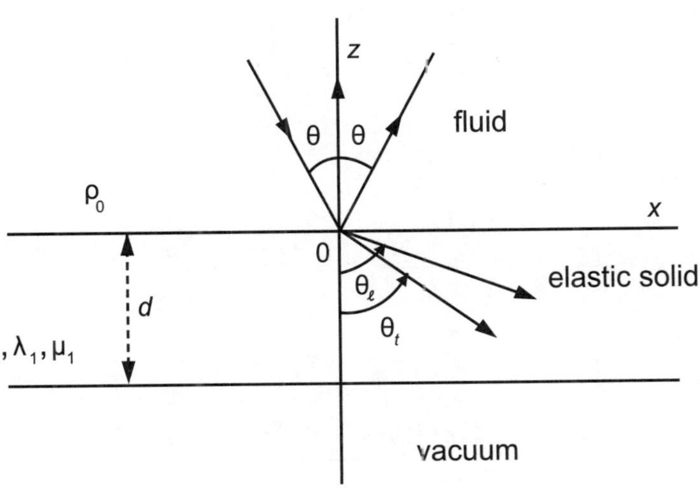

Figure B.2. Coordinate system and geometrical and acoustic parameters.

where ρ_0 is the density of the fluid.

The displacement in the elastic plate (medium 1) can be decomposed in the sum of the gradient of a scalar potential and the curl of a vector potential (see equation (B.25)):

$$\mathbf{u}_1 = \mathbf{grad}\ \phi_1 + \text{curl}\ \boldsymbol{\psi}_1. \tag{B.78}$$

Since the displacements of the incident and reflected fields are on the plane *xz*, the displacement of the transmitted field has no *y* component. The scalar potential ϕ_1 may therefore be written as

$$\phi_1 = \frac{A}{\rho_1 \omega^2}\left[D_L e^{i(\sigma x - \alpha z)} + U_L e^{i(\sigma x + \alpha z)}\right], \tag{B.79}$$

where ρ_1 is the density of the elastic medium.

Likewise, the term curl $\boldsymbol{\psi}_1$ in (B.78) has no *y* component. In order to simplify the derivation of the solution, we make the hypothesis that $\boldsymbol{\psi}_1 = \mathbf{y}\psi_1$ and write

$$\psi_1 = \frac{A}{\rho_1 w^2}\left[D_T\, e^{i(\sigma x - \beta z)} + U_T\, e^{i(\sigma x + \beta z)}\right]. \tag{B.80}$$

We verify afterward that this hypothesis is compatible with the boundary conditions.

Let k_ℓ and k_t be the propagation constants of the longitudinal and transverse waves, respectively. Then according to the Snell-Descartes law,

$$\sigma = k \sin\theta = k_\ell \sin\theta_\ell = k_t \sin\theta_t \tag{B.81}$$

and

$$\alpha = \sqrt{k_\ell^2 - \sigma^2}, \beta = \sqrt{k_t^2 - \sigma^2}, \tag{B.82}$$

$$k_\ell = \frac{\omega}{C_\ell}, C_\ell = \sqrt{\frac{\lambda_1 + 2\mu_1}{\rho_1}}, k_t = \frac{\omega}{C_t}, C_t = \sqrt{\frac{\mu_1}{\rho_1}}. \tag{B.83}$$

B.9.1. APPLICATION OF THE BOUNDARY CONDITIONS

We now apply the boundary conditions at $z = 0$ and $z = -d$. At $z = 0$, the following conditions must be verified:

1. The continuity of the normal component of the displacement $\mu_{1z} = \mu_0$ should be verified, which gives

$$\frac{\partial \phi_1}{\partial z} + \frac{\partial \psi_1}{\partial x} = \frac{1}{\rho_0 \omega^2} \frac{\partial}{\partial z}(P_i + P_r)$$

or equivalently

$$-\frac{\rho_1}{\rho_0} \cos\theta kR - \alpha D_L + \alpha U_L + \sigma D_T + \sigma U_T = -\frac{\rho_1}{\rho_0} k \cos\theta. \qquad \text{(B.84)}$$

2. The continuity of the normal stress should be verified:

$$\sigma_{zz} = (\lambda_1 + 2\mu_1)\frac{\partial u_{1z}}{\partial z} + \lambda_1 \frac{\partial u_{1x}}{\partial x} = -(P_i + P_r),$$

where $u_{1x} = \frac{\partial \phi_1}{\partial x} - \frac{\partial \psi_1}{\partial z}$ and

$$\lambda_1 + 2\mu_1 = \frac{\omega^2 \rho_1 (\beta^2 + \sigma^2)}{k_\ell^2 k_t^2}, \lambda_1 = \frac{\beta_1 \omega^2 (\beta^2 - 2\alpha^2 - \sigma^2)}{k_\ell^2 k_t^2} \qquad \text{(B.85)}$$

or, by replacing φ_1 and ψ_1 with their expressions (B.79) and (B.80),

$$-k_t^2 R + \left(\beta^2 - \sigma^2\right)\left(D_L + U_L\right) - 2\beta\sigma\left(D_T - U_T\right) = k_\ell^2. \qquad \text{(B.86)}$$

3. The tangential stress,

$$\sigma_{xz} = \mu_1 \left(\frac{\partial u_{1x}}{\partial z} + \lambda_1 \frac{\partial u_{1z}}{\partial x} \right) = 0,$$

vanishes and gives

$$2\alpha\sigma D_L - 2\alpha\sigma U_L + \left(\beta^2 - \sigma^2\right) D_T + \left(\beta^2 - \sigma^2\right) U_T = 0. \qquad \text{(B.87)}$$

At $z = -d$, the following boundary conditions must be verified:

1. The continuity of the normal stress should be verified:

$$\sigma_{zz} = \left(\beta^2 - \sigma^2\right) D_L e^{i\alpha d} + \left(\beta^2 - \sigma^2\right) U_L e^{-i\alpha d} - 2\beta\sigma D_T e^{i\beta d} + 2\beta\sigma U_T e^{-i\beta d} = 0. \qquad \text{(B.88)}$$

2. The tangential stress vanishes, giving

$$\sigma_{xz} = 2\alpha\sigma D_L e^{i\alpha d} - 2\alpha\sigma U_L e^{-i\alpha d} + \left(\beta^2 - \sigma^2\right) D_T e^{i\beta d} + \left(\beta^2 - \sigma^2\right) U_T e^{-i\beta d} = 0. \quad \text{(B.89)}$$

B.9.2. SOLUTION OF THE SYSTEM OF LINEAR EQUATIONS

The equations (B.84), (B.86), (B.87), (B.88), and (B.89) constitute a system of five linear equations of five unknowns. It can be readily verified that these equations are independent. The system has therefore a unique solution, which justifies our hypothesis on the vector potential. By choosing the new unknowns

$$U_L + D_L,\ U_L - D_L,\ U_T - D_T,\ U_T + D_T \text{ and } R,$$

we obtain the following result:

$$[F] \cdot [X] = [D],$$

where

$$[X] = \begin{bmatrix} R \\ U_L + D_L \\ U_L - D_L \\ U_T - D_T \\ U_T + D_T \end{bmatrix}, [D] = \begin{bmatrix} \frac{\rho_1}{\rho_0}\cos\theta k \\ k_t^2 \\ 0 \\ 0 \\ 0 \end{bmatrix} \quad \text{(B.91)}$$

and $[F]$ is the determinant,

$$[F] = \begin{bmatrix} k\frac{\rho_1}{\rho_0}\cos\theta & 0 & -\alpha & 0 & -\sigma \\ -k_t^2 & \beta^2 - \sigma^2 & 0 & 2\beta\sigma & 0 \\ 0 & 0 & -2\alpha\sigma & 0 & \beta^2 - \sigma^2 \\ 0 & (\beta^2 - \sigma^2)\cos\alpha d & -i(\beta^2 - \sigma^2)\sin\alpha d & 2\beta\sigma\cos\beta d & -2\beta\sigma i\sin\beta d \\ 0 & 2\alpha\sigma i\sin\alpha d & -2\alpha\sigma\cos\alpha d & -i(\beta^2 - \sigma^2)\sin\beta d & (\beta^2 - \sigma^2)\cos\beta d \end{bmatrix}. \quad \text{(B.92)}$$

It follows that

$$[F] = k\frac{\rho_1}{\rho_0}\cos\theta\ DET1 - k_t^2\ DET2,$$

$$R = \frac{k\dfrac{\rho_1}{\rho_0}\cos\theta\ DET1 - k_t^2\ DET2}{\dfrac{\rho_1}{\rho_0}k\cos\theta\ DET1 + k_t^2\ DET2}, \tag{B.93}$$

where $DET1$ and $DET2$ are the minors of $[F]$ with respect to the first column. Let $\varphi = \dfrac{\pi}{2} - \theta$ be the incidence angle measured with respect to the interface. Then the reflection coefficient can be written as

$$R = \frac{\sin\varphi - \eta}{\sin\varphi + \eta}, \tag{B.94}$$

where

$$\eta = \frac{\rho_0}{\rho_1}\frac{k_t^2}{k}\frac{DET2}{DET1}. \tag{B.95}$$

From (B.92), an explicit expression for η is given by

$$\eta = i\frac{\rho_0}{\rho_1}\frac{k_t^4}{k}\frac{\left[\left(\beta^2-\sigma^2\right)\cos\alpha d\sin\beta d + 4\alpha\beta\sigma^2\sin\alpha d\cos\beta d\right]\alpha}{8\beta\alpha\sigma^2\left(\beta^2-\sigma^2\right)\left(1-\cos\alpha d+\cos\beta d\right)+\sin\alpha d\sin\beta d\left[\left(\beta^2-\sigma^2\right)^4+16\alpha^2\beta^2\sigma^4\right]}. \tag{B.96}$$

This expression can be further reduced by stating

$$\tau_0 = \frac{\rho_0}{\rho_1}\frac{\alpha\ k_t^4}{k\left(\beta^2-\sigma^2\right)^2}, \quad \tau = \frac{4\alpha\beta\sigma^2}{\left(\beta^2-\sigma^2\right)^2}$$

and applying the trigonometric relation

$$\tan b\cot a + \tan a\cot b = 2\frac{1-\cos 2a\cos 2b}{\sin 2a\sin 2b}.$$

The final result is

$$\eta = i\tau_0 \frac{\cot \alpha d + \tau \cot \beta d}{1 + \left(\tan \frac{\alpha d}{2} \cot \frac{\beta d}{2} + \tan \frac{\beta d}{2} \cot \frac{\alpha d}{2} \right) \tau + \tau^2}. \tag{B.97}$$

Please refer to references [B.1] to [B.5] for additional information on linear elasticity.

REFERENCES

[B.1] Sokolnikoff, I. S. 1956. *Mathematical Theory of Elasticity.* New York: McGraw-Hill.

[B.2] Brekhowskikh, L. M., and O. A. Godin. 1990. *Acoustics of Layered Media I: Plane and Quasi-Plane Waves.* Berlin: Springer-Verlag.

[B.3] Bruneau, M. 1983. *Introduction aux Théories de L'acoustique.* Le Mans, France: Université du Maine.

[B.4] Morse, P. M., and H. Feshbach. 1953. *Methods of Theoretical Physics, Part 1.* New York: McGraw-Hill.

[B.5] Brilloin, L. 1960. *Les Tenseurs en Mécanique et en Élasticité,* Paris: Masson et Cie.

Appendix C

SURFACE WAVES

C.1. INTRODUCTION

Waves whose amplitudes decay rapidly with increasing distance from some surfaces are called surface waves. Various kinds of surface waves can exist near the boundaries of elastic solids. The surface wave near the free boundary of a solid was first described by Rayleigh [C.1] and is named after him.

Propagating disturbances conformed to the neighborhood of a surface occur not only in the vicinity of a free surface but also at the interface of two half-spaces filled with different materials. Such waves have components in both media, so they may be described as a set of different inhomogeneous plane waves in each medium that have a common velocity on the interface—the actual velocity of the interface wave. The surface wave at the interface of two elastic media has been found by Stoneley [C.2] and is called a Stoneley wave. But surface waves may also exist at the interface between a solid and a fluid.

Cagniard established the existence of an interface wave of complex velocity along the air-solid interface, the real part of which being very close to the velocity of the Rayleigh wave and the imaginary part of which tending toward zero with the density of air (see chapter 14, pages 233–46 in reference [C.3]). Later on, the geophysicist Scholte [C.4] found an interface wave of real velocity propagating along the fluid-solid interface, the components of which having an exponentially decreasing amplitude on both sides of the interface. Since this property is characteristic of Stoneley waves, Scholte called this wave a Stoneley wave. Today some call it "Scholte wave" while others call it a "Scholte-Stoneley wave," thus referring to both the one who found it and the characteristics it shares with Stoneley waves.

When two interfaces are close together, other types of guided waves exist. In a flat elastic plate with free boundaries, these waves are called Lamb waves, first described by Lamb [C.5]. Special cases are, for instance, an elastic plate with a boundary in contact with a fluid, the other boundary being free or in contact with the same fluid or a different fluid. All these waves are called generalized Lamb waves since they are a generalization of the case studied by Lamb.

C.2. RAYLEIGH WAVES

Let us begin by investigating the existence of Rayleigh waves near the free boundary of an elastic medium, which fill the region $x_2 > 0$ for the two-dimensional case of plane waves propagating in the x_1 direction.

If a surface wave exists near the free boundary of a solid, its displacement vector must be of the form $\mathbf{u} = \mathbf{u}_\ell + \mathbf{u}_t$ (ℓ = longitudinal, t = transverse) with

$$\begin{aligned} \mathbf{u}_\ell &= A\mathbf{d}^{(1)}\, e^{i\alpha x_2}\, e^{i\xi x_1 - i\omega t}, \\ \mathbf{u}_t &= B\mathbf{d}^{(2)}\, e^{i\beta x_2}\, e^{i\xi x_1 - i\omega t}, \end{aligned} \tag{C.1}$$

where $\mathrm{Im}(\alpha) > 0$, $\mathrm{Im}(\beta) > 0$, $\alpha^2 + \xi^2 = k_\ell^2$, and $\beta^2 + \xi^2 = k_t^2$ and where the unit vectors $\mathbf{d}^{(1)}$ and $\mathbf{d}^{(2)}$ are in the direction of the longitudinal and transverse displacements, respectively, which we suppose to be on the plane $x_1\, x_2$.

Since $\mathrm{curl}\,\mathbf{u}_\ell = 0$ and $\mathrm{div}\,\mathbf{u}_t = 0$, we have

$$d_2^{(1)} = \frac{\alpha}{\xi}\, d_1^{(1)}, \quad d_2^{(2)} = -\frac{\xi}{\beta}\, d_1^{(2)}, \tag{C.2}$$

where $d_1^{(q)}$ and $d_2^{(q)}$, (q = 1, 2) are the components of $\mathbf{d}^{(q)}$ along x_1 and x_2. The surface wave must satisfy the boundary conditions, which for a free boundary is

$$\sigma_{21} = \sigma_{22} = \sigma_{23} = 0. \tag{C.3}$$

According to (B19) and (C.1), the conditions (C.3) yield

$$\begin{aligned} \sigma_{21}^{(\ell)} + \sigma_{21}^{(t)} &= iA\mu\left(d_2^{(1)}\xi + d_1^{(1)}\alpha\right) + iB\mu\left(d_2^{(2)}\xi + d_1^{(2)}\beta\right) = 0, \\ \sigma_{22}^{(\ell)} + \sigma_{22}^{(t)} &= iA\left[\lambda\left(d_2^{(1)}\alpha + d_1^{(1)}\xi\right) + 2\mu\left(d_2^{(1)}\alpha\right)\right] \\ &\quad + iB\left[\lambda\left(d_2^{(2)}\beta + d_1^{(2)}\xi\right) + 2\mu\left(d_2^{(2)}\beta\right)\right] = 0. \end{aligned} \tag{C.4}$$

Since $\mathbf{u}_\ell$ and $\mathbf{u}_t$ do not depend on x_3, the condition $\sigma_{23} = 0$ is identically satisfied. Substitution of (C.2) into (C.4) gives

$$A_1 2\alpha + B_1\left(\beta - \frac{\xi^2}{\beta}\right) = 0,$$
$$A_1\left[(\lambda + 2\mu)\frac{\alpha^2}{\xi} + \lambda\xi\right] - 2\mu\xi B_1 = 0, \tag{C.5}$$

where $A_1 = Ad_1^{(1)}$ and $B_1 = Bd_1^{(2)}$.

The system of homogeneous equations (C.5) for the constants A_1 and B_1 has a nontrivial solution if the determinant of the coefficients vanishes, which leads to the following equation:

$$4\alpha\,\mu\,\xi + \left(\beta - \frac{\xi^2}{\beta}\right)\left[(\alpha + 2\mu)\frac{\alpha^2}{\xi} + \lambda\,\xi\right] = 0. \tag{C.6}$$

This relation is called the characteristic equation or dispersion equation and enables us to find the horizontal wave number ξ of the Rayleigh wave.

Since

$$\frac{\lambda + 2\mu}{\mu} = \frac{k_t^2}{k_\ell^2}, \quad \frac{\lambda}{\mu} = \frac{k_t^2}{k_\ell^2} - 2$$

and

$$\alpha = i\left(\xi^2 - k_\ell^2\right)^{1/2}, \quad \beta = i\left(\xi^2 - k_t^2\right)^{1/2}, \tag{C.7}$$

equation (C.6) reduces to

$$4\left(1 - \frac{k_\ell^2}{\xi^2}\right)^{1/2}\left(1 - \frac{k_t^2}{\xi^2}\right)^{1/2} = \left(\frac{k_t^2}{\xi^2} - 2\right)^2. \tag{C.8}$$

We would like to know the real solutions ξ_R of the characteristic equation. Using the notation

$$x = \frac{k_t^2}{\xi_R^2}, \quad \gamma = \frac{k_\ell^2}{k_t^2}, \tag{C.9}$$

where ξ_R is the velocity of the Rayleigh wave, we rewrite (C.8) in the following form:

$$4\sqrt{1-x}\,\sqrt{1-\gamma x} = (x-2)^2.$$

After squaring, the equation becomes algebraic:

$$f(x,\gamma) = x^3 - 8x^2 + 8(3-2\gamma)x - 16(1-\gamma) = 0. \qquad \text{(C.10)}$$

By virtue of the second equations of (B.29) and (B.30), since $\lambda > 0$ and $\mu > 0$, the parameter γ takes the values $0 \leq \gamma < \frac{1}{2}$. Therefore, denoting by C_{ph} the phase velocity of the surface wave, we have $C_{ph} = \omega / \xi_R$ and

$$x = \frac{k_t^2}{\xi_R^2} = \frac{\omega^2}{C_t^2 \xi_R^2} = \frac{C_{ph}^2}{C_t^2}. \qquad \text{(C.11)}$$

Surface waves are slow waves compared to volume waves: $C_{ph} < C_t < C_\ell$.

Thus according to (C.11) and since $f(o,\gamma) < 0$, $f(1,\gamma)$, and $\frac{\partial f}{\partial x}(x,\gamma) > 0$ at $x \leq 1$, all the roots of (C.10) are positive, and only one root of the equation is in the interval $0 < x < 1$. Hence the velocity of the Rayleigh wave is a single-valued function of the parameters of an elastic half-space. The cubic equation (C.10) has either one or three real roots depending on the value of γ. It can be shown that these roots correspond to a polarization exchange, which means that at the boundary the longitudinal wave transforms entirely into the transverse wave and vice versa—the reflection being absent.

C.3. SURFACE WAVES AT FLUID–SOLID INTERFACES

We investigate now the existence of surface waves at the interface of a fluid ($x_2 > 0$) and an elastic half-space ($x_2 < 0$). In the case of plane waves propagating in the x_1 direction, if a surface wave exists near the interface $x_2 = 0$ between an elastic solid and a fluid, its displacement vector must be of the form

$$\mathbf{u}^{(1)} = A\mathbf{d}^{(1)} e^{i\alpha_1 x_2} e^{i\xi x_1 - i\omega t} \qquad \text{(C.12)}$$

in the fluid, which we denote by medium 1 and

$$\mathbf{u}^{(2)} = \mathbf{u}_\ell^{(2)} + \mathbf{u}_t^{(2)} \qquad \text{(C.13)}$$

with

$$\mathbf{u}_{\ell}^{(2)} = B\,\mathbf{d}^{(2)}\, e^{i\alpha_2 x_2}\, e^{i\xi x_1 - i\omega t}, \tag{C.14}$$

$$\mathbf{u}_{t}^{(2)} = C\,\mathbf{d}^{(3)}\, e^{i\alpha_3 x_2}\, e^{i\xi x_1 - i\omega t} \tag{C.15}$$

in medium 2 corresponding to the solid, where

$$I_m(\alpha_1) > 0, \quad I_m(\alpha_2) < 0, \quad I_m(\alpha_3) < 0, \tag{C.16}$$

$$\alpha_1^2 + \xi^2 = k_1^2, \quad \alpha_2^2 + \xi^2 = k_\ell^2, \quad \alpha_3^2 - \xi^2 = k_t^2 \tag{C.17}$$

and where the unit vectors $\mathbf{d}^{(1)}$, $\mathbf{d}^{(2)}$, and $\mathbf{d}^{(3)}$ are in the direction of the longitudinal displacement in the fluid and in the longitudinal and transverse displacements in the solid, respectively, which we suppose to be the $x_1\, x_2$ plane. Since $\operatorname{curl}\mathbf{u}^{(1)} = 0$, $\operatorname{curl}\mathbf{u}_{\ell}^{(2)} = 0$, and $\operatorname{div}\mathbf{u}_{t}^{(2)} = 0$, the components $d_1^{(q)}$, $d_2^{(q)}$ of $\mathbf{d}^{(q)}$ ($q = 1, 2, 3$) are not independent.

They are related by

$$d_2^{(1)} = \frac{\alpha_1}{\xi}\, d_1^{(1)}, \quad d_2^{(2)} = \frac{\alpha_2}{\xi}\, d_1^{(2)}, \quad d_2^{(3)} = -\frac{\xi}{\alpha_3}\, d_1^{(3)}, \tag{C.18}$$

and since the vectors $\mathbf{d}^{(q)}$ are unit vectors, it follows from (C.18) that

$$d_1^{(1)} = \frac{\xi}{k_1}, \quad d_1^{(2)} = \frac{\xi}{k_\ell}, \quad d_1^{(3)} = \frac{\alpha_3}{k_t}. \tag{C.19}$$

The surface wave must satisfy the following boundary conditions (see equation (B.22)):

$$\mathbf{u}_2^{(1)} = \mathbf{u}_2^{(\ell)} + \mathbf{u}_2^{(t)}, \tag{C.20}$$

$$\sigma_{22}^{(1)} = \sigma_{22}^{(\ell)} + \sigma_{22}^{(t)}, \tag{C.21}$$

$$\sigma_{21}^{(\ell)} + \sigma_{21}^{(t)} = 0. \tag{C.22}$$

Since the displacements are all contained in the $x_1\, x_2$ plane, the other tangential components of the stress tensor are zero, which implies

$$\sigma_{23}^{(\ell)} = \sigma_{23}^{(t)} = 0.$$

Inserting (C.12) through (C.15) into the boundary conditions (C.20) through (C.22) and taking into account (C.18), (C.19), and (B.61), we obtain the following system of homogeneous equations for the constants *A, B, C*:

$$A\,\frac{\alpha_1}{k_1} - B\,\frac{\alpha_2}{k_\ell}\,d_1^{(2)} + C\,\frac{\xi}{k_t} = 0,$$

$$\rho_1\,\omega\,C_1\,A - \frac{\xi}{k_\ell}\,B\left[\langle\lambda + 2\mu\rangle\frac{\alpha_2^2}{\xi} + \lambda\xi\right] + 2C\,\mu\,\alpha_3\,\frac{\xi}{k_t} = 0, \qquad \text{(C.23)}$$

$$2\alpha_2 B\,\frac{\xi}{k_\ell} + C\,\frac{\alpha_3}{k_t}\left(\alpha_2 - \frac{\xi^2}{\alpha^2}\right) = 0,$$

where ρ_1 is the mass density of the fluid and λ and μ are the Lamé elastic constants of the solid.

The system (C.23) has a nontrivial solution if the determinant of the coefficients of A, B, C vanishes. Carrying out some simple computations, similar to those of section C.2, we obtain the following characteristic equation determining the wave number ξ:

$$4\xi^2\,\alpha_2\,\alpha_3 + \left(\xi^2 - \alpha_3^2\right)^2 = \frac{\rho_1}{\rho_2}\,\frac{\alpha_2\,k_t^4}{\sqrt{k_1^2 - \xi^2}}, \qquad \text{(C.24)}$$

where ρ_2 is the mass density of the solid. It is readily perceived that this equation differs from equation (C.8) for a Rayleigh wave in the presence of an additional term on the right-hand side to account for the effect of the fluid. Using the notations

$$x = \frac{k_t^2}{\xi^2}, \quad \eta = \frac{k_1^2}{k_t^2} \qquad \text{(C.25)}$$

and taking into account (C.16), equation (C.24) reduces to

$$4\left(1 - \gamma x\right)^{1/2}\left(1 - x\right)^{1/2} - \left(x - 2\right)^2 = \frac{\rho_1}{\rho_2}\sqrt{\left(1 - \gamma x\right)\left(1 - \eta x\right)}\;x^2. \qquad \text{(C.26)}$$

It is shown in reference [C.4] that for α_1, α_2, α_3 verifying (C.16), equation (C.26) always has one real root corresponding to a surface wave traveling along the boundary, with a phase velocity less than the velocity C_1 of the wave in the fluid and the velocities C_ℓ and C_t of the longitudinal and transverse waves in the solid. The indicated wave consists of an inhomogeneous wave in the fluid and two inhomogeneous waves in the solid. All three waves decay exponentially with distance, while leaving the boundary at $x_2 > 0$ and $x_2 < 0$.

It is possible to find the solution explicitly in the case where $\rho_1/\rho_2 <<1$ and $C_1^2/C_t^2 <<1$, corresponding to a rarefied medium (a gas) above an elastic half-space. Since we are seeking for the root $C_{ph} < C_1$, where $C_{ph} = \omega/\xi$, then $x = C_{ph}^2/C_t^2 << 1$. Following [C.6] and dividing both sides of (C.26) by $(1 - \gamma x)^{1/2}$ and expending the left-hand side in powers of x limited to the first term, we obtain

$$1-\eta x = x^2 / 4m^2\left(1-\gamma\right)^2. \tag{C.27}$$

where $m = \dfrac{\rho_2}{\rho_1}$.

In the first approximation by neglecting the second term of (C.27), we obtain $x = 1/\eta$. In the next approximation, we set this value of x on the right-hand side of (C.27) and then obtain

$$\eta x = 1 - \frac{1}{\left[2m\eta\left(1-\gamma\right)\right]^2}. \tag{C.28}$$

Since the velocity of the surface wave is

$$C_{ph} = \frac{\omega}{\xi} = \frac{\omega}{k_t}\sqrt{x} = C_t\sqrt{x}, \tag{C.29}$$

we obtain the following by noting that $1/\sqrt{\eta} = C_1/C_t$:

$$C_{ph} = C_1\left[1-\frac{1}{4m^2\eta^2\left(1-\gamma\right)^2}\right]^{1/2} \approx C_1\left[1-\frac{1}{8m^2\eta^2\left(1-\gamma\right)^2}\right].$$

We see that the velocity of the surface wave is somewhat smaller than the sound velocity in the gas. Furthermore, we have

$$\alpha_1 = i\sqrt{\xi^2 - k_1^2} = i\xi\sqrt{1-\eta x} = i\frac{1}{\sqrt{\eta}}\frac{k_1}{\sqrt{x}}\sqrt{1-\eta x} \cong i k_1\left[2m\eta\left(1-\gamma\right)\right]^{-1}$$

and

$$\alpha_2 = i\sqrt{\xi^2 - k_\ell^2} = i\xi\sqrt{1-\gamma x} \cong i k_1,$$

$$\alpha_3 = i\sqrt{\xi^2 - k_t^2} = i\xi\sqrt{1-x} \cong i k_1.$$

Thus since $m\eta = \rho_2 C_t^2 / \rho_1 C_1^2 >> 1$, the amplitude decreases very slowly in the gas, away from the boundary; while in the elastic solid, the entire wave field is concentrated in a layer with thickness of the order of a wavelength. We see that the surface wave has a different structure and velocity than a Rayleigh wave and does not transform into a Rayleigh wave in the solid as the density of the liquid tends to zero. As mentioned in the introduction of this appendix, the wave described by the positive root of the characteristic equation (C.26) is a particular case of the surface wave at the interface of two elastic media, found by Stoneley (but since it was found by Scholte, we call it the Scholte-Stoneley wave).

C.4. LEAKY WAVES

If we relax the condition (C.16) by letting $I_m \alpha_1$ be negative, then the characteristic equation (C.26) also has, in addition to the real root, a complex root corresponding to a system of three waves: one in the fluid and two in the solid, which transforms to a Rayleigh wave in the solid as the density of the fluid tends to zero. The fact that the root is complex has a simple physical interpretation: the wave will be partially radiated into the fluid, and it belongs to the class of leaky waves. The phase velocity of the leaky wave, in contrast to an ordinary surface wave, is not parallel to the boundary, and the amplitude of such a wave will fall off as it travels along the boundary. Moreover, as one goes away from the boundary into the fluid, along the normal to the boundary, one observes an increase in the amplitude of the wave.

The wave described by the complex root of the characteristic equation (C.26) is a type of Rayleigh wave whose velocity is somewhat altered because of the reaction of the fluid. We call it, therefore, a generalized Rayleigh wave. Different types of leaky waves can propagate along the interface of a solid and a fluid and also along the boundaries of solids. Although their amplitude falls off exponentially with distance, taking these waves into account is sometimes important in addressing the field generated by a line source or point source or the field diffracted by the edge of an elastic wedge or the tip of an elastic cone.

The leaky wave cannot exist alone since its field is unbounded at infinite distance from the interface. However, such a type of wave can be isolated, for instance, in the field diffracted by the edge of an elastic wedge immersed in a fluid. Here, as an observation point leaves the boundary, the increase in the wave's amplitude ceases at a certain distance, and it begins to decrease.

In order to classify the leaky waves corresponding to complex roots of the characteristic equation, it is important to introduce the concept of inhomogeneous plane waves. Let us look for the solution to the wave equation in homogeneous media (fluid or elastic solid), in the form

$$\mathbf{d} = \mathbf{d}_o \exp\left[i\left(\mathbf{K} \cdot \mathbf{r} - \omega t\right) \right], \tag{C.30}$$

where $\mathbf{d}$ is the complex displacement vector, $\mathbf{r}$ is a real vector associated to the observation point, $\mathbf{d}_o$ is a complex vector, and $\mathbf{K}$ is a complex wave vector

$$\mathbf{K} = \mathbf{K}' + i\mathbf{K}''. \tag{C.31}$$

Substituting this expression into the Helmholtz equation yields

$$\mathbf{K} \cdot \mathbf{K} = \frac{\omega^2}{C^2} = k^2, \tag{C.32}$$

which leads to

$$\mathbf{K}' \cdot \mathbf{K}'' = 0 \tag{C.33}$$

and

$$\mathbf{K}'^2 - \mathbf{K}''^2 = k^2. \tag{C.34}$$

The direction of propagation of the wave is given by $\mathbf{K}'$ and the direction corresponding to an exponential decrease of the wave amplitude is given by $\mathbf{K}''$. The expression

$$\mathbf{d} = \mathbf{d}_o \exp\left[i \left(\mathbf{K}' \cdot \mathbf{r} - \omega t\right) \right] \exp\left(-\mathbf{K}'' \cdot \mathbf{r}\right), \tag{C.35}$$

with $\mathbf{K}'$ and $\mathbf{K}''$ satisfying (C.33) and (C.34), describes an inhomogeneous plane wave. Its fronts (plane of constant phase) are orthogonal to $\mathbf{K}'$. The amplitude is constant at a plane orthogonal to the vector $\mathbf{K}''$. Hence constant phase planes and constant amplitude planes are orthogonal to each other. Expression (C.35) is the most general solution of the wave equation in an unbounded, nonviscous homogeneous medium.

Relation (C.34) is the most fundamental equation in the evanescent plane wave theory. Let us denote by C_{ph} the phase velocity of the inhomogeneous plane wave and denote by C the phase velocity of the homogeneous plane wave, both propagating in the same medium. We have

$$C_{ph} = \frac{\omega}{K'}, \quad C = \frac{\omega}{k}. \tag{C.36}$$

Then according to (C.34),

$$C_{ph} = \omega\left(k^2 + K''^2\right)^{-1/2} < C. \tag{C.37}$$

We see that the phase velocity of the inhomogeneous wave decreases as K'' increases and is always less than the phase velocity of the homogeneous plane wave in the same medium. Figure C.1 gives an intrinsic representation of the wave bivector $\mathbf{K} = (\mathbf{K}', \mathbf{K}'')$ of an inhomogeneous plane wave.

Consider now the case of the Scholte-Stoneley wave propagating at the interface $x_2 = 0$ between a fluid ($x_2 > 0$) and an elastic solid ($x_2 < 0$). This wave can be decomposed into a set of

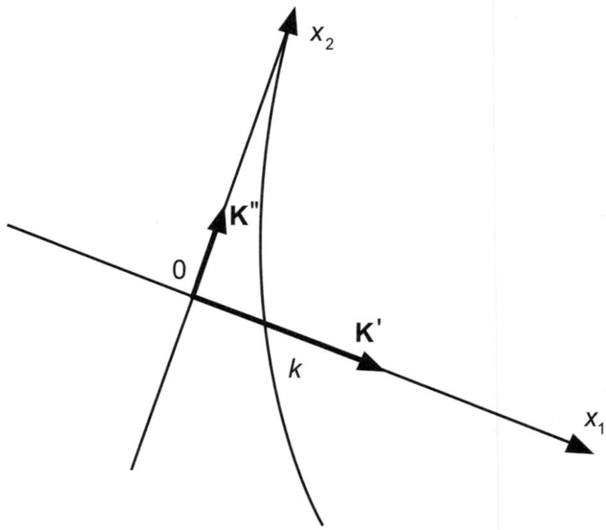

Figure C.1. Intrinsic representation of an inhomogeneous plane wave.

three inhomogeneous plane waves: a lamellar wave in the liquid of wave vector $\mathbf{K}_1$, a lamellar wave in the solid of wave vector $\mathbf{K}_L$, and a torsion wave in the solid of wave vector $\mathbf{K}_T$. Figure C.2 shows the decomposition of the Scholte-Stoneley wave. Each inhomogeneous plane wave is pictured by the real part and the imaginary part of its wave vector:

$$\mathbf{K}_1 = \mathbf{K}_1' + i\mathbf{K}_1'', \quad \mathbf{K}_L = \mathbf{K}_L' + i\mathbf{K}_L'', \quad \mathbf{K}_T = \mathbf{K}_T' + i\mathbf{K}_T''. \tag{C.38}$$

The three inhomogeneous plane waves are linked together on the interface by the Snell-Descartes laws:

$$\mathbf{K}_1 \cdot \mathbf{x}_1 = \mathbf{K}_L \cdot \mathbf{x}_1 = \mathbf{K}_T \cdot \mathbf{x}_1 = \xi, \tag{C.39}$$

where $\mathbf{x}_1$ is the unit vector along the x_1 axis. By separating the real and imaginary parts, (C.39) gives

$$\mathbf{K}_1' \cdot \mathbf{x}_1 = \mathbf{K}_L' \cdot \mathbf{x}_1 = \mathbf{K}_T' \cdot \mathbf{x}_1 = \xi_R, \tag{C.40}$$

$$\mathbf{K}_1'' \cdot \mathbf{x}_1 = \mathbf{K}_L'' \cdot \mathbf{x}_1 = \mathbf{K}_T'' \cdot \mathbf{x}_1 = \xi_I, \tag{C.41}$$

where ξ_R and ξ_I are, respectively, the real and the imaginary parts of ξ.

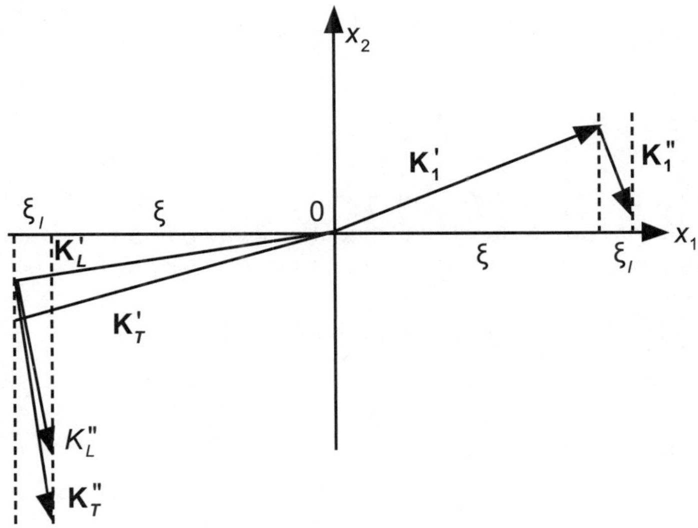

Figure C.2. The generalized Rayleigh wave as a set of three inhomogeneous plane waves.

Hence the projections on the interface of the real and imaginary parts of the wave vectors are equal. Relations (C.40) and (C.41) define the velocity C_W of the interface wave, as well as its damping coefficient α in the propagation direction

$$C_W = \frac{\omega}{\xi_R}, \quad \alpha = \xi_I. \tag{C.42}$$

Since the phase velocity of the inhomogeneous plane wave in the fluid is given by $C_{ph} = \omega / K_1'$ and since $K_1' > \xi_R$, we have

$$C_W > C_{ph1}, \tag{C.43}$$

which shows that the velocity of the interface wave is larger than its phase velocity in the fluid. The wave will therefore be partially radiated into the fluid and belongs to the class of leaky waves.

According to the geometrical property (see fig. C.2),

$$\frac{\xi_R}{K'} = \frac{\sqrt{K_1''^2 - \xi_I^2}}{K_1''}, \tag{C.44}$$

and for the relation (C.34), we obtain the following for a given value of $\xi = \xi_R + i\xi_I$:

$$K_1' = \left(k_1^2 + K_1''^2\right)^{1/2}, \quad K_1'' = \frac{1}{\sqrt{2}}\left(\xi_R^2 + \xi_I^2 - k_1^2 + \sqrt{\Delta}\right)^{1/2}, \tag{C.45}$$
$$\Delta = \left(\xi_R^2 + \xi_I^2 - k_1^2\right)^2 + 4k_1^2\,\xi_I^2.$$

Similar relations are verified by the corresponding inhomogeneous plane waves in the solid. However, these relations do not uniquely define the inhomogeneous plane waves corresponding to a given value of ξ. In the fluid, for instance, we have two inhomogeneous plane waves for $\xi_R > 0$, $\xi_I > 0$ (see fig. C.3).

These waves differ by the projection of the bivector ($\mathbf{K}_1'$, $\mathbf{K}_1''$) on the x_2 axis. For the wave (*a*), we have

$$\mathbf{K}_1' \cdot \mathbf{x}_2 = \sqrt{K_1'^2 - \xi_R^2}, \qquad \mathbf{K}_1'' \cdot \mathbf{x}_2 = -\sqrt{K_1''^2 - \xi_I^2}, \tag{C.46}$$

and for the wave (*b*), we have

$$\mathbf{K}_1' \cdot \mathbf{x}_2 = -\sqrt{K_1'^2 - \xi_R^2}, \qquad \mathbf{K}_1'' \cdot \mathbf{x}_2 = \sqrt{K_1''^2 - \xi_I^2}. \tag{C.47}$$

For $\xi_R > 0$ and $\xi_I > 0$, we can therefore define a pseudo-generalized Rayleigh wave represented in figure C.4.

We now come back to the characteristic equation (C.26). After squaring two times, the equation becomes algebraic in the unknown x defined by (C.25). Eliminating the trivial solution $x = 0$, we obtain the following:

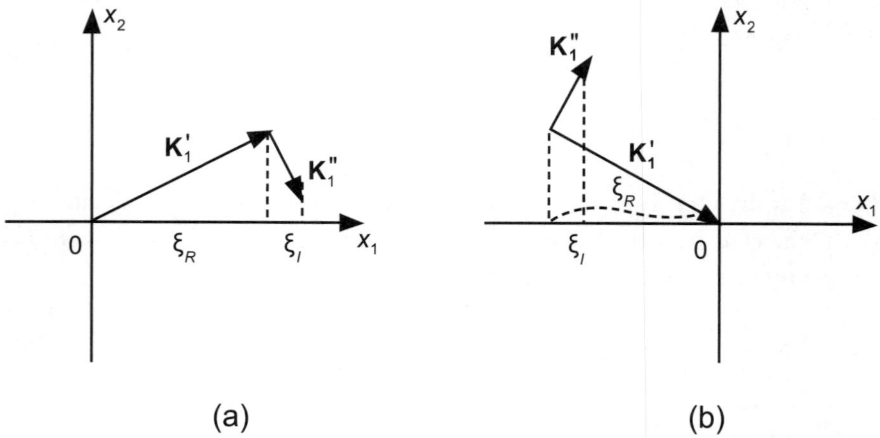

Figure C.3. Geometrical representation of two inhomogeneous plane waves in the fluid corresponding to the same value of ξ.

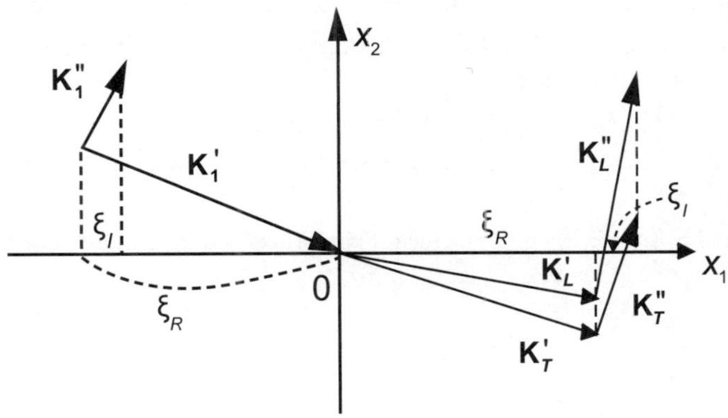

Figure C.4. Pseudo-generalized Rayleigh wave.

$$\left\{(1-\eta x)\left[x^3-8x^2+8(3-2\gamma)x-(16-\gamma)\right]+(1-\gamma x)x^3\right\}^2$$
$$=4(x-2)^4(1-\gamma x)(1-\eta x)x^2. \tag{C.48}$$

Equation (C.48) is of degree eight and therefore has eight roots, real or complex. Since each root has two corresponding interface waves, we obtained 16 interface waves for the fluid-solid interface, including the Scholte-Stoneley wave and the generalized Rayleigh wave.

A classification of the 16 interface waves has been established by Sebbag [C.7]. Similarly, at the free boundary of an elastic solid, the characteristic equation (C.10) has three roots, real or complex. They correspond to six waves that propagate along the surface, including the classical Rayleigh wave.

Additional information on Lamb and Rayleigh waves, inhomogeneous plane waves, and interface surface or leaky waves can be found in references [C.6], [C.8], [C.9], and [C.10].

REFERENCES

[C.1] Rayleigh, Lord. 1885. "On Waves Propagated along the Plane Surfaces of an Elastic Solid." *Proceedings of the London Mathematical Society* 17: 4–11.

[C.2] Stoneley, R. 1924. "Elastic Waves at the Interface of Separation of Two Solids." *Proceedings of the London Mathematical Society* 106: 416–28.

[C.3] Cagniard, L. 1939. *Reflexion et Refraction des Ondes Séismiques Progressives.* Paris: Gauthier-Villars.

[C.4] Scholte, J. G. 1949. "On True and Pseudo Rayleigh Waves." *Proc. Kon. Ned. Akad. Wet.* 52: 652–53.

[C.5] Lamb, H. 1917. "On Waves in an Elastic Plate." *Proceedings of the Royal Society* A: 93–114.

[C.6] Brekhowskikh, L. M., and O. A. Godin. 1990. *Acoustics of Layered Media I*. Berlin: Springer-Verlag.

[C.7] Sebbag, L. 1987. "Les Lois de la Réflexion-Réfraction des Ondes Planes Évanescentes et les Ondes D'interface." Paris: Université Paris.

[C.8] Viktorov, I. A. 1967. *Rayleigh and Lamb Waves*. New York: Plenum Press.

[C.9] Poirée, B. 1989. "Les Ondes Planes Évanescentes Dans les Fluides Parfaits et les Solides Élastiques." *Journal Acoustique* 2: 205–16.

[C.10] Poirée, B., and F. Luppé. 1991. "Evanescent Plane Waves and the Scholte-Stoneley Interface Wave." *Journal Acoustique* 4: 575–88.

Appendix D

GENERAL FORMULAS FOR THE PRINCIPAL RADII OF CURVATURE OF THE REFLECTED WAVE FRONT ON A THREE-DIMENSIONAL SURFACE

In this appendix, we give the formulas without demonstration. The interested reader is referred to reference [D.1] for an outline of the author's demonstration. The presentation of the formulas follows the presentation given in reference [D.2]. Examples of the application of these formulas are given in [D.3].

The principal radii of curvature of the reflected wave front at the point of reflection Q is given by

$$\frac{1}{\rho_1^r} = \frac{1}{2}\left(\frac{1}{\rho_1^i} + \frac{1}{\rho_2^i}\right) + \frac{1}{f_1},$$

$$\frac{1}{\rho_2^r} = \frac{1}{2}\left(\frac{1}{\rho_1^i} + \frac{1}{\rho_2^i}\right) + \frac{1}{f_2},$$

where

$$\frac{1}{f_{1,2}} = \frac{\cos\theta^i}{\left[\det(\Gamma)\right]^2}\left[\frac{(\Gamma_{22})^2 + (\Gamma_{12})^2}{a_1} + \frac{(\Gamma_{21})^2 + (\Gamma_{11})^2}{a_2}\right] \pm \frac{1}{2}\left\{\left(\frac{1}{\rho_1^i} - \frac{1}{\rho_2^i}\right)\right.$$

$$+\frac{4\cos\theta^i}{\left[\det(\Gamma)\right]^2}\left(\frac{1}{\rho_1^i}-\frac{1}{\rho_2^i}\right)\left[\frac{(\Gamma_{22})^2-(\Gamma_{12})^2}{a_1}+\frac{(\Gamma_{21})^2-(\Gamma_{11})^2}{a_2}\right]$$

$$+\frac{4\cos^2\theta^i}{\left[\det(\Gamma)\right]^2}\left[\frac{(\Gamma_{22})^2+(\Gamma_{12})^2}{a_1}+\frac{(\Gamma_{21})^2+(\Gamma_{11})^2}{a_2}\right]-\frac{4\left[\det(\Gamma)\right]^2}{a_1\, a_2}\Bigg\}.$$

Here, a_1,a_2 are the principal radii of curvature at Q of the surface, with principal directions defined by the unit vectors $\mathbf{u}_1$ and $\mathbf{u}_2$, and ρ_1^i, ρ_2^i are the principal radii of curvature of the incident wave front at Q with principal directions defined by the unit vectors $\mathbf{x}_1^i$ and $\mathbf{x}_2^i$. The Γ_{ij} are defined by

$$[\Gamma]=\begin{bmatrix}\Gamma_{11} & \Gamma_{12}\\ \Gamma_{21} & \Gamma_{22}\end{bmatrix}=\begin{bmatrix}\mathbf{x}_1^i\cdot\mathbf{u}_1 & \mathbf{x}_1^i\cdot\mathbf{u}_2\\ \mathbf{x}_2^i\cdot\mathbf{u}_1 & \mathbf{x}_2^i\cdot\mathbf{u}_2\end{bmatrix}.$$

REFERENCES

[D.1] Deschamps, G. A. 1972. "Ray Techniques in Electromagnetics." *Proceedings of the IEEE* 60 (9): 1022–35.

[D.2] Kouyoumjian, R. G., and P. H. Pathak. 1974. "A Uniform Geometrical Theory of Diffraction for an Edge in a Perfectly Conducting Surface." *Proceedings of the IEEE* 62 (11): 1448–61.

[D.3] McNamara, D. A., C. W. I. Pistorius, and J. A. G. Malherbe. 1990. *Introduction to the Uniform Geometrical Theory of Diffraction.* Norwood, MA: Artech House.

Appendix E

SYMMETRIC FORM OF THE MALIUZHINETS DIFFRACTION COEFFICIENT

According to the reciprocity theorem, the diffraction coefficient D in (2.193) or (2.204) must be invariant when we interchange ϕ and ϕ':

$$D(\phi, \phi'; k) = D(\phi', \phi; k). \tag{E.1}$$

In order to establish this property, which is not obvious in the formulas (2.193) and (2.204), we consider first the following quotients of ψ functions appearing in the diffraction coefficient:

$$\frac{\psi\left(\pi + \frac{n\pi}{2} - \phi\right)}{\psi\left(\frac{n\pi}{2} - \phi'\right)}, \frac{\psi\left(-\pi + \frac{n\pi}{2} - \phi\right)}{\psi\left(\frac{n\pi}{2} - \phi'\right)}. \tag{E.2}$$

According to (2.137) and (2.198), we have

$$\frac{\psi\left(\pi + \frac{n\pi}{2} - \phi\right)}{\psi\left(\frac{n\pi}{2} - \phi'\right)} = \frac{\psi_n\left(n\pi + \frac{3\pi}{2} - \phi - \theta_0\right)\psi_n\left(n\pi + \frac{\pi}{2} - \phi + \theta_0\right)\psi_n\left(\frac{\pi}{2} - \phi + \theta_n\right)\psi_n\left(\frac{3\pi}{2} - \phi - \theta_n\right)}{\psi_n\left(n\pi + \frac{\pi}{2} - \phi' - \theta_0\right)\psi_n\left(n\pi - \frac{\pi}{2} - \phi' + \theta_0\right)\psi_n\left(-\phi' - \frac{\pi}{2} + \theta_n\right)\psi_n\left(-\phi' + \frac{\pi}{2} - \theta_n\right)}. \tag{E.3}$$

For a passive system, we have $0 \le R_e \theta_{0,n} \le \pi/2$. If we consider an acute wedge angle $n\pi > \pi$, then the functions $\psi_n(u)$ on the right-hand side of (E.3), having their argument in the fundamental domain $-n\pi \le R_e u \le n\pi$ when $0 \le \phi \le n\pi$ and $0 \le \phi' \le n\pi$, are

$$\psi_n\left(\frac{\pi}{2} - \phi + \theta_n\right),\ \psi_n\left(n\pi - \frac{\pi}{2} - \phi' + \theta_0\right),\ \psi_n\left(-\phi' + \frac{\pi}{2} - \theta_n\right).$$

The arguments of all other functions fall outside the fundamental domain and can therefore not be calculated by making use of the Maliuzhinets function (2.139). However, owing to the Maliuzhinets relations (2.140), it is possible to make a translation of the arguments and bring them in the required fundamental domain. Applying the second relation (2.140) yields the following:

$$\psi_n\left(n\pi + \frac{3\pi}{2} - \phi - \theta_0\right) = \psi_n^2\left(\frac{\pi}{2}\right)\frac{\cos\left[\frac{1}{2n}\left(n\pi + \pi - \phi - \theta_0\right)\right]}{\psi_n\left(n\pi + \frac{\pi}{2} - \phi - \theta_0\right)}$$

$$= \frac{\cos\left[\frac{1}{2n}\left(n\pi + \pi - \phi - \theta_0\right)\right]}{\cos\left[\frac{1}{2n}\left(n\pi - \phi - \theta_0\right)\right]}\psi_n\left(n\pi - \frac{\pi}{2} - \phi - \theta_0\right),$$

$$\psi_n\left(n\pi + \frac{\pi}{2} - \phi + \theta_0\right) = \psi_n^2\left(\frac{\pi}{2}\right)\frac{\cos\left[\frac{1}{2n}\left(n\pi - \phi + \theta_0\right)\right]}{\psi_n\left(n\pi - \frac{\pi}{2} - \phi + \theta_0\right)},$$

$$\psi_n\left(\frac{3\pi}{2} - \phi - \theta_n\right) = \psi_n^2\left(\frac{\pi}{2}\right)\frac{\cos\left[\frac{1}{2n}\left(\pi - \phi - \theta_n\right)\right]}{\psi_n\left(\frac{\pi}{2} - \phi - \theta_n\right)},$$

$$\psi_n\left(n\pi + \frac{\pi}{2} - \phi' - \theta_0\right) = \psi_n^2\left(\frac{\pi}{2}\right)\frac{\cos\left[\frac{1}{2n}\left(n\pi - \phi' - \theta_0\right)\right]}{\psi_n\left(n\pi - \frac{\pi}{2} - \phi' - \theta_0\right)},$$

$$\psi_n\left(-\phi' - \frac{\pi}{2} + \theta_n\right) = \psi_n^2\left(\frac{\pi}{2}\right) \frac{\cos\left[\frac{1}{2n}\left(-\phi' + \theta_n\right)\right]}{\psi_n\left(-\phi' + \frac{\pi}{2} + \theta_n\right)}.$$

Inserting these relations into (E.2) yields

$$\frac{\psi\left(\pi + \frac{\pi}{2} - \phi\right)}{\psi\left(\frac{n\pi}{2} - \phi'\right)} = C_n^{(1)}\left(\phi,\phi';\theta_0,\theta_n\right) M_n\left(\phi,\phi';\theta_0,\theta_n\right), \tag{E.4}$$

where

$$C_n^{(1)}\left(\phi,\phi';\theta_0\theta_n\right) = \frac{\sin\left[\frac{1}{2n}\left(-\pi + \phi + \theta_0\right)\right] \sin\left[\frac{1}{2n}\left(\phi - \theta_0\right)\right] \cos\left[\frac{1}{2n}\left(\pi - \phi - \theta_n\right)\right]}{\sin\left[\frac{1}{2n}\left(\phi + \theta_0\right)\right] \sin\left[\frac{1}{2n}\left(\phi' + \theta_0\right)\right] \cos\left[\frac{1}{2n}\left(-\phi' + \theta_n\right)\right]}, \tag{E.5}$$

$$M_n\left(\phi,\phi';\theta_0,\theta_n\right) = \frac{\psi_n\left(n\pi - \frac{\pi}{2} - \phi - \theta_0\right) \psi_n\left(\frac{\pi}{2} - \phi + \theta_n\right) \psi_n\left(n\pi - \frac{\pi}{2} - \phi' - \theta_0\right) \psi_n\left(-\phi' + \frac{\pi}{2} + \theta_n\right)}{\psi_n\left(n\pi - \frac{\pi}{2} - \phi + \theta_0\right) \psi_n\left(\frac{\pi}{2} - \phi + \theta_n\right) \psi_n\left(n\pi - \frac{\pi}{2} - \phi' + \theta_0\right) \psi_n\left(-\phi' + \frac{\pi}{2} - \theta_n\right)}. \tag{E.6}$$

Likewise, the second term in (E.2) is given by

$$\frac{\psi\left(-\pi + \frac{n\pi}{2} - \phi\right)}{\psi\left(\frac{n\pi}{2} - \phi'\right)} = \frac{\psi_n\left(n\pi - \frac{\pi}{2} - \phi - \theta_0\right) \psi_n\left(n\pi - \frac{3\pi}{2} - \phi + \theta_0\right) \psi_n\left(-\frac{3\pi}{2} - \phi + \theta_n\right) \psi_n\left(-\frac{\pi}{2} - \phi - \theta_n\right)}{\psi_n\left(n\pi + \frac{\pi}{2} - \phi' - \theta_0\right) \psi_n\left(n\pi - \frac{\pi}{2} - \phi' + \theta_0\right) \psi_n\left(-\phi' - \frac{\pi}{2} + \theta_n\right) \psi_n\left(-\phi' + \frac{\pi}{2} + \theta_n\right)}. \tag{E.7}$$

On the right-hand side of (E.6), the functions $\psi_n(u)$ that verify $-n\pi \le R_e u \le n\pi$ are

$$\psi_n\left(n\pi - \frac{\pi}{2} - \phi - \theta_0\right), \psi_n\left(n\pi - \frac{\pi}{2} - \phi' + \theta_0\right), \psi_n\left(-\phi' + \frac{\pi}{2} - \theta_n\right). \tag{E.8}$$

Since the denominator in (E.7) is the same as in (E.3), the arguments of the functions that do not belong to the list (E.8) have already been transposed in the fundamental interval. It remains to transpose the arguments of the functions in the numerator of (E.7), which do not belong to (E.8). Applying the second relation (2.140) to these functions yields

$$\psi_n\left(n\pi - \frac{3\pi}{2} - \phi + \theta_0\right) = \psi_n^2\left(\frac{\pi}{2}\right)\frac{\cos\left[\frac{1}{2n}\left(n\pi - \phi + \theta_0\right)\right]}{\psi_n\left(n\pi - \frac{\pi}{2} + \phi + \theta_0\right)},$$

$$\psi_n\left(-\frac{3\pi}{2} - \phi + \theta_n\right) = \psi_n^2\left(\frac{\pi}{2}\right)\frac{\cos\left[\frac{1}{2n}\left(-\pi - \phi + \theta_n\right)\right]}{\psi_n\left(-\frac{\pi}{2} - \phi + \theta_n\right)} = \psi_n\left(\frac{\pi}{2} - \phi + \theta_n\right)\frac{\cos\left[\frac{1}{2n}\left(-\pi - \phi + \theta_n\right)\right]}{\cos\left[\frac{1}{2n}\left(-\phi + \theta_n\right)\right]},$$

$$\psi_n\left(-\frac{\pi}{2} - \phi - \theta_n\right) = \psi_n^2\left(\frac{\pi}{2}\right)\frac{\cos\left[\frac{1}{2n}\left(-\phi - \theta_n\right)\right]}{\psi_n\left(\frac{\pi}{2} - \phi - \theta_n\right)}.$$

By inserting these relations into (E.7), we obtain the final result:

$$\frac{\psi\left(-\pi + \frac{n\pi}{2} - \phi\right)}{\psi\left(\frac{n\pi}{2} - \phi'\right)} = C_n^{(2)}\left(\phi,\phi';\theta_0,\theta_n\right)M_n\left(\phi,\phi';\theta_0,\theta_n\right), \tag{E.9}$$

where $M_n(\phi, \phi'; \theta_0, \theta_n)$ is given by (E.6) and where

$$C_n^{(2)}\left(\phi,\phi';\theta_0,\theta_n\right) = \frac{\sin\left[\frac{1}{2n}\left(\pi + \phi - \theta_0\right)\right]\cos\left[\frac{1}{2n}\left(-\pi - \phi + \theta_n\right)\right]\cos\left[\frac{1}{2n}\left(\phi + \theta_n\right)\right]}{\sin\left[\frac{1}{2n}\left(\phi' + \theta_0\right)\right]\cos\left[\frac{1}{2n}\left(-\phi + \theta_n\right)\right]\cos\left[\frac{1}{2n}\left(-\phi' + \theta_n\right)\right]}. \tag{E.10}$$

We observe that $M_n(\phi, \phi'; \theta_0, \theta_n)$, given by (E.6), is symmetric with respect to ϕ and ϕ'. Moreover, by multiplying the numerator and the denominator of $C_n^{(1)}$ by $\cos\left[\frac{1}{2n}\left(-\phi + \theta_n\right)\right]$ and the numerator and denominator of $C_n^{(2)}$ by $\sin\left[\frac{1}{2n}\left(\phi + \theta_0\right)\right]$, we obtain

$$C_n^{(1)} = \frac{N_1}{Q_n}, \; C_n^{(2)} = \frac{N_2}{Q_n},$$

where the common denominator

$$Q_n = \sin\left[\frac{1}{2n}(\phi+\theta_0)\right]\sin\left[\frac{1}{2n}(\phi'+\theta_0)\right]\cos\left[\frac{1}{2n}(\phi'-\theta_n)\right]\cos\left[\frac{1}{2n}(\phi-\theta_n)\right] \quad \text{(E.11)}$$

is symmetric with respect to ϕ and ϕ' and where

$$N_1 = \sin\left[\frac{1}{2n}(-\pi+\phi+\theta_0)\right]\sin\left[\frac{1}{2n}(\phi-\theta_0)\right]\cos\left[\frac{1}{2n}(\pi-\phi-\theta_n)\right]\cos\left[\frac{1}{2n}(-\phi+\theta_n)\right], \quad \text{(E.12)}$$

$$N_2 = \sin\left[\frac{1}{2n}(\pi+\phi-\theta_0)\right]\cos\left[\frac{1}{2n}(-\pi-\phi+\theta_n)\right]\cos\left[\frac{1}{2n}(\phi+\theta_n)\right]\sin\left[\frac{1}{2n}(\phi+\theta_0)\right]. \quad \text{(E.13)}$$

According to (E.4) and (E.9), the diffraction coefficient given by (2.204) can be written in the form

$$D(\phi,\phi';k) = \frac{2}{n}\frac{M_n(\phi,\phi';\theta_0,\theta_n)}{Q_n(\phi,\phi';\theta_0,\theta_n)}N_n(\phi,\phi';\theta_0,\theta_n), \quad \text{(E.14)}$$

where

$$N_n(\phi,\phi';\theta_0,\theta_n) = \sin\frac{\phi'}{n}\left[\frac{N_2}{\cos\left(\frac{\phi+\pi}{n}\right)-\cos\frac{\phi'}{n}} - \frac{N_1}{\cos\left(\frac{\phi-\pi}{n}\right)-\cos\frac{\phi'}{n}}\right]. \quad \text{(E.15)}$$

By decomposing the cosine functions on the right-hand side of (E.15), we observe that the common denominator of the term in brackets is a symmetric function of ϕ and ϕ'. It remains to show that the numerator N given by

$$N = \sin\frac{\phi'}{n}\left\{N_2\left[\cos\left(\frac{\phi-\pi}{n}\right)-\cos\frac{\phi'}{n}\right] - N_1\left[\cos\left(\frac{\phi+\pi}{n}\right)-\cos\frac{\phi'}{n}\right]\right\}$$

is also symmetric. After some lengthy but elementary trigonometric calculations (not reproduced here) consisting of separating the trigonometric functions of ϕ and ϕ' from those depending on θ_0 and θ_n, we obtain

$$N = \left\{ \frac{1}{4} 2\sin\frac{\phi'}{n}\sin\frac{\phi}{n}\sin\frac{\pi}{n}\left[\cos\left(\frac{\theta_0+\theta_n}{2n}\right)\cos\left(\frac{\pi}{n}-\frac{\theta_0+\theta_n}{2n}\right)-\sin\left(\frac{\theta_n-\theta_0}{2n}\right)\right]\right.$$
$$-\frac{1}{2}\sin\frac{2\phi'}{n}\sin\frac{2\phi}{n}\sin\frac{\pi}{n}$$
$$\left.+\left(\sin\frac{\phi'}{n}\sin\frac{2\phi}{n}+\sin\frac{2\phi'}{n}\sin\frac{\phi}{n}\right)\sin\left(\frac{\theta_n-\theta_0}{2n}\right)\left[\cos\left(\frac{\pi}{n}-\frac{\theta_0+\theta_n}{2n}\right)-\cos\left(\frac{\theta_0+\theta_n}{2n}\right)\right]\right\},$$

which is obviously symmetric in ϕ and ϕ'. Hence N_n given by (E.15) is symmetric, and consequently $D(\phi,\phi'; k)$ is symmetric in ϕ and ϕ'.

Appendix F

ELEMENTS OF THE DETERMINANT OF THE BOUNDARY CONDITIONS FOR A CIRCULAR ELASTIC SHELL IN A FLUID

F.1. NORMAL INCIDENCE

In this appendix, we derive the elements of the determinant (2.385) corresponding to the linear systems (2.380) to (2.384). In polar coordinates, the displacements $\mathbf{u}_1$ and $\mathbf{u}_2$ in media 1 and 2 are related to the pressure p and the scalar and vector potentials ψ and $\mathbf{A} = \hat{\mathbf{z}} A_z$, respectively, by

$$\mathbf{u}_1 = \frac{1}{\rho_1 \omega^2} \nabla p = \frac{1}{\rho_1 \omega^2} \left(\hat{r} \frac{\partial p}{\partial r} + \frac{\hat{\theta}}{r} \frac{\partial p}{\partial \theta} \right), \tag{F.1}$$

$$\mathbf{u}_2 = \operatorname{grad} \psi + \operatorname{curl} \mathbf{A} = \hat{r} \frac{\partial \psi}{\partial r} + \frac{\hat{\theta}}{r} \frac{\partial \psi}{\partial \theta} + \frac{\hat{r}}{r} \frac{\partial A_z}{\partial \theta} - \hat{\theta} \frac{\partial A_z}{\partial r}. \tag{F.2}$$

Applying (F.1) and (F.2) to the equations (2.371), (2.372), (2.378), and (2.379), we directly obtain the following by imposing (2.381):

$$x_1 J_n'(x_1) = -x_1 H_n^{(1)'}(x_1) b_n + \rho_1 \omega^2 \left[x_\ell J_n'(x_\ell) c_n + x_\ell H_n^{(1)'}(x_\ell) d_n + n J_n(x_t) e_n + n H_n^{(1)}(x_t) f_n \right], \tag{F.3}$$

where

$$x_1 = k_1 a\,, \quad x_\ell = k_\ell a, \quad x_t = k_t a. \tag{F.4}$$

It follows immediately that

$$\begin{aligned}
&\beta_2 = x_1\, J_n'(x_1), \quad \alpha_{21} = -x_1\, H_n^{(1)'}(x_1), \quad \alpha_{22} = \rho_1\, \omega^2\, x_\ell\, J_n'(x_\ell),\\
&\alpha_{23} = \rho_1\, \omega^2\, x_\ell\, H_n^{(1)'}(x_\ell), \quad \alpha_{24} = \rho_1\, \omega^2\, n\, J_n(x_t), \quad \alpha_{25} = \rho_1\, \omega^2\, n\, H_n^{(1)}(x_t).
\end{aligned} \tag{F.5}$$

For applying the condition (2.380) on the continuity of the stress, we need to calculate the normal component of the stress tensor. When $i = j$, (B.19) takes the form

$$\sigma_{ii} = \lambda\, \mathrm{div}\, \mathbf{u} + 2\mu \frac{\partial u_i}{\partial x_i} = -\lambda\, \Delta\psi + 2\mu \frac{\partial u_i}{\partial x_i} = -\lambda\, k_\ell^2\, \psi + 2\mu \frac{\partial u_i}{\partial x_i}. \tag{F.6}$$

In polar coordinates, we have

$$\sigma_{rr} = -\lambda\, k_\ell^2\, \psi + 2\mu \left[\frac{\partial^2 \psi}{\partial r^2} + \frac{\partial}{\partial r}\left(\frac{1}{r} \frac{\partial A_z}{\partial \theta} \right) \right]. \tag{F.7}$$

Taking (F.7) into account in (2.380), we obtain

$$\begin{aligned}
\left(\frac{x_1}{k_1}\right)^2 J_n(x_1) = &-\left(\frac{x_1}{k_1}\right)^2 H_n^{(1)}(x_1) b_n + \left[x_\ell^2\, \lambda\, J_n(x_\ell) - 2\mu\, x_\ell^2\, J_n''(x_\ell) \right] c_n\\
&+ \left[x_\ell^2\, \lambda\, H_n^{(1)}(x_\ell) - 2\mu\, x_\ell^2\, H_n^{(1)\,''}(x_\ell) \right] d_n + 2\mu n \left[J_n(x_t) - x_t\, J_n'(x_t) \right] e_n\\
&+ 2\mu n \left[H_n^{(1)}(x_t) - x_t\, H_n^{(1)'}(x_t) \right] f_n.
\end{aligned} \tag{F.8}$$

From (F.8), it follows that

$$\begin{aligned}
&\beta_1 = \left(\frac{x_1}{k_1}\right)^2 J_n(x_1), \quad \alpha_{11} = -\left(\frac{x_1}{k_1}\right)^2 H_n^{(1)}(x_1), \quad \alpha_{12} = -2\mu\, x_\ell^2\, J_n''(x_\ell) + \lambda\, x_\ell^2\, J_n(x_\ell),\\
&\alpha_{13} = -2\mu\, x_\ell^2\, H_n^{(1)\,''}(x_\ell) + \lambda\, x_\ell^2\, H_n^{(1)}(x_\ell), \quad \alpha_{14} = 2\mu n \left[J_n(x_t) - x_t\, J_n'(x_t) \right],\\
&\alpha_{15} = 2\mu n \left[H_n^{(1)}(x_t) - x_t\, H_n^{(1)'}(x_t) \right].
\end{aligned} \tag{F.9}$$

The next equation in the linear system involves the tangential component of the stress σ_{ij} that, in cylindrical coordinates, reads

$$\sigma_{r\theta} = \sigma_{r\theta}^{(\ell)} + \sigma_{r\theta}^{(t)} \tag{F.10}$$

with

$$\sigma_{r\theta}^{(\ell)} = 2\mu\left[\frac{\partial}{\partial r}\left(\frac{1}{r}\frac{\partial\psi}{\partial\theta}\right) + \frac{1}{r}\frac{\partial^2\psi}{\partial r\partial\theta} - \frac{1}{r^2}\frac{\partial\psi}{\partial\theta}\right] = 2\mu\left[\frac{1}{r}\frac{\partial^2\psi}{\partial r\partial\theta} - \frac{1}{r^2}\frac{\partial\psi}{\partial\theta}\right], \tag{F.11}$$

$$\sigma_{r\theta}^{(t)} = \mu\left(\frac{1}{r^2}\frac{\partial^2 A_z}{\partial\theta^2} - \frac{\partial^2 A_z}{\partial r^2} + \frac{1}{r}\frac{\partial A_z}{\partial r}\right). \tag{F.12}$$

Taking (F.11) and (F.12) into account in (2.382) yields

$$\alpha_{32} = 2n\left[J_n(x_\ell) - x_\ell J_n'(x_\ell)\right], \ \alpha_{33} = 2n\left[H_n^{(1)}(x_\ell) - x_\ell H_n^{(1)'}(x_\ell)\right],$$
$$\alpha_{34} = -\left[x_t^2 J_n''(x_t) + n^2 J_n(x_t) - x_t J_n'(x_t)\right], \ \alpha_{35} = -\left[x_t^2 H_n^{(1)''}(x_t) + n^2 H_n^{(1)}(x_t) - x_t H_n^{(1)'}(x_t)\right]. \tag{F.13}$$

Using (F.7), (F.11), and (F.12), the coefficients of the unknowns in (2.383) and (2.384) are derived immediately and read

$$\alpha_{42} = 2\mu\, y_\ell^2\left[J_n''(y_\ell) - \lambda\, y_\ell^2 J_n(y_\ell)\right], \quad \alpha_{43} = 2\mu\, y_\ell^2\left[H_n^{(1)''}(y_\ell) - \lambda\, y_\ell^2 H_n^{(1)}(y_\ell)\right],$$
$$\alpha_{44} = -2\mu n\left[J_n(y_t) - y_t J_n'(y_t)\right], \quad \alpha_{45} = -2\mu n\left[H_n^{(1)}(y_t) - y_t H_n^{(1)'}(y_t)\right],$$
$$\alpha_{52} = 2n\left[J_n(y_\ell) - y_\ell J_n'(y_\ell)\right], \quad \alpha_{53} = 2n\left[H_n^{(1)}(y_\ell) - y_\ell H_n^{(1)'}(y_\ell)\right],$$
$$\alpha_{54} = -\left[y_t^2 J_n''(y_t) + n^2 J_n(y_t) - y_t J_n'(y_t)\right], \quad \alpha_{55} = -\left[y_t^2 H_n^{(1)''}(y_t) + n^2 H_n^{(1)}(y_t) - y_t H_n^{(1)'}(y_t)\right],$$

where $y_\ell = k_\ell b$ and $y_t = k_t b$. Note that the Hankel functions $H_n^{(1)}(x)$ can be replaced by the Neumann functions $Y_n(x)$ in the expressions of the elements $\alpha_{ij}\ (j \geq 2)$ of the determinant of the system without changing the solution.

F.2. OBLIQUE INCIDENCE

At oblique incidence, the formulas (2.371) and (2.372), which give the pressure in the fluid and the displacements in the elastic medium, particularly the scalar and vector potentials, are multiplied by $\exp(ik_z z)$. Moreover, among the three components of the vector potential A, only two are necessary. Applying the boundary conditions leads to seven linear equations of seven unknowns. We do not explicitly write these equations but give the elements of the determinant, directly written in the form (2.385):

$$\alpha_{11} = -\frac{\rho_1}{\rho_2} x_t^2 H_n^{(1)}\left(x_{1\perp}\right), \quad \alpha_{12} = 2\left[x_{\ell\perp} J_n'\left(x_{\ell\perp}\right) + \left(\frac{\lambda}{2\mu} x_\ell^2 - n^2 + x_{\ell\perp}^2\right) J_n\left(x_{\ell\perp}\right)\right],$$

$$\alpha_{13} = 2\left[x_{\ell\perp} Y_n'\left(x_{\ell\perp}\right) + \left(\frac{\lambda}{2\mu} x_\ell^2 - n^2 + x_{\ell\perp}^2\right) Y_n\left(x_{\ell\perp}\right)\right], \quad \alpha_{14} = -2n\left[x_{t\perp} J_n'\left(x_{t\perp}\right) - J_n\left(x_{t\perp}\right)\right],$$

$$\alpha_{15} = -2n\left[x_{t\perp} Y_n'\left(x_{t\perp}\right) - Y_n\left(x_{t\perp}\right)\right], \quad \alpha_{16} = 2\left[x_{t\perp} J_n'\left(x_{t\perp}\right) - \left(n^2 - x_{t\perp}\right) J_n\left(x_{t\perp}\right)\right],$$

$$\alpha_{17} = 2\left[x_{t\perp} Y_n'\left(x_{t\perp}\right) - \left(n^2 - x_{t\perp}\right) J_n\left(x_{t\perp}\right)\right],$$

$$\alpha_{21} = -x_{1\perp} H_n^{(1)'}\left(x_{1\perp}\right), \quad \alpha_{22} = x_{\ell\perp} J_n'\left(x_{\ell\perp}\right), \quad \alpha_{23} = x_{\ell\perp} Y_n'\left(x_{\ell\perp}\right), \quad \alpha_{24} = n\, J_n\left(x_{t\perp}\right),$$

$$\alpha_{25} = n\, Y_n\left(x_{t\perp}\right), \quad \alpha_{26} = x_{t\perp} J_n'\left(x_{t\perp}\right), \quad \alpha_{27} = x_{t\perp} Y_n'\left(x_{t\perp}\right),$$

$$\alpha_{31} = 0, \quad \alpha_{32} = -2n\left[x_{\ell\perp} J_n'\left(x_{\ell\perp}\right) - J_n\left(x_{\ell\perp}\right)\right], \quad \alpha_{33} = -2n\left[x_{\ell\perp} Y_n'\left(x_{\ell\perp}\right) - Y_n\left(x_{\ell\perp}\right)\right],$$

$$\alpha_{34} = \left(x_{t\perp}^2 - 2n^2\right) J_n\left(x_{t\perp}\right) + 2x_{t\perp} J_n'\left(x_{t\perp}\right), \quad \alpha_{35} = \left(x_{t\perp}^2 - 2n^2\right) Y_n\left(x_{t\perp}\right) + 2x_{t\perp} Y_n'\left(x_{t\perp}\right),$$

$$\alpha_{36} = 2n\left[J_n\left(x_{t\perp}\right) - x_{t\perp} J_n'\left(x_{t\perp}\right)\right], \quad \alpha_{37} = 2n\left[Y_n\left(x_{t\perp}\right) - x_{t\perp} Y_n'\left(x_{t\perp}\right)\right],$$

$$\alpha_{41} = 0\,,\ \alpha_{42} = -2\left[y_{\ell t} J_n'\left(y_{\ell t}\right) + \left(\frac{\lambda}{2\mu} y_\ell^2 - n^2 + y_{\ell\perp}^2 \right) J_n\left(y_{\ell\perp}\right)\right],$$

$$\alpha_{43} = -2\left[y_{\ell\perp} Y_n'\left(y_{\ell\perp}\right) + \left(\frac{\lambda}{2\mu} y_\ell^2 - n^2 + y_{\ell\perp}^2 \right) J_n\left(y_{\ell\perp}\right)\right],$$

$$\alpha_{44} = 2\left[y_{t\perp} J_n'\left(y_{t\perp}\right) - J_n\left(y_{t\perp}\right)\right],\ \ \alpha_{45} = 2\left[y_{t\perp} Y_n'\left(y_{t\perp}\right) - Y_n\left(y_{t\perp}\right)\right],$$

$$\alpha_{46} = -2\left[y_{t\perp} J_n'\left(y_{t\perp}\right) - \left(n^2 - y_{t\perp}\right) J_n\left(y_{t\perp}\right)\right],\ \ \alpha_{47} = -2\left[y_{t\perp} Y_n'\left(y_{t\perp}\right) - \left(n^2 - y_{t\perp}\right) Y_n\left(y_{t\perp}\right)\right],$$

$$\alpha_{51} = 0\,,\ \alpha_{52} = -2n\left[y_{\ell\perp} J_n'\left(y_{\ell\perp}\right) - J_n\left(y_{\ell\perp}\right)\right],\ \ \alpha_{53} = -2n\left[y_{\ell\perp} Y_n'\left(y_{\ell\perp}\right) - Y_n\left(y_{\ell\perp}\right)\right],$$

$$\alpha_{54} = \left(y_{t\perp}^2 - 2n^2\right) J_n\left(y_{t\perp}\right) + 2 y_{t\perp} J_n'\left(y_{t\perp}\right),\ \ \alpha_{55} = \left(y_{t\perp}^2 - 2n^2\right) Y_n\left(y_{t\perp}\right) + 2 y_{t\perp} Y_n'\left(y_{t\perp}\right),$$

$$\alpha_{56} = 2n\left[J_n\left(x_{t\perp}\right) - y_{t\perp} J_n'\left(y_{t\perp}\right)\right],\ \ \alpha_{57} = 2n\left[Y_n\left(y_{t\perp}\right) - y_{t\perp} J_n'\left(y_{t\perp}\right)\right],$$

$$\alpha_{61} = 0\,,\ \alpha_{62} = 2x_{1z}^2\, x_{\ell\perp} J_n'\left(x_{\ell\perp}\right),\ \alpha_{63} = 2x_{1z}^2\, x_{\ell\perp} Y_n'\left(x_{\ell\perp}\right),$$

$$\alpha_{64} = x_{1z}^2\, n\, J_n\left(x_{t\perp}\right),\ \ \alpha_{65} = x_{1z}^2\, n Y_n\left(x_{t\perp}\right),$$

$$\alpha_{66} = \left(x_{1z}^2 - x_{t\perp}\right) x_{t\perp} J_n'\left(x_{t\perp}\right),\ \ \alpha_{67} = \left(x_{1z}^2 - x_{t\perp}\right) x_{t\perp} Y_n'\left(x_{t\perp}\right),$$

$$\alpha_{71} = 0\,,\ \ \alpha_{72} = 2y_{1z}^2\, y_{\ell\perp} J_n'\left(y_{\ell\perp}\right),\ \ \alpha_{73} = 2y_{1z}^2\, y_{\ell\perp} Y_n'\left(y_{\ell\perp}\right),$$

$$\alpha_{74} = y_{1z}^2\, n\, J_n\left(\mathrm{x}_{t\perp}\right),\ \ \alpha_{75} = y_{1z}^2\, n Y_n\left(y_{t\perp}\right),$$

$$\alpha_{76} = \left(y_{1z}^2 - y_{t\perp}^2\right) y_{t\perp} J_n'\left(y_{t\perp}\right), \quad \alpha_{77} = \left(y_{1z}^2 - y_{t\perp}^2\right) y_{t\perp} Y_n'\left(y_{t\perp}\right),$$

where

$$x_{1\perp} = k_\perp a = k_1 a \sin\beta\,, \; y_{1\perp} = k_\perp b = k_1 b \sin\beta\,, \; x_{\ell\perp}^2 = \left(k_\ell a\right)^2 - \left(k_1 a\right)^2 \cos^2\beta,$$

$$x_{t\perp}^2 = \left(k_t a\right)^2 - \left(k_1 a\right)^2 \cos^2\beta, \; y_{\ell\perp}^2 = \left(k_\ell b\right)^2 - \left(k_1 b\right)^2 \cos^2\beta, \; y_{t\perp}^2 = \left(k_t b\right)^2 - \left(k_1 b\right)^2 \cos^2\beta,$$

$$x_t = k_t a\,, \quad x_\ell = k_\ell a, \quad y_\ell = k_\ell b, \quad y_t = k_t b.$$

We also obtain the following for the coefficients corresponding to the incident field:

$$\beta_1 = \frac{\rho_1}{\rho_2} x_{\mathrm{t}}^2 \, J_n\left(x_{1\perp}\right), \quad \beta_2 = x_{1\perp} \, J_n'\left(x_{1\perp}\right).$$

In the previously stated formulas for the elements α_{ij}, we have used the Bessel equation to express the second-order derivatives of the Bessel and Neumann functions in terms of the first- and zero-order derivatives of these functions.

INDEX

absolutely rigid boundary
- fluid waves, 400
- sound wave reflection, 403

absolutely soft boundary
- fluid waves, 401
- sound wave reflection, 403

acoustic field scattering
- convex impedance wedge, 272–93
- edge diffraction, impenetrable wedge, 53–62
- impedance surface characteristics, 40–44
- smooth convex object, shadow zone, 63–69

acoustic wave diffraction
- circular cylinder, 140–211
- elastic wedge-shaped shell, 125–40
- impedance wedge, Maliuzhinets's solution, 92–125
- straight wedge, normal and oblique incidence, 72–92

adiabatic motion
- ideal fluids, 398
- sound waves, 398–400

Airy function
- caustic canonical problem, 355–58
- circular impedance cylinder, 299–312
- diffraction coefficient, circular cylinder, 164–65
- grazing incidence and observation, 382–84
- pole path v-plane position, circular cylinder, 169–71

angular modes
- asymptotic expansion, circular cylinder, 149–62
- boundary conditions, circular cylinder, 142–47
- surface waves, uniform asymptotic solution, 283–89

ansatz
- geometric theory of diffraction and, 2
- uniform asymptotic solution, curved wedge, 271

astigmatic pencil of rays
- edge diffraction, acoustic field, 54–62
- transport equation resolution, 39–40

asymptotic expansion (evaluation)
- caustic canonical problem, 354–58
- creeping ray caustics, 363–65
- edge diffraction, ray caustics, 362–63
- elastic wedge-shaped shell, 132–37
- fluid-elastic interface, wave reflection and transmission, 45–53
- fundamentals, 22–28
- hollow elastic circular cylinder, 186–92
- impedance surface characteristics, 40–44
- impedance wedge, Maliuzhnets's solution, 110–22
- properties, 27–28
- three-dimensional convex wedge, 241–71
- Van der Waerden method, 255–60
- wave equation, source-free, unbounded medium, 28–40
- wedge diffraction, line source excitation, 83–89

asymptotic sequence
compatible sequence, 25–27
fundamentals, 23–24
axial caustics, observation point, 349–52

Bessel function
caustic canonical problem, 354–58
circular elastic shell, boundary conditions, 458
circular impedance cylinder, 298–312
concave circular cylinder, 214–33
creeping waves, 192–94
field diffraction in central region, 192–94
geometric acoustics extraction, 220–28
hollow elastic circular cylinder, 185–86, 192–94
integral representation, wedge diffraction, 78–82
line source canonical problem, 74–89, 214–20
remainder integral expression, 229–33
Watson transformation, 185–86
wedge diffraction, 74–89
Wronskian expression, 77–78, 215–20
boundary conditions
circular cylinder, 140–76
circular elastic shell, 453–58
concave circular cylinder, 212–35
creeping waves, 195–97
elastic plate, 426–28
elastic wedge-shaped shell, 126–32
fluid wave processes, 400–401, 426–28
hollow elastic circular cylinder, 179–82
linearized elasticity, 415
line source canonical problem, 212–33
Maliuzhinets's solution, 101–10
oblique incidence, 172–76, 195–97
oblique incidence, creeping waves, 195–97
Rayleigh waves, 432–34
reflection, fluid wave processes, 402–4
three-dimensional convex surface, 176–78
two-dimensional smooth convex surface, 297–98
wedge diffraction, 74–78
branch point singularities
concave circular cylinder, 217–20
Pauli-Clemmow method, 247–55
wedge diffraction, asymptotic approximation, 84–89

canonical problems
caustics, 354–58
concave circular cylinder, 212–33
edge diffraction, 53–58, 61–63
geometric theory of diffraction, 2–3
hollow elastic circular cylinder, 178–209
nonuniform asymptotic acoustic wave diffraction theory, 71–72
smooth convex object, acoustic field, 63–69
surface ray field, 67–69
three-dimensional convex wedge, 245–67
Cauchy's theorem
caustic canonical problem, 356–58
elastic wave excitation, 414
elastic wedge-shaped shell, 131–32
linear momentum and stress tensor, 411–12
wedge diffraction, plane wave incidence, 89–92
caustics
canonical problems, 354–58
circular caustic, tangential rays, 356–58
circular impedance cylinder, 309–12
concave circular cylinder, 227–28
curved wedge, 270–71
degenerated caustics, 349–52
elastic surface waves, 203
envelope, characteristic curves, 34–35
line cusp observation points, 366–67
observation points close to, 351–66
regular caustic solutions, 365–66

smooth surface ray reflection, 361–65
three-dimensional caustic, 358–61
three-dimensional smooth convex surface, 316–19
wave field calculation techniques, 349–67

circular cylinders
asymptotic expansion, 148–62
caustics, 349–51
concave circular cylinder, 212–33
curved wedge-shaped shell, 391–92
diffraction coefficients, 162–65, 391–92
hard, soft, or impedance boundary conditions, 140–76
hollow elastic circular cylinder, 178–209, 320–36
impedance cylinder, 298–312
line source excitation, 72–89
numerical results, 341–44
oblique incidence, 171–76
observation point location, 166–69
pole position, 169–71

circumferential waves, 213–20

coated wedge
asymptotic evaluation, 132–37
Lamb wave excitation, 139–40
oblique incidence, 137–38
physical parameters, 138
reflection coefficients, 138–39
solution derivation, 125–32

completeness relation
concave circular cylinder, line source canonical problem, 213–20
wedge diffraction, line source excitation, 77–78

complex v-plane
concave circular cylinder, 216–20, 230–33
generalized Lamb waves, 142
remainder integral expression, 230–33

complex ξ-plane
concave circular cylinder, 230–33
elastic wedge-shaped shell, 388–90
hybrid diffraction coefficients, 385–90
impedance wedge, 385–88
wedge diffraction, 83–89

concave surface
basic properties, 211–12
circular cylinder, 212–33
geometric acoustics, 220–28, 233–35
remainder integral, 228–33
whispering gallery modes, 393–94

contour deformation
circular impedance cylinder, 302–8
concave circular cylinder, 217–20, 222–28

convergence condition, 83–89

convex-concave rotational reflectors, 349–51

convex impedance wedge
circular cylinder, 175–76
planar interface, 273–89
uniform asymptotic solutions, 272–93

convex surface
smooth convex cylindrical surface, 371–73
three dimensional, boundary conditions, 176–78

convex three-dimensional shell
creeping wave solution, 209–10
elastic surface wave solution, 211
uniform asymptotic solution, 319–39

convolution product, elastic wedge-shaped shell, 127–32

coordinate system
elastic plate, fluid-vacuum separation, 425–30
elastic wedge-shaped shell, 126–32

creeping rays and waves
circular cylinder, 149–62, 168–76
curved wedge, 373–84
degenerated caustics, 349–52

creeping rays and waves (*continued*)
hollow elastic cylinder, 192–94, 320–36, 322–24
hybrid diffraction coefficients, 369–84
normal incidence, 192–94
oblique incidence, 171–76, 192–97
observation point location, 168–69
shadow zone, smooth convex object, 64–69
smooth surface caustics, 363–65
space wave diffraction, 373–78
three-dimensional caustic, 359–61
three-dimensional convex shell, 209–10, 336–39
three-dimensional convex surface, 176–78
three-dimensional smooth convex surface, 316–19

curl of vector
elastic plate, fluid-vacuum separation, 425–30
homogeneous isotropic solids, elastic wave excitation, 415–17
surface waves, fluid-solid interfaces, 435–38

curvature radii
circular cylinder, asymptotic expansion, 161–62
three-dimensional convex wedge, 241–71
three-dimensional surface wave front, 445–46

curved interface
impedance wedge, oblique incidence, 124–25
surface waves, 20–22
wave reflection and transmission, 48–53
wedge diffraction, 267–71

curved wedge
creeping waves, 373–84
edge diffraction, acoustic field, 53–62
elastic surface wave, 390–92
hybrid diffractions coefficients, 373–84, 390–92
Pauli-Clemmow method, 267–71
uniform asymptotic theory solution, 270–71
uniform theory of diffraction, 268–69
Van der Waerden method, 267–71

curves
eikonal equation resolution, 33–34
properties of, 34–35

curvilinear coordinates
concave circular cylinder, geometric acoustics extraction, 227–28
transport equation resolution, 36–40

Debye asymptotic expansion
caustic canonical problem, 354–58
circular cylinder, 148–62, 166–69
circular impedance cylinder, 298–312
concave circular cylinder, 221–28, 228–33
creeping waves, field diffraction in shadow region, 192–94
elastic surface waves, 197–203
hollow elastic circular cylinder, 187–94

degenerated caustics, 349–52

detour parameter
curved wedge, uniform asymptotic theory, 270–71
Van der Waerden uniform asymptotic solution, wedge diffraction, 265–67

diffraction coefficient
circular cylinder, 162–65, 175–76
convex impedance wedge, 272–93
hybrid diffraction coefficients, 369–94
impedance wedge, 92, 122–25, 279–80
Maliuzhinets's function, symmetric form, 447–52
noncircular convex cylinder, 165–66
oblique incidence, 92, 122–25, 175–76

Pauli-Clemmow uniform asymptotic solution, 263–64
planar interface, 279–80
surface waves, 286–89
three-dimensional convex surface, 177–78
three-dimensional convex wedge, 240–71
three-dimensional impedance wedge, 290–93
wedge diffraction, 88–89, 263–64

Dirac delta function, 73–89

Dirichlet boundary conditions, 72–92

dispersion equation, Rayleigh waves, 433–34

divergence factor
circular cylinder, 161–62
elastic surface waves, 203, 206–7
three-dimensional convex surface, 176–78
three-dimensional wedge, 382

doubly deflected ray, concave circular cylinder, 226–28

edge diffraction
constraints, 4
creeping waves, 369–84
homogeneous fluid, 11
hybrid diffraction coefficient, 362–63, 369–94
impenetrable wedge, 53–62
ray caustics, 362–63
surface waves, 384–92
whispering gallery waves, 392–94

edge excitation
surface waves, 384–92
whispering gallery waves, 392–94

edge excitation, hybrid diffraction coefficient
creeping waves, 369–84

eigenfunctions, 214–20

eikonal equation
creeping ray caustics, 363–65
resolution, 33–34
smooth convex object, 63–64
source-free, unbounded medium, 30–33
three-dimensional caustic, 358–61
wave field calculations, 349–51

elastic hollow cylinder shell, 178–209
asymptotic expansion and physical interpretation, 186–92
creeping waves, 192–94, 195–97
elastic surface waves, 197–203, 204–7
noncircular convex cylindrical shell, 195
noncircular cylindrical shell, 207–9
normal incidence, 178–82
Watson transformation, 182–86

elastic plate, fluid-vacuum separation, 425–30

elastic surface waves
hollow elastic circular cylinder, 197–203, 320–36
linear equations, 409–30
oblique incidence, 204–7
three-dimensional convex shell, 209–10, 336–39
uniform asymptotic solution, 324–34, 338–39

elastic wave excitation
fluid-elastic interface, 45–53
homogeneous isotropic solids, 415–17
linear equations, 414
smooth interface transmission, 9–10

elastic wedge-shaped shell
asymptotic evaluation, 132–37
hybrid diffraction coefficients, 388–90
Lamb wave excitation, 139–41
oblique incidence, 137–38
physical parameters, 138
reflection coefficient, 138–39
solution derivation, 125–32
uniform asymptotic solution, 294–96

Euler's theorem
circular cylinder, 173–76
sound waves, 398–400

excitation coefficient, noncircular cylindrical shell, 207–9

Fermat's theorem
 circular impedance cylinder, 308–12
 elastic surface waves, 204–7
 geometrical theory of diffraction, 3–12
 impedance wedge, 386–88
 surface waves, 12–22
 three-dimensional convex shell, 211
 three-dimensional smooth convex surface, 316–19
 three-dimensional wedge, 381–82
fluid-elastic interface
 vacuum separation, 425–30
 wave reflection and transmission, 45–53
fluid-solid interface, surface waves at, 434–38
fluid wave processes
 boundary conditions, 400–401
 constant sound velocity, 6
 harmonic waves, 401–2
 ideal fluids, wave propagation, 397–98
 line source excitation, wedge diffraction, 73–89
 sound waves, 398–400
Fock parameter
 circular cylinder, 173–76
 concave circular cylinder, 222–28
 smooth convex cylindrical surface, 371–73
 three-dimensional caustic, 360–61
 two-dimensional wedge, 375–78
 whispering gallery modes, 392–94
Fourier transform
 elastic wedge-shaped shell, 127–32
 impedance wedge, 101–10
Franz waves
 curved interface, 22
 hollow elastic circular cylinder, 184–86
 planar interface, 12
Frenet's formula, surface ray propagation, 12
Friedlander Keller expansion, 63–64

Gauss's theorem, 411–12
geodesic parameters
 circular cylinder, 171–76
 concave circular cylinder, 233–35
 creeping waves, 196–97
 elastic surface waves, 204–7, 338–39
 hybrid diffraction coefficients, 382
 noncircular cylindrical shell, 207–9
 three-dimensional convex shell, 209–11
 three-dimensional convex surface, 176–78
 three-dimensional smooth convex surface, 315–19
 three-dimensional wedge, 378–82
geometrical acoustics (GA)
 caustic field calculations, 351–67
 circular impedance cylinder, 308–12
 concave circular cylinder, 219–35
 concave surface, 211–12
 creeping waves, 193–94, 337–38
 curved wedge, 267–71
 elastic surface waves, 199–203, 329–34
 Fermat principle, 3–22
 hollow elastic circular cylinder, 189–94, 321–24
 smooth surface ray caustics, 361–62
 three-dimensional caustic, 361
 three-dimensional convex shell, 336–39
 three-dimensional smooth convex surface, 315–19
 Van der Waerden method, 266–67
 wedge diffraction, 86–89, 266–67
geometrical optics (GO)
 caustic field calculations, 351–67
 Fermat principle, 3–22
 source-free, unbounded medium, 30–33
 three-dimensional caustic, 361
 wedge diffraction, 86–89
geometrical theory of diffraction (GTD)
 acoustic field surface, 40–44
 asymptotic expansions, 22–40
 creeping ray caustics, 363–65
 curved interface, 20–22
 elastic wave excitation, 9–10
 Fermat principle, 3–22

fluid-elastic interface, 45–53
fluid segments, 6
homogeneous fluid, 8–9, 11
impedance wedge, 92–125
impenetrable wedge, 53–62
overview and basic principles, 1–3
path to ray configuration, 4–6
planar interface, 13–19
ray caustics, 362–63
smooth convex object, 63–69
smooth surface reflection, 7–8
source-free, unbounded medium wave equation, 28–40
surface rays and waves, 12–22
three-dimensional caustic, 361
two-dimensional wedge, 377–78

Green function
concave circular cylinder, 212–20, 222–28, 230–35
wedge diffraction, 72–89

Gudermann function, 136–37

half-space solution
impedance wedge, 94–101
planar interface reflection and refraction, 419–25

Hamilton-Jacobi equation, 33–34

Hankel function
circular cylinder, 144–47, 148–65
circular impedance cylinder, 298–312
concave circular cylinder, 215–33
creeping waves, 192–94
elastic surface waves, 197–203, 328–34
hollow elastic circular cylinder, 185–94, 321–24
impedance wedge, 384–88
line source excitation, 74–89
Watson transformation, 185–86

hard boundary conditions
circular cylinder, 140–76
circular impedance cylinder, 306–8
concave circular cylinder, 231–33
curved-face wedge, 267–71
fluid waves, 400
impenetrable wedge diffraction, 72–92
sound wave reflection at, 403
three-dimensional convex surface, 176–78
three-dimensional convex wedge, 240–67
wedge diffraction, 77–78

harmonic waves
governing equations, 401–2
hollow elastic circular cylinder, 178–82
linearized elasticity, 418–19

Heaviside function
convex impedance wedge, 272–93
elastic surface waves, 327–34
impedance wedge, 386–88
Pauli-Clemmow method, 249–55
three-dimensional convex wedge, 241–71, 249–55
wedge diffraction, 86–89

Helmholtz equation
circular cylinder, 140–76
concave surface, 212
creeping waves, 192–94
elastic wedge-shaped shell, 126–32
fluid-elastic interface, 45–53
geometric theory of diffraction, 2
harmonic waves, 402, 418–19
hollow elastic circular cylinder, 178–82
impedance wedge, 92–125
impenetrable wedge, 72–89
leaky waves, 439–43
line source excitation, 72–89
oblique incidence, 90–92, 192–94, 334–35
smooth convex object, 63–64
smooth convex shell, 334–35
source-free, unbounded medium, 29–33
three-dimensional caustic, 358–61
wedge diffraction, 74–78

Hilbert solution, 383–84
hollow elastic circular cylinder, 178–209
  asymptotic expansion and physical interpretation, 186–92
  creeping waves, 192–97
  elastic surface waves, 197–207, 324–34
  noncircular convex cylindrical shell, 195
  noncircular cylindrical shell, 207–9
  normal incidence, 178–82
  oblique incidence, 195–97, 204–7
  uniform asymptotic solution, 320–36
  Watson transformation, 182–86
holomorphic function
  elastic surface waves, 326–34
  Van der Waerden method, 255–60
homogeneous fluids
  edge diffraction, 11
  elastic wave excitation, 9–10
  reflection and refraction at interface, 404–6
  smooth interface transmission, 8–9
  smooth surface reflection, 7–8
Hooke's law, 413–14
hybrid diffraction coefficients, 369–94
  creeping waves, 369–84
  curved wedge, 373–84
  elastic wedge-shaped shell, 388–90
  grazing incidence and observation, 382–84
  impedance wedge, 384–88
  smooth convex cylindrical surface, 371–73
  surface waves, 384–92
  three-dimensional wedge, 378–82
  two-dimensional wedge, 373–78
  whispering gallery waves, 392–94

ideal fluids, wave propagation, 397–98
Imai transformation
  circular cylinder, 150–62
  creeping waves, 194
  hollow elastic circular cylinder, 186
impedance boundary conditions
  circular cylinder, 140–76
  diffraction coefficient, 165
  elastic wedge-shaped shell, 125–40
  oblique incidence, 175–76
  smooth convex impedance cylinder, 314–15
  sound wave reflection at, 404
  three-dimensional convex surface, 176–78
  three-dimensional wedge, 381–82
  two-dimensional wedge, 375–78
impedance cylinder, uniform asymptotic solution
  circular cylinders, 298–312
  smooth convex cylinder, 312–13
  smooth convex impedance cylinder, 314–15
impedance wedge
  convex impedance wedge, 272–89
  hybrid edge diffraction-excitation coefficients, 384–88
  Maliuzhinets's solution, 92–125
  three-dimensional wedge, 290–93
incident field
  circular impedance cylinder, 308–12
  elastic plate, 425–30
  impedance wedge, 41–44, 273–79
  impenetrable wedge, 53–62
  planar interface reflection and refraction, 273–79, 419–25
  three-dimensional convex wedge, 242–67
  three-dimensional impedance wedge, 292–93
  three-dimensional smooth convex surface, 316–19
inhomogeneous local plane waves
  elastic surface waves shading, 202–3
  impedance wedge, 387–88
  incident waves, 17–19

leaky waves, 439–43
planar interface, 17–19

integral representation
circular cylinder, 152–62
concave circular cylinder, 215–20
hollow elastic circular cylinder, 183–86
impedance wedge, 95–101
three-dimensional convex wedge, 244–46
wedge diffraction, 78–82

integration path
circular impedance cylinder, 299–312
concave circular cylinder, 214–28
elastic wedge-shaped shell, 130–32

inverse Fourier transform
elastic wedge-shaped shell, 127–32
harmonic waves, 402, 418–19
impedance wedge, 105–10

isotropic homogeneous elastic body
elastic wave excitation, 9–10
Hooke's law and, 413–14

jump discontinuity
hollow elastic circular cylinder, 323–24
Pauli-Clemmow method, 248–55
three-dimensional impedance wedge, 290–93
three-dimensional smooth convex surface, 248–55, 317–19
uniform theory of diffraction, 268–69

Keller's cone
edge diffraction, 11, 55–62
homogeneous fluids, 11
three-dimensional convex wedge, 242–67

Keller's nonuniform diffraction coefficient
asymptotic approximation, 89
impenetrable wedge, 72
Pauli-Clemmow uniform asymptotic solution, 264
two-dimensional wedge, 374–78

Kirchhoff approximation, 351

Kouyoumjian transition function
circular impedance cylinder, 304–8
elastic surface waves, 327–34
hollow elastic circular cylinder, 322–24
three-dimensional convex wedge, 247–55

Kronecker's delta, 413–14

Lamb waves
development of, 431–32
elastic wedge-shaped shell, 133–37
experimental parameters, 139–41
oblique incidence, 137–38
wedge diffraction, 340–41

Lamé constants
Hooke's law, 413–14
surface waves, 436–38

Laplace operator, wedge diffraction
line source excitation, 73–89
oblique plane wave incidence, 90–92

Laurent series, 108–10

law of reflection, 7–8

leaky waves
elastic surface waves as, 202–3
elastic wedge-shaped shell, 136–37
surface wave equations, 438–43

linearized elasticity
deformation, 409–10
fluid-vacuum separation, 425–30
homogeneous isotropic solids, 415–19
Hooke's law, 413–14
interfacial boundary conditions, 415
linear momentum and stress tensor, 410–12
planar interface reflection and refraction, 419–25
wave motion, 414–19

lines of existence, 136–37

line source excitation
concave circular cylinder, 212–33

line source excitation (*continued*)
impedance wedge, 384–88
wedge diffraction, 72–89
local cylindrical approximation, 210–11
logarithmic function
elastic wedge-shaped shell, 128–32
impedance wedge diffraction, 108–10
Luneburg-Kline expansion
geometrical theory of diffraction, 1
three-dimensional caustic, 358–61
transport equation resolution, 38–40

Maliuzhinets's function
convex impedance wedge, 272–73, 279–80
curved wedge-shaped shell, 391–92
impedance wedge, 92–125
surface waves, 281–89
symmetric diffraction coefficient, 447–52
two-dimensional wedge, 375–78
wedge diffraction, 107–10, 340–41
Meixner edge condition
elastic wedge-shaped shell, 126–32
impedance wedge, 102–10
oblique incidence, 91–92
wedge diffraction, 73–89
meromorphic function
impedance wedge diffraction, 107–10
surface waves, 281–89
Miller-type Airy function, 300–312
monochromatic waves
equations, 402
harmonic waves, 418–19
reflection at boundary, 402–4

Neumann boundary conditions
circular elastic shell, 455–58
impenetrable wedge diffraction, 72–92
noncircular convex cylinder
creeping wave solution, 192
diffraction coefficients, 165–66
noncircular cylindrical shell
normal and oblique incidence, 207–9
uniform asymptotic solution, 335–36
nonuniform asymptotic solution, acoustic wave diffraction
convex impedance wedge, 272–93
curved-face wedge, 267–71
Pauli-Clemmow method, 249–55
theory, 71–72
three-dimensional convex wedge, 240–42, 249–55
wedge diffraction, 72–140
normal incidence
creeping waves, field diffraction, 192–94
elastic surface waves, 197–203
hollow elastic circular cylinder, 179–82
impedance wedge, 94–110
Maliuzhinets's solution, 101–10
noncircular cylindrical shell, 207–9
straight wedge, 72–92
numerical results, uniform asymptotic solutions
circular cylinder, 341–45
convex impedance wedge, 279–80
wedge diffraction, 339–41

oblique incidence
circular cylinder, 171–76
circular elastic shell, 456–58
creeping waves, 194–97
elastic surface waves, 204–7
elastic wedge-shaped shell, 137–38
impedance wedge, 92–125, 289
noncircular cylindrical shell, 207–9
smooth convex shell, 334–35
three-dimensional impedance wedge, 290–93
two-dimensional wedge plane wave, 378–82
wedge diffraction, 90–92
observation points
caustic field calculations, 351–66

circular cylinder location, 166–69
concave circular cylinder, 231–33
elastic surface waves, 332–34
impedance wedge, 273–79
three-dimensional caustic, 359–61

Olver asymptotic expansion, 300–312

open waveguide theory, 142–47

oscillatory regime
concave circular cylinder, 231–33
Pauli-Clemmow method, 250–55

parallel rays, caustics, 349–51

path to ray configuration
geometrical theory of diffraction, 4–6
surface ray, 13–19

Pauli-Clemmow method
circular impedance cylinder, 304–8
curved-face wedge, 267–71
elastic surface waves, 325–34
impedance wedge, 273–79
integral evaluation and isolated pole singularity, 243–45
surface waves, 286–89
three-dimensional convex wedge, 245–55, 258–60
three-dimensional impedance wedge, 290–93
wedge diffraction, 260–64, 340–41

Pearcy function, 366–67

Pekeris functions
circular impedance cylinder, 305–8
creeping waves, 337–38
hollow elastic circular cylinder, 322–24
three-dimensional smooth convex surface, 315–19

perturbation theory, 28–40

planar interface
convex impedance wedge, 273–89
reflection and refraction at, 419–25
surface waves, 13–19
three-dimensional convex wedge, 242–67
wedge-shaped shell, 295–96

plane wave incidence
circular cylinder, 171–76, 341–44
convex impedance wedge, 273–89
creeping waves, 193–94
elastic wedge-shaped shell, 126–32, 389–90
hard and soft boundary conditions, 89–92
harmonic waves, 402
hollow elastic circular cylinder, 193–94, 320–36
homogeneous fluid interface, 404–6
homogeneous isotropic solids, 417
hybrid diffraction coefficients, 389–90
impedance wedge, 92–93, 123–25
noncircular cylindrical shell, 208–9
Pauli-Clemmow method, 255
three-dimensional convex wedge, 242–67
two-dimensional wedge, 378–79

point caustics, 349–52

pole path
circular cylinder, ν-plane position, 169–71
elastic surface waves, 197–203
elastic wedge-shaped shell, 130–32
surface waves, 281–89
three-dimensional convex wedge, 244–46

pseudo-ray path
circular impedance cylinder, 308–12
elastic surface waves, 332–34
hollow elastic circular cylinder, 322–24
Rayleigh waves, 442–43
two-dimensional wedge, 376–78

quasi-optics ansatz, 29–33

radiated inhomogeneous local plane waves, 17–19

radiation coefficient
circular cylinder, 140–41
noncircular cylindrical shell, 207–9
ray coordinate system
concave circular cylinder, 227–28
wave diffraction, 90–92
Rayleigh waves
curved interface, 21–22
elastic surface waves, 197–203
fluid-solid interfaces, 436–38
leaky waves as, 438–43
surface waves as, 432–34
ray theory (ray method)
curved interface, 20–22
overview, 1–2
planar interface, 13–19
transport equation resolution, 38–40
real ray path
curved interface, 21–22
planar interface, 15–19
reciprocity theorem
concave circular cylinder, 214–20
impedance wedge, 124–25
Maliuzhinets's solution, 447–52
noncircular convex cylinder, 165–66
noncircular cylindrical shell, 207–9
two-dimensional wedge, 376–78
reflected rays and waves
boundary conditions, 402–4
circular cylinder, 159–62
circular impedance cylinder, 308–12
elastic plate, fluid-vacuum separation, 425–30
elastic surface wave field, 324–34
fluid-elastic interface, 45–53
hollow elastic circular cylinder, 190–92, 320–36
homogeneous fluid interface, 404–6
impedance wedge, 273–79
planar interface, 273–79, 419–25
smooth body interference, 41–44
smooth surface caustics, 361–62
three-dimensional impedance wedge, 292–93
wedge diffraction, 339–41
reflection coefficient
elastic surface waves, 325–34
elastic wedge-shaped shell, 134–37
homogeneous fluid interface, 404–6
planar wedge-shaped shell, 295–96
smooth convex cylindrical surface, 373
three-dimensional smooth convex surface, 317–19
refraction
homogeneous fluid interface, 404–6
planar interface, 419–25
regions of validity, circular cylinder, 145–47
residue series
circular cylinder, 150–62
concave circular cylinder, 218–20
creeping waves, 192–94
elastic surface waves, 198–203
residue theorem
creeping waves, 337–38
hollow elastic circular cylinder, 322–24
impedance wedge, planar interface, 273–89
plane wave incidence, 89–92
wedge diffraction, 85–89
Riemann sheet
concave circular cylinder, 215–20
elastic surface waves, 330–34
Pauli-Clemmow method, 247–55
right-angle soft wedge, 339–41

saddle point conditions
circular cylinder, 152–62
circular impedance cylinder, 302–12
concave circular cylinder, 224–28
elastic surface waves, 325–34
impedance wedge, 385–88

Pauli-Clemmow uniform asymptotic solution, 260–64
three-dimensional convex wedge, 243–67
wedge diffraction, 82–89, 89–92

scalar function
elastic plate, 425–30
homogeneous isotropic solids, 415–17

Scholte-Stoneley wave
curved interface, 22
development of, 431–32
elastic wedge-shaped shell, 133–37, 390–91
hybrid diffraction coefficients, 390–91
leaky waves, 439–43
planar interface, 19

shadow boundaries
circular cylinder, 150–69
circular impedance cylinder, 298–312
creeping waves, 337–38
hollow elastic circular cylinder, 184–86, 189–94, 320–36
impedance wedge, 273–79
smooth convex object, 63–69
surface waves, 281–89
three-dimensional convex wedge, 242–67
three-dimensional impedance wedge, 290–93
three-dimensional smooth convex surface, 315–19
two-dimensional smooth convex surface, 297–315
wedge diffraction, 89, 339–41

smooth body interference, incident ray field, 41–44

smooth convex impedance cylinder, boundary layer field, 314–15

smooth convex shell, 319–39

smooth convex surface
acoustic field, 63–69
hybrid edge diffraction and excitation, 371–73
uniform asymptotic solutions, 297–319

smooth surface reflection
caustic rays, 361–65
homogeneous fluids, 7–8

Snell-Descartes law, 8–10, 17
elastic plate, 425–30
fluid-elastic interface, 47–53
homogeneous fluid interface, 406
leaky waves, 440–43

soft boundary conditions
circular cylinder, 140–76
circular impedance cylinder, 306–8
concave circular cylinder, 219–20
curved-face wedge, 267–71
fluid waves, 401
Helmholtz equation, 74–78
impenetrable wedge diffraction, 72–92
sound wave reflection at, 404
three-dimensional convex surface, 176–78
three-dimensional convex wedge, 240–67

Sommerfeld contour
normal and oblique incidence, 93–125
planar interface, 273–89

Sommerfeld radiation condition
hard and soft boundary conditions, 76–78
integral representation, 81–82
line source excitation, 73–89
oblique incidence, 91–92

sound rays and waves
elastic wave excitation, 10
governing equations, 398–400
path conditions, 4–6
reflection at boundary, 402–4

spectral theory of diffraction (STD)
creeping waves, 369–84
whispering gallery modes, 393–94

stationary points deformation
concave circular cylinder, 218–20, 222–28
Pauli-Clemmow method, 249–55
steepest descent paths (SDP)
caustic canonical problem, 357–58
circular cylinder, 152–62
circular impedance cylinder, 302–12
concave circular cylinder, 222–28
elastic surface waves, 325–34
elastic wedge-shaped shell, 132–37
hollow elastic circular cylinder, 189–92
impedance wedge, 273–89, 385–88
Pauli-Clemmow method, 248–55
plane wave incidence, 90–92
surface waves, 281–89
three-dimensional convex wedge, 242–67
wedge diffraction, 83–92, 260–67
Stokes lines
boundary conditions, 144–47
diffraction coefficient, 163–65
straight wedge, oblique incidence on, 122–25
stress components
elastic plate, 426–28
hollow elastic circular cylinder, 180–82
linearized elasticity, 410–12
surface rays and waves
convex impedance wedge, 281–89
curved interface, 20–22
elastic wedge-shaped shell, 132–37
Fermat's principle and, 12–22
fluid-solid interfaces, 434–38
hybrid edge diffraction-excitation coefficients, 384–92
leaky waves, 438–43
overview, 431–32
planar interface, 13–19
propagation, 12
Rayleigh waves, 432–34
smooth convex object, 67–69
three-dimensional impedance wedge, 293
tangent plane approximation, smooth convex cylindrical surface, 373
Taylor series
circular cylinder, 169–71
circular impedance cylinder, 308–12
elastic surface waves, 326–34
smooth convex cylindrical surface, 371–73
wedge diffraction, 86–89
three-dimensional caustic, 358–61
three-dimensional convex shell, uniform asymptotic solution, 336–39
three-dimensional convex wedge
hard or soft wedge, 240–71
Pauli-Clemmow method, 245–55
uniform asymptotic solution, 240–41, 290–93
three-dimensional convex wedge-shaped shell
boundary conditions, 176–78
creeping wave solution, 209–10
elastic surface wave solution, 211
uniform asymptotic solution, 296
three-dimensional curved wedge, 268–69
three-dimensional impenetrable wedge, edge diffraction, 53–62
three-dimensional smooth convex shell, 319–39
three-dimensional smooth convex surface, uniform asymptotic solutions, 297, 315–19
three-dimensional surface, curvature radii, 445–46
three-dimensional wedge, hybrid diffraction coefficients, 378–82
Tiberio method, impedance wedge, planar interface, 274–79
torsion, geodesic formula
three-dimensional convex shell, 209–11
three-dimensional convex surface, 177–78

three-dimensional smooth convex surface, 318–19
transition regions
circular impedance cylinder, 302–12
concave circular cylinder, 231–33
creeping waves, 336–38
hollow elastic circular cylinder, 321–24
shadow boundaries, 302–8
transmission
elastic wave excitation, 9–10
fluid-elastic interface, 45–53
homogeneous fluid interface, 404–6
planar interface, 421–25
smooth interface between two homogeneous fluids, 8–9
transport equation
asymptotic expansion derivation, 28–33
characteristic curve properties, 34–35
eikonal equation resolution, 33–34
resolution, 35–40
transverse curvature
harmonic waves, 418–19
three-dimensional convex shell, 211
three-dimensional convex surface, 176–78
two-dimensional caustic, uniform asymptotic expansion, 359–61
two-dimensional smooth convex surface, uniform asymptotic solutions, 297–315
two-dimensional wedge, hybrid diffraction coefficients, 373–78

Ufimtsev's correction, 384
uniform asymptotic solutions
basic principles, 239–40
caustic canonical problem, 355–58
circular cylinder, 341–45
circular impedance cylinder, 298–312
convex impedance wedge, 272–93
convex three-dimensional hard or soft wedge, 240–71
convex wedge-shaped elastic shell, 294–96
creeping ray caustics, 363–65
creeping waves, 336–38
curved-face hard or soft wedge, 267–71
diffraction coefficient, 279–80
hollow elastic cylinder, 320–36
impedance wedge, 273–89
incident and reflected fields, 273–79
noncircular cylindrical shell, 335–36
smooth convex three-dimensional surface, 315–19
smooth convex two-dimensional surface, 297–315
surface wave shadow boundaries, 281–89
three-dimensional caustic, 358–61
three-dimensional convex shell, 336–39
three-dimensional convex wedge, 240–41
three-dimensional impedance wedge, 290–93
three-dimensional smooth convex shell, 319–39
three-dimensional smooth convex surface, 297, 315–19
three-dimensional wedge-shaped shell, 296
two-dimensional smooth convex surface, 297–315
wedge diffraction, 259–67, 339–41
wedge-shaped shell, 295–96
uniform asymptotic theory (UAT), 270–71
uniform theory of diffraction (UTD)
curved-face wedge, 268–69
three-dimensional impedance wedge, 290–93
unit vectors
edge diffraction, 55–62
sound waves, 399–400

Van der Waerden method
circular cylinders, 341–44

Van der Waerden method (*continued*)
- curved-face wedge, 267–71
- elastic surface waves, 325–34
- impedance wedge, 273–79
- three-dimensional convex wedge, 244–45, 255–60
- wedge diffraction, 264–67

variables separation, 74–78

velocity of sound, 400

Wasylkiwskyj's solution
- geometric acoustics extraction, 223–28
- line source solution, 218–20

Watson transformation
- circular cylinder, 142–47
- circular impedance cylinder, 298–312
- hollow elastic circular cylinder, 182–86

wave field calculations
- caustic boundary layer, 349–67
- source-free, unbounded medium, 28–40

wave front
- characteristic curves, 34–35
- transport equation resolution, 35–40

wedge diffraction
- convex impedance wedge, 272–93
- curved wedge, 373–84
- elastic wedge-shaped shell, 125–40, 294–96, 388–90
- hard and soft boundaries, 72–92
- impedance wedge, 92–125
- line source excitation, 72–89
- nonuniform asymptotic acoustic wave diffraction theory, 72–140
- numerical results, 339–41
- Pauli-Clemmow uniform asymptotic solution, 260–64
- plane wave incidence, 89–92
- three-dimensional wedge, 378–82
- two-dimensional wedge, 373–78
- uniform asymptotic solutions, 259–67
- uniform theory of diffraction, 268–69
- Van der Waerden method, 264–67
- whispering gallery modes, 393–94

whispering gallery modes
- concave circular cylinder, 218–20
- hybrid diffraction coefficients, 392–94

Wiener-Hopf solution, 383–84

Wronksian expression
- circular cylinder, 163–65
- circular impedance cylinder, 307–8
- concave circular cylinder, 215–20
- wedge diffraction, 77–78
- whispering gallery modes, 393–94

zeroes of equation, trajectories
- circular cylinder, 169–71
- hollow elastic circular cylinder, 183–86

zero-order terms
- elastic surface waves, 287–89, 330–31
- Pauli-Clemmow method, 246–55
- three-dimensional convex wedge, 246–60
- Van der Waerden method, 256–60